AF660812

# *Introduction to*
# MICROPROCESSORS
# 8085 *and* 8086

# *Introduction to*
# MICROPROCESSORS
# 8085 *and* 8086

**Architecture, Interfacing and Programming**

***for***

**(All Indian Universities, Engineering and Technical Colleges and Institutions)**


**Rachhpal Singh**
HEAD
P.G. Deptt. of Computer Science & Applications
Khalsa College,
Amritsar

**Ikvinderpal Singh**
Lecturer
P.G. Deptt. of Computer Science & Applications
Khalsa College
Amritsar

**Jagbir Singh Virk**
Lecturer
P.G. Deptt. of Computer Science & Applications
Khalsa College
Amritsar



**KHANNA BOOK PUBLISHING CO. (P) LTD.**
Publisher, of Science, Technology and Engineering Books
4C/4344, Ansari Road, Darya Ganj, New Delhi-110002
**Phone:** 011-23244447-48 **Mobile:** +91-99109 09320
**E-mail:** contact@khannabooks.com
**Website:** www.khannabooks.com

# INTRODUCTION TO MICROPROCESSORS 8085 AND 8086

**Rachhpal Singh, Ikvinderpal Singh, Jagbir Singh Virk**

**Copyright © Khanna Book Publishing Co. (P) Ltd.**

This book is sold subject to the condition that it shall not, by way of trade or otherwise, be lent, resold, hired out, or otherwise circulated without the publisher's prior consent in any form of binding or cover other than that in which it is published and without a similar condition including this condition being imposed on the subsequent purchaser and without limiting the rights under copyright reserved above, no part of this publication may be reproduced, stored in or introduced into retrieval system, or transmitted any form or by any means (electronic, mechanical, photocopying, recording or otherwise), without the prior written permission of both, the copyright owner and the above mentioned publisher of this book.

**ISBN :** 978-93-80016-57-3

**First Edition:** 2010
**Reprint:** 2023

*Published by:*
**KHANNA BOOK PUBLISHING CO. (P) LTD.**
4C/4344, Ansari Road, Daryaganj, New Delhi-110002
**Phone:** 011-23244447/48 **Mobile:** 91-9910909320
**E-Mail:** contact@khannabooks.com

***Printed in India by:***
Vaibhav Products

***Dedicated to God***
***and***
***Our Parents***

# Preface

If you decided to read this book, then it means you have decided to learn about one of the most challenging, exciting and rapidly advancing fields in future technology. As a student you will find that this book is easily understood. It contains fundamental concepts and techniques of microprocessor.

**Chapter 1 and 2** covers the basics of microprocessor evolution and types. It highlights the architectural advancement of microprocessor and typical microprocessor development schemes.

Thereafter, internal architecture of microprocessor and operation of its components are explained using dynamic approach.

**Chapter 3 and 4** significantly expands the student's knowledge about the 8085 microprocessor. It contains lucid presentation of the Instruction set. Each and every instruction can be classified using different operations like data transfer, arithmetic operations, logical operations etc.

**Chapter 5 and 6** covers the basic Architecture of 8086 microprocessor, its various components and its different operating modes. Instruction set can be explained using different types of instructions.

**Chapter 7 and 8** introduces the concept of Assembly language programming based on intel 8085/8086 and its various Instructions like data transfer, arithmetic, logic etc.

Due to importance of peripheral devices **last chapters** has been devoted to the peripheral interfacing devices includes 8237 DMA Controller, 8255 programmable peripheral interface, 8253/8254 programmable timer/counter, 8259 programmable interrupt controller, 8251 USART and RS232C.

Each and every microprocessor described with the help of internal architecture, its pin diagram and description, its addressing modes, interrupts and its interfacing with different peripheral devices are given to have clear ideas.

Suggestions for improvement of the book will be gratefully accepted for future editions.

**Rachhpal Singh**
**Ikvinderpal Singh**
**Jagbir Singh Virk**

# Acknowledgement

First we would like to thanks almighty for giving us courage to bring up this book. We duly acknowledge the contribution of our colleagues Prof. Harbhajan Singh, Mr. Navjeet Singh, Mr. Avtar Singh Riar, Mr. Avtar Singh Sandhu, Mr. Mandeep Singh, Mr. Baljinder Singh, Mr. Sukhwant Singh, Mr. Kawaljeet Singh, Mr. Kawarpal Singh, Mr. Opinder Singh, Mr. Rajkaran Singh, Mr. Harmeet Singh, and our colleagues for their valuable co-operation and guidance. We are grateful to the entire team of M/s Khanna Book Publishing Co. (P) Ltd. for their encouragement and motivation during this project.

## Special Thanks

Our special thanks go to Dr. Daljit Singh, Principal, Khalsa College, Amritsar for their valuable guidance and timely support during the compilation of this book.

**Rachhpal Singh**
**Ikvinderpal Singh**
**Jagbir Singh Virk**

# Syllabus

## EEC-406: Introduction To Microprocessor

### Unit-I Introduction

Microprocessor evolution and types, microprocessor architecture and operation of its components, addressing modes, interrupts, data transfer schemes, instruction and data flow, timer and timing diagram. Interfacing devices. Architectural advancement of microprocessor. Typical microprocessor development schemes.

### Unit-II *-bit Microprocessors

Pin diagram and internal architecture of 8085 microprocessor, registers, ALU, Control & status, interrupt and machine cycle.

Instruction sets, Addressing modes. Instruction formats

Instruction Classification: data transfer, arithmetic operations, logical operations, branching operations, machine control and assembler directives.

### Unit-III 16-bit Microprocessor

Architecture of 8086 microprocessor: register organization, bus interface unit, execution unit, memory addressing, memory segmentation.

Operating modes. Instruction sets, instruction format, Types of instructions.

Interrupts: hardware and software interrupts.

### Unit-IV Programming

Assembly language programming based on intel 8085/8086.

Instructions, data transfer, arithmetic, logic, branch operations, looping, counting, indexing, programming techniques, counters and time delays, stacks and subroutines, conditional call and return instructions

### Unit-V Peripheral Interfacing

Peripheral Devices: 8237 DMA Controller, 8255 programmable peripheral interface, 8253/ 8254programmable timer/counter, 8259 programmable interrupt controller, 8251 USART and RS232C.

# Contents

CHAPTER 1

# Introduction to Microprocessors

## 1.1 INTRODUCTION

A microprocessor is a programmable integrated circuit that is used for executing instructions to process digital data or exercise digital control over other devices. It is employed primarily as the central processing unit (CPU) of a computer system. The complexity of present-day microprocessors makes even a modest description of how they work beyond the scope of this page. Thus, what is presented below is the architecture of a typical microprocessor from a couple of decades ago. The following discussion, simple as it is, nonetheless gives a reasonable understanding of how microprocessors in general work.

As mentioned, A microprocessor is a computer processor on a microchip. It's sometimes called a logic chip. It is the "engine" that goes into motion when you turn your computer on. A microprocessor is designed to perform arithmetic and logic operations that make use of small number-holding areas called registers. Typical microprocessor operations include adding, subtracting, comparing two numbers, and fetching numbers from one area to another. These operations are the result of a set of instructions that are part of the microprocessor design. When the computer is turned on, the microprocessor is designed to get the first instruction from the basic input/output system (BIOS) that comes with the computer as part of its memory. After that, either the BIOS, or the operating system that BIOS loads into computer memory, or an application program is "driving" the microprocessor, giving it instructions to perform.

## 1.2 COMPONENTS OF MICROPROCESSOR

1. **Software:** The **Software** is a set of instructions or commands needed for performing a specific task by a programmable device or a computing machine.
2. **Hardware:** The **Hardware** refers to the components or devices used to form computing machine in which the software can be run and tested.

Without software the Hardware is an idle machine. The microprocessors applications are classified primarily in two categories: reprogrammable systems and embedded systems. In reprogrammable systems, such as microcomputers, the microprocessor is used for computing and data processing, a Personal Computer (PC) is a typical illustration. In an embedded system, the microprocessor is a part of a final product and is not available for reprogramming to the end user.

The microprocessor operates in binary digits, 0 and 1, also known as bits. Bit is an abbreviation for the term binary digit. These digits are represented in terms of electrical voltages.

Each microprocessor recognizes and processes a group of bits called the word, and microprocessor are classified according to their word length. The fact that the microprocessor is programmable means it can be instructed to perform given tasks within its capability. The instructions are entered or stored in a storage device called memory, which can be read by the microprocessor.

Memory is like the pages of a notebook with space for a fixed number of binary numbers on each line.

### 1.2.1 Three Basic Characteristics Differentiate Microprocessors

1. **Instruction set:** The set of instructions that the microprocessor can execute.
2. **Bandwidth:** The number of bits processed in a single instruction.
3. **Clock speed:** Given in megahertz (MHz), the clock speed determines how many instructions per second the processor can execute.

## 1.3 BLOCK DESCRIPTION OF A MICROPROCESSOR

The basic units or blocks of a microprocessor are ALU, an array of registers and control unit.

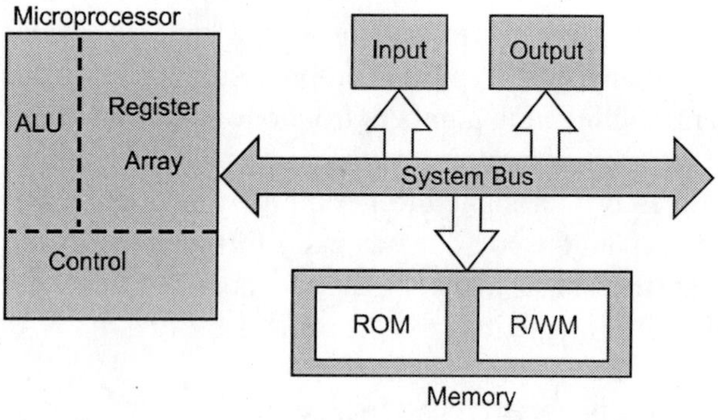


**Figure 1.1:** Basic Units or Blocks of a Microprocessor

### 1.3.1 ALU (Arithmetic/Logic Unit)

It performs such arithmetic operations as addition and subtraction, and such logic operations as AND, OR, and XOR. Results are stored either in registers or in memory.

### 1.3.2 Register Array

It consists of various registers identified by letter such as B, C, D, E, H, L, IX, and IY. These registers are used to store data and addresses temporarily during the execution of a program.

### 1.3.3 Control Unit

The control unit provides the necessary timing and control signals to all the operations in the microcomputer. It controls the flow of data between the microprocessor and memory and peripherals.

### 1.3.4 Input

The input section transfers data and instructions in binary from the outside world to the microprocessor. It includes such devices as a keyboard, switches, a scanner, and an analog-to-digital converter.

### 1.3.5 Output

The output section transfers data from the microprocessor to such output devices as LED, CRT, printer, magnetic tape, or another computer.

### 1.3.6 Memory

It stores such binary information as instructions and data, and provides that information to the microprocessor. To execute programs, the microprocessor reads instructions and data from memory and performs the computing operations in its ALU section. Results are either transferred to the output section for display or stored in memory for later use.

### 1.3.7 System Bus

It is a communication path between the microprocessor and peripherals. The microprocessor communicates with only one peripheral at a time. The timing is provided by the control unit of the microprocessor.

### 1.3.8 Four Operations Commonly Performed by MPU

1. **Memory Read:** Reads data (or instructions) from memory.
2. **Memory Write:** Writes Data (or instructions) into memory.
2. **I/O Read:** Accepts data from input devices.
3. **I/O Write:** Sends data to output devices.

### 1.3.9 Microprocessor Instruction Set and Computer Languages

The word (or word length), is the number of bits the microprocessor recognizes and processes at a time. The word length ranges from 4 bits for small microprocessor, to 64 bits for high-end microcomputers.

The byte is defined as a group of eight bits. The term "nibble", which stands for a group of four bits, is also found in popular computer magazines and books.

The instruction is defined as a complete task that microprocessor can perform. Instructions must be written in binary language, also know as machine language.

Assemble language is the program language that programmer can easily programs like as English words.

High-level languages (HLL) are the general-purpose languages as Visual BASIC, PASCAL, JAVA, C and C++. A program written in these languages can be machine independent.

## 1.4 MICROPROCESSOR EVOLUTION AND TYPES

From the earliest times the need to carry out calculations at huge speed and great accuracy has been developing. The first steps involved the development of counting and calculation aids such as the counting board and the abacus. Pascal (1623-62) was the son of a tax collector and a mathematical genius. He designed the first mechanical calculator (Pascaline) based on gears. It performed addition and subtraction.

Leibnitz (1646-1716) was a German mathematician and built the calculator (Stepped Reckoner) to do multiplication and division using a stepped cylindrical gear. It was not reliable due to the required accuracy of contemporary parts. Babbage (1792-1872) was a British inventor who designed an 'analytical engine' incorporating the ideas of a memory and card input/output for data and instructions. Again the current technology did not permit the complete construction of the machine. Babbage is largely remembered because of the work of Augusta Ada (Countess of Lovelace) who is probably the first computer programmer.

Burroughs (1855-98) introduced the first commercially successful mechanical adding machine of which a million were sold by 1926.

Hollerith (in 1889) developed an electromechanical punched-card tabulator to tabulate the data for 1890 U.S. census. Data was entered on punched cards and could be sorted according to the census requirements. The machine was powered by electricity. He formed the Tabulating Machine Company which became International Business Machines (IBM). IBM is still one of the largest computer companies in the world. Aiken (1900-73) a Harvard professor with the backing of IBM built the Harvard Mark I computer (51ft long) in 1944. It is based on relays (operate in milliseconds) as opposed to the use of gears. It required 3 seconds for a multiplication. Eckert and Mauchly designed and built the ENIAC (Electronic Numerical Integrator and Computer) in 1946 for military computations. It used vacuum tubes (valves) which were completely electronic (operated in microseconds) as opposed to the relay which is electromechanical. It weighed 30 tons, used 18000 valves, and required 140 kwatts of power. It is 1000 times faster than the Mark I multiplying in 3 milliseconds. ENIAC is a decimal machine and could not be programmed without altering its setup manually.

Atanasoff had built a specialized computer in 1941 and was visited by Mauchly before the construction of the ENIAC. He sued Mauchly in a case which was decided in his favor in 1974!

Von Neumann was a scientific genius and was a consultant on the ENIAC project. He formulated plans with Mauchly and Eckert for a new computer (EDVAC-Electronic Discrete Variable Automatic Computer) which was to store programs as well as data. This is called the stored program concept (SPC) and Von Neumann is credited with it. Almost all modern computers are based on this idea and are referred to as Von Neumann machines. He also concluded that the binary system was more suitable for computers since switches have only two values. He went on to design his own computer at Princeton which was a general purpose machine.

Alan Turing was a British mathematician who also made significant contributions to the early development of computing, especially to the theory of computation. He developed an abstract theoretical model of a computer called a Turing machine which is used to capture the notion of computable i.e. what problems can and what problems cannot be computed. Not all problems can be solved on a computer.

From the 1950's, the computer age took off in full force because of the invention of transistor. The development of the transistor revolutionized the development of computers. Invented at Bell Labs in 1947, transistors were much smaller, more rugged, cheaper to make and far more reliable than valves. Core memory was introduced and disk storage was also used. The hardware became smaller and more reliable, a trend that still continues.

In 1957 Japan's Electrotechnical Laboratory develops a transistor computer, the ETL Mark III that uses 130 transistors and 1,700 diodes. In 1959 on July 30, Robert Noyce and Grdon Moore file a patent application for integrated circuit technology on behalf of the Fairchild Semiconductor Corp and in the same year Jack Kilby at Texas Instruments designs a flip-flop IC.

In 1964 with a speed of 9 megaflops, Control Data Corp.'s CDC 6600 designed by Seymour Cray, claims the title of first commercially successful supercomputer. In 1968 the first computers to incorporate integrated circuits the B2500 and B3500 are introduced by Burroughs.

### 1.4.1 Intel 4004

Finally in 1971 the team of Ted Hoff, S. Mazor and F. Fagin develops the Intel 4004 microprocessor a "computer on a chip". This 4004, is the world's first commercially available microprocessor. This breakthrough invention powered the Busicom calculator and paved the way for embedding intelligence in inanimate objects as well as the personal computer. Just four years later, in 1975, Fortune magazine said, "The microprocessor is one of those rare innovations that simultaneously cuts manufacturing costs and ads to the value and capabilities of the product.

As a result, the microprocessor has invaded a host of existing products and created new products never before possible." This single invention revolutionized the way computers are designed and applied. It put intelligence into "dumb" machines and distributed processing capability into previously undreamed applications. The advent of intelligent machines based on microprocessors changed how we gather information, how we communicate, and how and where we work.

In mid-1969 Busicom, a now-defunct Japanese calculator manufacturer, asked Intel to design a set of chips for a family of high-performance programmable calculators. Maracian E. "Ted" Hoff, an engineer who had joined Intel the previous year was assigned to the project. In its original design, the calculator required twelve chips, which Hoff considered to complex to be cost-effective. Furthermore, Intel's small MOS staff was fully occupied with the 1101 (MOS type of static semiconductor memory) so the design resources were not available. Hoff came up with a novel alternative: by reducing the complexity of the instructions and providing a supporting memory device, he could create a general-purpose information processor. The processor, he reasoned, could find a wide array of uses for which it could be modified by programs stored in memory. "Instead of making their device act like a calculator," he recalled, "I wanted to make it function as a general purpose computer programmed to be a calculator."

To this end, Hoff and fellow engineers Federico Faggin and Stan Mazor came up with a design that involved four chips: a central processing unit (CPU) chip, a read-only memory (ROM) chip for the custom application programs, a random access memory (RAM) chip for processing data, and a shift register chip for input/output (I/O) port. The CPU chip, though it then had no name, would eventually be called a microprocessor. Measuring one-eighth of an inch wide by one-sixth of an inch long and made up of 2,300 MOS transistors, Intel's first microprocessor is equal in computing power to the first electronic computer, ENIAC, which filled 3000 cubic feet with 18,000 vacuum tubes.

The 4004, as it is to be called, would execute 60,000 operations a second, with by today's standards is primitive. It works at a clock rate of 108 KHz.

### 1.4.2 Intel 8008

The 8-bit 8008 microprocessor had been developed in tandem with the 4004 and was introduced in April 1972. It was intended to be a custom chip for Computer Terminals Corp. of Texas. But as it developed, CTC rejected the 8008 because it was too slow for the company's purpose and required too many supporting chips. However, Intel offered the 8008 on the open market, where its orientation to data/character manipulation versus the 4004's arithmetic orientation caught the eye of a new group of users. The 8008 is made up of 3,500 MOS transistors and could work at a clock rate of 200 KHz.

### 1.4.3 Intel 8080

It soon became obvious to Intel and its competitors that there were almost limitless number of applications for microprocessors. A big advance came in 1974 withIntel's 8080 chip, the first true general purpose microprocessor. It is much more highly integrated chip than its predecessors, with about 10 times the performance. It could execute about 290,000 operations a second and could address 64K bytes of memory. Both the 4004 and 8008 utilized the P-channel MOS technology, whereas the 8080 used the innovative N-channel MOS process yielding vast gains in speed, power, capacity and density. What's more the 8080 required only 6 support chips for operation as opposed to 20 with the 8008. It consisted of 60,000 transistors and worked at clock rate of 2 MHz.

### 1.4.4 Intel 8085

Two years later, in 1976 Intel came up with an improved version of 8080 which was called 8085. It use to work on a single 5v-power supply, is faster and integrated more functions. It consisted of 6,500 MOS transistor and could work at clock rates of 3 to 5MHz.

### 1.4.5 Intel 8086

In 1976 Intel decided to embark on a 16-bit project and the result was 8086 which came-in to market in 1978. 8086 is a 16-bit device with 10 times the performance of the 8080. It is build as an extension of the 8080's architectural concepts, making it easier for customers to use and for Intel market. In this microcontroller first time pipelining was introduced which furthermore increased the execution speed. Segmentation of memory i.e. dividing the memory space in to different segments for example code , data and stack segment was introduced for the first time. This segmentation of memory enabled the microprocessor to address the memory even though it had no register which

could hold 20 bit address. This was done using two registers base register and offset register to which a 0H had been hard wired in to the register as the lower nibble of base register and the higher nibble of offset register.

Thus when we add both of the register we get a 20-bit address. It had a pre-fetch queue of 6 instructions where in the instructions which were to be executed were fetching during execution of an instruction. Its architecture is designed to support parallel processing. It is made up of 29,000 MOS transistors and could work at clock rates of 5 to 10 MHz. It has a 16-bit ALU (Arithmetic and Logic Unit) and data bus but the width of address bus is 20-bit (could address only 1M byte of memory).

### 1.4.6 Intel 8088

Soon after this in 1979 Intel introduced another microprocessor 8088 which is similar in architecture to 8086 but the difference is in the available number of data bits of the data bus, they were limited to 8-bits even though the ALU is of 16-bits. This is for 8- bit applications requiring higher computational power. It had a pre-fetch queue of 4 instructions. A pivotal sale to IBM made the Intel 8088 the brains of IBM's new hit product—the IBM PC.

### 1.4.7 Intel 80186

By 1982 Intel came up with 80186 and 80286, two products compatible with the 8086 and 8088. The 80186, designed by a team under the leadership of Dave Stamm, integrated onto the CPU a number of functions previously implemented in peripheral chips, producing higher reliability and faster operations speeds at less cost. It had a prefetch queue of 6 instructions. It is suitable for high

volume applications such as computer workstations, word processor and personal computers. When compared with the previous microprocessors on chip, clock oscillator, interrupt controller, two DMA channels (with all support), chip select logic with operating modes (iRMX186 for master mode and Non iRMX186 for slave mode this is similar to min and max modes in 8086), and three timers.

Moreover there were ten extra instructions added to this microprocessor which are, PUSHA and POPA (to push and pop all the registers), IMUL destination, source, #immediate (an instruction of this type didn't exist in the previous μp), SHIFT / ROTATE destination register, #immediate (this instruction could shift or rotate a register contents given number of times), INS / OUTS (for input and output of a string, ex- INS DS:DI and OUTS DS DS:SI), other three instructions were required for operating system (a concept which became very prominent and still is from the starting days of microprocessors) ENTER, LEAVE and BOUND. It is made up of 134,000 MOS transistor to form 16-bit microprocessors with a 16-bit data bus, 20-bit address bus (could address only 1M byte of memory) and could work at clock rates of 4 and 6 MHz. This processor is called the first generation of microprocessors.

### 1.4.8 Intel 80286

The 80286 is called the second generation of microprocessor, it is more advanced to the 80186. This is the first Intel microprocessor offering multitasking and virtual memory. It is a 16-bit processor capable of addressing up to 16 MB (as it had an address but of 24-bits) of RAM and could also work with virtual memory (1GB). It had a prefetch queue of 6 instructions. The 286 is the first "real" processor. It introduced the concept of protected mode and real mode. To ensure proper operation, we must protect the operating system and all other programs and the shared resources.

The approach taken by many operating systems provide hardware support that allows us to differentiate among various modes of operation. A bit called the mode bit is added to the hardware of the computer to indicate the current mode.

With the mode bit, we are able to distinguish between a task that is executed on behalf of the operating system and one that is executed on behalf of the user. The dual mode of operation provides us with the means for protecting the operating system from errant users. This is accomplished by designing some of the machine instructions that may cause harm as "privileged instructions", these instructions are executed only in monitor mode. The 286 had an extra register called the Machine Status Register (MSW) whose lower nibble (containing D3 D2 D1 D0) defined the mode of operation and moreover it uses a four level memory protection which is an extension of the user/supervisory (protected/real) mode concept. It also had on-chip Memory Management Unit (MMU). This is also the first Intel processor that could run all of the software written for its predecessor. It has 134,000 transistors and could run at 6 to 12.5MHz

### 1.4.9 Intel 80386

The 80386 is the first popular 32-bit microprocessor. IA-32 first appeared with the 80386 processor, but the architecture was by no means completely new. IA-32's 8-bit predecessor first appeared in the Datapoint 2200 programmable terminal, released in 1971.It wasn't just an evolutionary product in Intel's growing family of microprocessors; it was revolutionary. It is a 32-bit chip that contained 275,000 transistors, could process five million instructions per second, and could run all popular operating systems, including Windows. It is also "multitasking," meaning it could run multiple programs at the same time. It has a pre-fetch queue length of 16 bytes. It has extensive memory management capabilities. It incorporates a sophisticated technique known as paging, in addition to the segmentation technique, for achieving virtual memory. The 80386 provided a new mode, virtual 8086 mode, in which real-mode programs could run while the processor was in protected mode. To support the concept of virtual memory to a grater extend it also has on-chip address translation unit.

This, combined with a moreflexible segmentation scheme and a larger addressable memory space (32 bits rather than 24, bring the total addressable memory to 4GB from 16MB and a virtual address space of 64 TB) has made 80386 protected mode the mode of choice for all modern operating systems. Later IA-32 implementations have not made significant changes or enhancements to protected mode. IA-32 adds the extended registers EAX, ECX, EDX, EBX, EBP, ESP, ESI, EDI,

EIP, and EFLAGS, as well as two additional, segment registers FS and GS. Originally all registers were special-purpose. For example, AX is originally an accumulator and could only be used as such. IA-32 lifted many of the restrictions on register usage, but some remain. For example, some instructions assume that a pointer in the EBX register is relative to the segment indexed by DS. In practice, 6 registers are available for generalpurpose use, far fewer than the number available in the ARM or IA-64 architectures.

The practical result of this register pressure is that IA-32 programs tend to make more frequent use of the stack for temporary storage. The 80386 has automatic self-test this feature is known as 'Built-In-Self-Test' (BIST). The BIST tests approximately one-half of the 80386 which includes the internal control ROM. After successful completion of BIST, the 80386 forms reset sequence after which it will start from the reset vector.

### 1.4.10 Intel 80486

The 80486s were not groundbreaking in terms of a radically different design philosophy, like the 80386. It did have four new features that made the 80486 about twice as fast as the fastest 80386. The most talked about new features were a built-in cache, (when the processor speeds reached the 20-25 MHz vicinity, reasonably priced DRAM memory could no longer be accessed with zero wait bus cycles) and a built-in math coprocessor (increased the throughput for floating point operations).

On average, the math coprocessor built into the 80486 yielded three times the greater performance than external 80387 Numeric Processing Unit (NPU). The speed difference between the 80386 and the 80486 made the Graphical User Interface (GUI) practical for everyday use. A concept which is known as, Unified Internal Code/Data Cache is used in 80486. It incorporates the advantage of the external cache coupled with the fact that every time a memory request is fulfilled by the internal cache, one less bus cycle results.

In addition, the access of data is much faster since the only delay would be to look up the data/code and to deliver it to the internal requester. This also frees up the bus for other bus masters. The address bus in 486 is bi-directional because of the presence of cache memory inside 486 (to enable cache invalidation). It also supports burst type of bus cycle which saves time during floating point operand fetch as well as cache memory fill. Internal data conversion logic for both 8-bit subsystem and 16-bit subsystems; dynamic bus sizing supporting 8, 16 and 32 bit cycles.

The 386 does not support direct interfacing of 8 bit subsystem. An external logic is needed for this purpose. The 486 incorporates several features in order to simplify the debugging process. The on-chip debugging aids of 486 are of three types: Breakpoint instruction, Single-set capability by Trap and Code and data breakpoint capability by means of debug register. The 486 also has a parity generator and parity checker inside it, providing parity logic for Data bus one parity bit for each data byte. This offers better reliability. It consists of 1.2 million transistors and could run at clock rates of 50MHz.

### 1.4.11 Intel Pentium

Cyrix and AMD are out in the marketplace selling their CPUs and math co processors, calling them 80386, 80387, just like Intel's. As any one can guess, Intel is not happy about this. The firm was so mad; they went to court to stop their competitors but one cannot copyright or patent numbers, the judge says. So Intel runs a contest to come up with a name for the 80586 that isn't a number. Penta means five, and on the 19th of October 1992, the name Pentium was announced. Thus the Pentium began as the fifth generation of the Intel x86 architecture. The Pentium had an L2 cache from 256KB to 1MB, used a 50, 60 or 66MHz system bus and contained from 3.1 to 3.3 million transistors.

As usual, the Pentium was backward compatible, while offering new features. The revolutionary step in this CPU was twin data pipelines. This enabled the CPU to execute two instructions at the same time. This is known as super scalar technology, typically found in RISC based CPUs.

The Pentium uses a 32-bit expansion bus, however the data bus is 64-bits which means the system memory is addressed at 64 bits at a time. This is an important distinction to remember when working with some types of RAM packaging,

### 1.4.12 Intel Pentium Pro

Towards the end of 1995 the Pentium Pro was announced. This Pentium introduced a new socket (Socket 8), utilizing 387 pins. The Pro series included ability to run multiple instructions in one cycle, could execute instructions out of order, and had dynamic branch prediction, as well as speculative execution. Also included was an impressive cache arrangement. For programmers, the Pro looks like a classic CISC CPU, while internally the CPU is very RISC oriented in design. This 3.3 Volt CPU (3.1V at 150 MHz) was designed with a 32-bit operating system (OS) such as Windows NT in mind.

While the Pro had Level 1 cache in the CPU, its real forte was the integrated Level 2 cache which allowed upwards of 1MB of cache to reside inside the CPU packaging to run at processor speed. This really improved performance in SMP based system boards.

The Pro chip was the first chip to be offered in the AT or the ATX format. The ATX format was preferred, as the Pro consumed more than 25 W of power, which generated a fair amount of heat. There were several major improvements of Pentium pro over Pentium, for example it had a superscalar architecture (microprocessor architecture containing more than one execution unit), 12-stage super pipeline, internal micro-ops similar to RISC like instructions and internal thermal protection. This microprocessor could be clocked to 200.00 MHz and consisted of 5.5 million transistors.

### 1.4.13 Intel Pentium II

This CPU had remarkable performance. The challenge Intel faced was the cost of production creating the Pro chip. The built-in L2 cache had a high failure rate at Intel fabrication plants. They came up with the Pentium II (P2). Intel began by separating the processor, and cache of the Pentium Pro, mounting them together on the circuit board with a big heat sink. Then by dropping the whole assembly onto the system board, using a **Single Edge Contact** (SEC) with 242 pins in the slot, and adding the 57 **MMX** (Multimedia extension) micro-code instructions, then Intel had the Pentium II. This way, defective cache modules don't force throwing out of a perfectly good CPU, because of a bad cache. And to further improve cache yields, the Pentium II ran cache at half the speed of the CPU.

Pentium II uses the **Dynamic Execution Technology**, which consists of three different facilities; Multiple branch prediction predicts program execution through several branches, accelerating the flow of work to the processor, Dataflow Analysis creates an optimized, reordered schedule f instructions by analyzing data dependencies between instructions and Speculative Execution carries out instructions speculatively thereby ensures that the multiple execution unit remains busy, boosting overall performance.

Pentium II includes data integrity and reliability features such as Error Correction Code (ECC), Fault Analysis, Recovery and Functional Redundancy Checking for both system and L2 cache buses. The pipelined Floating-Point Unit (FPU) supports the 32-bit and 64-bit formats specified in IEEE standard 754, as well as an 80-bit format.

Parityprotected address/request and response system bus signals with a retry mechanism for high data integrity and reliability. An on-die diode monitors the die temperature. A thermal sensor located on the motherboard can monitor the die temperature of the Pentium II processor for thermal management purposes. This microprocessor could work at clock rates of 300MHz and is made up of 7.5 million transistors.

### 1.4.14 Intel Pentium III

Similar to Pentium II, the Pentium III processor also uses a Dynamic Execution micro-architecture: a unique combination of multiple branch prediction, data flow analysis, and speculative execution. The Pentium III has two major differences with Pentium II: Improved MMX and Processor serial number feature. The improved MMX has totally 70 instructions enabling advanced imaging, 3D streaming audio and video, and speech recognition for enhanced Internet Experience: technology instructions for enhanced media and communication performance.

Additionally, Streaming SIMD (single-instruction, multiple data) Extensions for enhanced floating point and 3-D application performance. It also consisted of **Internet Streaming SIMD Extensions.** which consisted of 70 instructions and includes single instruction, multiple data for floating-point, additional SIMD integer and cacheability control instructions. Data Pre-fetch Logic anticipates the data needed by the application programs and pre-loads into the Advanced Transfer Cache increasing performance. The processor has multiple low power states such as Sleep, and Deep to conserve power during idle times. The system bus runs at 100MHz and 133MHzallowing for a 33% increase in available bandwidth to the processor. The Processor Serial Number extends the concept of processor identification by providing a 96-bit software accessible processor number that may be used by applications to identify a system. Applications include membership authentication, data backup/restore protection, removable storage data protection, and managed access to files.

### 1.4.15 Intel Pentium 4

The Pentium 4 processor is Intel's microprocessor that was introduced at 1.5GHz in November of 2000. It implements the new **Intel NetBurst micro-architecture** that features significantly higher clock rates and world-class performance. It includes several important new features and innovations that will allow the Intel Pentium 4 processor to deliver industry-leading performance for the next several years.

The Pentium 4 processor is designed to deliver performance across applications where end users can truly appreciate and experience its performance. For example, it allows a much better user experience in areas such as Internet audio and streaming video, image processing, video content creation, speech recognition, 3D applications and games, multi-media and multi-tasking user environments. The Pentium 4 processor enables realtime MPEG2 video encoding and near real-time MPEG4 encoding, allowing efficient video editing and video conferencing. It delivers world-class performance on 3D applications and games. It adds 144 new 128-bit Single Instruction Multiple Data (SIMD) instructions called **SSE2 (Streaming SIMD Extension 2)** that improves performance for multi-media, content creation, scientific, and engineering applications. Intel NetBurst micro-architecture of the Pentium 4 processor has four main sections: the in-order front end, the out-of-order execution engine, the integer and floating-point execution units, and the memory subsystem.

The Pentium 4 processor has a **20-stage misprediction pipeline** while the P6 micro-architecture has a 10-stage misprediction (This pipeline covers the cycles it takes a processor to recover from a branch that went a different direction than the early fetch hardware predicted at the beginning of the machine pipeline) pipeline. By dividing the pipeline into smaller pieces, doing less work during each pipeline stage (fewer gates of logic), the clock rate can be a lot higher. The Pentium 4 processor has a system bus with 3.2 G-bytes per second of bandwidth. This high bandwidth is a key enabler for applications that stream data from memory.

This bandwidth is achieved with a 64-bit wide bus capable of transferring data at a rate of 400MHz. It uses a source-synchronous protocol that quad-pumps the 100MHz bus to give 400 million data transfers per second. It has a split-transaction, deeply pipelined protocol to allow the memory subsystem to overlap many simultaneous requests to actually deliver high memory bandwidths in a real system. The bus protocol has a 64- byte access length. The Pentium 4 processor has 42 million transistors implemented on Intel's 0.18u CMOS process, with six levels of aluminum interconnect.

### 1.4.16 Intel Itanium

Intel, with partner Hewlett-Packard, developed a next generation 64-bit processor architecture called IA-64 (the 80x86 design was renamed IA-32) - the first implementation was named Itanium. Itanium core processor is not binary compatible with X86 processors, instead it has a separate compatibility unit in hardware to provide IA32 compatibility. Itanium only allow memory operands in load and store operations. As Itanium was a 64-bit processor so could address memory up to 4 GByte of RAM. The processor uses Explicitly Parallel Instruction Computing (EPIC), in which the compiler lines up instructions for parallel execution. Features were added to ensure compatibility with both Intel x86 and HP UNIX applications.

During development, it was widely expected that this would become the dominant processor architecture for servers, workstations, and perhaps even desktops, bumping the ubiquitous x86 architecture, and providing an industry-standard architecture across an unprecedented range of computing platforms, but it didn't happen so. The Itanium processor was specifically designed to provide a very high level of parallel processing, to enable high performance without requiring very high clock frequencies (which can lead to excessive power consumption and heat generation). Key strengths of the Itanium architecture include, Up to 6 instructions/cycle: The Itanium processor can handle up to 6 simultaneous 64-bit instructions per clock cycle, and the dual-core version can support up to two software threads per core, Extensive execution resources per core: 256 application registers (128 general purpose, 128 floating point) and 64 predicate registers, Large cache: 24MB in the dual-core version (12MB per core), providing data to each core at up to 48GB/s, Large address space: 50-bit physical / 64-bit virtual, Small, energy-efficient core: Since Itanium relies on the compiler for scheduling instructions for parallel throughput (other architectures rely on runtime optimization within the processor itself), it has fewer transistors in each core. This may be an advantage in current and future multi-core designs.

### 1.4.17 Intel Itanium 2

The Itanium 2 is an IA-64 microprocessor developed jointly by Hewlett- Packard (HP) and Intel, and introduced on July 8, 2002. The first Itanium 2 processor (code-named McKinley) was substantially more powerful than the original Itanium processor, roughly doubling performance, and providing competitive performance across a range of data- and compute-intensive workloads. Several generations of Itanium 2 processors have followed.

The Itanium 2 processor architecture is, dubbed **Explicitly Parallel Instruction Computing (EPIC)**. It is theoretically capable of performing roughly 8 times more work per clock cycle than other CISC and RISC architectures due to its Parallel Computing Micro-architecture. However, performance is heavily dependent on software compilers and their ability to generate code which efficiently uses the available execution units of the processor.

All Itanium 2 processors to date share a common cache hierarchy. They have 16 KB of Level 1 instruction cache and 16 KB of Level 1 data cache. The L2 cache is unified (both instruction and data) and is 256 KB. The Level 3 cache is also unified and varies in size from 1.5 MB to 24 MB. In an interesting design choice, the L2 cache contains sufficient logic to handle semaphore operations without disturbing the main ALU. The latest Itanium processor, however, features a split L2 cache, adding a dedicated 1MB L2 cache for instructions and thereby effectively growing the original 256 KB L2 cache, which becomes a dedicated data cache.

Most systems sold by enterprise server vendors that contain 4 or more processor sockets use proprietary **Non-Uniform Memory Access (NUMA)** architectures that supersede the more limited front side bus of 1 and 2 CPU socket servers. The Itanium 2 bus is occasionally referred to as the Scalability Port, but much more frequently as the McKinley bus. It is a 200 MHz, 128-bit wide, double pumped bus capable of 6.4 GB/s — more than three times the bandwidth of the original Itanium bus, known as the Merced bus. In 2004, Intel released processors with a 266 MHz bus, increasing bandwidth to 8.5 GB/s. In early 2005, processors with a 10.6 GB/s, 333 MHz bus were released.

### 1.4.18 Intel Pentium D

The Pentium D is a series of microprocessors that was introduced by Intel at the spring 2005 Intel Developer Forum. A 9xx-series Pentium D package contains two Pentium 4 dies, unlike other multi-core processors (including the Pentium D 8xx-series) that place both cores on a single die. The Pentium D was the first announced multi-core CPU (along with its more expensive twin, the Pentium Extreme Edition) from any manufacturer intended for desktop computers. Intel underscored the significance of this introduction by predicting that by the end of 2006 over 70% of its shipping desktop CPUs would be multi-core.

Historically, processor manufacturers have responded to the demand for more processing power primarily by delivering faster processor speeds. However, the challenge of managing power and cooling requirements for today's powerful processors has prompted a reevaluation of this approach to processor design. With heat rising incrementally faster than the rate at which signals move through the processor, known as clock speed, an increase in performance can create an even larger increase in heat. The answer is multi-core microprocessor.

For example, by moving from a single high-speed core, which generates a corresponding increase in heat, to multiple slower cores, which produce a corresponding reduction in heat, enterprises can potentially improve application performance while reducing their thermal output. A multi-core microprocessor is one which combines two or more independent processors into a single package, often a single integrated circuit (IC); to be more specific it has more than one execution unit with in a single integrated circuit. A dual-core device contains only two independent microprocessor execution units, as shown in the figure below. In general, multi-core microprocessors allow a computing device to exhibit some form of thread-level parallelism (TLP) without including multiple microprocessors in separate physical packages. This form of TLP is often known as chip-level multiprocessing, or CMP.

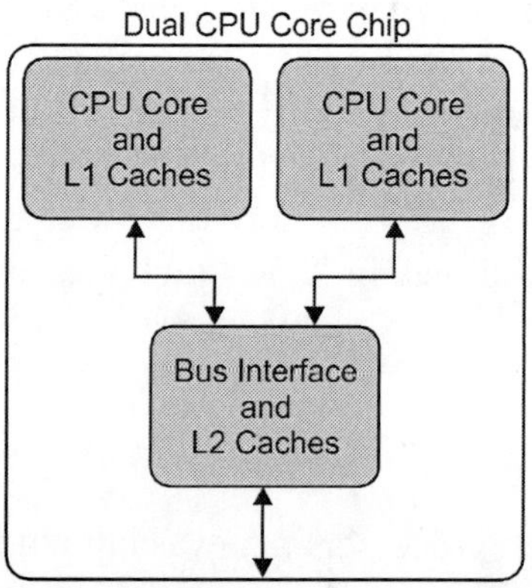


The Pentium D 820 runs in at 2.8GHz, is dual-core, its highlights are; it features two 16KB data caches in addition to data cache, each core includes an Execution Trace Cache that stores up to 12 K decoded micro-ops in the order of program execution, Streaming SIMD Extensions 3(SSE3) significantly accelerates performance of digital media applications and includes additional integer and cache ability instructions that may improve other aspects of performance, Execute Disable Bit feature combined with a supported operating system, allows memory to be marked as executable and nonexecutable and if code attempts to run in non-executable memory the processor raises an error to the operating system, it also has internal performance counters for performance monitoring and event counting and it also includes a thermal monitor feature that allows motherboards to be more cost effective.

Analysts have speculated that the clock rate race between Intel and AMD is largely over, with no more exponential gains in clock rate likely. Instead, as long as Moore's Law holds true, it is expected that the increasing number of transistors that chipmakers can incorporate into their CPUs will be used to increase CPU throughput through other methods, such as adding cores.

### 1.4.19 Intel Core 2 DUO

Core 2 Duo was the first family of desktop-class microprocessors based on Core microarchitecture. While the first Core 2 Duo processors had much lower core frequency and approximately the same FSB frequency and level 2 cache size as Pentium D microprocessors, they had better performance than the fastest Pentium D 960 due to much more efficient microarchitecture. The only exception to this were the slowest (less than 2 GHz) Core 2 Duo CPUs, that could perform slightly worse in some benchmarks. Newer dual-core CPUs have such improvements as higher core and FSB frequency, larger level 2 cache size, and lower power consumption. All Core 2 Duo processors use the same socket 775 package as many Pentium 4 and all Pentium D microprocessors, and can work in a number of Pentium 4 and Pentium D motherboards.

### 1.4.20 Intel Xeon

The Xeon is a brand of multiprocessing- or multi-package-capable x86 microprocessors from Intel Corporation targeted at the non-consumer server, workstation and embedded system markets. The Xeon brand has been maintained over several generations of x86 and x86-64 processors. Older models added the Xeon moniker to the end of the name of their corresponding desktop processor, but more recent models used the name Xeon on its own. The Xeon CPUs generally have more cache than their desktop counterparts in addition to multiprocessing capabilities.

**CURRENT TECHNOLOGY** allows for one processor socket to provide access to one logical core. But this approach is expected to change, enabling one processor socket to provide access to two, four, or more processor cores. Future processors will be designed to allow multiple processor cores to be contained inside a single processor module. For example, a tightly coupled set of dual processor cores could be designed to compute independently of each other—allowing applications to interact with the processor cores as two separate processors even though they share a single socket. This design would allow the OS to "thread" the application across the multiple processor cores and could help improve processing efficiency

A multi-core structure would also include cache modules. These modules could either be shared or independent. Actual implementations of multi-core processors would vary depending on manufacturer and product development over time. Variations may include shared or independent cache modules, bus implementations, and additional threading capabilities such as Intel Hyper-Threading (HT) Technology. A multi-core arrangement that provides two or more low-clock-speed cores could be designed to provide excellent performance while minimizing power consumption and delivering lower heat output than configurations that rely on a single high-clock-speed core. The above stated example shows how multi-core technology could manifest in a standard server configuration and how multiple low-clock-speed cores could deliver greater performance than a single high-clock-speed core for networked applications.

## 1.5 ARCHITECTURAL ADVANCEMENTS OF MICROPROCESSORS

A number of advancements that had taken place in the computer have migrated to microprocessor field with the continual advancement of microelectronics technology. The concepts of multitasking, pipelining, and multiprocessing are already there in the latest microprocessors. The pipelined microprocessors have a number of smaller independent units connected in series like a pipeline. Each unit is able to perform its tasks independently and the system as a whole is able to give much more throughput than the single microprocessor, Multitasking provides the environment in which the microprocessor can execute multiple tasks simultaneously by cycle stealing. The concept of virtual memory increases the memory capacity beyond the physical memory space possible through width of the address bus. The Memory Management Unity (MMU) is now integrated on microprocessor chips. Thus microprocessors of the 2000s are equivalent to supercomputers of 1990s and this process of growth is going to continue in the following years as well,

We shall now describe some of the concepts like pipelining, cache memory and virtual memory.

### 1.5.1 Pipelining

Pipelining in computer is used to enhance the execution speed. A number of processing elements or processors are used in pipelining. To explain the concept, let us assume that a task T is to be executed on a computer. This task T can be further divided into N sub-tasks as T1 to TN. These sub-tasks are independent of each other and the output of sub-task T1 is the input of sub-task T2 and thus the output of sub-task TN is the final output of task T. If we have N processors, P1 to PN, which are connected in the same way, i.e. the output of processor P1 is the input to processor P2 and so on (Figure 1.2) and if all sub-tasks take equal time. this connection will then achieve N times enhancement in the execution speed.

Let us consider the example of four processing elements P1 to **P4** The task T has to be subdivided into four sub-tasks T I to T4. Each processor takes one clock cycle 10 execute its sub- task as depicted below.

**Cycle No.** **P1 → P2 → P3 → P4**

1. ----------------T1
2. --------------------T2
3. ------------------------T3
4. ----------------------------T4

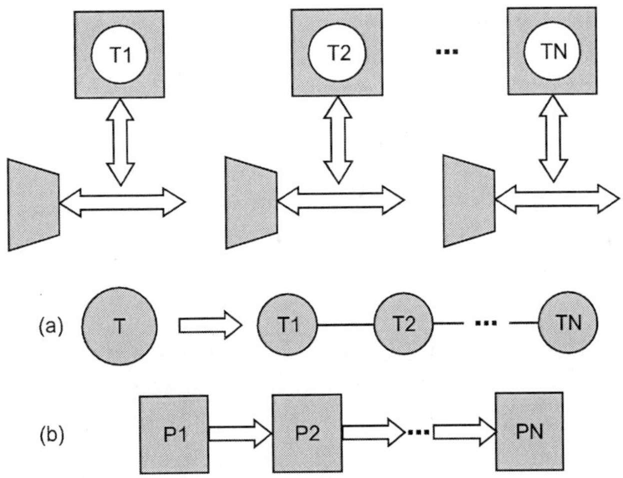


**Figure 1.2:** Independent Task Allocation

Thus a single task T will take four clock cycles which is same as a non-pipelined processor. Therefore, conclusions can he drawn that, for a single task there is no advantage derived from a pipelined processor. However, let us consider an example in which there is continuous flow of tasks T(1), T(2), T(3), T(4), ... each of which can be divided into four sub- tasks [e.g. task T(l) will have sub-tasks T1(1), T2(1), T3(1) and T4(1)] as before. Now the pipeline will look like:

| Cycle | P1 | P2 | P3 | P4 |
|---|---|---|---|---|
| 1. | T1(1) | | | |
| 2. | T1(2) | T1(1) | | |
| 3. | T1(3) | T2(2) | T2(1) | |
| 4. | T1(4) | T2(3) | T3(2) | T4(1) |
| 5. | T1(5) | T2(4) | T3(3) | T4(2) |
| 6. | T1(6) | T295) | T3(4) | T4(3) |

Now:

- The task T (1) consisting of sub-tasks T1 (1), T2 (1), T3 (1) and T4 (1) will get completed at the end of cycle 4.
- The task T (2) consisting of sub-tasks T1 (2), T2 (2), T3 (2) and T4 (2) will get completed at the end of cycle 5.
- Similarly the task T (3) will get completed at the end of cycle 6.

Thus all tasks after task T (1) take only one cycle to get completed. Therefore for subsequent tasks, four times enhancement in speed has been achieved. if the number of tasks is very large, the total speed enhancement (including task T(1)) approximates to four times.

The pipelining technique has been used to increase the instruction's execution speed in a number of computers. The technique is called 'instruction Pipeline'. An instruction execution contains the following four independent sub-tasks.

1. Instruction Fetch
2. Instruction Decode
3. Operand Fetch
4. Execute.

if four hardware modules or processing elements are designed to execute these four sub- tasks, four times enhancement in the execution speed is possible. This however is the maximum theoretical enhancement possible.

Unfortunately; not all instructions, are independent. If instruction I2 has to work on the result of the previous instruction I1, i.e. the result of instruction I1 is the operand for instruction then the operand fetch for I2 has to wait for I1 to get completed. The control logic inserts stall or wasted clock cycles into pipeline until such dependencies are resolved.

Another problem frequently encountered is that related to branch instructions. As an example, if instruction I2 is a branch instruction. the execution of instructions I3 and I4 in the pipeline becomes meaningless as program will branch to a new location. In such cases, the entire pipeline must be flushed. The problem increases if the number of stages in the pipeline is more (e.g. Intel Pentium 4 has a 20-stage pipeline).

The problem could be alleviated by incorporating branch prediction. Suppose as part of the instruction decode, it is possible to predict

- whether a branch will take place (in case of conditional branch),
- the location to which the program control will be transferred,

then by the time the instruction is executed, the pipeline can be flushed and the execution of the instruction from the new address may be initiated, However, a poor branch prediction may further exacerbate the problem.

The 8086, 80186, 80286 and 80386 have a two-stage pipeline. i.e. Bus Interface Unit (Instruction Fetch) and Execution Unit (Instruction Execution).

The 80486 has a five-stage instruction pipeline whereas the Pentium processor has a 20-stage instruction pipeline with branch prediction.

### 1.5.2 Cache Memory

Increasing the memory throughput will increase the execution speed of the processor. To achieve this, one or more fast buffers named cache between the processor and the main memory are provided. The cache memory has cycle time compatible with the processor speed. This facilitates the data and instruction to be supplied to the processor fast enough for processing. The working of a cache memory system is explained in Figure 1.3.

The cache memory consists of a few kilobytes of high speed Static RAM (SRAM). Whereas the main memory consists of a few megabytes to gigabytes of slower but cheaper Dynamic RAM (DRAM). Now let us consider the operation.

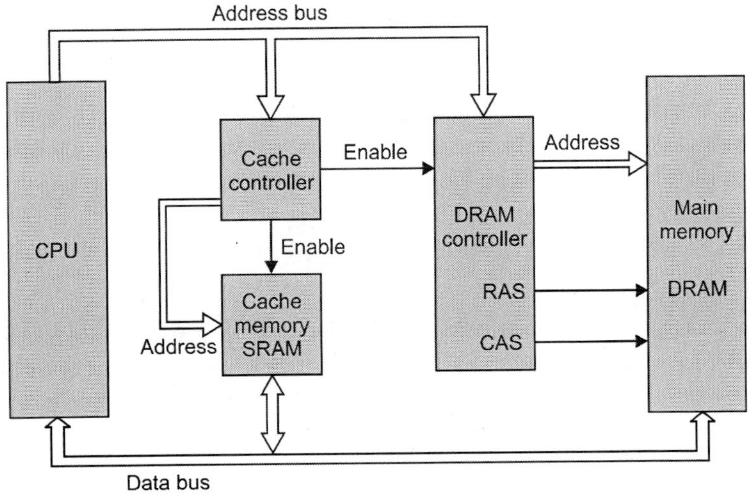


RAS = Row Address Strobe, and
CAS = Column Address Strobe and required for read/write operation in DRAM

**Figure 1.3:** Cache memory-principle of operation

- When the CPU wants to read a byte or word. it outputs address on the address bus.
- The cache controller checks whether the required contents are available in cache memory.
- If the addressed byte/word is available in cache memory, the cache controller enables the cache memory to output the addressed byte/word on the data bus.
- If the addressed byte/word is not present in cache memory, the cache controller enables the DRAM controller.
- The DRAM controller sends the address to main memory to get the byte/word. Since DRAM is slower, this access will require some wait states to be inserted,
- The byte/word which is read is transferred. to CPU as well as to cache memory. if needs this byte/word again, it can be accessed without any wait state.

The percentage of accesses where the CPU finds the desired byte/word in cache is called hit rate. For write operation, if the cache controller finds that the addressed byte/Word is present in the cache memory then the controller will write the new byte/word to the cache memory without any wait State, It intimates the CPU that the write operation is complete, The cache controller then writes the byte/word to main memory.

The above write operation methodology is called the Posted Write Through Method. The write to main memory is transparent to the CPU unless the main memory is involved in the previous write operation.

To keep track of which main memory locations are currently present in cache memory, the cache controller uses a cache directory. Each location in the cache is represented by an entry in the directory. The exact format of the cache directory depends on the cache scheme used. The basic cache schemes are—Direct mapped $cache_s$ Two way set associative cache and Fully associative cache, We shall not describe these here. Interested readers may consult (Douglas, 1991) and (Tannenbaum, 2005) on this topic.

Modern processors have incorporated both specialized caches as well as cache hierarchy. Pipelined CPUs access memory from multiple points in pipeline, i.e. different processing elements in pipeline access memory for different information. Different physical caches for each of these points are used in the architecture. The idea is to allocate one physical resource (cache) to service one point in pipeline. Thus three separate caches—Instruction, TLB (Translation Look Aside Huffer) and Data are incorporated to serve instruction Fetch, Virtual to physical address translation and Data fetch.

A victim cache is used to hold blocks evicted from a CPU cache due to conflict or capacity miss. It is put between the main cache and its refill path. The blocks that were evicted from main cache on a miss are stored in victim cache. The technique reduces the penalty incurred by cache on a miss. The microprocessors AMD K7 and K8 have large secondary cache as victim cache to avoid duplicate storage of contents of large primary cache.

Trace cache is a specialized cache incorporated to store traces of instructions that have already been fetched. A trace cache stores instructions either after decode or after their retirement. Generally, instructions are added to trace caches in groups representing either basic blocks or dynamic instruction traces. A basic block contains a group of non-branching instructions ending with a branch. On the other hand, a dynamic trace consists of instructions whose results are actually used, and eliminates instructions taken from following branches since they are not executed, The trace cache mechanism is used to increase the instruction fetch bandwidth. The microprocessor Pentium 4 uses trace caches to store already decoded micro operations so that the instructions, if needed next time, can be taken frost.) trace caches without going for decode.

The simulation studies of cache behaviour vis-a-vis cache size have indicated that capacity miss (i.e misses which a cache of given size will have regardless of its associativity or block size) falls steeply between 32 KB and 64 KB. Thus a CPU designer will like to keep the cache size around 64 KB.

Another consideration is the trade-off between cache latency and hit rate. Large caches have better hit rates but are slower. To overcome this, many computers use multiple levels of cache memory. Figure 1.4 shows three levels of a cache memory system. In a multilevel cache system, the small but fast cache is used first, followed by large and slower caches which are ultimately followed by even slower main memory.

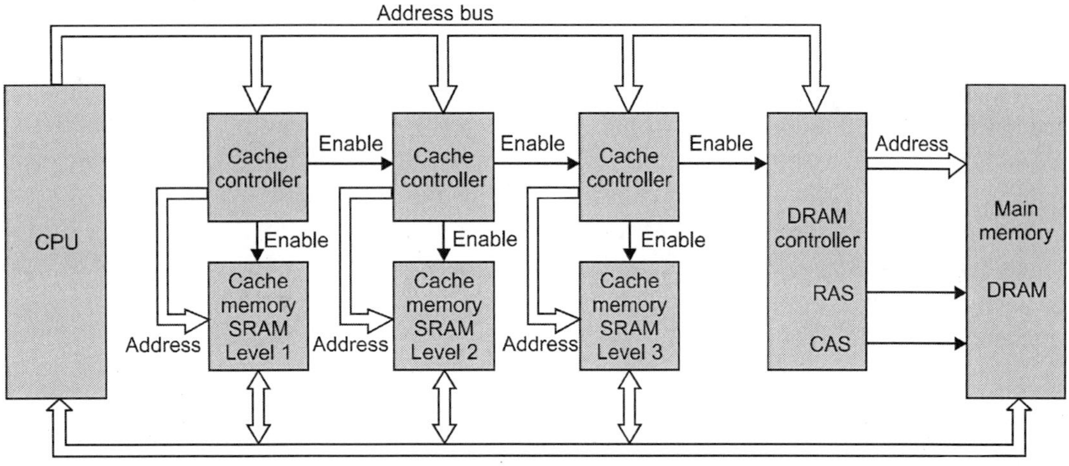


**Figure 1.4:** Multilevel cache system

As explained in Figure 1.4, multilevel caches operate by checking the smallest level 1 cache first. If the addressed byte/word is found, i.e. it hits, then the processor proceeds with high speed, otherwise the next higher cache is checked, and so on. At the end, main memory is checked. Multilevel caches can be either inclusive or exclusive. In art inclusive cache design the data in L1 cache will also be in L2 cache. The microprocessors Pentium 2, 3 and 4 have the inclusive multilevel cache system.

In exclusive caches, data is present either in L1 or L2 caches. The AMD microprocessor Athlon follows the exclusive multilevel cache system. The advantage of exclusive caches is that they store more data. When L1 misses and L2 hits on an access, the hitting cache line in the L2 is exchanged with a line in L1.

The advantage of inclusive cache is that when external devices or other processors in a multiprocessor system wish to remove a cache line from the processor, the processor needs to check only the L2 cache. In the other designs both L2 and L1 cache will required to be checked.

Though considered very small, but at the extreme case, the 6-byte instruction queue of the 8086 processor can be considered an example of on-chip cache. Other examples include Pentium 2, 3, 4 using the inclusive multilevel cache system and the AMD Athlon 64 (whose core design is known as K8) using the exclusive multilevel cache.

The Itanium II processor uses 6 MB unified level 3 cache on-chip. The IBM Power 4 series has 256 MB level 3 cache off-chip, shared among several processors.

### 1.5.3 Memory Management

The simplest model of memory management in computers in the beginning was Linear Addressing or Linear Memory organization. Here, only a single program occupied the complete memory space. The systems designer/programmer made a plan to use the memory space according to application and then developed the software. The earliest microprocessors like Intel 8080, Z80, and Intel 8085 used this concept. Even embedded processors like Intel 8051 /, Intel 8096 use this model as the processors are dedicated to specific applications.

With the passage of time, as computers grew in hardware complexity, they were required to be used in timesharing multitasking environments. This involved allocating a fixed amount of time to programs of different users- Relocatability of programs became the main criteria. In order to achieve relocatabi!ity, segmented addressing was used. In this mode of addressing the memory space is divided into a number of segments. The segment registers hold the memory address of the beginning of the segment. All addresses in programs are in relation to the memory addresses stored in segment registers. Figure 1.5 shows that the complete memory segment of 64 KB is divided into three segments, A, B and C. Thus,

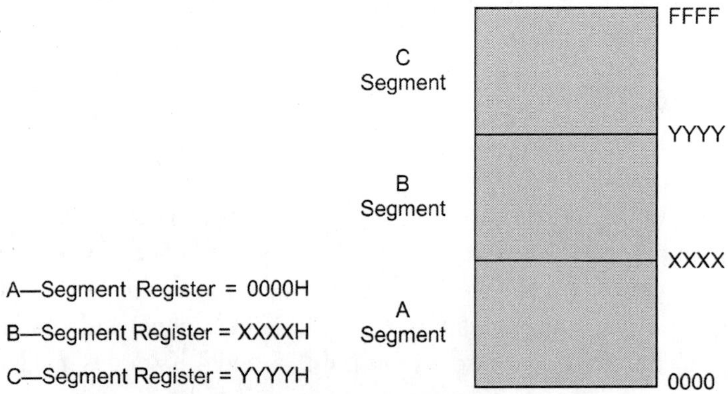


**Figure 1.5:** Segmented memory system

Segment Registers A, B and C store the starting address, i.e. Base Address 0000H, XXXXH and YYYYH of segments in memory. Now each memory address, whether for instruction or for data is defined as

$$\text{Memory Address} = \text{Base Address} + \text{Offset}$$

The offset for any information stored in memory is the distance in memory location from the start of the segment.

Now consider that a program P is loaded in segment A starting from address WWWW. Relocating this program in another segment will require changing only the base address, Thus in multitasking systems, different user programs could be located as per the availability of memory space, The 8086 incorporated a segmented memory system with four segments—Code Segment, Data Segment, Extra Segment, and Stack Segment

### 1.5.4 Virtual Memory System

As computer applications grew in complexity and consequently computer programs became complex and lengthy, it became evident that physical memory size would become a bottleneck. The virtual memory system was evolved as a compromise in which the complete program was divided into fixed/variable size pages and stored in hard disk. The main memory was also divided into pages of the same size. The basic idea is that pages can be swapped between the disk and the main memory as and when needed. This task is performed by the operating system, Thus, for a programmer, the main memory size is much more than the physical memory available which is virtually correct.

The software designer decides the virtual address space of his program and data. The virtual address space is divided into pages. The physical memory is also divided into corresponding units of the same size called frames. Figure 1.6 shows an example where the page size is 4 KB, the physical memory is 64 KB and the virtual memory is 256 KB.

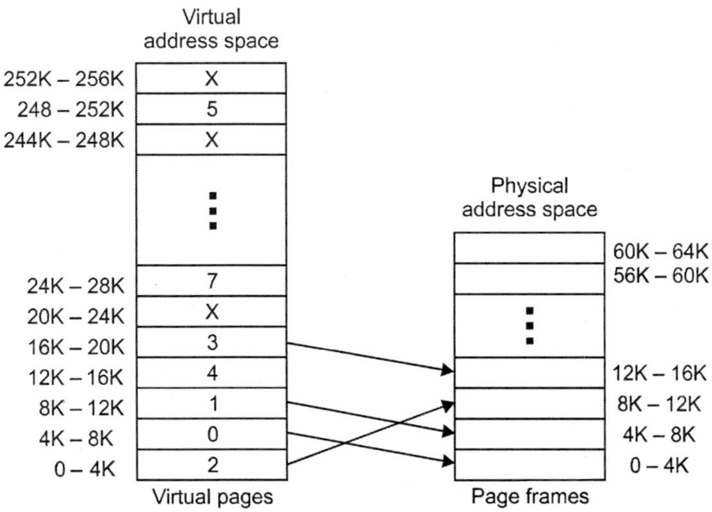


**Figure 1.6:** Mapping of virtual address space and physical address space

Number of pages in virtual space= 64 (0 to 63)

Number of page frames in physical memory = 16 (0 to 15)

Let us assume that a virtual address of g292 (8K + LOU) is generated (for instruction fetch or memory read/write as a result of instruction execution) by the CPU. This occurs at virtual page no, 2. Here 100 is the offset within the page. This virtual address is mapped to page frame no.1 in the physical address space which starts at 4K. Thus the physical address will be 4K + offSet = 4096 + 100 = 4196.

The task of translation of virtual address to physical address is carried out by the memory management unit (MMU). The MMU is a hardware device, functionally positioned between the CPU and the actual memory. However, many processors like Intel 80286, 80386, 80486, Pentium and Motorola MC 68031 and MC 68040 have the MMU integrated on the chip.

The memory management unit uses the page table to map virtual pages to page frames. Figure 1.7 shows the example of how the virtual address 8292 is translated to physical address using the page table. Since page size is 4K, i.e. 212, 12 bits of the address whether physical or virtual are assigned to offset. Since the virtual memory is 256 KB and the physical memory is 64 KB, the upper 6 bits in ease of virtual address will be assigned to virtual page number and the upper 4 bits will be assigned to a page frame number_ For each page, the page table keeps information on the present/absent, i.e. whether the page is present in main memory or not.

The physical address may be loaded to bus, The page frame numbers in page table are the most significant bits of the physical memory address. If the page is present in memory, memory fetch is effected. Otherwise the MMU raises a page fault interrupt. The operating system selects a page frame where the required page may be loaded from the disk. The contents of the selected page frame are saved into disk, and the referenced page is fetched from the disk and is placed in the selected page frame, The operating system changes the page table and restarts the instruction that caused the page fault.

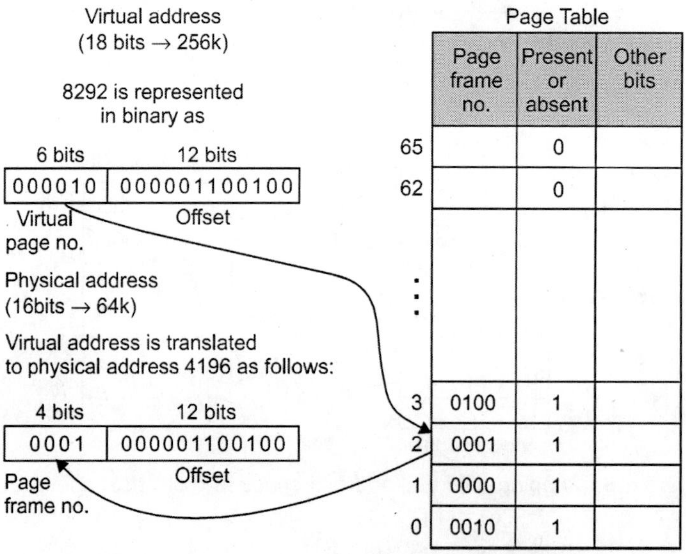


**Figure 1.7:** Translation of virtual address to physical address

In addition to the Present/Absent, the page frame may contain the following information:

**Dirty bit:** If bit = 1, the page in the memory has been modified. While page swapping between pages in disk and memory; the dirty bit is examined. If dirty bit = 0, the page in physical memory has not been modified, and thus need not be written hack to memory, A new page can be written over this page,

**Referenced bit:** If bit 1, the program has made reference to the page, meaning that the page may be currently in use. The operating system normally likes to select a page frame which has not been referenced for page replacement,

**Protection bits:** These bits specify whether the data in the page can be read/written/ executed.

The mapping from virtual to physical address must be done for every memory reference. Every instruction fetch requires a memory reference. Also, many instructions have a memory operand. Therefore, the mapping must be extremely fast lest it becomes the bottleneck.

For fast operations, a number of page table design alternatives like Multi Level Page Tables, Translation Lookaside Buffers (TLBs) etc. have been suggested. Translation Lookaside Buffers is a fast look-up table to correlate the virtual page number (through key) to the page frame number. The TLB is put as hardware cache in microprocessor to make the mapping very fast. The Intel Pentium Processor uses 32-bit virtual address with 4K page size.

The MMUs in the 80286, the 80386 and the 80486 manage the segment-based virtual memory through the Segment Selector and Descriptor Table. Each address in the 80286 has the following two components.

- 16-bit segment selector
- 16-hit offset

Using the segment selector, the MMU accesses the descriptor for the desired segment in the Descriptor Table. The descriptor contains the base address of the segment, the privilege level of the segment and some other control bits. The selector has 14 address bits and 2 privilege level bits, meaning that the total number of entries in Descriptor Table (Figure 1.8) and therefore the total number of segments is 16384. Since offset is represented in 16 bits, the total number of memory location, i.e. the size of each segment is 64 KB. Thus the total virtual space available in the 80286 is 1GB, compared to physical memory of 16 MB has 24 address lines).

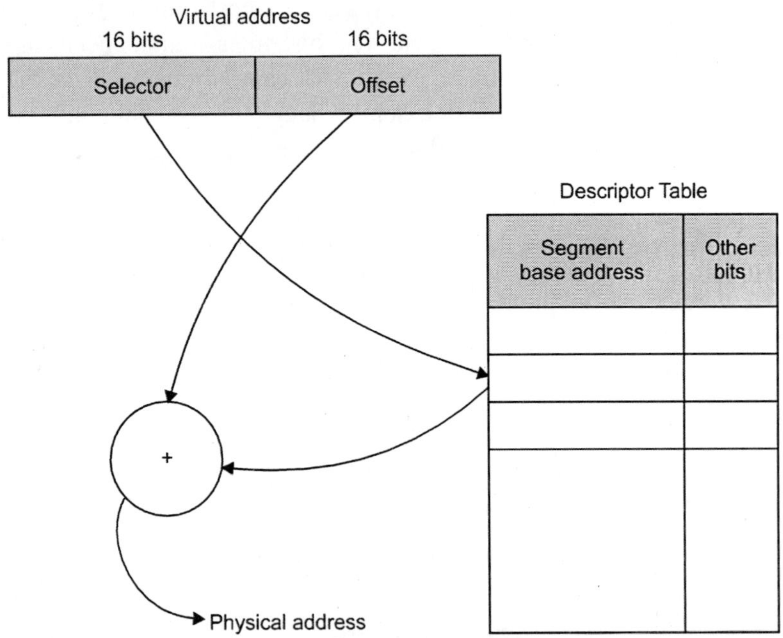


**Figure 1.8:** Segment-based virtual memory in intel 80286, 80386, 80486

You would notice that the basic concept is the same, as described earlier. The descriptor table is basically the page table and the segment selector is the virtual page address.

It is often opined that the virtual memory concept is the extension of cache memory, a concept where the disk has replaced the main memory and the main memory has replaced the cache memory.

## 1.6 TYPICAL MICROPROCESSOR DEVELOPMENT SCHEMES

A number of approaches can be used to simplify the development of microprocessor-based systems. Each of these depends upon the availability of suitable development tools. The four main approaches used are:

1. Microcomputer kit
2. Dedicated microprocessor development system
3. Universal microprocessor development system
4. Software package/simulator

### 1.6.1 Microcomputer Kit

The main purpose of a microcomputer kit is to impart training on the microprocessors. The use of a microcomputer kit to develop application software is perhaps the most crude and error- prone of all the methods. A typical microcomputer kit essentially consists of the central processing unit of the target machine on which the application program in binary code is executed, a small amount of the random access memory, serial input/output ports for interfacing to the peripherals, interrupt facility, and a quartz crystal controlled clock, etc. The system program consists of a simple monitor, an assembler in some cases and a debugger; Such a kit can only be used for the development of simple programs written in machine code or assembly language. The debugging facilities provided are very limited. So it is very difficult to develop an application program by using the microcomputer kit.

### 1.6.2 Dedicated Microprocessor Development System

A dedicated microprocessor development system (MDS) is a microcomputer system specially designed to aid in the development of a prototype system incorporating a specific microprocessor. it incorporates various user ails and test programs that enable a prototype system to be fully analyzed. As implied by the term dedicated, such systems can only be used to develop software for one or a limited range of microprocessors manufactured by the same manufacturer. The main drawback of an MDS system is that the user will have to base all his future systems on one particular microprocessor.

A typical microprocessor development system will have the following configuration.

- A central processing unit
- Random access memory upwards from 16 KB
- A visual display monitor

- An alphanumeric key board
- A printer
- A secondary mass storage device, i.e. either a floppy disk or a hard disk
- Ports for interfacing other peripherals
- Hardware debugging aids
- PROW EPROM Programmer, UV eraser
- An in-circuit emulator
- Slots for additional hoards.

The heart of the system which actually facilitates system integration and debugging is the in-circuit emulator (ICE), The hardware debugging aids include Logic State Analyzer, Signature Analyser, etc. We shall describe these facilities in brief in the following sections.

### *1. Logic State Analyzer*

When errors occur during data transfers **or** interrupt servicing, it is often necessary to examine a number of the streams of data which have been handled. If we think of an 8-bit microprocessor, there are 16 address lines and 8 data lines in addition to the control lines to investigate. So a single- or twin-beam oscilloscope is not a suitable tool for this purpose. An instrument developed particularly for this situation is the Logic State Analyzer. It has 16-48 data probes which can be attached to the bus lines to capture data on the rising edge of each clock pulse. Generally, a block of 250 successive bits are captured by each probe and stored in a semiconductor memory. The sequence is started by a trigger pulse which can be either an electrical signal such as an interrupt or a liming pulse or a derived form of a particular bit pattern which appears on the line under test. Once captured, the data can be displayed either as a sequence of words coded in hexadecimal, octal or binary format or as a set of waveforms.

Where the program appears to jump out of the intended sequence, the probes can be put on the address lines with the sampling enabled by the Valid Memory Address (VMA) signal. The data captured will then be a list of addresses from which the microprocessor has fetched successive instructions and data. If any interface package appears to be working incorrectly, its address can be used as the trigger signal and the data lines can be examined for the correct signals.

Although this procedure gives a useful record of the data flow and addresses used, it is essentially a sampled system. Any short-duration transients which occur between the dock pulses could cause false data to be received, though not perceivable by the display. In order to check for this, it is usually possible to operate the analyzer asynchronously. In this mode, the data sampling block is derived in the analyzer and is not locked to the microprocessor clock. By running the sampling clock 5-10 times faster than the microprocessor clock, transients which are long enough to cause a logical error will normally be sampled and stored and also seen on the display.

### *2. Signature Analyzer*

The signature analyser is an important tool for hardware debugging. This was originally developed as a quick method of checking the complex digital instruments but is equally applicable to the microprocessors as well. This allows a stream of data to be checked easily fur any single-bit error. The process **is** similar lo the one that is used to generate the check character, normally sent after each block in many kinds of digital transmission, and uses a shift register **with** feedback connections.

By itself the signature conveys no meaning; it is used in a comparison process by first measuring the signature of a suitable set of nodes for a set of standard input signals in a system known to be working correctly. These form a standard signal for the system. By measuring the signature at successive nodes of **a** faulty system, it is possible to locate the fault within some particular section of the equipment, It should be noted that however long the train of data signal may be, if only one bit is in error, the signature obtained will differ markedly from the correct one. The test thus has a high discrimination capability.

### *3. Prom/Eprom Programmer and UV Eraser*

Once the application software has been developed and checked in the prototype hardware module, the next task is to store the software. It can he either in a disk, floppy, or in semiconductor memory. In an application where the software does not require frequent alteration or modification, it is better to load the program in the semiconductor memory (PROM/EPROM), PROM is a programmable ROM and once it is programmed, i.e. data is stored in it, the same cannot he erased. But in an EPROM, though the data is stored in the same way, it is erasable and we can store the new data after erasing. The ultraviolet rays are allowed to fall on the small window of the EPROM chip to erase the previously stored data, The UV erasers and EPROM programmers are available commercially.

### *4. IN-Circuit Emulator (ICE)*

The ICE is used to combine software testing with hardware testing. The MDS with the ICE option contains at least two microprocessors. The MDS processor supervises the system resources. executes the system monitor commands. and drives the system peripherals, The second processor, i.e., the ICE processor, interfaces directly to the designer's prototype or production system via an external cable, In-circuit emulation is carried out by pulling out the socket-mounted microprocessor chip from the prototype microcomputer system for which the software is being developed and replacing it by the in-circuit emulator. Thus the hardware system to be tested has access to all the resources of the MDS and can be tested in real time, **As** the system operates in real-time, the response of the input/output system to particular events can be tested by interrupting program execution and by checking the contents of memory locations and address registers,

Sometimes it may be required to emulate the system more precisely. In real-time emulation we may not get the detailed trace data, The single step facility is usually available to get the detailed trace data.

This single stepping differs from what is available in a microprocessor kit, Here, the number of steps to be emulated can be programmed. We can also specify the particular conditions for halting. At the end of emulation, the debugged code can be stored as well. This facility of combined hardware and software testing of the target system is extremely useful and can considerably shorten the system development time.

The MDS also includes considerable software facility. Typical software modules included **in** MDS constitute a single or multitask disk operating system which contains

- Assemblers
- High-level language compiler/interpreters
- Loaders
- Linkers
- Test editors
- Debug monitors
- File handling/management routines.

An operating system provides interface between the hardware and the user. It consists of the set of program modules, as mentioned above, for allocating and controlling the resources available on the system automatically.

### 1.6.3 Universal Microprocessor Development System

A dedicated MDS can only be used for developing the microcomputer hardware/software for a single microprocessor or a few of the same series manufactured by the same company. Though it is true that majority of the applications can be implemented using any of the available microprocessors, efficiency wise, some microprocessors may be superior to the others. So the MDS manufacturers have come forward with a universal MDS, which can be used to develop and test the hardware/ software for a number of different systems. The term " universal" is rather misleading because no single development system can be used to develop and test the hardware/ software for all the available microprocessors.

However, manufacturers do provide an attractive method of developing the hardware/software For a wide range of microprocessors provided by a number of different suppliers. In addition to all the normal Facilities available in a dedicated system, the universal system has the facility to plug in different modules which result in the development system emulating the behaviour of a particular microprocessor.

### 1.6.4 Simulators

Simulator is an alternative to the MDS for the development and testing of the microcomputer software. As the term implies, simulators are used to simulate the features, such as memory and registers of a target microprocessor on a host computer (personal computer). The application program object code for the target microprocessor, generated by a cross compiler or a cross assembler is executed on the host computer system. As the execution proceeds, the contents of the simulated registers and memory locations are altered as they would be in the real system. These values can then be printed out and analyzed. Here the execution takes place in simulated time and not in real time. The distinct advantages of simulators are given below.

- As there are no real-time execution constraints, a more thorough check can be carried out on the program output thereby helping with the process of debugging the program.
- It is also possible to evaluate the likely performance of a microprocessor for a particular application.

The main drawback lies in the difficulty in simulating the effect of input and output values. Testing and debugging should be carried out in real time, being not possible on a simulator.

## QUESTION BANK

Q1. What is Microprocessor? Explain different components of microprocessor.

Q2. Explain the block description of microprocessor with the help of block diagram?

Q3. Explain different types of microprocessors.

Q4. Write a detailed description about the evolution of microprocessors.

Q5. Discuss the various architectural advancements of microprocessors.

Q6. Briefly explain the concept of pipelining.

Q7. Write a short note on cache memory and virtual memory.

Q8. Explain the typical microprocessor development schemes in detail.

CHAPTER 2

# Microprocessor Architecture

## 2.1 ARCHITECTURE OF MICROPROCESSOR

The microprocessor uses the same type of logic that is used in a digital computer's central processing unit (CPU). Because it is similar to the CPU and it is constructed with microcircuit (IC) technology. The microprocessor has digital circuit for data handling and computation under program control. (The microprocessor is a data processing unit) Data processing is the microprocessor's main function. Data processing includes both computation and data handling. Computation is performed by logic circuits that make up what is usually called the arithmetic logic unit (ALU). These logic circuits enable us to use functions that cause data changes. Among these functions are Add, Subtract, AND, OR, XOR, Compare, Increment, and Decrement. The ALU cannot perform any of these functions with out data operation on. In order to process data, the microprocessor must have control logic which tells the microprocessor how to decode and execute the program.

The control logic steps the microprocessors through the stored program steps (instructions) in memory. It calls (fetches) them one at a time. After the instruction is fetched, the microprocessor's control logic decodes the instruction. Then the control logic carries out (executes) the decoded instruction. Because the instructions are stored in memory, you can change them when you want to.

**Review:** The microprocessor's purpose is to process data. To do this, it must have logic to process and handle data, and control logic. The processing logic moves data from place and performs operations on the data.

Microprocessor is a multipurpose, programmable, clock-driven, register-base, electronic device that reads binary instructions from a storage device called memory, accepts binary data as input and processes data according to those instructions, and provides results as output.

## 2.2 OVERVIEW OF MICROCOMPUTER STRUCTURE AND OPERATION

Figure 2.1 Shows a block diagram for a simple microcomputer. The major parts are the central processing unit or CPU, Memory and the input and output circuitry or I/O. Connecting these parts are three sets of parallel lines called buses. The three buses are the address bus, the data bus, and the control bus. Let's take a brief look at each of these parts.

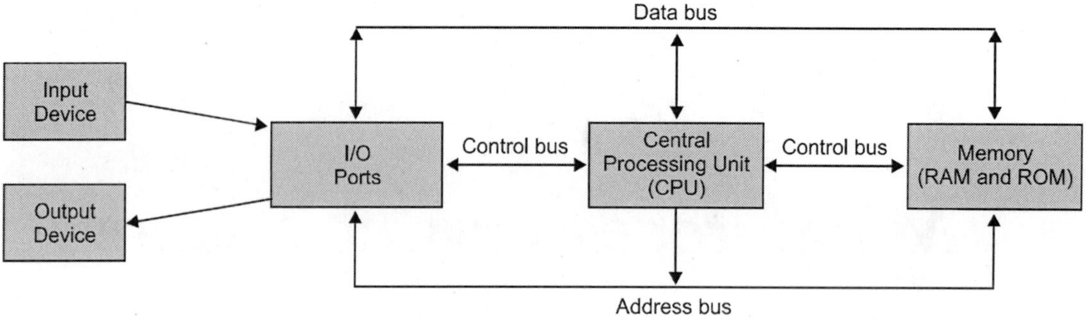


**Figure 2.1:** Block diagram of a simple microcomputer

### 2.2.1 Memory

The memory section consists of a mixture of RAM and ROM. It may also have magnetic hard disks, or optical disks. Memory has two purposes. The first purpose is to store the binary codes for the sequences of instructions you want the computer to carry out. When you write a computer program, you are really writing a sequential list of instructions for the computer. The second purpose of the memory is to store the binary-coded data with which the computer is going to be working. This data might be the inventory records of a supermarket, for example.

### 2.2.2 Input/Output

The input/output section allows the computer to take data from the outside world or send data to the outside world. Peripherals such as keyboards, video display terminals, printers, and modems are connected to the I/O section. These allow the user and the computer to communicate with each other. The actual physical devices used to interface the computer buses to external systems are often called ports. Ports in a computer function just as shipping ports do for a country. An input port allows data from a keyboard, an A/D converter, or some other source to be read into the computer under control of the CPU. An output port is used to send data from the computer to some peripheral, such as a video display terminal , a printer, or a D/A converter,

Physically, the simplest type of input or output port is just a set of parallel D flip-flops. If they are being used as an input port, the D inputs are connected to the external device, and the Q outputs are connected to the data bus which runs to the CPU. Data will then be transferred through the latches when they are enabled by a control signal from the CPU. In a system where they are being used as an output port, the D inputs of the latches are connected to the data bus, and the Q outputs are connected to some external device. Data sent out on the data bus by the CPU will be transferred to the external device when the latches are enabled by a control signal from the CPU.

### 2.2.3 Central Processing Unit

The central processing unit or CPU controls the operation of the computer. In a microcomputer the

CPU is a microprocessor. The CPU fetches binary-coded instructions from memory, decodes the instructions into a series of simple actions, and carries out these actions in a sequence of steps.

The CPU also contains an address counter or instruction pointer register, which holds the address of the next instruction or data item to be fetched from memory; general-purpose registers, which are used for temporary storage of binary data; and circuitry, which generates the control bus signals.

### 2.2.4 Address Bus

The address bus consists of 16, 20, 24, or 32 parallel signal lines. On these lines the CPU sends out the address of the memory location that is to be written to or read from. The number of memory locations that the CPU can address is determined by the number of address lines. If the CPU has N address lines, then it can directly address $2^N$ memory locations.

For example, a CPU with 16 address lines can address $2^{16}$ or 65,536 memory locations, a CPU with 20 address lines can address $2^{20}$ or 1,048,576 locations, and a CPU with 24 address lines can address $2^{24}$ or 16,777,216 locations. When the CPU reads data from or writes data to a port, it sends the port address out on the address bus.

### 2.2.5 Data Bus

The data bus consists of 8, 16, or 32 parallel signal lines. As indicated by the double-ended arrows on the data bus line in Figure 2.1, the data bus lines are bidirectional. This means that the CPU can read data in from memory or to a port on these lines, or it can send data out to memory or to a port on these lines, or it can send data out to memory or to a port on these lines. Many devices in a system will have their outputs connected to the data bus, but only one device at a time will have its outputs enabled. Any device connected on the data bus must have three-state outputs so that its outputs can be disabled when it is not being used to put data on the bus.

### 2.2.6 Control Bus

The control bus consists of 4 to 10 parallel signal lines. The CPU sends out signals on the control bus to enable the outputs of addressed memory devices or port devices. Typical control bus signals are Memory Read, Memory Write, I/O Read, and I/O Write. To read a byte of data from a memory location, for example, the CPU sends our the memory address of the desired byte on the address bys and then sends out a Memory Read signal on the control bus. The Memory Read signal enables the addressed memory device to output a data word onto the data bus. The data word from memory travels along the data bus to the CPU.

### 2.2.7 Hardware, Software and Firmware

When working around computers, you hear the terms hardware, software, and firmware almost constantly. Hardware is the name given to the physical devices and circuitry of the computer. Software refers to the programs written for the computer. Firmware is the term given to programs stored in ROMs or in other devices which permanently keep their stored information.

## 2.3 ADDRESSING MODES

The operation field of an instruction tells the operation to be performed by CPU . This operation must be executed on some data stored in computer registers or memory words. The way operands are chosen during program execution is dependent on the addressing mode of the instruction. The addressing mode species a rule for interpreting or modifying the address field of the instruction before the operand is actually referenced. Computers use addressing mode techniques for the purpose of accommodating one or both of the following provisions:

1. To give programming versatility to the user.
2. To reduce the number of bits in the addressing field of the instruction.

To understand the various addressing modes to be presented in this section, it is imperative that we understand the basic operation cycle of the computer. The control unit of a computer is designed to go through an instruction cycle that is divided into three major phases:

1. Fetch the instruction from memory.
2. Decode the instruction.
3. Execute the instruction.

Program counter (PC) There is one register in the computer called the program counter or PC that keeps track of the instruction in the program stored in memory. PC holds the address of the instruction of be executed next and is incremented each time an instruction is fetched from memory. The decoding done in step 2 determines the operation to be performed, the addressing mode of the instruction, and the location of the operands. The computer then executes the instruction and returns to step 1 to fetch the next instruction in sequence.

In some computers the addressing mode of the instruction is specified with a district binary code, just like the operation code is specified. Other computers use a single binary code that designates both the operation and the mode of the instruction, Instruction may be defined with a variety of addressing modes, and sometimes, two or more addressing modes are combined in one instruction.

An example of an instructions format with a addressing mode field is shown in Fig. 2.2. The operation code specifies the operation to be performed. The mode field is used to locate the operands needed for the operation. There may or may not be an address field in the instruction. If there is an address field, it may designate a memory address or a processor register. Although most addressing modes modify the address field of the instruction, there are two modes that need no address field at all. These are the implied and immediate modes.

### 1. *Implied Mode*

In this mode, the operands are specified implicitly in the definition of the instruction. For example, The instruction "increment accumulator" is an implied-mode instruction because the operand in the accumulator register is implied in the definition of the instruction. In fact, all register reference instructions that use an accumulator are implied-mode instructions.

| Opcode | Mode | Address |
|---|---|---|

**Figure 2.2:** Instruction format with mode field.

Zero-address instructions in a stack-organized computer are implied-mode instruction since the operands are implied to be to be on top of the stack.

### 2. *Immediate Mode*

In this mode, the operand is specified in the instruction itself. In other words, an immediate-mode instruction has an operand field rather than an address field. The operand field contains the actual operand to be used in conjunction with the operation specified in the instruction. Immediate-mode instructions are useful for initializing registers to a constant value.

### 3. *Register Mode*

In this mode the operands are in registers that reside within the CPU. The particular register is selected from a register field in the instruction. A k-bit field can specify any one of $2^k$ register.

### 4. *Register Indirect Mode*

In this mode the instruction specifies a register in the CPU whose contents give the address of the operand in memory. In other words, the selected register contains the address of the operand rather than the operand itself. Before using a register indirect mode instruction, the programmer must ensure that the memory address of the operand is placed in the processor register with a previous instruction. A reference to the register is then equivalent to specifying a memory address.

The advantage of a register indirect mode instruction is that the address field of the instruction uses fever bits to select a register than would have been required to specify a memory address directly.

### 5. *Autoincrement or Autodecrement Mode*

This is similar to the register indirect mode except that the register is incremented or decremented after (or before) its value is used to access memory. When the address stored in the register refers to a table of data in memory. It is necessary to increment or decrement the register after every access to the table. This can be achieved by using the increment or decrement instruction. However, because it is such a common requirement, some computers incorporate a special mode that automatically increments or decrements the content of the register after data access.

The address field of an instruction is used by the control unit in the CPU of obtain the operand from memory. Something the value given in the address field is the address of the operand, but sometimes it is just an address from which the address of the operand is calculated. To differentiate among the various addressing modes it is necessary to distinguish between the address part of the instruction and the effective address used by the control when part of the instruction .(The effective address is defined to be the memory address obtained from the computation dictated by the given addressing mode. The effective address is the address of the obtained in a computational-type instruction. It is the address where control branches in response to a branch-type instruction.

### 6. *Direct Address Mode*

In this mode the effective address is equal to the address part of the instruction. The operand resides in memory and its address is given directly by the address by the address field of the instruction. In a branch-type instruction the address field specifies the actual branch address.

### 7. *Indirect Address Mode*

In this mode the address field of the instruction gives the address where the effective address is stored in memory. Control fetches the instruction from memory and uses its address part to access memory again to read the effective address.

A few addressing modes require that the address field of the instruction be added to the content of a specifies register in the CPU. The effective address in these modes is obtained from the following computation:

Effective address = address part of instruction + content of CPU register

The CPU register used in the computation may by the program counter, an index register, or a base register. In either case we have a different addressing mode which is used for a different application.

### 8. *Relative Address Mode*

In this mode the content of the program counter is added to the address part of the instruction in order to obtain the effective address. The address part of the instruction is usually a signed number (in 2's complement representation) which can be either positive or negative. When this number is added to the program counter, the result produces an effective address whose position in memory is relative to the address of the next instruction.

To clarify with an example, assume that the program counter contains the number 825 and the address part of the instruction during to fetch phase and the program counter is then incremented to 626. The effective address computation for the relative address mode is 626 + 74 = 700. This is 74 memory locations forward from the address of the next instruction. Relative addressing is often used with branch-type instruction when the branch address is in the area surrounding the instruction word itself. It result in a shorter address field in the instruction format since the relative address can be specified with a smaller number of bits compared to the number of bits required to designate the entire memory address.

### 9. *Indexed Addressing Mode*

In this mode the content of an index register is added to the part of the instruction to obtain the effective address. The index register is a special CPU register that contains an index value. The address field of the instruction defines the beginning address of a data array in memory. Each operand in the array is stored in memory relative to the beginning address. The distance between the beginning address and the address of the operand is the index value stored in the index register. Any operand in the array can be accessed with the same instruction provided that the index register contains the correct index value. The index register can be incremented to facilitate access to consecutive operands. Note that if an index type instruction does not include an address field in its format, the instruction converts to the register indirect mode of operation.

Some computers dedicate one CPU register to function solely as an index register. This register is involved implicitly when the index-mode instruction is used. In computers with many processor register, any one of the CPU registers can contain the index number. In such a case the register must be specified explicitly in a register field within the instruction format.

### *10. Base Register Addressing Mode*

In this mode the content of a register is added to the address part of the instruction to obtain the effective address. This is similar to the indexed addressing mode register. The difference between the two modes is in the way they are used rather than in the way that they are computed. An index register is assumed to hold an index number that is relative to the address part of the instruction. A base register is assumed to hold a base address and the address field of the instruction gives a displacement relative to this base address.

The base register addressing mode is used in computers to facilitate the relocation of programs in memory. When programs and data are moved from one segment of memory to another, as required in multiprogramming systems, the address values of instructions must reflect this change of position. With a base register, the displacement values of instructions do not have change. Only the value of the base register requires updating to reflect the beginning of a new memory segment.

## 2.4 INTERRUPTS

An interrupt is a condition that causes the microprocessor to temporarily work on a different task, and then later return to it's previous task. Interrupts can be internal or external interrupts. Internal interrupts, or "software interrupts" are triggered by a software instruction, and operate similarly to a jump or branch instruction. An external interrupt, or a "hardware interrupt" is caused by an external hardware module. As an example, many computer systems use interrupt driven I/O, a process where pressing a key on the keyboard, or clicking a button on the mouse triggers an interrupt. The processor stops what it is doing, it reads the input from the keyboard or mouse, and then it returns to the current program.

The image below shows conceptually how an interrupt happens:

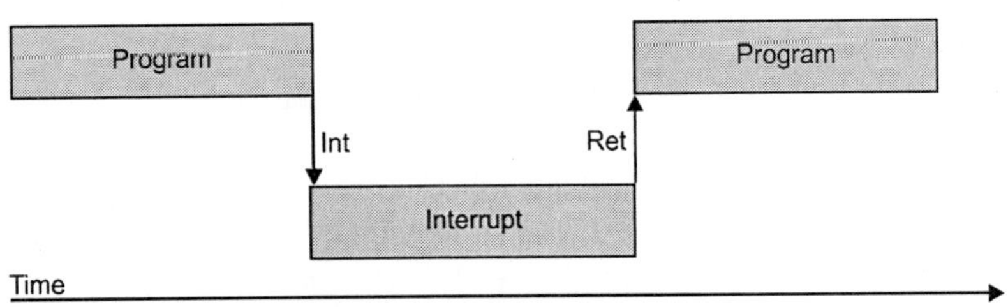

The grey bars represent the control flow. The top line is the program that is currently running, and the bottom bar is the interrupt service routine (ISR). Notice that when the interrupt (Int) occurs, the program stops executing, and the microcontroller begins to execute the ISR. Once the ISR is complete, the microcontroller returns to processing the program where it left off.

What happens when external hardware requests another interrupt while the processor is already in the middle of executing the ISR for a previous interrupt request?

When the first interrupt was requested, hardware in the processor causes it to finish the current instruction, disable further interrupts, and jump to the interrupt handler.

The processor ignores further interrupts until it gets to the part of the interrupt handler that has the "return from interrupt" instruction, which re-enables interrupts.

Some processors, if an interrupt occurs while interrupts were turned off, will immediately jump to that interrupt handler as soon as interrupts are turned back on. With this sort of processor, an interrupt storm "starves" the main loop background task. Other processors execute at least one instruction of the main loop before handling the interrupt, so the main loop may execute extremely slowly, but at least it never "starves"

### 2.4.1 Inerrupt as Hadware Event

An interrupt is an event in hardware that triggers the processor to jump from its current program counter to a specific point in the code. Interrupts are designed to be special events whose occurrence cannot be predicted precisely (or at all). The MSP has many different kinds of events that can trigger interrupts, and for each one the processor will send the execution to a unique, specific point in memory. Each interrupt is assigned a word long segment at the upper end of memory. This is enough memory for a jump to the location in memory where the interrupt will actually be handled.

Interrupts in general can be divided into two kinds- **maskable and non-maskable.** A **maskable** interrupt is an interrupt whose trigger event is not always important, so the programmer can decide that the event should not cause the program to jump. A **non-maskable** interrupt (like the reset button) is so important that it should never be ignored. The processor will always jump to this interrupt when it happens. Often, maskable interrupts are turned off by default to simplify the default behavior of the device. Special control registers allow non-maskable and specific non-maskable interrupts to be turned on. Interrupts generally have a "priority;" when two interrupts happen at the same time, the higher priority interrupt will take precedence over the lower priority one. Thus if a peripheral timer goes off at the same time as the reset button is pushed, the processor will ignore the peripheral timer because the reset is more important (higher priority).

The function that is called or the particular assembly code that is executed when the interrupt happens is called the Interrupt Service Routine (ISR). Other terms of note are: An interrupt flag (IFG) this is the bit that is set that triggers the interrupt, leaving the interrupt resets this flag to the normal state. An interrupt enable (IE) is the control bit that tells the processor that a particular maskable interrupt should or should not be ignored. There is usually one such bit per interrupt, and they are often found together in a register with other interrupt enable bits. The most important interrupt on MSP430 is the reset interrupt. When the processor detects a reset or powers up for the first time, it jumps to the beginning of memory and executes the instructions there. The highest priority interrupt vector begins at the address 0xfffe. The lowest priority interrupt begins at 0xFFE0. The complete set of interrupts is ranked by priority:

15 non-maskable: External reset, power up, watchdog timer reset, invalid flash memory activation

14 non-maskable: oscillator fault, flash memory access violation, NMI

13 maskable: timer B capture compare register 0

12 maskable: timer B capture compare registers 1-6, timer B interrupt

11 maskable: comparator A interrupt

10 maskable: watchdog timer interrupt

9 maskable: USART0 receive interrupt, I2C interrupt

8 maskable: USART0 transmit interrupt

7 maskable: A/D converter interrupt

6 maskable: timer A capture compare register 0 interrupt

5 maskable: timer A capture compare registers 1-2 interrupt

4 maskable: port 1 interrupts

3 maskable: USART1 receive interrupt

2 maskable: USART1 transmit interrupt

1 maskable: port 2 interrupts

0 maskable: D/A converter interrupt

When the interrupt first occurs on the MSP there is a precise order of events that will occur. This process takes 6 instruction cycles to occur.

1. The current instruction completes.
2. The program counter as it is after the above instruction is pushed onto the stack. The stack is memory whose contents are kept in last in first out order. The stack pointer is always updated to point to the most recent element added to the stack. This allows the processor to call functions and track interrupts. When something is pushed onto the stack, the stack pointer is incremented and the pushed data is written to that location. When you copy out of the stack and decrement the stack pointer, this is called popping something off the stack.
3. The status register is pushed onto the stack.
4. The highest priority interrupt waiting to occur is selected.
5. Single source interrupts have their interrupt request flags reset automatically. Multiple source interrupt flags do not do this so that the interrupt service routine can determine what the precise cause was.
6. The status register with the exception of the SCG0 bit is cleared. This will bring the processor out of any low-power modes. This also disables interrupts (the GIE bit) during the interrupt.

7. The content of the interrupt vector is loaded into the program counter. Specifically the processor executes the instruction at the particular memory location (the interrupt vector) for that interrupt. This should always be a jump to the interrupt service routine.

The interrupt service routine is the code that the programmer writes to take care of the work that needs to be done when a particular interrupt happens. This can be anything you need it to be. Because entering the interrupt turned off the GIE bit, you will not receive any interrupts that happen while you are still in the interrupt service routine. You can turn the interrupts back on if you need to receive interrupts during your interrupt, but usually it is a better idea to make interrupt service routines shorter instead. In C interrupts are simply functions with special declarations. You never call these functions; the compiler simply sets up the interrupt vector table to call your function when the particular interrupt occurs.

### 2.4.2 Inerrupt as Software Event

The normal program execution of a microprocessor can be interrupted by some external device or by inserting a special instruction in a program. An interrupt caused by an external device is called hardware interrupt. A microprocessor has interrupt input lines to receive interrupt signals. When an external device wants to interrupt the microprocessor, it sends an interrupt signal to the microprocessor through the interrupt line to which it is connected. Slow I/O devices are connected to a microprocessor through an interrupt line.

When data are ready, an input device interrupts CPU. After completing the current instruction at hand the CPU responds to the input device. The CPU executes a subroutine to transfer data from the interrupting device. This subroutine is called interrupt service subroutine (ISS). When data transfer is completed the CPU returns beck to the main program which it was executing before the interrupt occurred.

When the normal program execution of a microprocessor is interrupted by a special instruction, it is called **software interrupt.**

1. **Enabling, Disabling and Masking of Interrupts:** When interrupts are to be used they are enabled (made effective) by software. This is known as enabling of an interrupt.

   When a microprocessor is executing certain important program or subroutine, it may be desired that interrupt should not occur during that period. Then can be done by disabling (made infective) interrupts. An interrupt can be disabled by the software.

   In certain situations it may be desired that some of the several interrupts should not be entertained, and hence they are masked, I.e. made ineffective by software. On this basis the interrupts are classified into two categories, maskable interrupts and nonmaskable interrupts.

2. **Exceptions:** The interruption in normal program execution of a microprocessor by abnormal internal conditions is called an exception. On the other hand, an interruption of microprocessor caused by external events is known as an interrupt. Software interrupts caused by INT N instruction are also treated as exceptions by the microprocessor. Exceptions include faults, traps and aborts.

3. **Faults:** Program faults come under this category of exception. They are detected and corrected before the execution of the faulty instruction. Examples of faults are: attempts to executes illegal or undefined instructions, access nonexistent or protected memory, etc.
4. **Traps:** The exceptions detected immediately after the execution of an instruction that caused the problem are called traps. Examples of traps are: exceptional conditions such as division by zero, overflow, user defined interrupts, etc.
5. **Aborts:** Abnormal conditions caused by hardware faults are called aborts.

Definitions of interrupts and exceptions differ from book to book. Sometimes all sorts of internal abnormal conditions and external events are dealt under the single heading of interrupts or sometimes under exceptions.

## 2.5 DATA TRANSFER SCHEMES

Binary information received from an external device is stored in memory for further processing. Information transferred from the central computer into an external device originates in the memory unit. The CPU merely executes the I/O instructions and may accept the data temporarily, but the ultimate source or destination is the memory unit.

Data transfer between the central computer and I/O devices may be handled in a variety of modes. Some modes use the CPU as an intermediate path; others transfer the data directly to and from the memory unit. Data transfer to and from peripherals may be handled in three possible modes:

1. Programmed I/O
2. Interrupt-initiated I/O
3. Direct memory access (DMA)

### 2.5.1 Programmed I/O

Programmed I/O operation are the result of I/O instruction written in the computer program. Each data item transfer is initiated by an instruction in the program. Usually, the transfer is to and from a CPU register and peripheral. Other instructions are needed to transfer the data to and from CPU and memory. Transferring data under program control requires constant monitoring of the peripheral by the CPU. Once a data transfer is initiated, the CPU is required to monitor the interface to see when a transfer can again be made. It is up to the programmed instructions executed in the CPU to keep close tabs on everything that is taking place in the interface unit and the I/O device.

#### *Example of Programmed I/O*

In the programmed I/O method, the I/O device does not have direct access to memory. A transfer from an I/O device to memory requires the execution of several instructions by the CPU, including an input instruction to transfer the data from the device to the CPU and a store instruction to transfer the data from the CPU to memory. Other instructions may be needed to verify that the data are available from the device and to count the numbers of words transferred.

An example of data transfer from an I/O device through an interface into the CPU is shown in Figure 2.3. The device transfers bytes of data one at a time as they are available. When a byte of data is available, the device places it in the I/O bus and enables its data valid line. The interface accepts the byte into its data register and enables the data accepted line. The interface sets a bit in the status register that we will refer to as an F or "flag" bit. The device can now disable the data valid line, but it will not transfer another byte until the data accepted line is disabled by the interface.

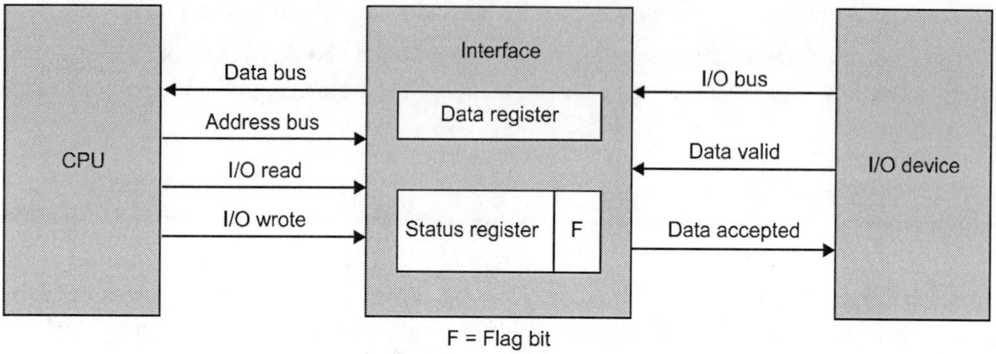


**Figure 2.3:** Data transfer from I/O device to CPU

A program is written for the compute to check the flag in the status register to determine if a byte has been placed in the data register by the I/O device. This is done by reading the status register into a CPU register and checking the value the flag bit. If the flag is equal to 1, the CPU reads the data from the data register. The flag bit is then cleared to 0 by either the CPU or the interface, depending on how the interface circuits are designed. Once the flag is cleared, the interface disables the data accepted line and the device can then transfer the next data byte.

A flowchart of the program that must be written for the CPU is shown in Figure 2.4. It is assumed that the device is sending a sequence of bytes that must be stores in memory. The transfer of each byte requires three instructions:

1. Read the status register.
2. Check the status of the flag bit and branch to step 1 if not set or to step 3 if set.
3. Read the data register.

Each byte is read into a CPU register and then transferred to memory with a store instruction. A common I/O programming task is to transfer a block of words from an I/O device and store them in a memory buffer.

The programmed I/O method is particularly useful in small low-speed computers or in systems that are dedicated to monitor a device continuously. The difference in information transfer rate between the CPU and the I/O device makes this type of transfer inefficient. To see why this is inefficient, consider a typical computer that can execute the two instructions that read the status register and check the flag in 1 ms. Assume hat the input device transfers its data at an average rate of 100 bytes per second. This is equivalent to one byte every 10,000 ms. This means that the CPU is wasting time while checking the flag instead of doing some other useful processing task.

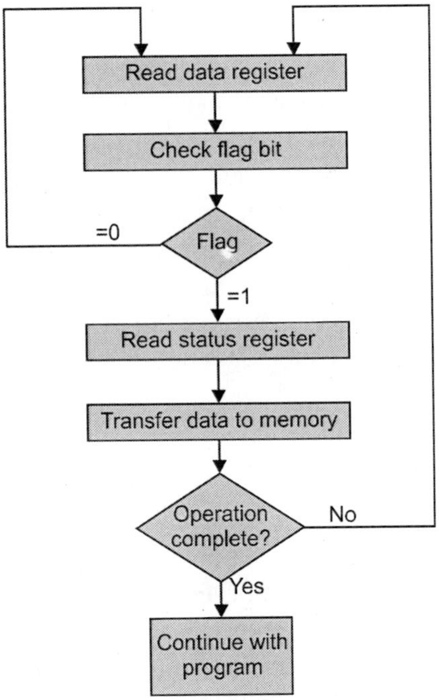


**Figure 2.4:** Flowchart for CPU program to input data

### 2.5.2 Interrupt-initiated I/O

In the programmed I/O method, the CPU stays in a program loop until the I/O unit indicates that it is ready for data transfer. This is a time-consuming process since it keeps the processor busy needlessly. It can be avoided by using an interrupt facility and special commands to inform the interface to issue an interrupt request signal when the data are available from the device. In the meantime the CPU can proceed to execute another program. The interface meanwhile keeps monitoring the device. When the interface determines that the device is ready for data transfer, it generates an interrupt request to the computer. Upon detecting the external interrupt signal, the CPU momentarily stops the task it is processing, branches to a service program to process the I/O transfer, and then returns to the task it was originally performing.

An alternative to the CPU constantly monitoring the flag is to let the interface inform the computer when it is ready to transfer data. This mode of transfer uses the interrupt facility. While the CPU is running a program, it does not check the flag. However, when the flag is set, the computer is momentarily interrupted from proceeding with the current program and is informed of the fact that the flag has been set. The CPU deviates from what it is doing to take care of the input or output transfer. After the transfer is completed, the computer returns to the previous program to continue what it was doing before the interrupt.

### *Vectored Interrupt*

The CPU responds to the interrupt signal by storing the return address from the program counter into a memory stack and then control branches to a service routine that processes the required I/O transfer. The way that the processor chooses the branch address of the service routine varies from one unit to another. In principle, there are two methods for accomplishing this. One is called vectored interrupt and the other, nonvectored interrupt. In a nonvectored interrupt, the branch address is assigned to a fixed location in memory. In a vectored interrupt, the source that interrupts supplies the branch information to the computer. This information is called the interrupt vector. In some computers the interrupt vector is the first address of the I/O service routine. In other computers the interrupt vector is an address that points to a location in memory where the beginning address of the I/O service routine is stored.

## 2.5.3 Direct Memory Access (DMA)

Transfer of data under programmed I/O is between CPU and peripheral. In direct memory access (DMA) the interface transfers data into and out of the memory unit through the memory bus. The CPU initiates the transfer by supplying the interface with the starting address and the number of words needed to be transferred and then proceeds to execute other tasks. When the transfer is made, the DMA request memory cycles through the memory bus. When the request is granted by the memory controller, the DMA transfers the data directly into memory. The CPU merely delays its memory access operation to allow the direct memory I/O transfer. Sicne peripheral speed is usually slower that processor speed, I/O-memory transfers are infrequent compared to processor access to memory.

The transfer of data between a fast storage device such as magnetic disk and memory is often limited by the speed of the CPU. Removing the CPU from the path and letting the peripheral device manage the memory buses directly would improve the speed of transfer. This transfer technique is called direct memory access (DMA). During DMA transfer, the CPU idle and has no control of the memory buses. A DMA controller takes over the buses to manage the transfer directly between the I/O device and memory.

### *Bus Request*

The CPU may be placed in an idle state in a variety of ways. One common method extensively used in microprocessors is t disable the buses through special control signals. Figure 2.5 shows two control signals in the CPU that facilitate the DMA transfer. The bus request (BR) input is used by the DMA controller to request the CPU to relinquish control of the buses. When this input is active, the CPU terminates the execution of the current instruction and places the address bus, the data bus, and the read and write lines into a high-impedance state. The high-impedance state behaves like an open circuit, which means that the output is disconnected and does not have a logic significance. The CPU activates the bus grant (BG) output to inform the external DMA that the buses are in the high-impedance state. The DMA that originated the bus request can now take control of the buses to conduct memory transfers without processor intervention. When the DMA terminates the transfer, it disables the bus request line. The CPU disables the bus grant, takes control of the buses, and returns to its normal operation.

### Burst Transfer

When the DMA takes control of the bus system, it communicates directly with the memory. The transfer can be made in several ways. In DMA burst transfer, a block sequence consisting of a number of memory words is transferred in a continuous burst while the DMA controller is master of the memory buses. This mode of transfer is needed for fast devices such as magnetic disks, where data transmission cannot be stopped or slowed down until an entire block is transferred. An alternative technique called cycle stealing allows the DMA control of the buses to the CPU. The CPU merely delays its operation for one memory cycle to allow the direct memory I/O transfer to "steal" one memory cycle.

### DMA Controller

The DMA controller needs the usual circuits of an interface to communicate with the CPU and I/O device. In addition, it needs an address register, a word count register, and a set of address lines. The address register and address lines are used for direct communication with the memory. The word count register specifies the number of words that must be transferred. The data transfer may be done directly between the device and memory under control of the DMA.

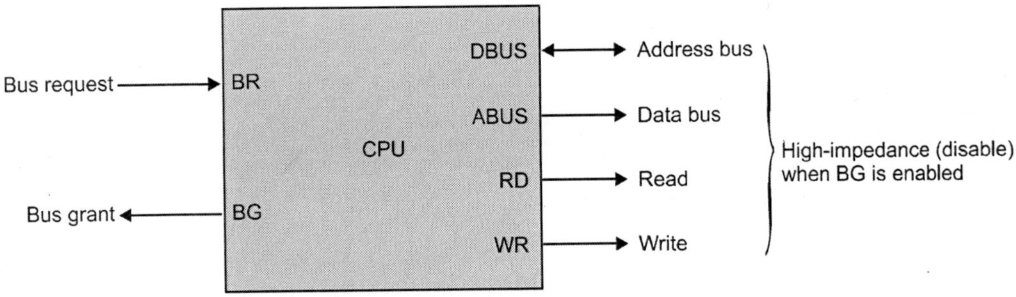


**Figure 2.5:** CPU bus signals for DMA transfer.

Figure 2.6 shows the block diagram of a typical DMA controller. The unit communicates with the CPU via the data bus and control lines. The registers in the DMA are selected by the CPU through the address bus by enabling the DS (DMA select) and RS (register select) inputs. The RD (read) and WR (write) inputs are bidirectional. When the BG (bus grant) input is 0, the CPU can communicate with the DMA registers through the data bus to read from or write to the DMA registers. When BG = 1, the CPU has relinquished the buses and the DMA can communicate directly with the memory by specifying an address in the address bus and activating the RD or WR control. The DMA communicates with the external peripheral through the request and acknowledge lines by using a prescribed handshaking procedure.

The DMA controller has three registers: an address register, a word count register, and a control register. The address register contains an address to specify the desired location in memory.

The address bits go through bus buffers into the address bus. The address register is incremented after each word that is transferred t memory. The word count register holds the number of word to be transferred. This register is decremented by one after each word transfer and internally tested for zero. The control register specifies the mode of transfer. All registers in the DMA appear to the CPU as I/O interface registers. Thus the CPU can read from or write into the DMA registers under program control via the data bus.

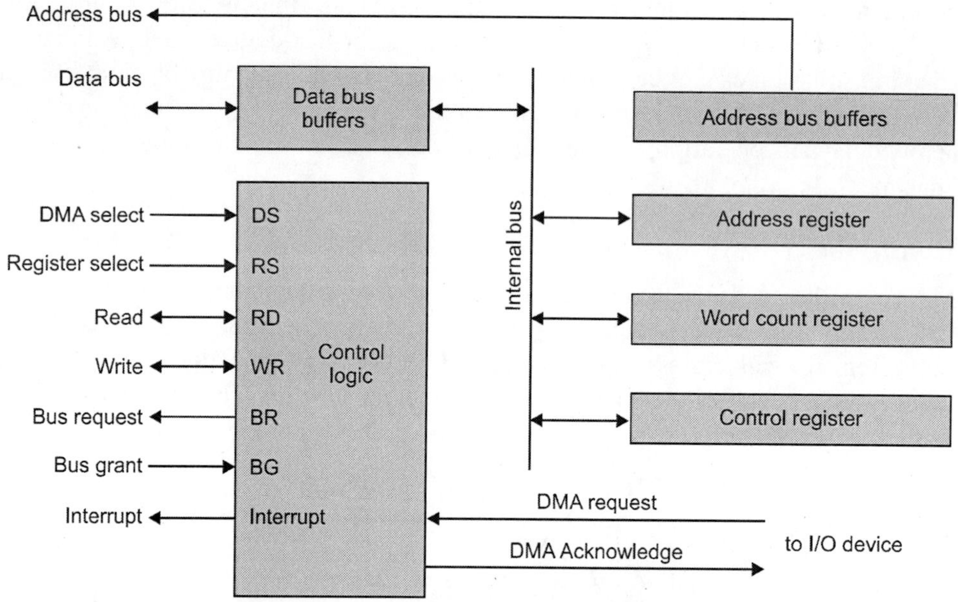


**Figure 2.6:** Block diagram of DMA controller.

The DMA is first initialized by the CPU. After that, the DMA starts and continues to transfer data between memory and peripheral unit until an entire block is transferred. The initialization process is essentially a program consisting of I/O instructions that include the address for selecting particular DMA registers. The CPU initializes the DMA by sending the following information through the data bus:

1. The starting address of the memory block where data are available (for read) or where data are to be stored (for write).
2. The word count, which is the number of words in the memory block.
3. Control to specify the mode of transfer such as read or write.
4. A control to start the DMA transfer.

The starting address is stored in the address register. The word count is stored in the word count register, and the control information in the control register. Once the DMA is initialized, the CPU stops communicating with the DMA unless it receives an interrupt signal or if it wants to check how many words have been transferred.

### *DMA Transfer*

The position of the DMA controller among the other components in a computer system is illustrated in Figure 2.7. The CPU communicates with the DMA through the address and data buses as with any interface unit. The DMA has its own address, which activates the DS and RS lines. The CPU initializes the DMA through the data bus. Once the DMA receives the start control command, it can start the transfer between the peripheral device and the memory.

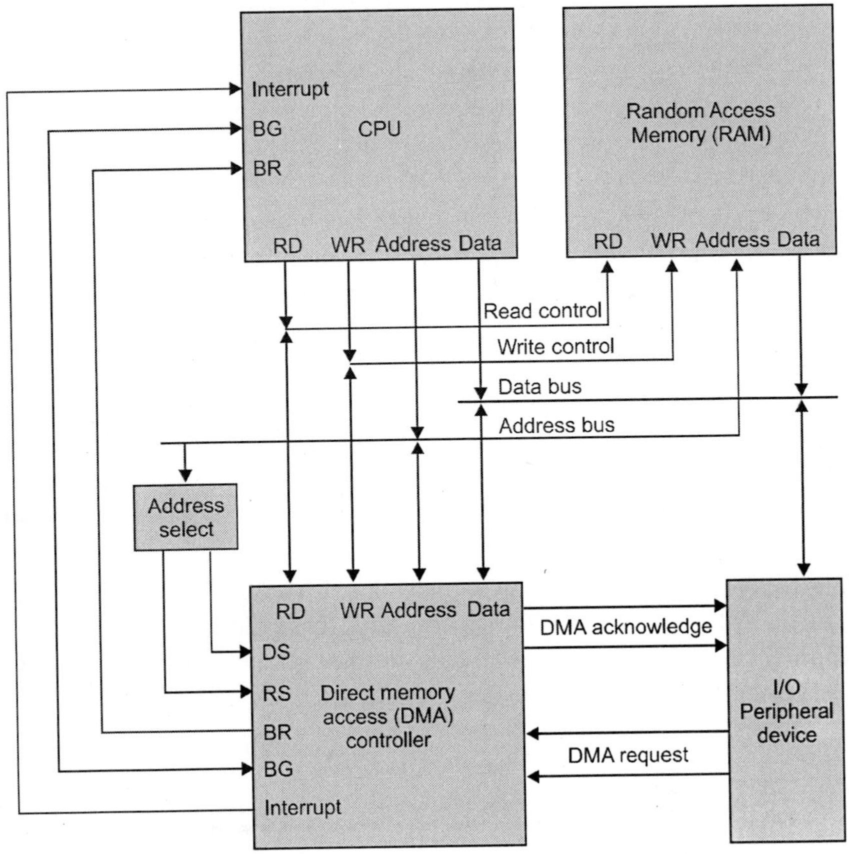


**Figure 2.7:** DMA transfer in a computer system.

When the peripheral device sends a DMA request, the DMA controller activates the BR line, informing the CPU to relinquish the buses. The CPU responds with its BG line, informing the DMA that its buses are disabled. The DMA then puts the current value of its address register into the address bus, initiates the RD or WR signal, and sends a DMA acknowledge to the peripheral device. Note that the RD and WR lines in the DMA controller are bidirectional. The direction of transfer depends on the status of the BG line. When BG = 0, the RD and WR are input lines allowing the CPU to communicate with the internal DMA registers. When BG = 1, the RD and WR are output lines from the DMA registers. When BG = 1, the RD and WR are output lines from the DMA controller to the random-access memory to specify the read or write operation for the data.

When the peripheral device receives a DMA acknowledge, it puts a word in the data bus (for write) or receives a word from the data bus (for read). Thus the DMA controls the read or write operations and supplies the address for the memory. The peripheral unit can then communicate with memory through the data bus for direct transfer between the two while the CPU is momentarily disabled.

For each word that is transferred, the DMA increments its address register and decrements its word count register. If the word count does not reach zero, the DMA checks the request line coming from the peripheral. For a high-speed device, the line will be active as soon as the previous transfer is completed. A second transfer is then initiated, and the process continues until the entire block is transferred. If the peripheral speed is slower, the DMA request line may come somewhat later. In this case the DMA disables the bus request line so that CPU an continue to execute its program. When the peripheral requests a transfer, the DMA request the buses again.

If the word count, register reaches zero, the DMA stops any further transfer and removes its bus request. It also informs the CPU of the termination by means of an interrupt. When the CPU responds to the interrupt, it reads the content of the word count register. The zero value of this register indicates that all words were transferred successfully. The CPU can read this register at any time to check the number of words already transferred.

A DMA controller may have more that one channel. In this case, each channel has a request and acknowledge pair of control signals which are connected to separate peripheral devices. Each channel also has its own address register and word count register within the DMA controller. A priority among the channels may be established so that channels with high priority are serviced before channels with lower priority.

DMA transfer is very useful in many applications. It is used for fast transfer of information between magnetic disks and memory. It is also useful for updating the display in an interactive terminal. Typically, an image of the screen display of the terminal is kept in memory which can be updated under program control. The contents of the memory can be transferred to the screen periodically by means of DMA transfer.

## 2.6 INSTRUCTION AND DATA FLOW

A microprocessor acts as the central processing unit (CPU) in a microcom-puter system. It is generally a single chip LSI circuit. It contains several registers, a timing and control unit and data paths. The registers are used during instruction execution. The timing and control unit is responsible for generating timing signals that are necessary for the proper execution of in-structions and the control of peripheral devices. Data and instructions are transferred within the microprocessor and peripheral devices on the data paths.

A variety of register types and organizations exists amongst various mi-croprocessors. The timing used for instruction execution and peripheral control and the data paths also differ from one microprocessor to another While selecting a microprocessor for a particular application, a design en-gineer faces the tough task of identifying one amongst a large number of microprocessors with differing organizations. An appreciation of these organizations and a knowledge of their functions will aid the engineer try microprocessor selection.

Knowledge of timing and control signals generated by a microprocessor is equally important for the evaluation of a microprocessor- The timing structure and register organization of a microprocessor affect the system design and its cost. A poorly chosen microprocessor may result in a costly and inefficient design.

### 2.6.1 Timing and Control Unit

The primary task of any microprocessor is to execute programs. These pro-grams consist of a sequence of instructions stored in the memory interfaced to the microprocessor. As soon as the microprocessor is turned on and the input voltage Stabilizes, it enters a loop that may best be described by the following algorithm:

```
while power-is-on do
begin
fetch the next instruction;
execute this instruction;                    (2.1)
end
```

As is evident from (2.1), the two tasks of fetching and executing in-structions are repetitively executed by the microprocessor until the power is switched off. It should be noted that, as described in (2.1), these two tasks are strictly sequential in nature— only one instruction is dealt with during a certain time slot. Many microprocessors follow (2.1) though some do not. An example of the latter family is Intel's 8086.

Each of the two operations— fetch and execute— is performed in time slots of fixed or varying length. The fetch operation is performed in a time' slot of fixed length (say, FC) whereas the execute operation may require variable length (say, EC) time slots, the length being dependent on the instruction to be executed. The total time required to perform these two operations once, that is to go around the loop in (2.1) once, is known as the instruction cycle time (X). As is evident, it is given as:

$$IC = FC + EC$$

The fetch and execute operations are carried out in time slots also known as the fetch cycle and the execute cycle. It is from these names that we have derived the abbreviations Fe and EC, respectively. Figure 2.8 illustrates these notions.

An examination of Figure 2.8 reveals that the fetch and execute operations are carried out in synchronism with a clock. Furthermore, several dock cycles elapse before a fetch or execute operation is completed. This is typical of all microprocessors. The clock itself may be generated internally or supplied from an external source.

### 2.6.2 The Fetch Operation

The fetch operation causes an instruction byte containing the operation code, to be brought into the microprocessor from the memory. Note that the instruction itself may be more than 1 byte long, the other byte con-taining data or operand address. If the operation code is two nibbles long (1 nibble = 4 bits), which may be the case for 4-bit microprocessors, then too, according to our convention, the fetch operation fetches the complete operation code in one fetch cycle.

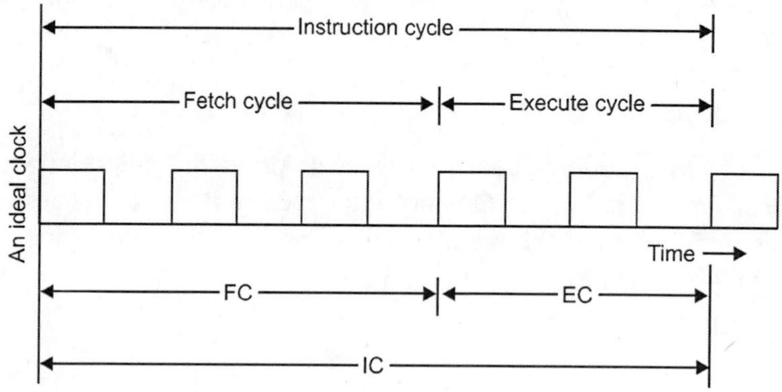


**Figure 2.8:** Fetch, execute and instruction cycles.

The fetch operation itself may be -described by the following simple algorithm.

begin
send the address of the next
instruction to memory; (2.2)
receive the instruction from the memory;
end

We already know that a program counter register (PC) within the micro-processor keeps track of the address of the next instruction to be executed. Thus, at the beginning of the fetch cycle, the contents of PC are sent to the memory. The memory responds by reading the contents of the received address and transferring them to the microprocessor. Figure 2.9 exhibits the fetch operation. Note that it requires one clock cycle for transferring the address. In the next two clock cycles, two operations are performed. The first of these is the reading of the instruction by the memory from the location whose address it has received in the previous clock cycle. The second of these is the transfer of the instruction from the memory to the microprocessor. However, this distinction may not be so precise in a typical microprocessor,

It is important to note in Figure 2.9 that only two clock cycles have been used to read the instruction. However, the access time of the memory may be more than two clock cycles. In such a microcomputer, the micropro-cessor has to wait for more than two clock cycles before it can receive the instruction operation code. The extra clock cycles are known as wait clocks and most microprocessors available today have provisions for introducing these wait cycles within a fetch cycle in order to cope with slow memories.

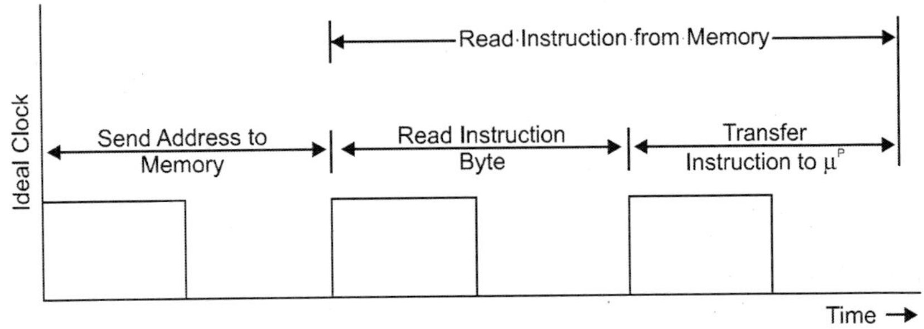


**Figure 2.9:** A typical fetch cycle.

### 2.6.3 The Execute Cycle

A simple algorithm describing the execute cycle is given below:

begin
decode the instruction fetched;
if operand in memory then (2.3)
fetch operand;
execute instruction;
end

Figure 2.10 exhibits a typical execute cycle. As is evident from (2.3), the time required for an execute cycle will depend on the instruction type. If the instruction is only one byte long, and the operands are in the general purpose registers, then only the decoding and execute operations are per-formed, Both these are, in many microprocessors, carried out in just one dock cycle. However, if the instruction contains data or address of data, then one or two extra memory retch cycles are required to fetch the operand and operand address. These fetch cycles are similar to the instruction fetch cycle except that the quantity fetched is an address or data.

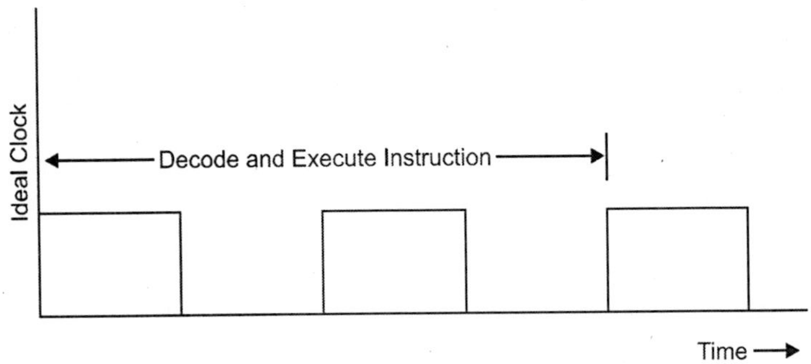


**Figure 2.10:** Decoding and execution

### 2.6.4 Machine Cycle and State

We have seen that an instruction cycle consists of an opcode fetch followed by an execute cycle. The execute cycle itself consists of zero or more fetch cycles which may be required to fetch the operand. All these operations are performed in different time slots known as machine cycles. An instruction cycle may consist of one or more machine cycles. The first of these is always the opcode fetch cycle. Some instructions, which require only register to register transfer inside the microprocessor, may be executed in one machine :yeti, whereas some which require data transfer between microprocessor and memory or an I/O device, may need more than one machine cycle.

One machine cycle itself consists of several clock cycles. Each of these clock cycles is known as a stale. The states within a machine cycle are generally referred to as T1, T2, ..... Figure 2.11 exhibits a typical instruction cycle and the machine cycles within it.

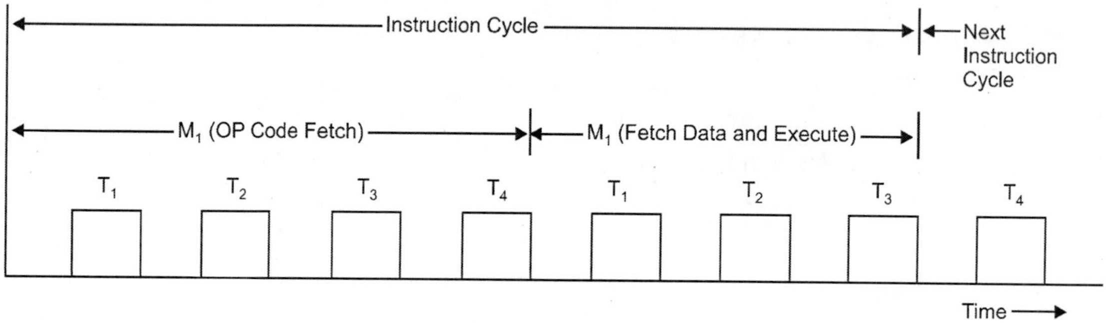


**Figure 2.11:** A typical instruction cycle

It shows two machine cycles MI and M2. MIA which the operation code is fetched and decoded, consists of four states. In M2, which consists of three states, the data contained in the next byte of the instruction is fetched and the instruction executed.

### 2.6.5 Address and Data Transfer

During a fetch cycle, an operation code, an operand or the operand address is fetched. Figure 2.12(b) exhibits the flow of instruction bytes. As mentioned earlier, at the beginning of the fetch cycle, the microprocessor transfers the next instruction address, stored in the PC, to a special register known as the memory address register (MAR) or simply address register (AR) from which it is transferred to the memory unit on the address bus. This process may require one or more clock cycles depending on the width of the address bus and the address size.

The decoder circuits within the memory are activated and the instruc-tion code from the addressed memory byte is accessed. This code is then transferred over the data bus into the memory buffer register (MBR) or data register (DR) within the microprocessor- As the data which is now in MBR, is an instruction, it is transferred to the instruction register which decodes the instruction and causes suitable timing and control signals to be generated with the aid of the timing and control unit.

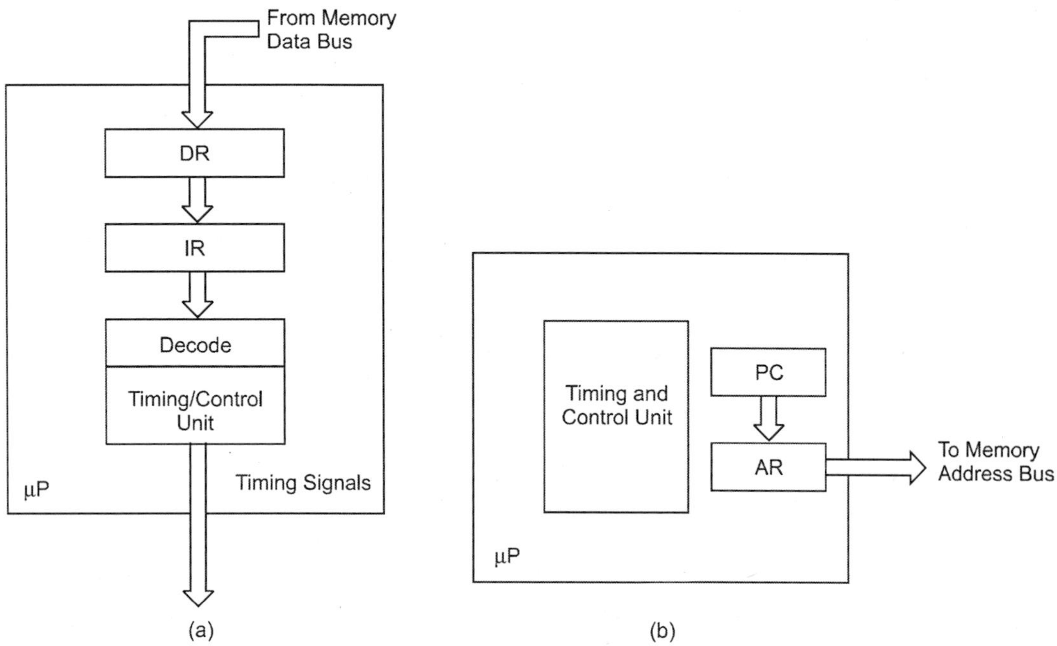


**Figure 2.12:** Address and data transfer to memory during an instruction cycle: (a) Address transfer (b) Data transfer

## 2.7 TIMER AND TIMING DIAGRAMS

### 2.7.1 Timers

It is common for microprocessor to have one or more timers. A timer is essentially just a counter that counts the number of microprocessor clock cycles and then causes an interrupt when the count expires. Here are a few features of the usual microprocessor timers:

- A pre—scaler divides the microprocessor dock signal by some constant, perhaps 20. before the signal gets to the timer,
- The counter can reset itself to its initial value when it expires and then continue to count, so that it can be the source of a regular, periodic interrupt.
- The timer can drive an output pin on the microprocessor, either causing a pulse whenever the timer expires or creating a square wave with an edge at every timer expiration.
- The timer has an input pin that enables or disable-5 counting, The timer circuit also nifty be able to function as a counter that alums pulses on dim input pin.

Most timers are set up by writing values into a small collection of registers, typically registers to hold the count a register with a collection of bits to enable the counter, to reset the interrupt, to control what the timer does to its output pin, if any, and so on.

### 2.7.2 Timing Diagrams

Timing diagrams are used to illustrate sequences of events and dependences among events, For the reader unfamiliar with timing diagrams, this appendix provides brief explanation. Communication among devices connected to a bus takes place along a set of lines capable of carrying signal. Two different signal levels (voltage levels), representing binary 0 and binary 1, may be transmitted. A timing diagram shows the signal level on a line as function of time (Figure 2.13a). By convention, the binary 1 signal level is depicted as a higher level than that of binary 0. Usually; binary 0 is the default value. That is, if no data or oilier signal is being transmitted. then the level on a line is that which represents binary 0.A signal transition front 0 to 1 is frequently referred to as the signal's leading, edge; a transition from 1 to 0 is referred to as a trailing edge. Such transitions are not instantaneous, but this transition time is usually small compared with the duration of a signal level.

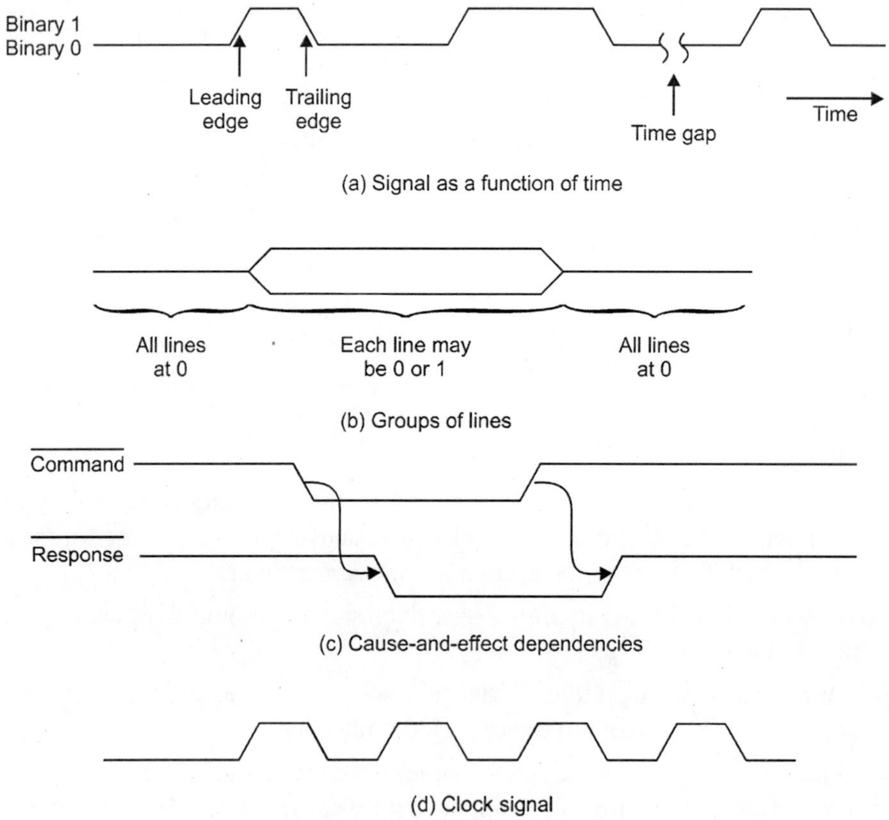


**Figure 2.13:** Timing Diagrams

For clarity, the transition is usually depicted as an angled line that exaggerates the relative amount of lime that the transition takes. Occasionally, you will see diagrams that use vertical lines, which incorrectly suggests that the transition is instantaneous, On a timing diagram., it may happen that a variable or at least irrelevant amount of time elapses between events of interest. This is depicted by a gap in the time line, Signals arc sometimes represented in groups (Figure 2.13b), For example, if data are transferred a byte at a time, then eight lines are required. Generally, it is not important to know the exact value being transferred on such a group, but rather whether signals are present or not.

A signal transition on one line may trigger an attached device to make signal changes on other lines, For example, if a memory module detects a read control signal (0 or 1 transition), it will place data signals on the data lines. Such cause-and--effect relationships produce sequences of events. Arrows arc used on timing diagrams to show these dependencies (Figure 2.13c),

In Figure 2.13c, the overbar over the signal name indicates that the signal is active low as shown. For example, $\overline{\text{Command}}$ is active, or asserted, at 0 volts, This means that $\overline{\text{Command}}$ = 0 is interpreted as logical 1, or true.

A clock line is often part of a system bus. An electronic clock is connected to the clock line and provides a repetitive, regular sequence of transitions (Figure 2.13d), Other events may be synchronized to the clock signal.

## 2.8 INTERFACING DEVICES

Any microprocessor-based system design involves interfacing of the proces-sor with one or more peripheral devices for the purpose or communication with the environment. In the next chapters we have discussed vari-ous interfacing techniques. During the early days of the microprocessor revolution, the implementation of these techniques required a considerable amount of hardware consisting essentially of MSI devices causing the sys-tem design and development to be a time consuming activity.

Over the past several years, manufacturers have developed a large num-ber of general and special purpose peripheral devices, most of them being single chip circuits. These devices prove to be of tremendous value to a sys-tem designer. Their use reduces the chip count of a system and simplifies the design process. They can be programmed to fit a variety of configurations.

The purpose of this chapter is to introduce devices commonly used to simplify the task of interfacing, Owing to space constraints, it is neither possible to describe all types of devices supplied by different manufacturers nor is it feasible to describe all versions of one type of device supplied by different manufacturers. We have therefore carefully chosen a few types of devices that are most likely to be used in system design and described the versions supplied by Intel. We describe specific devices from Intel so that we could illustrate their use in 8085 and 8086/8088 based systems.

Though gates, shift registers, decoders, latches, buffers, etc. also play an important role in interfacing, the devices described in this chapter are much more complex and deserve special study by any microcomputer designer.

### 2.8.1 Types of Interfacing Devices

Interfacing devices may be classified into two categories:

1. General purpose peripherals, and

2. Special purpose peripherals (also known as dedicated function periph-erals).

We use the term peripheral for these devices as well as for the other I/O devices, such as printers and floppy drives, which would be interfaced to the MP through them.

1. **General purpose peripherals** are devices that perform a specific task but may be used for interfacing a variety of I/O devices to the MP.

   Examples of general purpose peripherals are:

   (i) Programmable Peripheral Interface (PP1), also known as Peripheral Interface Adapter (PIA)

   (ii) Programmable Interrupt Controller (PIC)

   (iii) Programmable DMA Controller

   (iv) Programmable Communications Interface

   (v) Programmable Interval Timer.

   **(i) Programmable Peripheral Interface (PP1):** The Intel 8255A is a general purpose **programmable I/O device** designed for use with Intel microprocessors.

   The 8255A is a widely used, programmable, parallel I/O device. It can be programmed to transfer data under various conditions, from simple I/O to interrupt I/O. It is flexible, versatile, and economical (when multiple I/O ports are required), but somewhat complex. It is an important general-purpose I/O device that can be used with almost any microprocessor.

   **(ii) Programmable Interrupt Controller (PIC):** The 8259A is a **programmable interrupt controller** designed to work with Intel microprocessor 8085, 8086 and 8088. The 8259A interrupt controller can

   1. Manage eight interrupts according to the instructions written into its control registers. This is equivalent to providing eight interrupt pins on the processor in place of one INTR (8085) pin.
   2. Vector an interrupt request anywhere in the memory map. However, all eight interrupts are spaced at the interval of either four or eight locations. This eliminates the major drawback of the 8085 interrupts in which all interrupts are vectored to memory locations on page 00H.
   3. Resolve eight levels of interrupt priorities in a variety of modes, such as fully nested mode, automatic rotation mode, and specific rotation mode (to be explained later).
   4. Mask each interrupt request individually.
   5. Read the status of pending interrupts, in-service interrupts, and masked interrupts.
   6. Be set up to accept either the level-triggered or the edge-triggered interrupt request.
   7. Be expanded to 64 priority levels by cascading additional 8259As.
   8. Be set up to work with either the 8085 microprocessor mode or the 8086/8088 microprocessor mode.

(iii) **Programmable DMA Controller** - The 8237 is a **programmable Direct Memory Access controller (DMA)** houses in a 40-pin package. It has four independent channels with each channel capable of transferring 64K bytes. It must interface with two types of devices: the MPU and peripherals such as floppy disks. As mentioned earlier, the DMA plays two roles in a given system: It is an I/O to the microprocessor (slave mode) and it is a data transfer processor to peripherals such a floppy disks (master mode). Many of its signals that are input in the I/O mode become outputs in the processor mode. It also needs additional signal lines to communicate with the addresses of 64K data bytes, and these signals must be generated externally by the using latches and buffers. The 8237 is a complex device. To maintain clarity, the following discussion is divided into five segments: DMA channels and interfacing, DMA signals, system interface, programming DMA execution.

(iv) **Programmable Communications Interface** - The 8251A is a **USART (Universal Synchronous Asynchronous Receiver Transmitter)** for serial data communication. As a peripheral device of a microcomputer system, the 82C51A receives parallel data from the CPU and transmits serial data after conversion. This device also receives serial data from the outside and transmits parallel data to the CPU after conversion. The 82C51A configures a fully static circuit using silicon gate CMOS technology. Therefore, it operates on extremely low power at 100 mA (max) of standby current by suspending all operations.

1. Wide power supply voltage range from 3 V to 6 V
2. Wide temperature range from –40°C to 85°C
3. Synchronous communication upto 64 Kbaud
4. Asynchronous communication upto 38.4 Kbaud
5. Transmitting/receiving operations under double buffered configuration.
6. Error detection (parity, overrun and framing)
7. 28-pin Plastic DIP (DIP28-P-600-2.54): (Product name: MSM82C51A-2RS)
8. 28-pin Plastic QFJ (QFJ28-P-S450-1.27): (Product name: MSM82C51A-2JS)
9. 32-pin Plastic SSOP(SSOP32-P-430-1.00-K): (Product name: MSM82C51A-2GS-K)

The 825IA is a programmable chip designed for synchronous and asynchronous serial data communication, packaged in a 28-pin DIP. The 825I A is the enhanced version of its predecessor, the 8251, and is compatible with the 8251.

(v) **Programmable Interval Timer** - The 8253 is a **programmable Interval timer/counter** specifically designed for use with the Intel Micro computer systems. Its function is that of a general purpose, multi-timing element that can be treated as an array of I/O ports in the system software.

The 8253 solves one of the most common problems any microcomputer system. The generation of accurate time delay under software control. Instead of selling up timing loops in systems software. The programmer configures the 8253 to match his requirements. Initializes one of the counters of the 8253 with the desired quantity, then upon command the 8253 will count-out the delay and interrupt the CPU when it has completed its tasks. It is easy to see that the software overhead is minimal and that multiple delays can easily be maintained by assignment of priority levels.

2. **Special purpose peripherals** are devices that may be used for interfacing a MP to a specific type of I/O device. These peripherals are much more complex and therefore relatively more expensive than the general purpose peripherals.

   Some of the special purpose peripherals are:
   (i) Programmable CRT Controller
   (ii) Programmable Floppy Disk Controller
   (iii) Programmable Hard Disk Controller
   (iv) Programmable Keyboard and Display Interface.

As a meaningful description of these devices would require a description of the I/O devices they control, these peripheral chips are not considered further in this book. Several manufacturer application notes describe the above special purpose peripherals in depth.

All Examples in the subsequent sections assume that any peripheral device connected to the 8085 is addressed using the I/O mapped mode. We begin by describing a simple scheme for I/O address decoding using the 71LS138 decoder.

## QUESTION BANK

Q1. What do you mean by architecture of microprocessor?

Q2. Explain the overview of microcomputer structure and its operations.

Q3. Explain different types of addressing modes in detail.

Q4. What are interrupts? Explain different types of interrupts.

Q5. Discuss the various data transfer schemes.

Q6. What is DMA? Explain in detail description of DMA controller and DMA transfer with the help of diagrams.

Q7. Discuss the working of instruction and data flow in detail.

Q8. Explain the working of timing and control unit.

Q9. What are timers? Also explain the various timing diagrams.

Q10. What are the interfacing devices? Also explain its different types.

CHAPTER 3

# 8085 Microprocessor

## 3.1 MICROPROCESSOR ARCHITECTURE AND ITS OPERATIONS

Microprocessor is a multipurpose, programmable, clock-driven, register based electronic device that reads binary instructions from a storage device called memory, accepts binary data as input and processes data according to those instructions, and provides as output.

Microprocessor is also called programmable digital device because it can be instructed to perform given tasks within its capability. Microprocessor is designed to understand and execute many binary instructions.

"It is a program controlled semiconductor device (IC), which fetches, decodes and executes instructions."

A microprocessor is used to execute a series of steps or instructions, which collectively constitutes a 'program'.Every microprocessor has a unique set of instructions that it can execute.This set of instructions is known as its, well, instruction set. Every instruction on the instruction set does something unique, and has different requirements in terms of which part(s) of the microprocessor to utilize or what data to work on.

**Basic Microprocessor Architecture has the following parts:**

1. An Arithmetic and Logic Unit (ALU), which is where the arithmetic and logic operations of the microprocessor take place.
2. A data bus system where data that need to be processed are transported.
3. An address bus system that provides the address of the memory location being accessed.
4. A control unit for orchestrating the program execution of the microprocessor.
5. An instruction register/decoder where instructions are loaded one at a time and 'interpreted'.
6. A program counter that indicates the memory address where the next instruction will come from.
7. Various registers, flags, and pointers.

A microprocessor executes a program stored in memory by fetching the instructions of the program (and whatever data they require) one at a time and performing these instructions.Memory in this context basically refers to external memory devices that complement the microprocessor and the input/output devices of the computer system. The manner in which the next instruction will be executed depends on the results of the last operation.Thus, the output of the microprocessor depends on the instructions and the input data provided to it.

All the various functions performed by the microprocessor can be classified in three categories:

1. Microprocessor system bus structure
2. Internal Registers
3. External initiated signals

**Four Operations Commonly Performed By Mpu**

1. **Memory Read:** Reads data (or instructions) from memory.
2. **Memory Write:** Writes Data (or instructions) into memory.
2. **I/O Read:** Accepts data from input devices.
3. **I/O Write:** Sends data to output devices.

## 3.2 8085 SYSTEM BUS STRUCTURE

Typical system uses a number of busses, collection of wires, which transmit binary numbers, one bit per wire. A typical microprocessor communicates with memory and other devices (input and output) using three buses: Address Bus, Data Bus and Control Bus.

To communicate with a peripheral, the MPU needs to perform the following steps:

1. Identifying a peripheral or a memory location.
2. Transferring binary information.
3. Providing timing signals.

### 3.2.1 Address Bus

The address bus is a group of 16 lines generally identified as $A_0$ to $A_{15}$. The address bus is unidirectional: bits flow in one direction-from the MPU to peripheral devices. The MPU uses the address bus to perform the first function: identifying a peripheral or a memory location.

In computer system, each peripheral or memory location is identified by a binary number (0 or 1), called address, and the address bus is used to carry a 16-bit address. One wire for each bit, therefore 16 bits = 16 wires. Binary number carried alerts memory to 'open' the designated box. Data (binary) can then be put in or taken out. The Address Bus consists of 16 wires, therefore 16 bits. Its "width" is 16 bits. A 16 bit binary number allows $2^{16}$ different numbers, or 32000 different numbers, i.e. 0000000000000000 up to 1111111111111111. Because memory consists of boxes, each with a unique address, the size of the address bus determines the size of memory, which can be used. To communicate with memory the microprocessor sends an address on the address bus, e.g. 0000000000000011 (3 in decimal), to the memory. The memory the selects box number 3 for reading or writing data. Address bus is unidirectional, i.e. numbers only sent from microprocessor to memory, not other way.

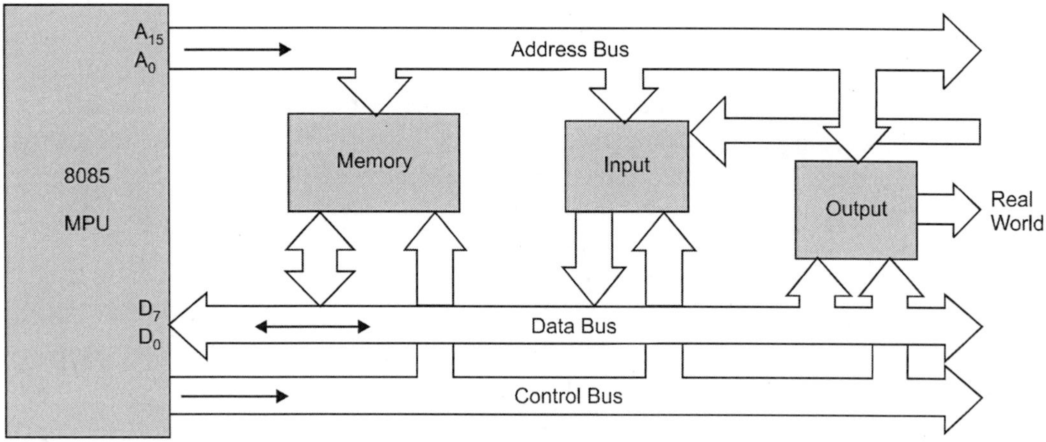


**Figure 3.1:** The 8085 Bus structure

### 3.2.2 Data Bus

Data Bus is used to carries 'data', in binary form, between µP and other external units, such as memory. Typical size is 8 or 16 bits. Size determined by size of boxes in memory and µP size helps determine performance of µP. The Data Bus typically consists of 8 wires. The data bus is a group of eight lines used for data flow.These lines are bi-directional - data flow in both directions between the MPU and memory and peripheral devices. The MPU uses the data bus to perform the second function: transferring binary information. The eight data lines enable the MPU to manipulate 8-bit data ranging from 00 to FF ($2^8$ = 256 numbers).Therefore, $2^8$ combinations of binary digits. Data bus used to transmit "data", i.e. information, results of arithmetic, etc, between memory and the microprocessor. Bus is bi-directional. Size of the data bus determines what arithmetic can be done. If only 8 bits wide then largest number is 11111111 (255 in decimal). Therefore, larger numbers have to be broken down into chunks of 255. This slows microprocessor. Data Bus also carries instructions from memory to the microprocessor. Size of the bus therefore limits the number of possible instructions to 256, each specified by a separate number.

### 3.2.3 Control Bus

The control bus carries synchronization signals and providing timing signals. The MPU generates specific control signals for every operation it performs. These signals are used to identify a device type with which the MPU wants to communicate. Control Bus are various lines which have specific functions for coordinating and controlling µP operations. E.g. Read/Not Write line, single binary digit. Control whether memory is being 'written to' (data stored in memory) or 'read from' (data taken out of memory) 1 = Read, 0 = Write. May also include clock line(s) for timing/synchronizing, 'interrupts', 'reset' etc. Typically µP has 10 control lines. Cannot function correctly without these vital control signals. The Control Bus carries control signals partly unidirectional, partly bidirectional. Control signals are things like "read or write". This tells memory that we are **reading from** a location, specified on the address bus, or **writing to** a location specified. Various other signals to control and coordinate the operation of the system. Modern day microprocessors, like 80386, 80486 have much larger busses. Typically 16 or 32 bit busses, which allow larger number of instructions, more memory location, and faster arithmetic. Microcontrollers organized along same lines, except: because microcontrollers have memory etc inside the chip, the busses may all be internal. In the microprocessor the three busses are external to the chip (except for the internal data bus). In case of external busses, the chip connects to the busses via buffers, which are simply an electronic connection between external bus and the internal data bus.

## 3.3 INTERNAL REGISTERS OF 8085

The 8085 have six general-purpose registers to store 8-bit data during program execution.

These registers are identified as B, C, D, E, H, and L.

They can be combined as register pairs-BC, DE, and HL-to perform some 16-bit operations. These registers are programmable, meaning that a programmer can use them to load or copy data from the registers by using instructions. For example, the instruction MOV A, C copies the data from register C to register A.

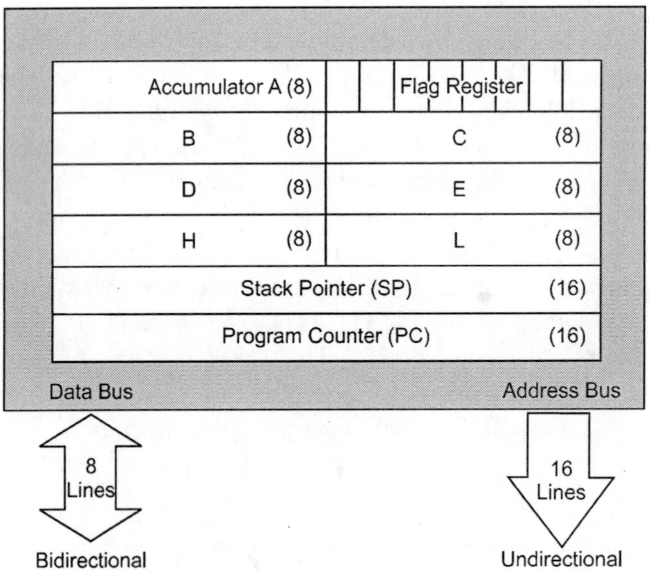


**Figure 3.2:** The 8085 Registers

### 1. Accumulator (A)

The accumulator is an 8-bit register that is part of the arithmetic/logic unit (ALU). This register is used to store 8-bit data and to perform arithmetic and logical operations. The result of an operation is stored in the accumulator.

## 2. Flags

The ALU includes five flip-flops that are set or reset according to the result of an operation. The microprocessor uses the flags for testing the data conditions.

They are Zero (Z), Carry (CY), Sign (S), Parity (P), and Auxiliary Carry (AC) flags. The most commonly used flags are Sign, Zero, and Carry.

The bit position for the flags in flag register is,

| D7 | D6 | D5 | D4 | D3 | D2 | D1 | D0 |
|---|---|---|---|---|---|---|---|
| S | Z | | AC | | P | | CY |

(i) **Sign Flag (S):** After execution of any arithmetic and logical operation, if D7 of the result is 1, the sign flag is set. Otherwise it is reset.

D7 is reserved for indicating the sign; the remaining is the magnitude of number.

If D7 is 1, the number will be viewed as negative number. If D7 is 0, the number will beviewed as positive number.

(ii) **Zero Flag (z):** If the result of arithmetic and logical operation is zero, then zero flag is set otherwise it is reset.

(iii) **Auxiliary Carry Flag (AC):** If D3 generates any carry when doing any arithmetic and logical operation, this flag is set.

Otherwise it is reset.

(iv) **Parity Flag (P):** If the result of arithmetic and logical operation contains even number of 1's then this flag will be set and if it is odd number of 1's it will be reset.

(v) **Carry Flag (CY):** If any arithmetic and logical operation result any carry then carry flag is set otherwise it is reset.

## 3. Arithmetic Logic Unit

The ALU unit performs such arithmetic operations as addition and subtraction, and such logic operations as AND, OR, and exclusive OR. It receives the data from accumulator and registers. According to the result it set or reset the flags.

## 4. Program Counter (PC)

The program counter is a 16-bit register. It is also called as memory pointer. Memory locations have 16-bit addresses. The function of the program counter is to point to the memory address of the next instruction to be fetched. When a byte is fetched, the program counter is incremented by one and which points the next memory address in the memory.

## 5. Stack Pointer (SP)

The stack pointer is also a 16-bit register, used as a memory pointer. It points to a memory location in R/W memory, called the stack. The beginning of the stack is defined by loading a 16-bit address in the stack pointer.

### 6. Externally Initiated Signals

External initiated signals are RESET, INTERRUPT, HOLD and READY signals.

I. **RESET:** When the signal on this pin is activated, all internal operations are suspended the program counter is set to zero, the MPU is reset.

II. **INTERRUPT:** When the signal on this pin is activated, the microprocessors interrupted from the normal execution of instruction and ask to execute other instructions.

III. **HOLD:** When the signal on this pin is activated, it indicates that a peripheral such as a DMA controller is requesting to use the address and data bus.

IV. **READY:** This signal is used to delay the microprocessor read or write cycles until a slow responding peripheral is ready to send or accept data.

## 3.4 GENERAL DESCRIPTION

The Intel 8085 microprocessor is an HMOS 8-bit device. Sixteen address bits provide access to 65,536 bytes of 8 bits each. Eight bi-directional data lines provide access to a system data bus. Control is provided by a variety of lines which support memory and I/O interfacing, and a flexible interrupt system. The 8085 provides an upward mobility in design from the 8080 by supporting all of the 8080's instruction set and interrupt capabilities. At the same time, it provides cleaner designs by virtue of a greater on-device component density, and by requiring only a 5 volt supply. In addition, the 8085 is available in two clock speeds.

The 8085 comes in two models, the 8085A and the 8085A-2. The 8085A expects a main clock frequency of 3 MHz, while the 8085A-2 expects a main clock frequency of 5 MHz. In both cases, the clock is a single phase square wave. This single clock is generated within the 8085 itself, requiring only a crystal externally. This eliminates the need for an external clock generator device. In all other respects, the A and A-2 devices are identical.

The 8085 supports the interrupt structure of the 8080, including the RST instruction and the eight vectors. It extends these by the addition of four more interrupts, each with their own pins, three of which are maskable, and which use vector areas between the existing ones of the 8080. The 8085 is adaptable for use with the 8259 Priority Interrupt Controller, a programmable device. It is possible, upon an interrupt from this device, to jam either a RST instruction onto the data lines, or a CALL instruction to any location in RAM directly.

The 8085 has two pins dedicated to the generation or reception of serial data. While these do not constitute a complete serial I/O system, they do allow the MP to send and receive serial bits, albeit with a large software overhead. The 8085 therefore finds itself useful as a complete control device for remote control applications.

The 8085 supports the entire 8080 instruction set. In addition, two new instructions are added. These instructions permit software control over the extended interrupt capabilities of the '85, by making the new interrupts both maskable and interrogatable. The masks can be set, examined, etc.. The same instructions also allow investigation of the serial input line, and generation of conditions on the serial output line.

Unlike the 8080 which had discrete pins for the address and data busses, the 8085 make use of multiplexing of the lower 8 bits of the address with the data bits on the same 8 pins. This requires that the external circuitry be able to catch and hold the A0-A7 lines for later use. The upper 8 bits of the address have their own pins, however. Three primary control bus lines allow the device to identify whether the cycle in progress is for RAM or I/O, and whether it is a Read or a Write. Two status pins are provided, to allow advance knowledge of certain events in multiprocessor applications.

The internal timing of the device makes use of machine cycles in which, in almost every case, a bus cycle is involved. Each machine cycle consists of several T-states, which are defined by the clock input signal. Thus, many clock cycles are needed to effect one complete instruction. The 8085 has many new support devices to ease design work. These include the 8259 Programmable Interrupt controller, the 8202 Dynamic RAM controller, plus several new I/O devices with various amounts of RAM, ROM, parallel I/O, and timer-counters. The general approach was to make the device as compatible with the Multibus architecture as possible.

### 3.4.1 Interfacing the 8085

A brief description of the signals between the 8085 and the outside world follows. 8085 pin-outs and its detailed description shown in Figure 3.3(a), 3.3(b), 3.3(c).

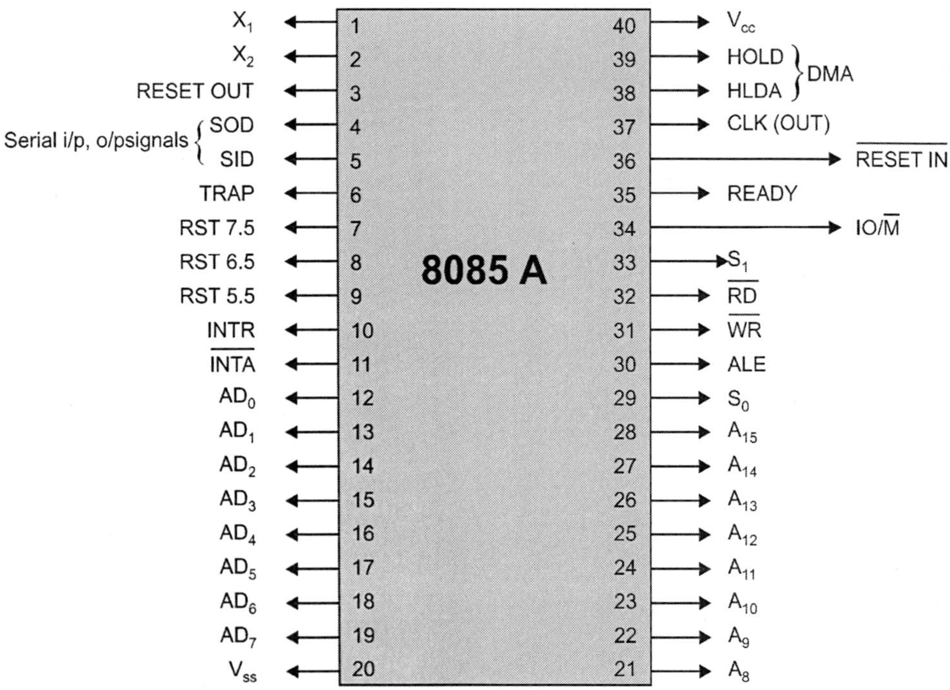


**Figure 3.3(a):**Pin Diagram of 8085

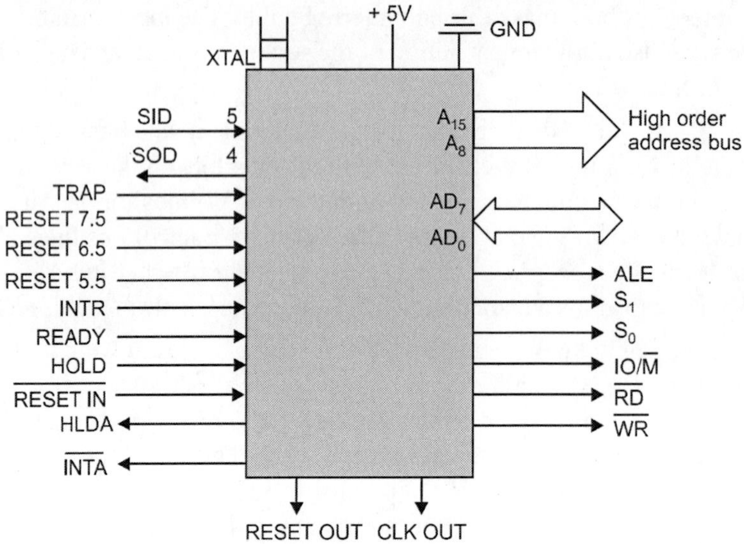


**Figure 3.3(b):** Signal Groups of 8085

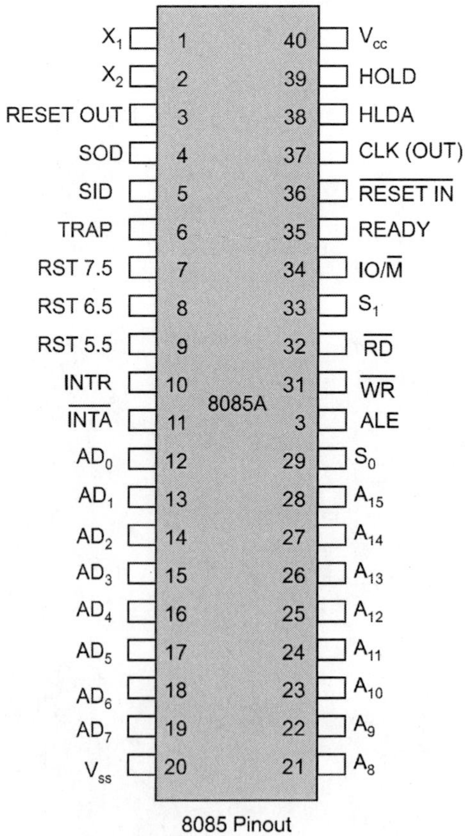


8085 Pinout

**Figure 3.3(c):**Pin Diagram of 8085

1. **ADDRESS LINES A8 - A15:** These tri-state lines are outbound only. They provide the upper 8 bits of the 16-bit-wide address which identifies one unique 8-bit byte within the MP's address space, or the 8-bit address of an I/O device. Sixteen address lines provide an address space of 65,536 locations.
2. **MULTIPLEXED ADDRESS-DATA LINES AD0 - AD7:** These tri-state lines may by either inbound or outbound. They provide a multiplexing between the lower 8 bits of the 16-bit-wide address early in a machine cycle and 8 data bits later in the cycle (shown in figure 3.4). When containing addresses, these lines are outbound only; when containing data, they may be either inbound or outbound, depending upon the nature of the machine cycle. They also will contain the 8 bits of an I/O device address during an I/O operation.

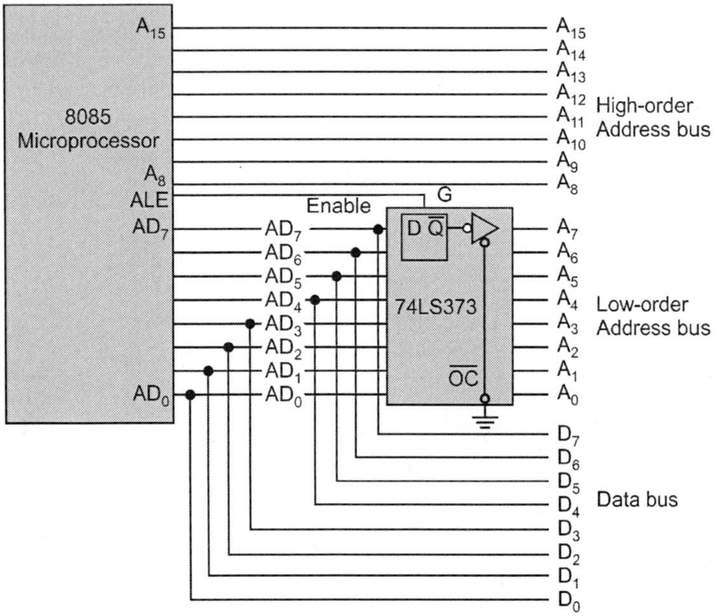


**Figure 3.4:**Multiplexed Address-Data Lines.

3. **ADDRESS LATCH ENABLE (ALE) (pin 30):** This signal appears outbound early in a machine cycle to advise the external circuitry that the AD0 - AD7 lines contain the lower 8 bits of a memory address. It should be used to clock a catch-and-hold circuit such as a 74LS245 or 74LS373 (shown in figure 3.4), so that the full address will be available to the system for the rest of the machine cycle. The documentation states that the falling edge of ALE is the point at which the signals on the AD lines, as well as the so, S1, and I-O/M lines (below) will be stable and may be taken by the external circuitry. It indicates whether bus functions as address bus or data bus.

   If ALE = 1 then Bus functions as address bus.

   If ALE = 0 then Bus functions as data bus.

4. **STATUS LINES S0, S1, & I-O/M (pin 29, 33 and 34):** These three status lines serve to indicate the general status of the processor with respect to what function the MP will perform during the machine cycle. The S0 and S1 lines are made available for circuits which need advanced warning of the ensuing operation, such as very slow RAM or other specialized devices. The system may not normally need to monitor these lines. The IO/M line approximates in one line what the S0 and S1 lines do in two. It indicates whether the operation will be directed toward memory (line is low), or toward I/O (line is high). This pin tells whether I/O or memory operation is being performed.

   If IO/M = 1 then I/O operation is being performed.

   If IO/M = 0 then Memory operation is being performed.

**Table 3.1: It Indicates The Current Operation Of The Microprocessor**

| Operations | IO/M | $S_0$ | $S_1$ |
|---|---|---|---|
| Opcode Fetch | 0 | 1 | 1 |
| Memory Read | 0 | 1 | 0 |
| Memory Write | 0 | 0 | 1 |
| I/O Read | 1 | 1 | 0 |
| I/O Write | 1 | 0 | 1 |
| Interrupt Ack. | 1 | 1 | 1 |
| Halt | High Impedance | 0 | 0 |

5. **READ & WRITE (/RD & /WR) (pin 32 and 31):** These lines indicate which direction the MP expects to pass data between itself and the external data bus. Read indicates that the MP is expecting data to be fed to it; Write indicates that the MP is going to send data away from itself. These lines also serve to time the event, as well as identify its direction.
6. **READY (pin 35):** This is an input line which may be used as a signal from external RAM that a wait state is needed, since the RAM is not able to provide the data or accept it in the time allowed by the MP. The negation of Ready, by being pulled low, will cause the 8085 to enter wait states. See the timing diagrams for critical timing.
7. **HOLD & HOLD ACKNOWLEDGE (HOLD & HLDA) (pin 38 and 39):** These lines provide the 8085 with a DMA capability by allowing another processor on the same system buses to request control of the buses. Upon receipt of HOLD, the '8085 will tri-state its address, data, and certain control lines, and then generate HLDA. This signals the other processor that it may proceed. The '8085 will remain off the buses until HOLD is negated.
8. **INTERRUPT & INTERRUPT ACKNOWLEDGE (INTR & INTA) (pin 6-10and 11):** These lines provide a vectored interrupt capability to the 8085. Upon receipt of INTR, the '85 will complete the instruction in process, and then generate INTA as it enters the next machine cycle. The interrupting device will jam a Restart (RST) instruction onto the data bus, which the '85 uses to locate an interrupt vector in low RAM.

9. **RST 5.5, 6.5, 7.5 (pin 9, 8 and 7):** These three lines are additional interrupt lines which generate an automatic Restart, without jamming, to vectors in low RAM which are between those used by the normal INTR instruction. The 5.5 line, for example, will cause an automatic restart to a 4-byte vector located between 5 and 6 of the normal vectors used by INTR. These lines have priority over the INTR line, and each other. They also have certain electrical characteristics for assertion, and may be masked off or on by software. More on the 8085's interrupts later.
10. **TRAP (pin 6):** This is an unmaskable interrupt with a fixed vector in low RAM. See the interrupt discussion later in the notes.
11. **RESET IN & RESET OUT (pin 36 and 3):** These lines provide for both MP and system reset. The reset circuitry in the 8224, used with the 8080, has been brought inside the MP. The RESET IN line is generated asynchronously by some sort of external circuit, such as an RC network or Reset switch. Upon receipt of this signal, the 8085 will internally synchronize the Reset with the clock of the processor, and then generate RESET OUT for other devices in the system. See the reference manual for details.
12. **X1 & X2(pin 1 and 2):** These two pins provide connection for an external frequency determining circuit to feed the 8085's clock. This is normally a crystal, although other resonant circuits may be used. X1 alone may be used as a single input from an external oscillator. The internal oscillator of the 8085 will divide the frequency by two for the system clock.
13. **CLOCK (CLK) (pin 37):** This line provides a system clock signal to external circuits which need to be in synchronization with the MP.
14. **SERIAL INPUT DATA & SERIAL OUTPUT DATA (SID & SOD) (pin 5 and 4):** These two lines provide for a single serial input or output line to/from the 8085. These lines are brought into the device as $D_7$, and may be tested or set by the Read Interrupt Mask (RIM) or Set Interrupt Mask (SIM) instructions. These two instructions also have control over the mask which controls the RST 5.5, 6.5, and 7.5, and TRAP, interrupts. The SID and SOD lines are simple single bit I/O lines; any timing required to provide external communication via them must be provided by the software.
15. **Vcc & Vss (pin 40 and 20):** These are the power connections for +5 volts and ground, respectively.

## 3.5 THE 8085 MICROPROCESSOR

The microprocessor uses the same type of logic that is used in a digital computer's central processing unit (CPU). Because it is similar to the CPU and it is constructed with microcircuit (IC) technology. The microprocessor has digital circuit for data handling and computation under program control. (The microprocessor is a data processing unit) Data processing is the microprocessor's main function.

Data processing includes both computation and data handling. Computation is performed by logic circuits that make up what is usually called the arithmetic logic unit (ALU). These logic circuits enable us to use functions that cause data changes. Among these functions are Add, Subtract, AND, OR, XOR, Compare, Increment, and Decrement. The ALU cannot perform any of these functions with out data operation on. In order to process data, the microprocessor must have control logic which tells the microprocessor how to decode and execute the program.

**Program is a set of instructions for processing the data**

The control logic steps the microprocessors through the stored program steps (instructions) in memory. It calls (fetches) them one at a time. After the instruction is fetched, the microprocessor's control logic decodes the instruction. Then the control logic carries out (executes) the decoded instruction. Because the instructions are stored in memory, you can change them when you want to.

The microprocessor is a clock-driven semiconductor device consisting of electronic logic circuits manufactured by using either a large-scale integration (LSI) or very-large-scale integration (VLSI) technique. The microprocessor is capable of performing various computing functions and making decisions to change the sequence of program execution.

In large computers, a CPU implemented on one or more circuit boards performs these computing functions. The microprocessor is in many ways similar to the CPU, but includes the logic circuitry, including the control unit, on one chip. The microprocessor can be divided into three segments for the sake clarity, arithmetic/logic unit (ALU), register array, and control unit.8085 is a 40 pin IC, DIP package. The signals from the pins can be grouped as follows

This figure shows the pin out of 8085 microprocessor. It is a 8-bit microprocessor, which can address 64K of memory. 8085 microprocessor has 40 pins and requires a +5V power supply and can operate with a 3 MHz single-phase clock. Each pin has an input or output signal.

These signals are classified as follows:

1. Address bus
2. Data bus
3. Control and status signal
4. Power supply and frequency signals
5. Externally initiated signals
6. Serial i/o ports

### 1. Address Bus

The 8085 microprocessor has a 16 bit address lines ($A_{15}$–$A_0$), These are the unidirectional bus, the address buses are divided in to two groups, higher order bus ($A_{15}$–$A_8$) and lower ordered bus ($A_7$–$A_0$). The lower order address bus $AD_7$–$AD_0$ are called as multiplexed bus. The lower ordered bus can carry the address and data.

### 2. Multiplexed Address-data Bus

These multiplexed set of lines used to carry the lower order 8 bit address as well as data bus. During the opcode fetch operation, in the first clock cycle, the lines deliver the lower order address

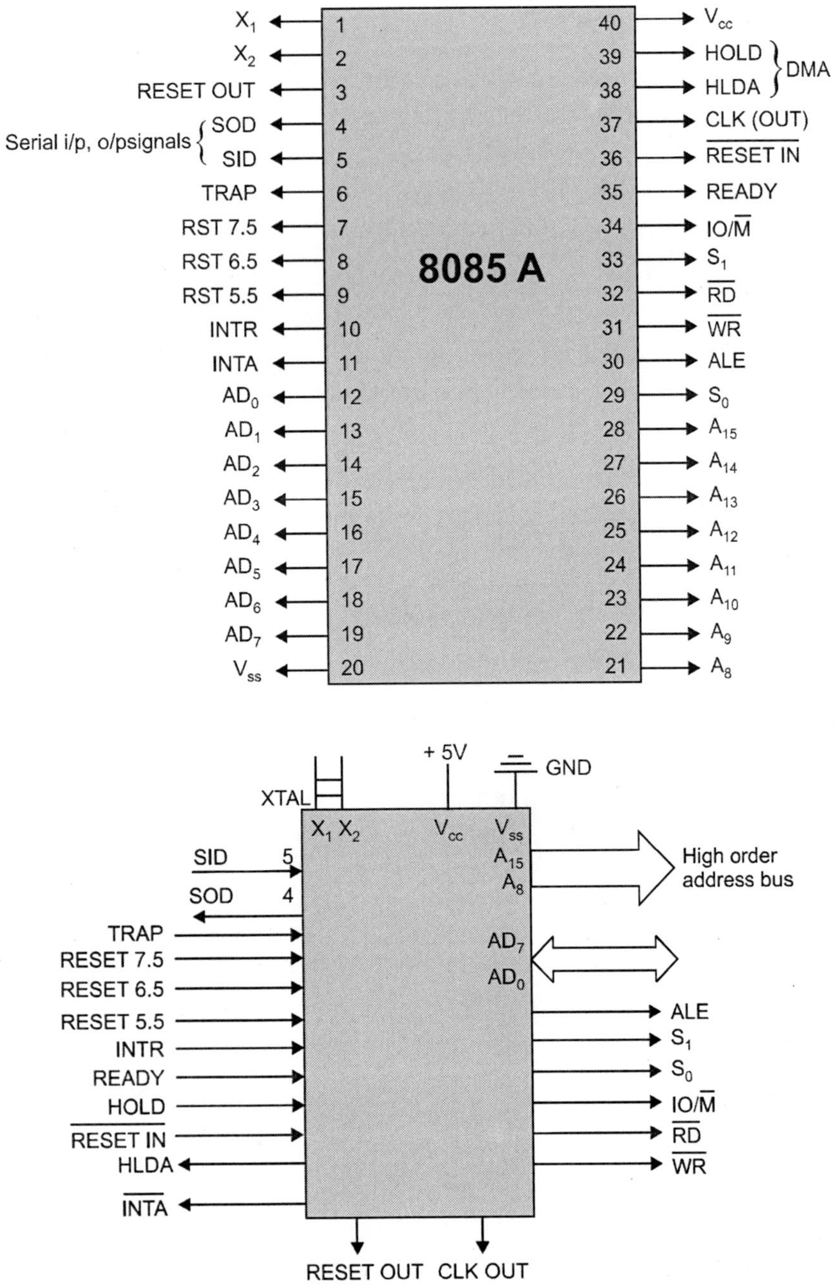


**Figure 3.5:** 8085 Pin Configuration and signals

A0–A7. In the subsequent IO/memory, read/write clock cycle the lines are used as data bus. The CPU may read or write out data through these lines. These buses can carry address and data, the address transfer and the data transfer is controlled by the control signal ALE.

### 3. Control and Status Signal

The group of signals includes two control signals ( $\overline{RD}$ and $\overline{WR}$ ), three status signals (IO/ $\overline{M}$ , $S_1$ and $S_0$) to identify the operation, and a special signal (ALE), to select the mode of multiplexed bus.

- ALE (Address Latch Enable): This is a positive pulse generated at the beginning of an operation. It indicates that AD7 to ADo are lower order address lines. This signal is used to enable the latch of the lower order address from the multiplexed bus.
- $\overline{RD}$ : Read control signal is an active low signal. This indicates that the microprocessor connected to I/O or memory device has to read the data available in the data bus.
- $\overline{WR}$ : Write control signal is an active low signal. This indicates that the data on the data bus are to be written in the selected I/O or memory device.
- IO/ $\overline{M}$ **:** This status signals indicates whether the operation is with I/O devices or memory when it is high, it indicates an I/O operation. When it is low, it indicates a memory operation.
- **$S_1$ and $S_0$:** These are status signals, to identify the operation such as opcode fetch, memory read, memory write, I/O read, I/O write, etc.

**Table 3.2: It is used To Know the Type of Current Operation Of The Microprocessor.**

| IO/M (Active Low) | S1 | S2 | Data Bus Status (Output) |
|---|---|---|---|
| 0 | 0 | 0 | Halt |
| 0 | 0 | 1 | Memory Write |
| 0 | 1 | 0 | Memory Read |
| 1 | 0 | 1 | IO Write |
| 1 | 1 | 0 | IO Read |
| 0 | 1 | 1 | Opcode Fetch |
| 1 | 1 | 1 | Interrupt acknowledge |

### 4. Power Supply and Clock Frequency

**(i) Power supply signals are:**

- Vcc: + 5V power supply
- Vss: Ground Reference.

**(ii) Clock signals are:**

- X1, X2: A crystal is connected at these two pins. Two internally divides the Frequency; therefore to operate a system at 3MHz, the crystal should have a frequency of 6MHz.
- CLK: clock signal, which controls the whole operations of the microprocessor.

### 5. Externally Initiated Signals

They are the signals initiated by an external device to request the microprocessor to do a particular task or work.

There are five hardware interrupts called,

TRAP<br>RST 7.5<br>RST 6.5 } (inputs)<br>RST 5.5<br>INTR<br>INTA (active low output)

Other External initiated signals are RESET, HOLD and READY signals.

### *8085 Interrupts*

- **INTR:** This signal is active high signal, which is used to indicate the interrupt.
- **INTA:** This signal is active low signal, which issued to acknowledge the interrupt.
- **RST:** Restart Interrupt signal, which are vectored interrupts that transfer the program control to specific memory location.
- **TRAP:** This is a non-maskable interrupt and has the highest priority.
- **HOLD:** This signal indicates that a peripheral such as a DMA controllers is requesting to use the address and data bus.
- **HLDA:** This is an acknowledge signal for the HOLD request signal given by the microprocessor.
- **READY:** This signal is used to delay the microprocessor read or write cycles until a slow responding peripheral is ready to send or accept data.
- **RESET IN:** When the signal on this pin goes low, the program counter is set to zero, the MPU is reset.
- **RESET OUT:** This signal indicates that the MPU is being reset.

### 6. Serial I/O Ports

The 8085 has two signals to perform the serial transmission through SID and SOD. (Serial input data and serial output data).

- **SID** – serial input data: The data on this line is loaded into accumulator bit – 7 whenever a RIM instruction is executed.
- **SOD** – Serial output data: This line is set or reset as specified by the SIM instruction.

These two signals are used to establish serial communication between the microprocessor and external serial I/O devices.

## 3.6 MICROPROCESSOR COMMUNICATION AND BUS TIMINGS

Now, we have to examine the process of communication between the microprocessor and memory to understand the functions of various signals.

The first step in the communication process is reading from memory or fetching an instruction. This process is called **fetch cycle**. We need to understand timings of various signals in relation to the system clock. It can be better understood through timing diagram as shown below.

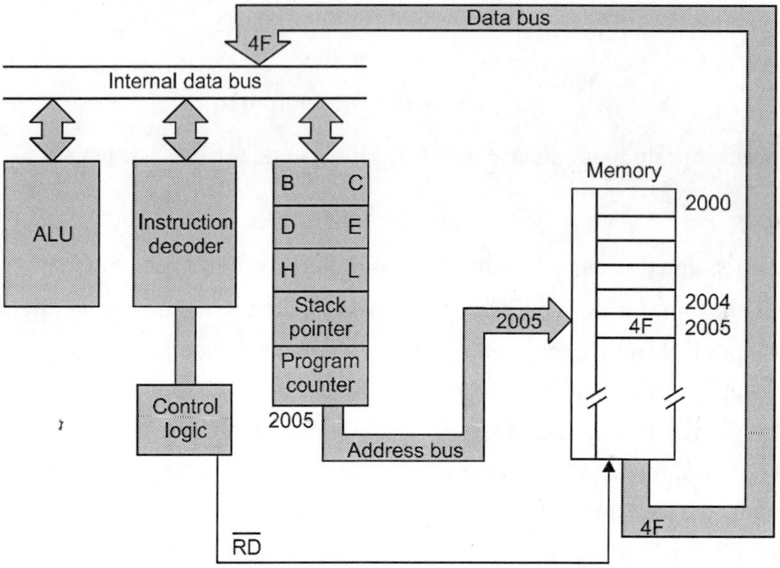


**Figure 3.6:** Data flow from memory to microprocessor

Consider the example of fetching the machine code of instruction MOV C, A (0100 1111 = 4F) stored at the address 2005.

***Step1:*** Program Counter places the 16-bit memory address on the address bus. In Timing diagram, during the T-state $T_1$ the higher order memory address 20H is placed on the address lines $A_{15}$–$A_8$, the lower order address 05H is placed on the bus $AD_7$–$AD_0$, and ALE signal goes high. The status signal IO/M goes low, indicating this is a memory-related operation.

***Step2:*** The control unit sends the control signal RD to enable the memory chip The control signal RD is sent out during the clock period $T_2$, thus enabling the memory chip. This signal is active for two clock periods.

***Step3:*** The byte from the memory location is placed on the data bus. When the memory is enabled, the instruction byte (4F) is placed on the bus $AD_7 - AD_0$ and transferred to the microprocessor. The RD signal causes 4F to be placed on the multiplexed bus and when RD goes high, it causes the bus to go into high impedance.

***Step4:*** The byte is placed in the instruction decoder of the microprocessor and the task is carried out according to the instruction. The instruction decoder decodes the machine code and the contents of the accumulator are copied into register C. This task is performed during period $T_4$.

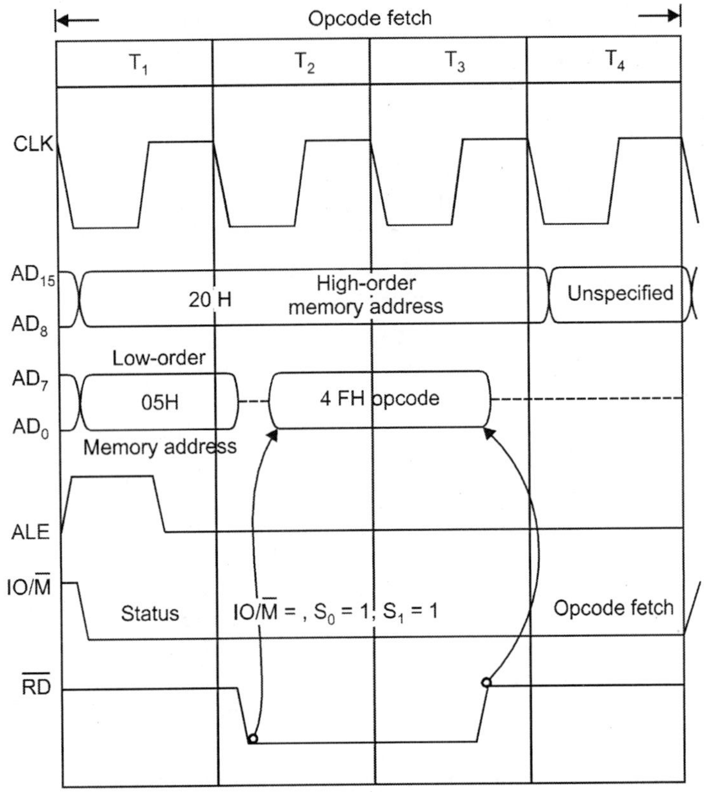


**Figure 3.7:** Timing diagram for transferring bytes from memory to microprocessor.

### 3.6.1 Demultiplexing The Bus $AD_7$–$AD_0$

In the above timing diagram, demultiplexing $AD_7$–$AD_0$ becomes apparent. It clearly shows that, the lower order address (05H) is lost after the first clock period. This address needs to be latched and used for identifying the memory address. If the bus $AD_7$–$AD_0$ is used to identify the memory location 2005H, the address will change to 204FH after the first clock period.

The Figure 3.8 shows a schematic that uses a latch and ALE signal to demultiplex the bus. The bus $AD_7$–$AD_0$ is connected as the input to the latch 74LS373. The ALE signal is connected to the enable pin of the latch and the output control signal of latch is grounded.

ALE goes high during $T_1$. When ALE is high, the latch is transparent and output changes according to input. During $T_1$, the output of the latch is 05H. When ALE goes low, the data byte 05H is latched until next ALE. And the output of latch represents the lower order address bus $A_7$–$A_0$.

After carefully observing *timing diagram* of instruction fetch shown above, we can make following observations:

1. The machine code 4FH is one-byte instruction that copies the contents of the accumulator into register C.

2. The 8085 microprocessor requires one external operation – fetching the machine code from memory location 2005h.
3. The entire operation, fetching, decoding and executing requires four clock periods.

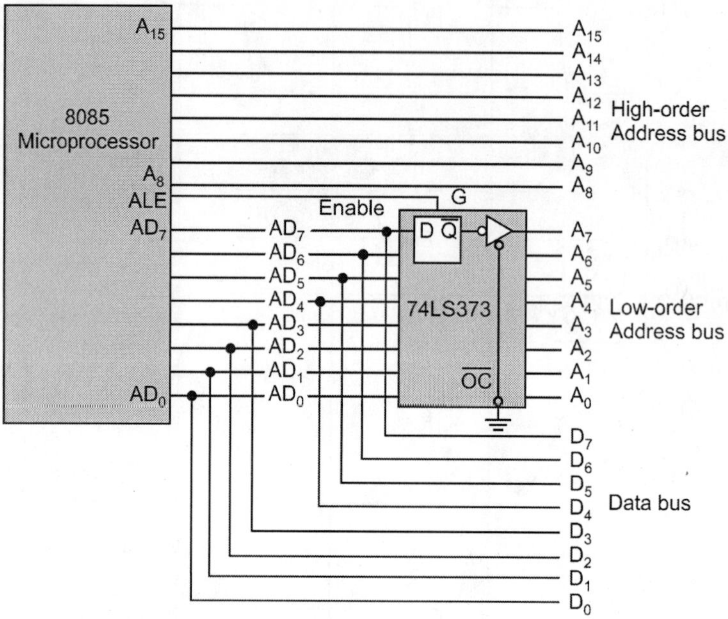


**Figure 3.8:** Schematic of latching low order address bus

Now, we can define following terms:

1. **Instruction cycle:** It is the time required to complete the execution of an instruction. 8085 has instructions which consume one to six machine cycles.
2. **Machine Cycle:** It is defined as the time required to complete one operation of accessing memory, I/O, or acknowledging an external request. This cycle may consist of three to six T-states.
3. **T-state:** It is defined as one subdivision of the operation performed in one clock period. These subdivisions are internal states synchronized with the system clock and each T-state is precisely equal to one clock period.

### 3.6.2 Generating the Control Signals

In Figure 3.7 shows the $\overline{RD}$ (Read) as a control signal. Because this signal is used both for reading memory and reading an input device, it is necessary to generate two different Read signals: one for memory and another for input. Similarly, two separate Write signals must be generated.

In Figure 3.9 shows that four different control signals are generated by combining the signals, $\overline{RD}$, $\overline{WR}$ and $IO/\overline{M}$. The signal $IO/\overline{M}$ goes low for the memory operation. This signal is ANDed with $\overline{RD}$ and $\overline{WR}$ signals by using the 74LS32 quadruple two-input OR gates, as shown in Figure 3.9. The OR gates are functionally connected as negative NAND gates. When both input signals go low, the outputs of the gates go low and generate $\overline{MEMR}$ (Memory Read) and $\overline{MEMW}$ (Memory Write) control signals. When the IO/ signals goes high, it indicates the peripheral I/O operation. Figure 3.9 shows that this signals is complemented using the Hex inverter 74LS04 and ANDed with the $\overline{RD}$ and $\overline{WR}$ signals to generate $\overline{IOR}$ (I/O Read) and $\overline{IOW}$ (I/O Write) control signals.

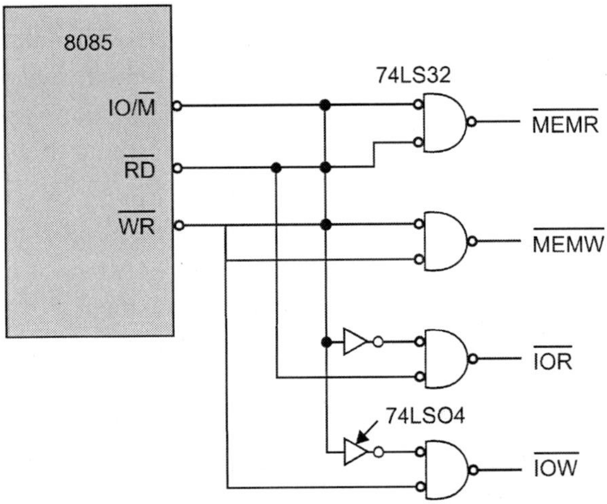


**Figure 3.9:** Schematic to Generate Read/Write Control Signals for Memory and I/O

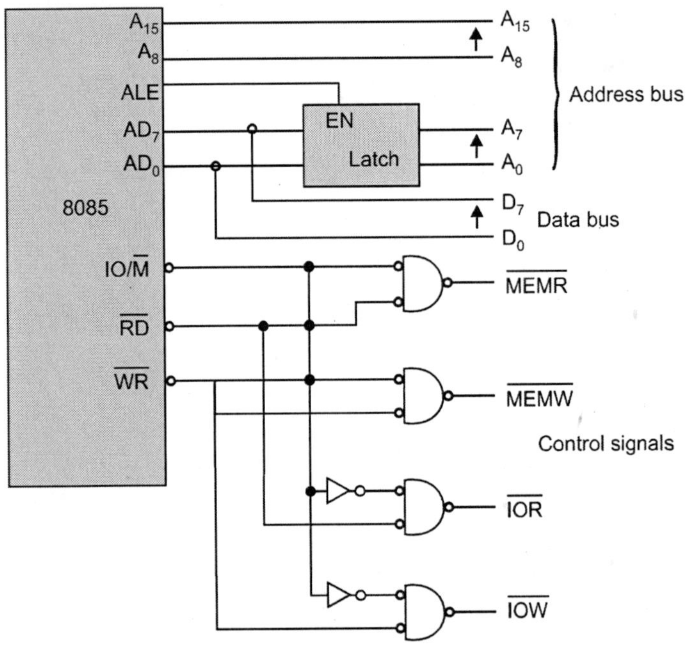


**Figure 3.10:** 8085 Demultiplexed Address and Data Bus with Control Signals.

To demultiplex the bus and to generate the necessary control signals, the 8085 microprocessor requires a latch and logic gates to build the MPU, as shown in Figure 3.10. This MPU can be interfaced with any memory or I/O.

## 3.7 MEMORY INTERFACING

Memory is an integral part of a microcomputer system. Our concept is how to interface a memory chip with the microprocessor. While executing a program, the microprocessor needs to access memory quite frequently to read instruction codes and data stored in memory; the interfacing circuit enables that access. Memory has certain signal requirements to write into and read from its registers. Similarly, the microprocessor initiates a set of signals when it wants to read from and write into memory. The interfacing process involves designing a circuit that will match the memory requirements with the microprocessor signals.

Here, we will examine memory structure and its requirements and the 8085 Memory Read and Write machine cycles. Then we will derive the basic steps necessary to interface memory with the 8085. In this topic, we will illustrate memory interfacing, using memory chips such as 2732 EPROM and 6116 static R/W memory, and will discuss address decoding and memory addresses.

### 3.7.1 Memory Structure and Its Requirements

Read/Write memory (R/WM) is a group of registers used to store binary information. Figure 3.11(a) shows a typical R/W memory chip; it has 2048 registers and each register can store eight bits indicated by eight input and eight output data lines. The chip has 11 address lines $A_{10}$–$A_0$, one Chip Select ($\overline{CS}$), and two control lines: Read ($\overline{RD}$) to enable the output buffer and Write ($\overline{WR}$) to enable the input buffer. Figure 3.11(a) also shows the internal decoder to decode the address lines. Figure 3.11(b) shows the logic diagram of a typical EPROM (Erasable Programmable Read-Only Memory) with 4096 (4K) registers. It has 12 address lines $A_{11}$-$A_0$, one Chip Select ($\overline{CS}$), and one Read control signal.

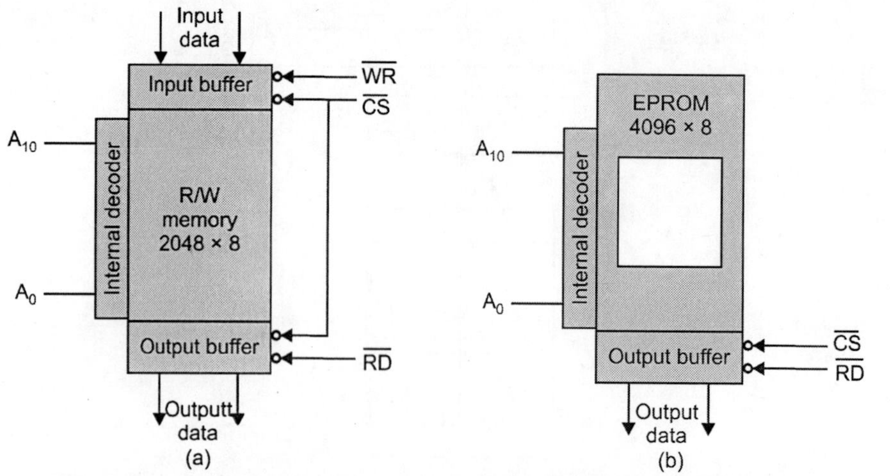


**Figure 3.11:** (a) Typical Memory Chips (b) R/W Static Memory and EPROM

This chip must be programmed (written into) before it can be used as a read-only memory. Figure 3.11(b) also shows a quartz window on the chip that is used to expose the chip to ultraviolet rays for erasing the program. Once the chip is programmed, the window is covered with opaque tape to avoid accidental erasing. For interfacing the R/W memory, Figure 3.11 (a), and EPROM, Figure 3.11 (b), the process is similar; the only difference is that the EPROM does not require the WR signal.

### 3.7.2 Basic Concepts in Memory Interfacing

The primary function of memory interfacing is that the microprocessor should be able to read from and write into a given register of a memory chip. To perform these operations, the microprocessor should

1. Be able to select the chip.
2. Identify the register.
3. Enable the appropriate buffer.

Let us examine the timing diagram of the Memory Read operation (Figure 3.12) to understand how the 8085 can read from memory. Figure 3.12 is used to illustrate memory interfacing concepts.

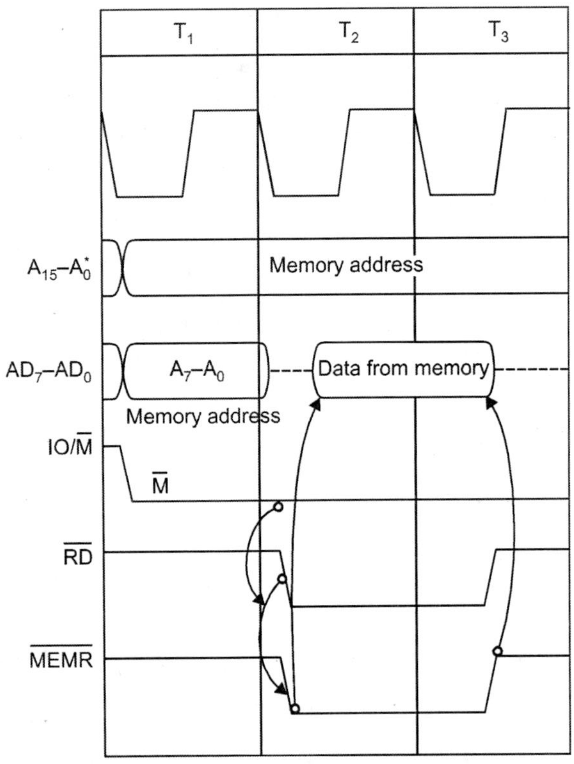


**Figure 3.12:** Timing of the Memory Read Cycle

1. The 8085 places a 16-bit address on the address bus, and with this address only one register should be selected. For the memory chip in Figure 3.11(a), only 11 address lines are required to identify 2048 registers. Therefore, we can connect the low-order address lines $A_{10}$–$A_0$ of the 8085 address bus to the memory chip. The internal decoder of the memory chip will identify and select the register for the EPROM, Figure 3.11(a).
2. The remaining 8085 address lines ($A_{15}$-$A_{11}$) should be decoded to generate a Chip Select ($\overline{CS}$) signal unique to that combination of address logic.
3. The 8085 provides two signals– $IO/\overline{M}$ and $\overline{RD}$ –to indicate that it is a memory real operation. The IO/ andcan be combined to generate the $\overline{MEMR}$ (Memory Read) control signal that can be used to enable the output buffer by connecting to the memory signal $\overline{RD}$.
4. Figure 3.12 also shows that memory places the data byte from the addressed register during $T_2$, and that is read by the microprocessor before the end of $T_3$.

To write into a register, the microprocessor performs similar steps as it reads from a register. Figure 3.13 shows the Memory Write cycle. In the Write operation, the 8085 places the address and data and asserts the $IO/\overline{M}$ signal. After allowing sufficient time for data to become stable, it asserts the Write ($\overline{WR}$) signal. The $IO/\overline{M}$ and $\overline{WR}$ signals can be combined to generate the $\overline{MEMR}$ control signal that enables the input buffer of the memory chip and stores the byte in the selected memory register.

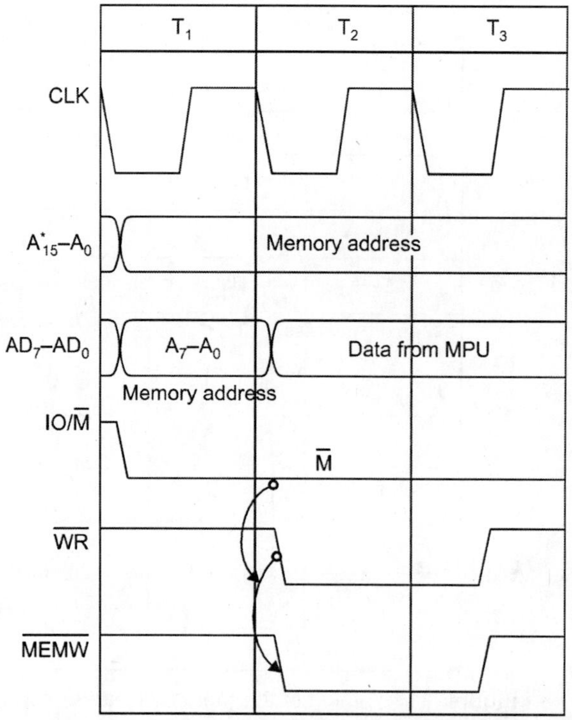


**Figure 3.13:** Timing of the Memory Write Cycle

To interface memory with the microprocessor, we can summarize the above steps as follows:

1. Connect the required address lines of the address bus to the address lines of the memory chip.
2. Decode the remaining address lines of the address bus to generate the Chip Select signal, as discussed in the section (3.7.3), and connect the signal to select the chip.
3. Generate control signal $\overline{MEMR}$ and $\overline{MEMW}$ by combining $\overline{RD}$ and $\overline{WR}$ signals with $IO/\overline{M}$, and use them to enable appropriate buffers.

### 3.7.3 Address Decoding

The process of address decoding should used to identifying a register for a given address. We should be able to generate a unique pulse for a given address. For example, in Figure 3.11(b), 12 address lines ($A_{11}$–$A_0$) are connected to the memory chip, and the remaining four address lines ($A_{15}$–$A_{12}$) of the 8085 microprocessor must be decoded. Figure 3.14 shows two methods of decoding these lines: one by using a NAND gate and the other by using a 3-to-8 decoder. The output of the NAND goes active and selects the chip only when all address lines $A_{15}$–$A_{12}$ are at logic 1. We can obtain the same results by using O–$_7$ of the 3-to-8 decoder, which is capable of decoding eight different input addresses. In the decoder circuit, three input lines can have eight different logic combinations from 000 to 111; each input combination can be identified by the corresponding output line if Enable lines are active. In this circuit, the Enables lines

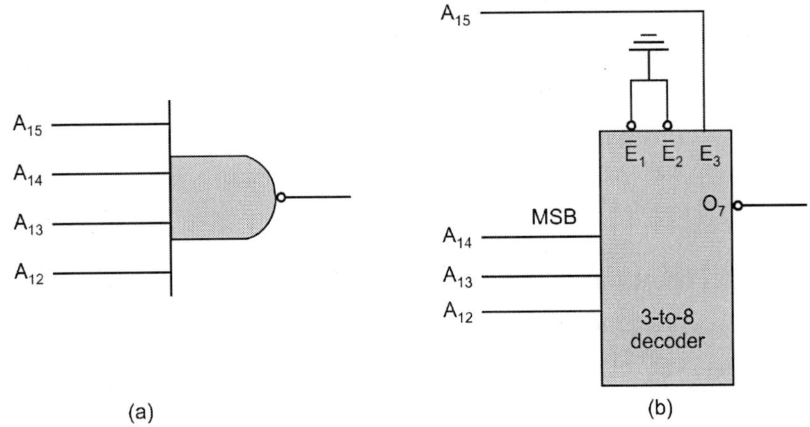


**Figure 3.14:** (a) Address Decoding Using NAND Gate (b) 3-to-8 Decoder

$\overline{E_1}$ and $\overline{E_2}$ are enabled by grounding, and $A_{15}$ must be at logic 1 to enable $E_3$. We will use this address decoding scheme to interface a 4K EPROM and a 2K R/W memory as illustrated in the next two examples.

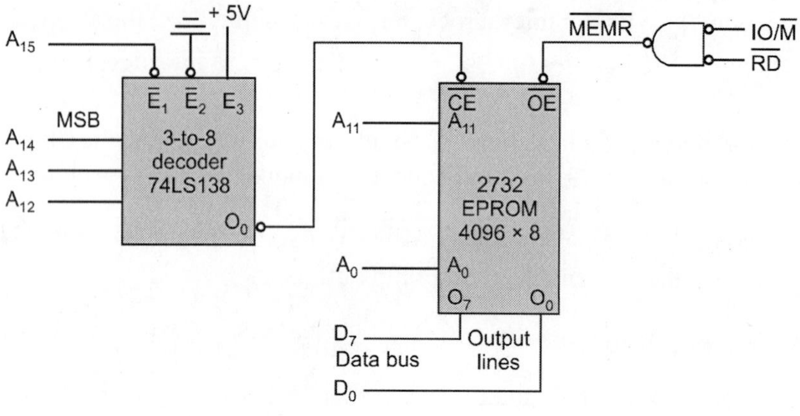


**Figure 3.15:** Interfacing the 2732 EPROM

### 3.7.4 Interfacing Circuit

In Figure 3.15 shows an interfacing circuit using a 3-to-8 decoder to interface the 2732 EPROM memory chip. It is assumed here that the chip has already been programmed, and we will analyze the interfacing circuit in terms of the same three steps outlined previously:

**Step 1:** The 8085 address lines $A_{11}$–$A_0$ are connected to pins $A_{11}$–$A_0$ of the memory chip to address 4096 registers.

**Step 2:** The decoder is used to decode four address lines $A_{15}$–$A_{12}$. The output $O_0$ of the decoder is connected to Chip Enable ($\overline{CE}$). The $\overline{CE}$ is asserted only when the address on $A_{15}$–$A_{12}$ is 0000; $A_{15}$ (low) enables the decoder and the input 000 asserts the output $O_0$.

**Step 3:** For this EPROM, we need one control signal: Memory Read ($\overline{MEMR}$), active low. The $\overline{MEMR}$ is connected toto enable the output buffer; $\overline{OE}$ is the same as $\overline{RD}$ in Figure 3.11.

### 3.7.5 Address Decoding and Memory Addresses

We can obtain the address range of this memory chip by analyzing the possible logic levels on the 16 address lines. The logic levels on the address lines $A_{15}$–$A_{12}$ must be 0000 to assert the Chip Enable, and the address lines $A_{11}$–$A_0$ can assume any combinations from all 0s to all ls. Therefore, the memory address of this chip ranges from 0000H to 0FFFH, as shown below.

| $A_{15}$ | $A_{14}$ | $A_{13}$ | $A_{12}$ | $A_{11}$ | $A_{10}$ | $A_9$ | $A_8$ | $A_7$ | $A_6$ | $A_5$ | $A_4$ | $A_3$ | $A_2$ | $A_1$ | $A_0$ | | |
|---|---|---|---|---|---|---|---|---|---|---|---|---|---|---|---|---|---|
| 0 | 0 | 0 | 0 | 0 | 0 | 0 | 0 | 0 | 0 | 0 | 0 | 0 | 0 | 0 | 0 | = | 0000H |
| | | | | ↓ | | | | | | | | | | | ↓ | | |
| 0 | 0 | 0 | 0 | 1 | 1 | 1 | 1 | 1 | 1 | 1 | 1 | 1 | 1 | 1 | 1 | = | 0FFFH |
| Chip Enable | | | | Register Select | | | | | | | | | | | | | |

We can verify the memory address range in terms of our analogy of page and line numbers, as discussed in Chapter 3, Section 3.22. The chip's 4096 bytes of memory can be viewed as 16 pages with 256 lines each. The high-order Hex digits range from 00 to 0F, indicating 16 pages–0000H to 00FFH and 100H to 01FFH, for example.

Now, to examine how an address is decoded and how the microprocessor reads from this memory, let us assume that the 8085 places the address 0FFHH on the address bus. The address 0000 (0H) goes to the decoder, and the output line $O_0$ of the decoder selects the chip. The remaining address FFFH goes on the address lines of the chip, and the internal decoder of the chip decodes the address and selects the register FFFH. Thus, the address 0FFFH selects the register as shown in Figure 3.16. When the 8085 asserts the RD signal, the output buffer is enabled and the contents of the register 0FFFH are placed on the data bus for the processor to read. Analyze the interfacing circuit in Figure 3.17 and find its memory address range.

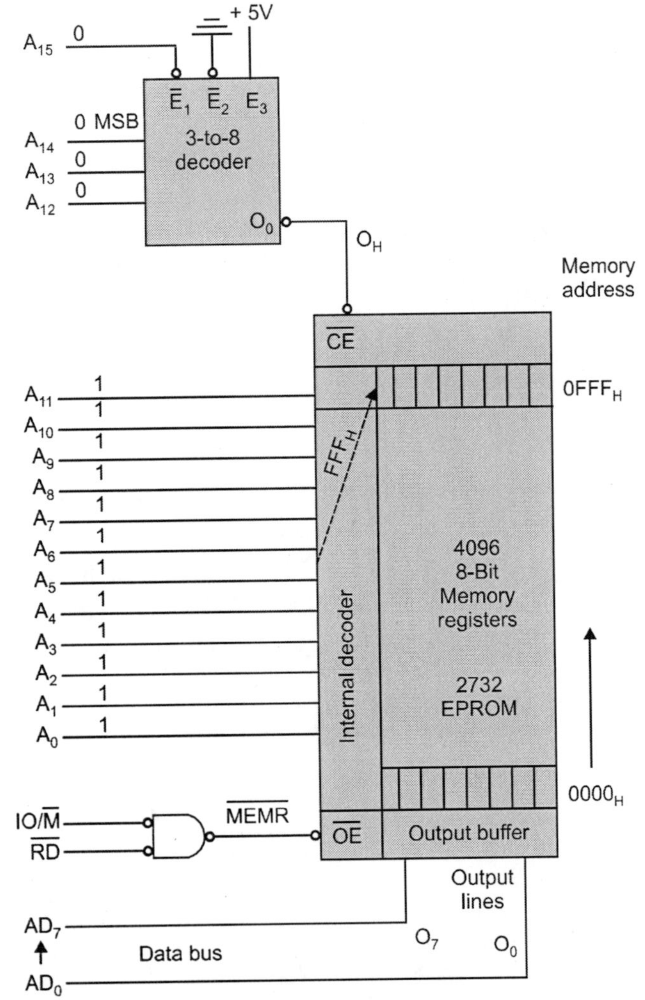


**Figure 3.16:** Address Decoding and Reading from Memory

In Figure 3.17 shows the interfacing of the 6116 memory chip with 2048 (2K) registers. The memory chip requires 11 address lines ($A_{10}$–$A_0$) to decode 2048 registers. The remaining address lines $A_{15}$–$A_{11}$ are connected to the decoder. However, in this circuit, the decoder is enabled by the IO/M signal in addition to the address lines $A_{15}$ and $A_{14}$, and the $\overline{RD}$ and $\overline{WR}$ signals of the 8085 are directly connected to the memory chip.

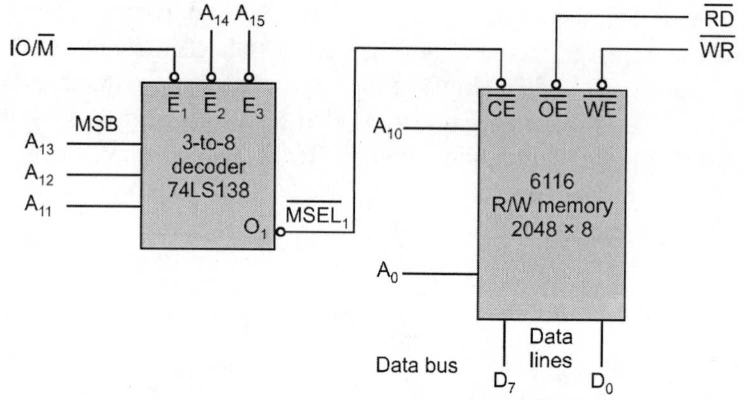


**Figure 3.17:** Interfacing R/W Memory

The signals $\overline{MEMR}$ and $\overline{MEMW}$ need not be generated separately; thus, technique saves two gates. The memory address of this chip ranges from 8800H to 8FFFH, as shown below.

| $A_{15}$ | $A_{14}$ | $A_{13}$ | $A_{12}$ | $A_{11}$ | $A_{10}$ | $A_9$ | $A_8$ | $A_7$ | $A_6$ | $A_5$ | $A_4$ | $A_3$ | $A_2$ | $A_1$ | $A_0$ | |
|---|---|---|---|---|---|---|---|---|---|---|---|---|---|---|---|---|
| 1 | 0 | 0 | 0 | 1 | 0 | 0 | 0 | 0 | 0 | 0 | 0 | 0 | 0 | 0 | 0 | = 8800H |
| ↓ | | | | | ↓ | | | | | | | | | | ↓ | |
| | | | | | 1 | 1 | 1 | 1 | 1 | 1 | 1 | 1 | 1 | 1 | 1 | = 8FFFH |
| $MSEL_1$ | | | | | | | | | | | | | | | | |

The output line $O_1$ of the decoder is connected to CE of the memory chip, and it is identified as $MSEL_1$ because it is asserted only when $IO/\overline{M}$ is low.

The address lines $A_{15}$ (high) and $A_{14}$ (low) enable the decoder, and the input lines to the decoder – $A_{13}$, $A_{12}$, and $A_{11}$ (001) – activate the output $O_1$ ($MSEL_1$) to select the memory chip. The address lines $A_{10}$–$A_0$ can assume any logic combination between all 0s to all 1s, as shown above.

## 3.8 DETAILED LOOK AT MICROPROCESSOR AND ARCHITECTURE

Architecture reveals the internal logic of a microprocessor. The 8085 architecture comprises of the following blocks:

1. ALU logic
2. Register logic
3. Timing and Execution logic
4. Interrupt logic
5. Serial I/O logic

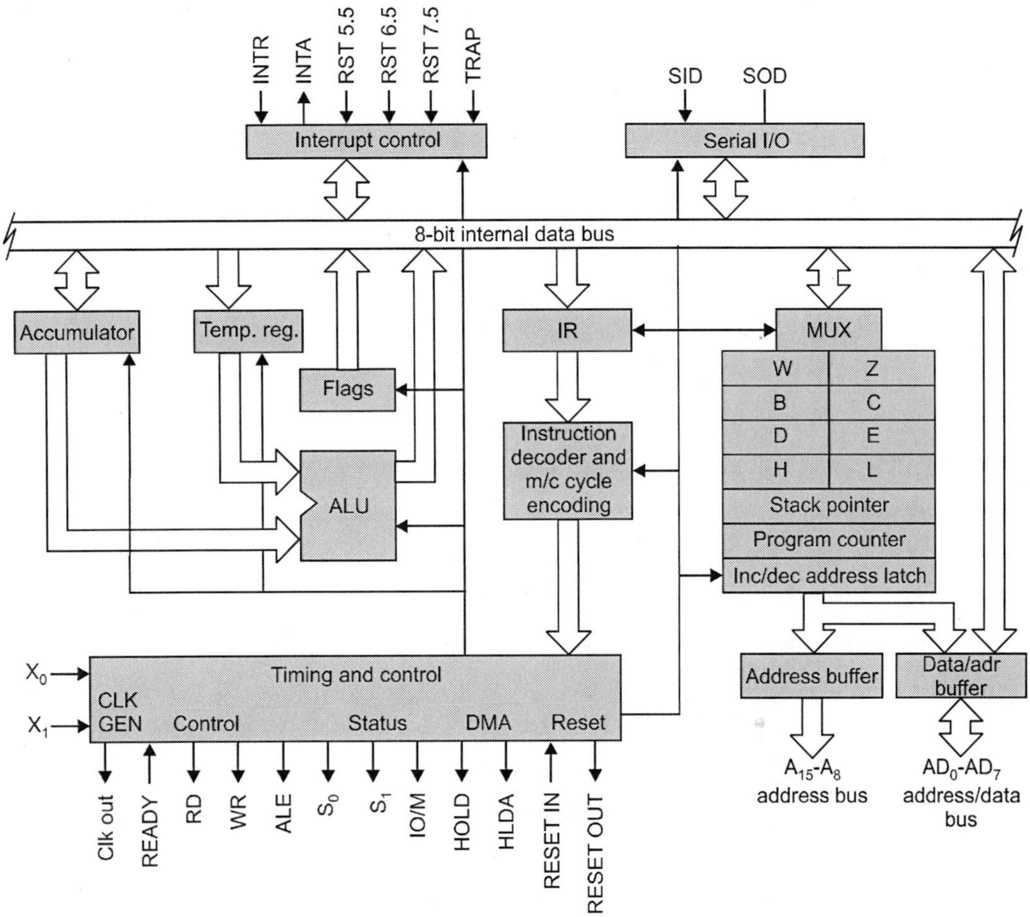


**Figure 3.18:** Detailed Look at 8085 Microprocessor

## 1. ALU Logic

The Arithmetic and Logic Unit performs arithmetic and logic operations. It comprises of the accumulator, temporary registers, flag register and arithmetic and logic circuits.

### *Accumulator*

- It is an 8-bit register
- Stores one of the operands during arithmetic and logic operations
- Stores result of the operation

### *Temporary Register*

- It is used to hold another operand during arithmetic and logic operations.
- It is not accessible to the programmer.

### *Flag Register*

- It is an 8-bit register
- It makes use of only five bits, each one is called a flag
- They reflect the result of an arithmetic or logic operation

| D7 | D6 | D5 | D4 | D3 | D2 | D1 | D0 |
|---|---|---|---|---|---|---|---|
| S | Z | | AC | | P | | CY |

**Sign(S) flag:** set/reset after the execution of an arithmetic or logic operation

1 (set) – if bit $d_7$ of result is 1 (negative number)

0 (reset) – if bit $d_7$ of result is 0 (positive number)

**Zero (Z) flag:** set/reset after the execution of an arithmetic or logic operation

1 – Result is zero

0 – result is non – zero

**Auxiliary Carry (AC) flag:** set/reset after the execution of an arithmetic or logic operation

1 – If carry is generated by bit $d_3$

0 – If there is no carry out of bit $d_3$

It is used internally for BCD operations and is not available to the programmer

**Parity (P) flag:** set/reset after the execution of an arithmetic or logic operation

1 – Even number of 1s in the result

0 – Odd number of 1s in the result

**Carry (CY) flag:** set/reset after the execution of an arithmetic or logic operation

1 – if an arithmetic operation results in carry/borrow

0 – if no carry/borrow

### *The ALU of 8085 provides*

1. **Arithmetic operations** – Addition, subtraction, incrementing and decrementing operations
2. **Logical operations** – AND, OR, EXOR & NOT

## 2. Register Logic

This logic provides a set of registers and the circuits for accessing these registers. They are broadly classified as –

(i) General purpose registers

(ii) Special function registers

**(i) General Purpose registers:** These are the ones that are accessible to the programmer. They are 8-bit registers – B, C, D, E, H, and L

Some of them are used in combination to form 16-bit register pairs. The following combinations are possible – BC, DE and HL

**(ii) Special Function registers:** Specific functions are assigned to the registers in this group.

1. **Program Counter (PC):** This is used for sequencing the execution of instructions. It is always pointing to the memory address from where the next byte is to be fetched
2. **Stack Pointer (SP):** This is a pointer register. Its always pointing to the top of the stack.
3. **Timing and Execution logic:** The instruction Register holds the instruction fetched from memory.

   The Instruction Decoder decodes the instruction in the Instruction Register and passed relevant information to the timing and control unit.

   Timing and Control Unit synchronizes all microprocessor operations with the clock and generates the control signals necessary for communication between microprocessor and peripherals.

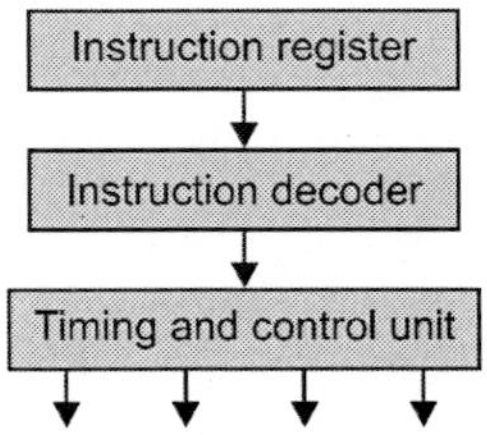


4. **Interrupt Logic:** This logic supports 5 interrupts with the following features:
   - Priority
   - Masking and Non-masking
   - Vectoring and Non-vectoring

   The five interrupt signals are: TRAP, RST 7.5, RST 6.5, RST 5.5, and INTR.

   One signal called Interrupt acknowledge (INTA) is an output signal. This is to acknowledge the acceptance to service the interrupt.
5. **Serial I/O Logic:** This logic supports serial communication with the help of 2 signals: SID (input) and SOD (output).

## 3.9 INSTRUCTION SET

An **instruction** is a binary pattern designed inside a microprocessor to perform a specific function. The entire group of instructions, called the **instruction set**, determines what functions the microprocessor can perform. These instructions can be classified into the following five functional categories: data transfer (copy) operations, arithmetic operations, logical operations, branching operations, and machine-control operations.

### 3.9.1 Data Transfer (Copy) Operations

This group of instructions copy data from a location called a source to another location called a destination, without modifying the contents of the source. In technical manuals, the term *data transfer* is used for this copying function. However, the term *transfer* is misleading; it creates the impression that the contents of the source are destroyed when, in fact, the contents are retained without any modification. The various types of data transfer (copy) are listed below together with examples of each type:

| Type | Examples |
|---|---|
| 1. Between registers. | 1. Copy the contents of the register B into register D. |
| 2. Specific data byte to a register or a memory location. | 2. Load register B with the data byte 32H. |
| 3. Between a memory location and a register. | 3. From a memory location 2000H to register B. |
| 4. Between an I/O device and the accumulator. | 4. From an input keyboard to the accumulator. |

### 3.9.2 Arithmetic Operations

These instructions perform arithmetic operations such as addition, subtraction, increment, and decrement.

1. **Addition:** Any 8-bit number, or the contents of a register or the contents of a memory location can be added to the contents of the accumulator and the sum is stored in the accumulator. No two other 8-bit registers can be added directly (e.g., the contents of register B cannot be added directly to the contents of the register C). The instruction DAD is an exception; it adds 16-bit data directly in register pairs.
2. **Subtraction:** Any 8-bit number, or the contents of a register, or the contents of a memory location can be subtracted from the contents of the accumulator and the results stored in the accumulator. The subtraction is performed in 2's compliment, and the results if negative, are expressed in 2's complement. No two other registers can be subtracted directly.
3. **Increment/Decrement:** The 8-bit contents of a register or a memory location can be incremented or decrement by 1. Similarly, the 16-bit contents of a register pair (such as BC) can be incremented or decrement by 1. These increment and decrement operations differ from addition and subtraction in an important way; i.e., they can be performed in any one of the registers or in a memory location.

### 3.9.3 Logical Operations

These instructions perform various logical operations with the contents of the accumulator.

1. **AND, OR Exclusive-OR:** Any 8-bit number, or the contents of a register, or of a memory location can be logically ANDed, ORed, or Exclusive-ORed with the contents of the accumulator. The results are stored in the accumulator.
2. **Rotate:** Each bit in the accumulator can be shifted either left or right to the next position.

3. **Compare:** Any 8-bit number, or the contents of a register, or a memory location can be compared for equality, greater than, or less than, with the contents of the accumulator.
4. **Complement:** The contents of the accumulator can be complemented. All 0s are replaced by 1s and all 1s are replaced by 0s.

### 3.9.4 Branching Operations

This group of instructions alters the sequence of program execution either conditionally or unconditionally.

1. **Jump:** Conditional jumps are an important aspect of the decision-making process in the programming. These instructions test for a certain conditions (e.g., Zero or Carry flag) and alter the program sequence when the condition is met. In addition, the instruction set includes an instruction called *unconditional jump.*
2. **Call, Return, and Restart:** These instructions change the sequence of a program either by calling a subroutine or returning from a subroutine. The conditional Call and Return instructions also can test condition flags.

### 3.9.5 Machine Control Operations

These instructions control machine functions such as Halt, Interrupt, or do nothing. The microprocessor operations related to data manipulation can be summarized in four functions:

1. Copying data
2. Performing arithmetic operations
3. Performing logical operations
4. Testing for a given condition and alerting the program sequence

Some important aspects of the instruction set are noted below:

1. **In data transfer,** the contents of the source are not destroyed; only the contents of the destination are changed. The data copy instructions do not affect the flags.
2. **Arithmetic and Logical operations** are performed with the contents of the accumulator, and the results are stored in the accumulator (with some expectations). The flags are affected according to the results.
3. Any register including the memory can be used for **increment and decrement.**
4. A program sequence can be changed either **conditionally or by testing** for a given data condition.

## 8085 INSTRUCTION SET

### 1. Data Transfer Instructions

| | |
|---|---|
| **MOV** | Copy from source to destination |

| | |
|---|---|
| **MVI** | Move immediate 8-bit |
| **LDA** | Load accumulator |
| **LDAX** | Load accumulator indirect |
| **LXI** | Load register pair immediate |
| **LHLD** | Load H and L registers direct |
| **STA** | Store accumulator direct |
| **STAX** | Store accumulator indirect |
| **SHLD** | Store H and L registers direct |
| **XCHG** | Exchange H and L with D and E |
| **SPHL** | Copy H and L registers to the stack pointer |
| **XTHL** | Exchange H and L with top of stack |
| **PUSH** | Push register pair onto stack |
| **POP** | Pop of stack to register pair |
| **OUT** | Output data from accumulator to a port with 8-bit address |
| **IN** | Input data to accumulator from a port with 8-bit address |

## 2. Arithmetic Instructions

| | |
|---|---|
| **ADD** | Add register or memory to accumulator |
| **ADC** | Add register to accumulator with carry |
| **ADI** | Add immediate to accumulator |
| **ACI** | Add immediate to accumulator with carry |
| **DAD** | Add register pair to H and L registers |
| **SUB** | Subtract register or memory from accumulator |
| **SBB** | Subtract source and borrow from accumulator |
| **SUI** | Subtract immediate from accumulator |
| **SBI** | Subtract immediate from accumulator with borrow |
| **INR** | Increment register or memory by 1 |
| **INX** | Increment register pair by 1 |
| **DCR** | Decrement register or memory by 1 |
| **DCX** | Decrement register pair by 1 |
| **DAA** | Decimal adjust accumulator |

## 3. Machine Control Instructions

| | |
|---|---|
| **NOP** | No operation |
| **HLT** | Halt |

| | |
|---|---|
| **DI** | Disable interrupts |
| **EI** | Enable interrupts |
| **RIM** | Read interrupt mask |
| **SIM** | Set interrupt mask |

### 4. Branching Instructions

| | |
|---|---|
| **JMP** | Jump unconditionally |
| **JC** | Jump on carry |
| **JNC** | Jump on no carry |
| **JP** | Jump on positive |
| **JM** | Jump on minus |
| **JZ** | Jump on zero |
| **JNZ** | Jump on no zero |
| **JPE** | Jump on parity even |
| **JPO** | Jump on parity odd |
| **CALL** | Call unconditionally |
| **CC** | Call on carry |
| **CNC** | Call on no carry |
| **CP** | Call on positive |
| **CM** | Call on minus |
| **CZ** | Call on zero |
| **CNZ** | Call on no zero |
| **CPE** | Call on parity even |
| **CPO** | Call on parity odd |
| **RET** | Return unconditionally |
| **RC** | Return on carry |
| **RNC** | Return on no carry |
| **RP** | Return on positive |
| **RM** | Return on minus |
| **RZ** | Return on zero |
| **RNZ** | Return on no zero |
| **RPE** | Return on parity even |
| **RPO** | Return on parity odd |
| **PCHL** | Load program counter with HL contents |
| **RST** | Restart |

## 5. Logical Instructions

| | |
|---|---|
| **CMP** | Compare register or memory with accumulator |
| **CPI** | Compare immediate with accumulator |
| **ANA** | Logical AND register or memory with accumulator |
| **ANI** | Logical AND immediate with accumulator |
| **XRA** | Exclusive OR register or memory with accumulator |
| **XRI** | Exclusive OR immediate with accumulator |
| **ORA** | Logical OR register or memory with accumulator |
| **ORI** | Logical OR immediate with accumulator |
| **RLC** | Rotate accumulator left |
| **RRC** | Rotate accumulator right |
| **RAL** | Rotate accumulator left through carry |
| **RAR** | Rotate accumulator right through carry |
| **CMA** | Complement accumulator |
| **CMC** | Complement carry |
| **STC** | Set carry |

## SUMMARY OF 8085 INSTRUCTION SET

### ACI data (8b)

**Description:** Add Immediate 8-bit data to the accumulator with carry.

**Bytes/M-Cycles/T-States:** 2/2/7

**Hex Code:** CE

**Flags:** All flags are affected based upon the result of the addition.

### ADC R

**Description:** Add register to accumulator with carry.

**Bytes/M-Cycles/T-States:** 1/1/4

| | | Register |
|---|---|---|
| **Hex Codes:** | 8F | A |
| | 88 | B |
| | 89 | C |
| | 8A | D |
| | 8B | E |

| | |
|---|---|
| 8C | H |
| 8D | L |

**Flags:** All flags are affected based upon the result of the addition.

### ADC M

**Description:** Add contents of memory location pointed to by HL register pair to the accumulator with carry.

**Bytes/M-Cycles/T-States:** 1/2/7

**Hex Codes:** 8E

**Flags:** All flags are affected based upon the result of the addition.

### ADD R

**Description:** Add register to accumulator.

**Bytes/M-Cycles/T-States:** 1/1/4

| | | Register |
|---|---|---|
| **Hex Codes:** | 87 | A |
| | 80 | B |
| | 81 | C |
| | 82 | D |
| | 83 | H |
| | 84 | E |
| | 85 | L |

**Flags:** All flags are affected based upon the result of the addition.

### ADD M

**Description:** Add contents of memory location pointed to by HL to the accumulator.

**Bytes/M-Cycles/T-States:** 1/2/7

**Hex Codes:** 86

**Flags:** All flags are affected based upon the result of the addition.

### ADI Data (8b)

**Description:** Add the immediate 8-bit data to the accumulator.

**Bytes/M-Cycles/T-States:** 2/2/7

**Hex Codes:** C6

**Flags:** All flags are affected based upon the result of the addition.

### ANA R

**Description:** The contents of the accumulator and the register are logically ANDed and the result is put in the accumulator.

**Bytes/M-Cycles/T-States:** 1/1/4

| | | Register |
|---|---|---|
| **Hex Codes:** | A7 | A |
| | A0 | B |
| | Al | C |
| | A2 | D |
| | A3 | E |
| | A4 | H |
| | A5 | L |

**Flags:** S, Z, and P are modified based upon the result of the operation. CY is reset, and AC is set.

### ANA M

**Description:** The contents of the accumulator and the contents of the memory location pointed to by HL are logically ANDed, and the result is put in the accumulator.

**Bytes/M-Cycles/T-States:** 1/2/7

**Hex Codes:** A6

**Flags:** S, Z, and P are modified based upon the result of the operation. CY is reset, and AC is set.

### ANI Data (8b)

**Description:** The contents of the accumulator and the 8-bit data are ANDed and the result is put in the accumulator.

**Bytes/M-Cycles/T-States:** 2/2/7

**Hex Codes:** E6

**Flags:** S, Z, and P are modified based upon the result of the operation. CY is reset, and AC is set.

### CALL Address (16b)

**Description:** The program sequence is transferred to the address specified by the 16-bit address. Before the program is transferred, the address of the instruction following the CALL instruction is pushed onto the stack.

**Bytes/M-Cycles/T-States:** 3/5/18

**Hex Codes:** CD

**Flags:** No flags are affected.

### CC Address (16b)

**Description:** The program sequence is transferred to the address specified by the 16-bit address if the CY flag is set. If CY = 0, no transfer takes place. If the transfer takes place, the address of the instruction following the CC instruction is pushed onto the stack.

**Bytes/M-Cycles/T-States:** 3/2/9 if transfer is not taken
3/5/18 if the transfer is taken

**Hex Codes:** DC

**Flags:** No flags are affected.

### CNC Address (16b)

**Description:** The program sequence is transferred to the address specified by the 16-bit address if the CY flag is not set. If CY = 1, no transfer takes place. If the transfer takes place, the address of the instruction following the CNC instruction is pushed onto the stack.

**Bytes/M-Cycles/T-States:** 3/2/9 if transfer is not taken
3/5/18 if the transfer is taken

**Hex Codes:** D4

**Flags:** No flags are affected.

### CP Address (16b)

**Description:** The program sequence is transferred to the address specified by the 16-bit address if positive, or if the S flag = 0. If S = 1, no transfer takes place. If the transfer takes place, the address of the instruction following the CP instruction is pushed onto the stack.

**Bytes/M-Cycles/T-States:** 3/2/9 if transfer is not taken
3/5/18 if the transfer is taken

**Hex Codes:** F4

**Flags:** No flags are affected.

### CM Address (16b)

**Description:** The program sequence is transferred to the address specified by the 16-bit address if minus, or if the S flag = 1. If S = 0, no transfer takes place. If the transfer takes place, the address of the instruction following the CM instruction is pushed onto the stack.

**Bytes/M-Cycles/T-States:** 3/2/9 if transfer is not taken
3/5/18 if the transfer is taken

**Hex Codes:** FC

**Flags:** No flags are affected.

### CPE Address (16b)

**Description:** The program sequence is transferred to the address speci-fied by the 16-bit address if parity is even, or the P flag = 1. If P = 0, no transfer takes place. If the transfer takes place, the address of the instruction following the CPE instruction is pushed onto the stack.

| **Bytes/M-Cycles/T-States:** | 3/2/9 | if transfer is not taken |
|---|---|---|
| | 3/5/18 | if the transfer is taken |

**Hex Codes:** EC

**Flags:** No flags affected.

### CPO Address (16b)

**Description:** The program sequence is transferred to the address speci-fied by the 16-bit address if parity is odd, or if the P flag = 0. If P = 1, no transfer takes place. If the transfer takes place, the address of the instruction following the CPO instruction is pushed onto the stack.

| **Bytes/M-Cycles/T-States:** | 3/2/9 | if transfer is not taken |
|---|---|---|
| | 3/5/18 | if the transfer is taken |

**Hex Codes:** E4

**Flags:** No flags are affected.

### CZ Address (16b)

**Description:** The program sequence is transferred to the address specified by the 16-bit address if zero, or if the Z flag = **1.** If Z = 0, no transfer takes place. If the transfer takes place, the address of the instruction following the CZ instruction is pushed onto the stack.

| **Bytes/M-Cycles/T-States:** | 3/2/9 | if transfer is not taken |
|---|---|---|
| | 3/5/18 | if the transfer is taken |

**Hex Codes:** CC

**Flags:** No flags are affected.

### CNZ Address (16b)

**Description:** The program sequence is transferred to the address speci-fied by the 16-bit address if not zero, or if the Z flag = 0. If Z = **1,** no transfer takes place. If the transfer takes place, the address of the instruction following the CNZ instruction is pushed onto the stack.

| **Bytes/M-Cycles/T-States:** | 3/2/9 | if transfer is not taken |
|---|---|---|
| | 3/5/18 | if the transfer is taken |

**Hex Codes:** C4

**Flags:** No flags are affected.

### CMA

**Description:** The contents of the accumulator are complemented.

**Bytes/M-Cycles/T-States:** 1/1/4

**Hex Codes:** 2F

**Flags:** No flags are affected.

### CMC

**Description:** The carry flag is complemented.

**Bytes/M-Cycles/T-States:** 1/1/4

**Hex Codes:** 3F

**Flags:** The CY flag is complemented. No other flags are affected.

### CMP R

**Description:** The contents of the register are compared to the contents of the accumulator. Both contents are unaffected, and the following flags are used to show the results of the compare:

If $A < R$ CY = 1 and Z = 0

If $A = R$ CY=O and Z= 1

If $A > R$ CY=O and Z = 0

**Bytes/M-Cycles/T-States:** 1/1/4

| | | Registers |
|---|---|---|
| **Hex Codes:** | BF | A |
| | B8 | B |
| | B9 | C |
| | BA | D |
| | BB | E |
| | BC | H |
| | BD | L |

**Flags:** S, P, and AC are also affected based upon the results of the operation, besides Z and CY.

### CMP M

**Description:** The contents of the memory location pointed to by HL are compared to the contents of the accumulator. Both contents are unaffected, and the following flags are used to show the results of the compare:

If A. < R CY = 1 and Z = 0

If A = R CY=0 and Z = 1

If A > R CY=O and Z = 0

**Bytes/M-Cycles/T-States:** 1/2/7

**Hex Codes:** BE

**Flags:** S, P, and AC are also affected based upon the results of the operation, besides Z and CY.

### CPI data (8b)

**Description:** The 8-bit data is compared with the contents of the accumulator. The contents of the accumulator arc unaffected. The following flags are used to show the results of the compare:

If A < R CY =I and Z = 0

If A = R CY = 0 and Z = 1

If A > R CY= 0 and Z = 0

**Bytes/M-Cycles/T-States:** 2/2/7

**Hex Codes:** FE

**Flags:** S, P, and AC are also affected based upon the results of the operation, besides Z and CY.

### DAA

**Description:** The contents of the accumulator are converted from a binary value to two 4-bit Binary Coded Decimal (BCD) digits.

**Bytes/M-Cycles/T-States:** 1/1/4

Hex Codes: 27

**Flags:** S, Z, AC, P, and CY flags are affected based upon the results of the operation.

### DAD Rp

**Description:** The contents of the register pair (Rp) are added to the contents of the register pair HL. The source register pair is unchanged, and the results are stored in **HL.**

Bytes/M-Cycles/T-States: 1/3/10

Register Pair

| **Hex Codes:** | |
|---|---|
| 09 | BC |
| 19 | DE |
| 29 | HL |
| 39 | SP |

**Flags:** If the result is larger than 16 bits, the CY flag is set, otherwise no flags are affected.

## DCR R

**Description:** The contents of the register are decremented by one. The result is stored in the register.

**Bytes/M-Cycles/T-States:** 1/1/4

| | | Register |
|---|---|---|
| **Hex Codes:** | 3D | A |
| | 05 | B |
| | 0D | C |
| | 15 | D |
| | 1D | E |
| | 25 | H |
| | 2D | L |

**Flags:** S, AC, Z, and P are affected by the results of the operation. The CY flag is not affected.

## DCR M

**Description:** The contents of the memory location pointed to by **HL** are decremented by one, and the results are stored in the memory location.

**Bytes/M-Cycles/T-States:** 1/3/10

**Hex Code:** 35

**Flags:** S, AC, Z, and P are affected by the results of the operation. The CY flag is not affected.

## DCX Rp

**Description:** The contents of the register pair are decremented by 1. The result is stored in the register pair. The register pair is treated as a 16-bit number.

**Bytes/M-Cycles/T-States:** 1/1/6

| Hex Codes | Register Pair |
|---|---|
| 0B | BC |
| 1B | DE |
| 2B | HL |
| 3B | SP |

**Flags:** No flags are affected.

## DI

**Description:** The Interrupt Enable flip-flop is reset, and all of the interrupts except the TRAP interrupt are disabled.

**Bytes/M-Cycles/T-States:** 1/1/4

**Hex Codes:** F3

**Flags:** No flags are affected.

## EI

**Description:** The interrupt Enable flip-flop is set and all interrupts are enabled.

**Bytes/M-Cycles/T-States:** 1/1/4

**Hex Codes:** FB

**Flags:** No flags are affected.

## HLT

**Description:** The MPU finishes executing the current instruction and halts any further execution. The MPU enters the Halt Acknowledge machine cycle, and Wait states are inserted in every clock period. It requires an interrupt or a reset to get the MPU out of the Halt state.

**Bytes/M-Cycles/T-States:** One / two or more / five or more

**Hex Codes:** 76

**Flags:** No flags are affected.

## IN Port Address (8b)

**Description:** The contents of the input port designated are read and loaded into the accumulator.

**Bytes/M-Cycles/T-States:** 2/3/10

**Hex Codes:** DB

**Flags:** No flags are affected.

## INR R

**Description:** The contents of the register are incremented by one and stored in the register.

**Bytes/M-Cycles/T-States:** 1/1/4

| | | Register |
|---|---|---|
| **Hex Codes:** | 3C | A |
| | 04 | B |
| | 0C | C |
| | 14 | D |
| | 1C | E |
| | 24 | H |
| | 2C | L |

**Flags:** S, Z, P, AC are affected by the results of the operation. CY is not modified.

## INR M

**Description:** The contents of the memory location pointed to by HL are incremented by one and the result is put in the memory location.

**Bytes/M-Cycles/T-States:** 1/3/10

**Hex Codes:** 34

**Flags:** S, Z, P, and AC are affected by the results of the operation. CY is not modified.

## INN X Rp

**Description:** The contents of the register pair are incremented by 1 and stored in the register pair. The instruction views the two registers as a 16-bit number.

**Bytes/M-Cycles/T-States:** 1/1/6

| | | Register Pair |
|---|---|---|
| **Hex Codes:** | 03 | BC |
| | 13 | DE |
| | 23 | HL |
| | 33 | SP |

**Flags:** No flags are affected.

## JMP Address (16b)

**Description:** The program execution is transferred to the memory address specified.

**Bytes/M-Cycles/T-States:** 3/3/10

**Hex Codes:** C3

**Flags:** No flags are affected.

## JC Address (16b)

**Description:** Program execution is transferred to the memory address specified if the carry flag is set, or CY = 1. If CY = 0, no transfer takes place.

| | | |
|---|---|---|
| **Bytes/M-Cycles/T-States:** | 3/2/7 | if condition is not true |
| | 3/3/10 | if condition is true |

**Hex Codes:** DA.

**Flags:** No flags are affected.

## JNC Address (16b)

**Description:** Program execution is transferred to the memory address specified if the carry flag is not set, or CY = 0. If CY = 1, no transfer takes place.

**Bytes/M-Cycles/T-States:** 3/2/7 if condition is not true

3/3/10 if condition is true

**Hex Codes:** D2

**Flags:** No flags are affected.

### JP Address (16b)

**Description:** Program execution is transferred to the memory address specified if positive, or S = 0. If S = 1, no transfer takes place.

| **Bytes/M-Cycles/T-States:** | 3/2/7 | if condition is not true |
|---|---|---|
| | 3/3/10 | if condition is true |

**Hex Codes:** F2

**Flags:** No flags are affected.

### JM Address (16b)

**Description:** Program execution is transferred to the memory address specified if minus, or S = 1. If S = 0, no transfer takes place.

| **Bytes/M-Cycles/T-States:** | 3/2/7 | if condition is not true |
|---|---|---|
| | 3/3/10 | if condition is true |

**Hex Codes:** FA

**Flags:** No flags are affected.

### JPE Address (16b)

**Description:** Program execution is transferred to the memory address specified if parity is even, or P = 1. If P = 0, no transfer takes place.

| **Bytes/M-Cycles/T-States:** | 3/2/7 | if condition is not true |
|---|---|---|
| | 3/3/10 | if condition is true |

**Hex Codes:** EA

**Flags:** No flags are affected.

### JPO Address (16b)

**Description:** Program execution is transferred to the memory address specified if parity is odd, or P = 0. If P = 1, no transfer takes place.

| **Bytes/M-Cycles/T-States:** | 3/2/7 | if condition is not true |
|---|---|---|
| | 3/3/10 | if condition is true |

**Hex Codes:** E2

**Flags:** No flags are affected.

### JZ Address (16b)

**Description:** Program execution is transferred to the memory address specified if zero, or Z = 1. If Z = 0, no transfer takes place.

**Bytes/M-Cycles/T-States:** 3/2/7 if condition is not true
3/3/10 if condition is true

**Hex Codes:** CA

**Flags:** No flags are affected.

### JNZ Address (16b)

**Description:** Program execution is transferred to the memory address specified if not zero, or Z = 0. If Z = 1, no transfer takes place.

**Bytes/M-Cycles/T-States:** 3/2/7 if condition is not true
3/3/10 if condition is true

**Hex Codes:** C2

**Flags:** No flags are affected.

### LDA Address (16b)

**Description:** The contents of the memory location specified are transferred to the accumulator.

**Bytes/M-Cycles/T-States:** 3/4/13

**Hex Codes:** 3A

**Flags:** No flags are affected.

### LDAX Rp

**Description:** The contents of the memory location pointed to by the register pair are loaded into the accumulator.

**Bytes/M-Cycles/T-States:** 1/2/7

| | | Register Pairs |
|---|---|---|
| **Hex Codes:** | 0A | BC |
| | 1A | DE |

**Flags:** No flags are affected.

### LHLD Address(16b)

**Description:** The contents of the memory location specified are loaded into register L and the contents of the next memory location are loaded into register H.

**Bytes/M-Cycles/T-States:** 3/5/16

**Hex Codes:** 2A

**Flags:** No flags are affected.

## LXI Rp, Sata (16b)

**Description:** The 16-bit data is loaded into the register pair.

**Bytes/M-Cycles/T-States:** 3/3/10

| | | Register Pair |
|---|---|---|
| **Hex Codes:** | 01 | BC |
| | 11 | DE |
| | 21 | HL |
| | 31 | SP |

**Flags:** No flags are affected.

## MOV Rd, Rs

**Description:** The contents of the source register Rs are transferred into the destination register Rd.

**Bytes/M-Cycles/T-States:** 1/1/4

**Hex Codes:** Source Register

| | A | B | C | D | E | H | L |
|---|---|---|---|---|---|---|---|
| A | 7F | 78 | 79 | 7A | 7B | 7C | 7D |
| B | 47 | 40 | 41 | 42 | 43 | 44 | 45 |
| C | 41 | 48 | 49 | 4A | 4B | 4C | 4D |
| D | 57 | 50 | 51 | 52 | 53 | 54 | 55 |
| E | 5F | 58 | 59 | 5A | 5B | 5C | 5D |
| H | 67 | 60 | 61 | 62 | 63 | 64 | 65 |
| L | 6F | 68 | 69 | 6A | 6B | 6C | 6D |

**Flags:** No flags are affected.

## MOV M, Rs

**Description:** The contents of the source register Rs are transferred to the memory location pointed to by **HL.**

**Bytes/M-Cycles/T-States:** 1/2/7

| **Hex Codes:** | **Source Register** |
|---|---|
| 77 | A |
| 70 | B |
| 71 | C |

| | |
|---|---|
| 72 | D |
| 73 | E |
| 74 | H |
| 75 | L |

**Flags:** No flags are affected.

## MOV Rd, M

**Description:** The contents of the memory location pointed to by HL are transferred to the destination register.

**Bytes/M-Cycles/T-States:** 1/2/7

| **Hex Codes:** | **Destination Register** |
|---|---|
| 7E | A |
| 46 | B |
| 4E | C |
| 56 | D |
| 5E | E |
| 66 | H |
| 6E | L |

**Flags:** No flags are affected.

## MVI R, Data(8b)

**Description:** The 8 bits of data are stored in the register.

**Bytes/M-Cycles/T-States:** 2/2/7

| **Hex Codes:** | Register |
|---|---|
| 3E | A |
| 06 | B |
| 0E | C |
| 16 | D |
| lE | E |
| 26 | H |
| 2E | L |

**Flags:** No flags are affected.

## MVI M, data (8b)

**Description:** The 8 bits of data are stored in the memory location pointed to by **FIL.**

**Bytes/M-Cycles/T-States:** 2/3/10

**Hex Codes:** 36

**Flags:** No flags are affected.

### NOP

**Description:** No operation is performed. The instruction is fetched and decoded, but no operation is executed.

**Bytes/M-Cycles/T-States:** 1/1/4

**Hex Codes:** 00

**Flags:** No flags are affected.

### ORA R

**Description:** The contents of the accumulator are logically OR'd with the contents of the register. The result is stored in the accumulator.

**Bytes/M-Cycles/T-States:** 1/1/4

| | | Register |
|---|---|---|
| **Hex Codes:** | B7 | A |
| | B0 | B |
| | B1 | C |
| | B2 | D |
| | B3 | E |
| | B4 | H |
| | B5 | L |

**Flags:** Z, S, and P are affected based upon the operation. AC and CY are reset.

### ORA M

**Description:** The contents of the accumulator are logically OR'd with the contents of the memory location pointed to by HL.

**Bytes/M-Cycles/T-States:** 1/2/7

**Hex Codes:** B6

**Flags:** Z, S, and P are affected based upon the operation. AC and CY are reset.

### ORI Data (8b)

**Description:** The contents of the accumulator are logically OR'd with the 8 bits of data. The result is stored in the accumulator.

**Bytes/M-Cycles/T-States:** 2/2/7

**Hex Codes:** F6

**Flags: Z,** S, and P are affected based upon the operation. AC and CY are reset.

## OUT Port Address (8b)

**Description:** The contents of the accumulator are copied out to the output port specified.

**Bytes/M-Cycles/T-States:** 2/3/10

**Hex Codes:** D3

**Flags:** No flags are affected.

## PCHL

**Description:** The contents of registers H and L are copied into the program counter. H is the high-order bits and L is the low-order bits.

**Bytes/M-Cycles/T-States:** 1/1/6

**Hex Codes:** E9

**Flags:** No flags are affected.

## POP Rp

**Description:** The contents of the memory location (stack) pointed to by the stack pointer are copied to the low-order register of the register pair. (C, E, L, and flags.) The stack pointer is then incremented and the contents of that memory location being pointed to are copied to the high-order register of the register pair.

**Bytes/M-Cycles/T-States:** 1/3/10

| | | Register |
|---|---|---|
| **Hex Codes:** | C1 | BC |
| | D1 | DE |
| | E1 | HL |
| | F1 | PSW |

**Flags:** No flags are affected.

## PUSH Rp

**Description:** The contents of the register pair are copied into the stack. The high-order register (B, D, H, A) is put on the stack first, then the contents of the low-order register (C, E, L, flags) are put onto the stack.

**Bytes/M-Cycles/T-States:** 1/3/12

Register Pair

**Hex Codes:** C5 B
D5 D
E5 H
F5 PSW

**Flags:** No flags are affected.

## RAL

**Description:** The contents of the accumulator are rotated left by one position through the carry flag.

**Bytes/M-Cycles/T-States:** 1/1/4

**Hex Codes:** 17

**Flags:** CY is modified according to hit D7. S, Z, P, and AC are not affected.

## RAR

**Description:** The contents of the accumulator are rotated right by one position through the carry flag.

Bytes/M-Cycles/T-States: 1/1/4

**Hex Codes:** 1F

**Flags:** CY is modified according to bit DO. S, Z, P, and AC are not affected.

## RLC

**Description:** The contents of the accumulator are rotated left one position. Bit D7 is placed in both DO and CY.

**Bytes/M-Cycles/T-States:** 1/1/4

**Hex Codes:** 07

**Flags:** CY is modified according to bit D7. S, Z, P, and AC are not affected.

## RRC

**Description:** The contents of the accumulator are rotated right by 1 bit. Bit DO is placed in both D7 and CY at the same time.

**Bytes/M-Cycles/T-States:** 1/1/4

**Hex Codes:** OF

**Flags:** CY is modified according to bit DO. S, Z, P, and AC are not affected.

## RET

**Description:** The program sequence is transferred from the subroutine to the calling program. The two bytes from the top of the stack are copied into the Program Counter and this is the address that program execution begins.

Bytes/M-Cycles/T-States: 1/3/10

**Hex Codes:** C9

**Flags:** No flags are affected.

## RC

**Description:** The program sequence is transferred from the subroutine to the calling program if the carry flag is set, or CY = 1. If CY = 0, no transfer takes place.

| **Bytes/M-Cycles/T-States:** | 1/1/6 | if condition is not true |
|---|---|---|
| | 1/3/12 | if condition is true |

**Hex Codes:** D8

**Flags:** No flags are affected.

## RNC

**Description:** The program sequence is transferred from the subroutine to the calling program if the carry flag is **not** set, or CY = 0. If CY=1, no transfer takes place.

| **Bytes/M-Cycles/T-States:** | 1/1/6 | if condition is not true |
|---|---|---|
| | 1/3/12 | if condition is true |

**Hex Codes:** D0

**Flags:** No flags are affected.

## RP

**Description:** the program sequence is transferred from the subroutine to the calling program if positive, or S = 0. If S = 1, no transfer takes place.

| **Bytes/M-Cycles/T-States:** | 1/1/6 | if condition is not true |
|---|---|---|
| | 1/3/12 | if condition is true |

**Hex Codes:** F0

**Flags:** No flags are affected.

## RM

**Description:** The program sequence is transferred from the subroutine to the calling program if minus, or S = 1. If S = 0, no transfer takes place.

**Bytes/M-Cycles/T-States:** 1/1/6 if condition is not true

| | 1/3/12 | if condition is true |
|---|---|---|

**Hex Codes:** F8

**Flags:** No flags are affected.

## RPE

**Description:** The program sequence is transferred from the subroutine to the calling program if the parity is even, or P = 1. If P = 0, no transfer takes place.

| **Bytes/M-Cycles/T-States:** | 1/1/6 | if condition is not true |
|---|---|---|
| | 1/3/12 | if condition is true |

**Hex Codes:** E8

**Flags:** No flags are affected

## RPO

**Description:** The program sequence is transferred from the subroutine to the calling program if the parity is odd, or P =0. If P=1, no transfer takes place.

| **Bytes/M-Cycles/T-States:** | 1/1/6 | if condition is not true |
|---|---|---|
| | 1/3/12 | if condition is true |

**Hex Codes:** E0

**Flags:** No flags are affected

## RZ

**Description:** The program sequence is transferred from the subroutine to the calling program if zero, or Z = 1. If Z = 0, no transfer takes place.

| **Bytes/M-Cycles/T-States:** | 1/1/6 | if condition is not true |
|---|---|---|
| | 1/3/12 | if condition is true |

**Hex Codes:** C8

**Flags:** No flags are affected.

## RNZ

**Description:** The program sequence is transferred from the subroutine to the calling program if not zero, or Z = 0. If Z = 1, no transfer takes place.

| **Bytes/M-Cycles/T-States:** | 1/1/6 | if condition is not true |
|---|---|---|
| | 1/3/12 | if condition is true |

**Hex Codes:** C0

**Flags:** No flags are affected.

## RIM

**Description:** This instruction is used to both read in the status of interrupts 7.5, 6.5, and 5.5, as well as to read in the serial input data bit. An 8-bit word is read in and stored in the accumulator. The layout of that word is shown in Figure A1-1.

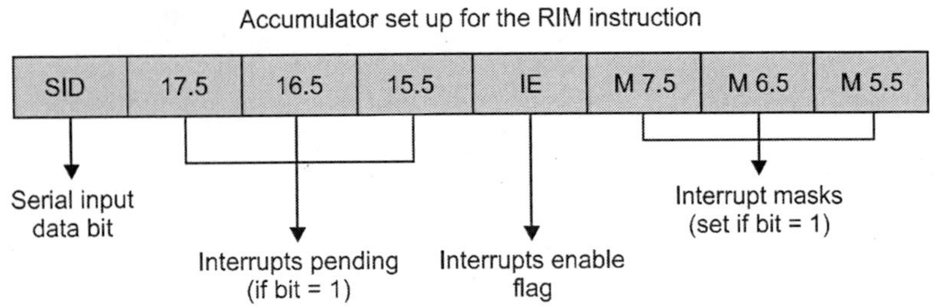


**Figure A1-1 n** The RIM instruction layout in the accumulator

**Bytes/M-Cycles/T-States:** 1/1/4

**Hex Codes:** 20

**Flags:** No flags are affected

## RST n (where n = 0 – 7)

**Description:** This instruction operates like a call instruction that goes to one of eight predetermined memory locations on page 0. Each instruction (RST 0-RST 7) goes to a specific address listed next.

| Instruction | Restart Address |
|---|---|
| RST 0 | 0000 |
| RST 1 | 0008 |
| RST 2 | 0010 |
| RST 3 | 0018 |
| RST 4 | 0020 |
| RST 5 | 0028 |
| RST 6 | 0030 |
| RST 7 | 0038 |

**Bytes/M-Cycles/T-States:** 1/3/12

**Hex Codes:**

| | |
|---|---|
| C7 | RST 0 |
| CF | RST 1 |
| D7 | RST 2 |
| DF | RST 3 |
| E7 | RST 4 |

| | |
|---|---|
| EF | RST 5 |
| F7 | RST 6 |
| FF | RST 7 |

**Flags:** No flags are accepted.

### SBB R

**Description:** The contents of the register and the borrow flag are subtracted from the contents of the accumulator and the results are stored in the accumulator.

**Bytes/M-Cycles/T-States:** 1/1/4

| | | Register |
|---|---|---|
| **Hex Codes:** | 9F | A |
| | 98 | B |
| | 99 | C |
| | 9A | D |
| | 9B | E |
| | 9C | H |
| | 9D | L |

**Flags:** All flags are affected based upon the results of the operation.

### SBB M

**Description:** The contents of the memory location pointed to by **HL** and the borrow flag are subtracted from the contents of the accumulator. The results are then stored in the accumulator.

**Bytes/M-Cycles/T-States:** 1/2/7

**Hex Codes:** 9E

**Flags:** All flags are affected based upon the results of the operation.

### SBI Data (8b)

**Description:** The 8 bits of data and the borrow flag are subtracted from the accumulator and the results are stored in the accumulator.

**Bytes/M-Cycles/T-States:** 2/2/7

**Hex Codes:** DE

**Flags:** All flags are affected based upon the results of the operation.

### SHLD Address (16b)

**Description:** The contents of register L are stored at the memory location specified and the contents of register H are stored at the next memory location by incrementing the operand by 1.

**Bytes/M-Cycles/T-States:** 3/5/16

**Hex Codes:** 22

**Flags:** No flags are affected.

## SIM

**Description:** This is an instruction that is used to set the interrupt masks as well as set the serial output data bit. The accumulator is laid out as shown in Figure A1-2.

Accumulator set up for the SIM instruction

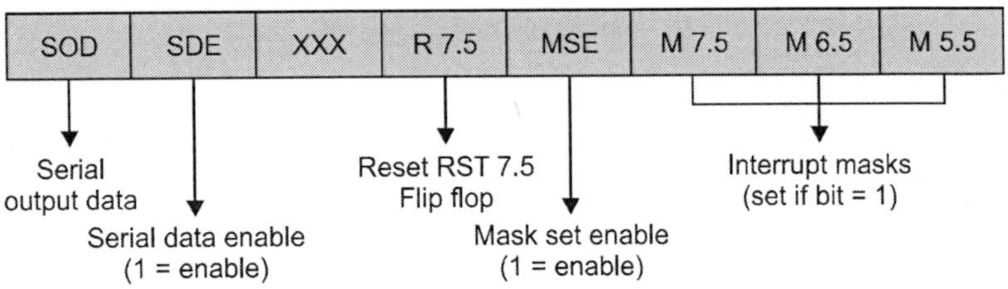


**The SIM instruction accumulator layout**

**Bytes/M-Cycles/T-States:** 1/1/4

**Hex Codes:** 30

**Flags:** No flags are affected.

## SPHL

**Description:** The contents of registers H and L are loaded into the Stack Pointer. H has the high-order part of the address, while L has the low-order portion.

**Bytes/M-Cycles/T-States:** 1/1/6

**Hex Codes:** F9

**Flags:** No flags are affected.

## STA Address (16b)

**Description:** The contents of the accumulator are stored at the memory location specified.

**Bytes/M-Cycles/T-States:** 3/4/13

**Hex Codes:** 32

**Flags:** No flags are affected.

## STAX Rp (only BC or DE)

**Description:** The contents of the accumulator are stored at the memory location pointed to by the register pair. The contents of the accumulator are not affected.

**Bytes/M-Cycles/T-States:** 1/2/7

| | Register Pair |
|---|---|
| **Hex Codes:** 02 | BC |
| 12 | DE |

**Flags:** No flags are affected.

### STC

**Description:** The carry flag, CY, is set to 1.

**Bytes/M-Cycles/T-States:** 1/1/4

**Hex Codes:** 37

**Flags:** Only the CY flag is affected.

### SUB R

**Description:** The contents of the register are subtracted from the accumulator and the result is stored in the accumulator.

**Bytes/M-Cycles/T-States:** 1/1/4

| | Register |
|---|---|
| **Hex Codes:** 97 | A |
| 90 | B |
| 91 | C |
| 92 | D |
| 93 | E |
| 94 | H |
| 95 | **L** |

**Flags:** All flags are affected by the result of the operation.

### SUB M

**Description:** The contents of the memory location pointed to by HL are subtracted from the accumulator and the result is stored in the accumulator.

**Bytes/M-Cycles/T-States:** 1/2/7

**Hex Codes:** 96

**Flags:** All flags are affected by the result of the operation.

### SUI Data (8b)

**Description:** The 8-bit data is subtracted from the contents of the accumulator. The result is stored in the accumulator.

**Bytes/M-Cycles/T-States:** 2/2/7

**Hex Codes:** D6

**Flags:** All flags are affected by the result of the operation.

### XCHG

**Description:** The contents of register H and register D are exchanged, and the contents of register L and register E are exchanged.

**Bytes/M-Cycles/T-States:** 1/1/4

**Hex Codes:** EB

**Flags:** No flags are affected.

### XRA R

**Description:** The contents of the register are Exclusive OR'd with the contents of the accumulator. The results are then stored in the accumulator.

**Bytes/M-Cycles/T-States:** 1/1/4

| | | Register |
|---|---|---|
| **Hex Codes:** | AF | A |
| | A8 | B |
| | A9 | C |
| | AA | D |
| | AB | E |
| | AC | H |
| | AD | L |

**Flags:** Z, S, and P are affected based upon the operation. CY and AC are reset.

### XRA M

**Description:** The contents of the memory location pointed to by HL are Exclusive OR'd with the contents of the accumulator. The results are stored in the accumulator.

**Bytes/M-Cycles/T-States:** 1/2/7

**Hex Codes:** AE

**Flags:** Z, S, and P are affected based upon the operation. CY and AC are reset.

### XRI Data (8b)

**Description:** The 8 bits of data are Exclusive OR'd with the contents of the accumulator. The results are then stored in the accumulator.

**Bytes/M-Cycles/T-States:** 2/2/7

**Hex Codes:** EE

**Flags:** Z, S, and P are affected based upon the operation. CY and AC are reset.

### XTHL

**Description:** The contents of the L register are exchanged with the stack location pointed to by the stack pointer. The contents of the H register are exchanged with the stack location pointed to by the stack pointer +1. The stack pointer remains unchanged.

**Bytes/M-Cycles/T-States:** 1/5/16

**Hex Codes:** E3

**Flags:** No flags are affected.

## 3.10 ADDRESSING MODES

Addressing modes are an aspect of the instruction set architecture in most central processing unit (CPU) designs. The various addressing modes that are defined in a given instruction set architecture define how machine language instructions in that architecture identify the operand (or operands) of each instruction. An addressing mode specifies how to calculate the effective memory address of an operand by using information held in registers and/or constants contained within a machine instruction or elsewhere.

The instructions MOV B, A or MVI A, 82H are to copy data from a source into a destination. In these instructions the source can be a register, an input port, or an 8-bit number (00H to FFH). Similarly, a destination can be a register or an output port. The sources and destination are operands. The various formats for specifying operands are called the ADDRESSING MODES.

**The 8085 Microprocessor has 5 Types of Addressing Modes**

1. Immediate Addressing mode
2. Register Addressing mode
3. Register Indirect Mode
4. Direct Addressing mode
5. Indirect Addressing mode
6. Implied Addressing mode

### 1. Immediate Addressing

This mode results in a two or three byte instruction. The first byte contains the instruction itself. The second and third bytes contain the immediate data, either as a single 8-bit value, or as a 16-bit value. If the 16-bit value is used, the bytes are reversed as discussed previously, with the second byte containing the low order byte, and the third byte containing the high order byte.

Data is present in the instruction. Load the immediate data to the destination provided. When the data is directly specified in the instruction, then it is called immediate addressing mode.

**Example:**

1. MVI R, data
2. MVI B, 72H
   LXI H, 5000H
3. MVI B, 3EH: Move the data 3EH given in the instruction to B register; LXI SP, 2700H.

## 2. Register Addressing

This mode results in single-byte instructions. The byte contains bits which specify a register or register pair in which the data is located.

Data is provided through the registers. When the data is stored in the register and if the register is specified in the instruction, then it is called register addressing mode

**Example:**

1. MOV Rd, Rs
2. MOV D, C: The contents of C are moved to register D.
3. MOV A, B - Move the content of B register to A register; SPHL; ADD C.

## 3. Register Indirect Mode

This mode results in single-byte instructions. The byte contains bits which specify a register pair, which in turn contains the address of the data in memory.

Thus, the instruction indirectly specifies the address of the data by referring to a register pair for the absolute address. Note that the high order byte of the address is stored in the leftmost register of the pair, while the low order byte of the address is stored in the rightmost register of the pair. The address 3000H, therefore, would be stored in the HL register pair as 30 in H, and 00 in L. EX. MOV A, M - The memory data addressed by H L pair is moved to A register. LDAX B.

## 4. Direct Addressing

This mode creates instructions three bytes long. The first byte contains the operation to be performed. The second and third bytes contain the address in memory where the data byte may be found. Thus, the instruction directly specifies the absolute location of the data.

Note that the second byte of the instruction contains the low order byte of the address, while the third byte contains the high order byte of the address. This illustrates the inverse addressing of the device.

When the memory Address specified with in the instruction, then it is called direct addressing mode. Used to accept data from outside devices to store in the accumulator or send the data stored in the accumulator to the outside device. Accept the data from the port 00H and store them into the accumulator or Send the data from the accumulator to the port 01H.

**Example:**

1. IN 00H or OUT 01H
2. LDA 5000H;

The 5000H memory content is stored in to the accumulator.

EX. LDA 1050H - Load the data available in memory location 1050H in to accumulator; SHLD 3000H

### 5. Indirect Addressing

This means that the Effective Address is calculated by the processor. And the content of the address (and the one following) is used to form a second address. The second address is where the data is stored.

Note that this requires several memory accesses; two accesses to retrieve the 16-bit address and a further access (or accesses) to retrieve the data which is to be loaded into the register.

**Example:**

MOV A, M

The data moved to accumulator from the memory, which is pointed by the memory pointer.

### 6. Implied Addressing Mode

When the instruction itself specifies the data to be operated, then it is called Implied addressing mode.

**Example:**

CMA - Complement the content of accumulator.

RRC - Rotate accumulator content with out carry.

CMA - Complement the content of accumulator; RAL

## 3.11 INSTRUCTION FORMAT

An **instruction** is a command to the microprocessor to perform a given task on a specified data. Each instruction has two parts: one is task to be performed, called the **operation code** (opcode), and the second is the data to be operated on, called the **operand.** The operand (or data) can be specified in various ways. It may include 8-bit (or 16-bit) data, an internal register, a memory location, or 8-bit (or 16-bit) address. In some instructions, the operand is implicit.

### 3.11.1 Instruction Word Size

The 8085 instruction set is classified into the following three groups according to word size:

1. One-word or 1-byte instructions
2. Two-word or 2-byte instructions

3. Three-word or 3-byte instructions

In the 8085, "byte" and "word" are synonymous because it is an 8-bit microprocessor. However, instructions are commonly referred to in terms of bytes rather than words.

### *1. One-byte Instructions*

A 1-byte instruction includes the opcode and operand in the same byte. Operand(s) are internal register and are coded into the instruction.

**For example:**

| Task | Op code | Operand | Binary | Hex |
|---|---|---|---|---|
| Copy the contents of the accumulator in the register C | MOV | C, A | 01001111 | 4FH |
| Add the contents of register B to the contents of the accumulator | ADD | B | 10000000 | 80H |
| Invert (compliment) each bit in the accumulator | CMA | | 00101111 | 2FH |

These instructions are 1-byte instructions performing three different tasks. In the first instruction, both operand registers are specified. In the second instruction, the operand B is specified and the accumulator is assumed. Similarly, in the third instruction, the accumulator is assumed to be the implicit operand. These instructions are stored in 8- bit binary format in memory; each requires one memory location.

### MOV rd, rs

rd <— rs copies contents of rs into rd.

Coded as 01 ddd sss where ddd is a code for one of the 7 general registers which is the destination of the data, sss is the code of the source register.

**Example:** MOV A,B

Coded as 01111000 = 78H = 170 octal (octal was used extensively in instruction design of such processors).

### ADD r

A <— A + r

### *2. Two-byte Instructions*

In a two-byte instruction, the first byte specifies the operation code and the second byte specifies the operand. Source operand is a data byte immediately following the opcode.

**For example:**

| Task | Op code | Operand | Binary | Hex | |
|---|---|---|---|---|---|
| Load on 8-bit data byte in the accumulator | MVI | A, Data | 00111110<br>Data | 3E<br>Data | First byte<br>Second byte |

Assume that the data byte is 32H. The assembly language instruction is written as

| Mnemonics | Hex code |
|---|---|
| MVI, A 32H | 3E 32H |

The instruction would require two memory locations to store in memory.

### MVI r, data

r <— data

Example: MVI A, 30H coded as 3EH 30H as two contiguous bytes. This is an example of immediate addressing.

### ADI data

A <— A + data

### OUT port

where port is an 8-bit device address. (Port) <— A. Since the byte is not the data but points directly to where it is located this is called direct addressing.

### *3. Three-byte Instructions*

In a three-byte instruction, the first byte specifies the opcode, and the following two bytes specify the 16-bit address. Note that the second byte is the low-order address and the third byte is the high-order address.

**opcode + data byte + data byte**

**For example:**

| Task | Opcode | Operand | Binary code | Hex Code | |
|---|---|---|---|---|---|
| Transfer the program sequence to the memory location 2085H | JMP | 2085H | 1100 0011<br>1000 0101<br>0010 0000 | C3<br>85<br>20 | First byte<br>Second byte<br>Third byte |

This instruction would require three memory locations to store in memory. Three byte instructions - opcode + data byte + data byte

### LXI rp, data16

rp is one of the pairs of registers BC, DE, HL used as 16-bit registers. The two data bytes are 16-bit data in L H order of significance.

rp <— data16

**Example:** LXI H,0520H coded as 21H 20H 50H in three bytes. This is also immediate addressing.

### LDA addr

A <— (addr) Addr is a 16-bit address in L H order. Example: LDA 2134H coded as 3AH 34H 21H. This is also an example of direct addressing.

### Programs Illustrate Concept of Addressing Modes

1. **Write an assembly program to add two numbers**

   Program

   MVI D, 8BH

   MVI C, 6FH

   MOV A, C

   1100 0011

   1000 0101

   0010 0000

   ADD D

   OUT PORT1

   HLT

2. **Write an assembly program to multiply a number by 8**

   Program

   MVI A, 30H

   RRC

   RRC

   RRC

   OUT PORT1

   HLT

3. **Write an assembly program to find greatest between two numbers**

   Program

   MVI B, 30H

   MVI C, 40H

```
MOV A, B
CMP C
JZ EQU
JC GRT
OUT PORT1
HLT
EQU: MVI A, 01H
OUT PORT1
HLT
GRT: MOV A, C
OUT PORT1
HLT
```

## 3.12 INTERRUPT STRUCTURE OF 8085

- Interrupt is signals send by an external device to the processor, to request the processor to perform a particular task or work.
- Mainly in the microprocessor based system the interrupts are used for data transfer between the peripheral and the microprocessor.
- The processor will check the interrupts always at the 2nd T-state of last machine cycle.
- If there is any interrupt it accept the interrupt and send the INTA (active low) signal to the peripheral.
- The vectored address of particular interrupt is stored in program counter.
- The processor executes an interrupt service routine (ISR) addressed in program counter.
- It returned to main program by RET instruction.

### 3.12.1 Types Of Interrupts

It supports two types of interrupts.

1. Hardware Interrupts
2. Software Interrupts

#### *1. Software Interrupts*

The software interrupts are program instructions. These instructions are inserted at desired locations in a program. The 8085 has eight software interrupts from RST 0 to RST 7. The vector address for these interrupts can be calculated as follows.

Interrupt number * 8 = vector address

For RST 5, 5 *8 = 40 = 28H

Vector address for interrupt RST 5 is 0028H

The Table shows the vector addresses of all interrupts.

| Intrrupt | Vector address |
|---|---|
| RST 0 | $0000_H$ |
| RST 1 | $0008_H$ |
| RST 2 | $0010_H$ |
| RST 3 | $0018_H$ |
| RST 4 | $0020_H$ |
| RST 5 | $0028_H$ |
| RST 6 | $0030_H$ |
| RST 7 | $0038_H$ |

### 2. *Hardware Interrupts*

An external device initiates the hardware interrupts and placing an appropriate signal at the interrupt pin of the processor.

If the interrupt is accepted then the processor executes an interrupt service routine.

The 8085 has five hardware interrupts:

(1) TRAP (2) RST 7.5 (3) RST 6.5 (4) RST 5.5 (5) INTR

| Intrrupt | Vector address |
|---|---|
| RST 7.5 | $003C_H$ |
| RST 6.5 | $0034_H$ |
| RST 5.5 | $002C_H$ |
| TRAP | $0024_H$ |

**(i) TRAP**

This interrupt is a non-maskable interrupt. It is unaffected by any mask or interrupt enable. TRAP bas the highest priority and vectored interrupt. TRAP interrupt is edge and level triggered. This means hat the TRAP must go high and remain high until it is acknowledged. In sudden power failure, it executes a ISR and send the data from main memory to backup memory. The signal, which overrides the TRAP, is HOLD signal. (i.e., If the processor receives HOLD and TRAP at the same time then HOLD is recognized first and then TRAP is recognized).

There are two ways to clear TRAP interrupt.

1. By resetting microprocessor (External signal)
2. By giving a high TRAP ACKNOWLEDGE (Internal signal)

### (ii) RST 7.5

The RST 7.5 interrupt is a maskable interrupt. It has the second highest priority. It is edge sensitive. i.e. Input goes to high and no need to maintain high state until it recognized.

**Maskable interrupt:** It is disabled by,

1. DI instruction
2. System or processor reset.
3. After reorganization of interrupt.

Enabled by EI instruction.

### (iii) RST 6.5 and 5.5

The RST 6.5 and RST 5.5 both are level triggered. i.e. Input goes to high and stay high until it recognized.

**Maskable interrupt:** It is disabled by,

1. DI, SIM instruction
2. System or processor reset.
3. After reorganization of interrupt.

Enabled by EI instruction.

The RST 6.5 has the third priority whereas RST 5.5 has the fourth priority.

### (iv) INTR

INTR is a maskable interrupt. It is disabled by,

1. DI, SIM instruction
2. System or processor reset.
3. After reorganization of interrupt.

Enabled by EI instruction.

**Non- vectored interrupt:** After receiving INTA (active low) signal, it has to supply the address of ISR. It has lowest priority. It is a level sensitive interrupts.i.e. Input goes to high and it is necessary to maintain high state until it recognized.

The following sequence of events occurs when INTR signal goes high.

1. The 8085 checks the status of INTR signal during execution of each instruction.
2. If INTR signal is high, then 8085 complete its current instruction and sends active lowinterrupt acknowledge signal, if the interrupt is enabled.
3. In response to the acknowledge signal, external logic places an instruction OPCODE on the data bus. In the case of multibyte instruction, additional interrupt acknowledge machine cycles are generated by the 8085 to transfer the additional bytes into the microprocessor.

4. On receiving the instruction, the 8085 save the address of next instruction on stack and executereceived instruction.

### (v) SIM and RIM for interrupts

The 8085 provide additional masking facility for RST 7.5, RST 6.5 and RST 5.5 using SIM instruction. The status of these interrupts can be read by executing RIM instruction. The masking or unmasking of RST 7.5, RST 6.5 and RST 5.5 interrupts can be performed by moving an 8-bit data to accumulator and then executing SIM instruction.

The format of the 8-bit data is shown below.

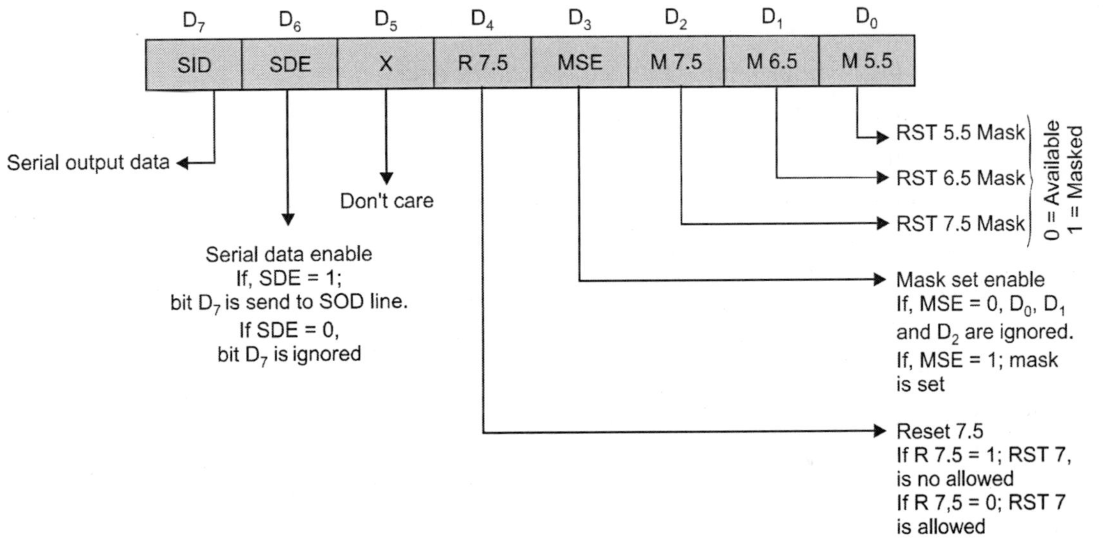


The status of pending interrupts can be read from accumulator after executing RIM instruction. When RIM instruction is executed an 8-bit data is loaded in accumulator, which can be interpreted as shown in Fig.

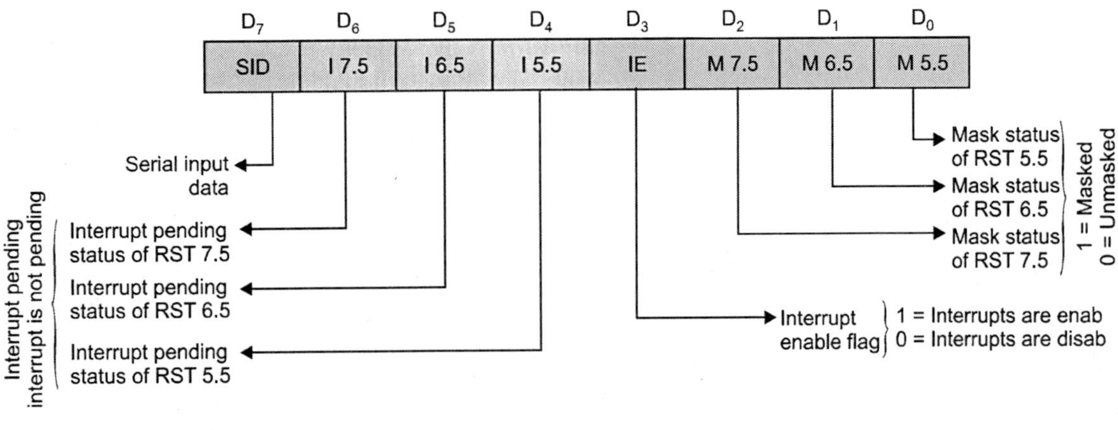

| Interrupt type | Trigger | Priority | Maskable | Vector address |
|---|---|---|---|---|
| TRAP | Edge and level | 1st | No | 0024H |
| RST 7.5 | Edge | 2nd | Yes | 003CH |
| RST 6.5 | Level | 3rd | Yes | 0034H |
| RST 5.5 | Level | 4th | Yes | 002CH |
| INTR | Level | 5th | Yes | – |

## QUESTION BANK

Q1. Explain the microprocessor architecture and its operations.

Q2. Explain the overview of 8085 system bus structure.

Q3. Explain different types of internal registers of 8085.

Q4. Explain the general description of microprocessor.

Q5. Discuss the 8085 microprocessor pin diagram and its pins description.

Q6. Explain the working of 8085 microprocessor in detail.

Q7. Discuss the microprocessor communication and bus timings.

Q8. Explain the memory interfacing and its structure with the help of diagrams.

Q9. Explain detailed description of 8085 microprocessor and its architecture.

Q10. What is instruction set? Explain different types of instructions.

Q11. Explain different types of addressing modes available in 8085 microprocessor.

Q12. What do you mean by instruction format? Also explain its different types.

Q13. What are interrupts? Explain different types of interrupts available in 8085 microprocessor.

Q14. Explain detail description of software and hardware interrupts.

CHAPTER 4

# 8085 Instruction Classification

## 4.1 INSTRUCTION ORGANIZATION

The 8085's instructions are made up of bytes. In microprocessor parlance, a byte is described as 8 contiguous binary bits treated as a unit. The least significant bit is on the right, and is labeled Bit 0. The most significant bit is on the left, and is Bit 7. Thus, the machine coding is "origin zero", unless noted otherwise. Note also that there is no parity bit, or provision for it, as would be found in larger systems.

The 8085's instructions are one, two, or three bytes long. In all cases, the first byte contains the essential information, such as the OP code. The second and third bytes, if included, provide operand information that won't fit in the first byte.

## 4.2 DATA TRANSFER GROUP

This group of instructions provides the 8085 with the ability to move data around inside the RAM, between the RAM and the registers of the MP, and between registers within the MP. They are important because a good deal of moving must be done to offset the fact that in an 8-bit byte, there is insufficient room to define the operands as specifically as is done. These instructions do not affect the condition codes.

**A few comments on the OP code groups follow:**

### *1. MOV Group*

These three instructions (MOV r1,r2, MOV r, M, and MOV M, r) are the general means of moving data between memory and registers. They move one byte with each execution. The second and third examples illustrate the use of the Register Indirect mode of addressing, in which the H&L registers of the MP contain an address, and the data is moved to or from that address. This saves space, in that the instruction is only one byte long. It requires, however, that the H&L registers be previously set up with the address required. The letter "M", when appearing as an operand in this description, specifies Register Indirect mode with H&L as the register to contain the address. No other register pair is used as such.

### 2. MVI Group

These two instructions (MVI r, data, and MVI M, data) provide a means of loading a byte immediately into a register or a memory address. Note that the Register Indirect mode again is evident. The immediate data is stored as a byte directly below the instruction byte.

### 3. LXI Instruction

This single instruction (LXI rp, data 16) provides a means of loading any register pair with a two-byte value. The second byte l of the instruction is loaded into the leftmost, or low-order, register of the pair, while the third byte is loaded into the rightmost or high order, register of the pair.

### 4. LDA & STA

The Load Accumulator Direct (LDA) and the Store Accumulator Direct (STA) instructions provide a means of moving a byte between the accumulator and a RAM address. This may also be done with the MOV OP code, but only indirectly, that is, with the help of the H&L registers. The address of the byte to be loaded or stored follows the instruction, again with the inverse addressing.

### 5. LHLD & SHLD

The Load H&L Direct (LHLD) and Store H&L Direct (SHLD) instructions provide a means of moving two bytes between the HL register pair and a RAM address. Since the H&L register pair is heavily used in the Register Indirect mode, these instructions provide a quick means of loading the two bytes of an address into the pair in only one instruction. The two bytes following the instruction contain an address in RAM, again low-order in byte 2, and high-order in byte 3. For the LHLD, this address points to a single byte, this is obtained and loaded into the L register. The second byte from RAM is obtained from the address one higher than the RAM byte, and loaded into the H register. The SHLD simply stores as above, instead of loads.

### 6. LDAX & STAX

The Load Accumulator Indirect (LDAX) and Store Accumulator Indirect (STAX) instructions provide a means of moving data between the U accumulator and a memory location indirectly, with the RAM address contained in either the BC or DE register pair. This is not the same as the MOV, which uses only the HL register pair. This instruction permits the accumulator to access groups of data bytes, as may be necessary with long precision arithmetic. Obviously, the BC or DE pair must be previously loaded with the address desired.

### 7. XCHG Instruction

The Exchange (XCHG) instruction permits the HL register pair's contents to be exchanged with the DE register pair's contents. This allows an address to be built in the DE pair, then, when ready, to be transferred at once to the HL pair. This would be advantageous in complex data handling.

## Data Transfer Instructions

| Opcode | Operand | Description |
|---|---|---|
| **Copy from source to destination** | | |
| MOV | Rd, Rs<br>M, Rs<br>Rd, M | This instruction copies the contents of the source register into the destination register; the contents of the source register are not altered. If one of the operands is a memory location, its location is specified by the contents of the HL registers.<br>**Example: MOV B, C or MOV B, M** |
| **Move immediate 8-bit** | | |
| MVI | Rd, data<br>M, data | The 8-bit data is stored in the destination register or memory. If the operand is a memory location, its location is specified by the contents of the HL registers.<br>**Example: MVI B, 57H or MVI M, 57H** |
| **Load accumulator** | | |
| LDA | 16-bit address | The contents of a memory location, specified by a 16-bit address in the operand, are copied to the accumulator. The contents of the source are not altered.<br>**Example: LDA 2034H** |
| **Load accumulator indirect** | | |
| LDAX | B/D Reg. pair | The contents of the designated register pair point to a memory location. This instruction copies the contents of that memory location into the accumulator. The contents of either the register pair or the memory location are not altered.<br>**Example: LDAX B** |
| **Load register pair immediate** | | |
| LXI | Reg. pair, 16-bit data | The instruction loads 16-bit data in the register pair designated in the operand.<br>**Example: LXI H, 2034H or LXI H, XYZ** |
| **Load H and L registers direct** | | |
| LHLD | 16-bit address | The instruction copies the contents of the memory location pointed out by the 16-bit address into register L and copies the contents of the next memory location into register H. The contents of source memory locations are not altered.<br>**Example: LHLD 2040H** |
| **Store accumulator direct** | | |
| STA | 16-bit address | The contents of the accumulator are copied into the memory location specified by the operand. This is a 3- byte instruction, the second byte specifies the low- |

order address and the third byte specifies the high-order address.

**Example: STA 4350H**

**Store accumulator indirect**

STAX Reg. pair

The contents of the accumulator are copied into the memory location specified by the contents of the operand (register pair). The contents of the accumulator are not altered.

**Example: STAX B**

**Store H and L registers direct**

SHLD 16-bit address

The contents of register L are stored into the memory location specified by the 16-bit address in the operand and the contents of H register are stored into the next memory location by incrementing the operand. The contents of registers HL are not altered. This is a 3-byte instruction, the second byte specifies the low-order address and the third byte specifies the high-order address.

**Example: SHLD 2470H**

**Exchange H and L with D and E**

XCHG none

The contents of register H are exchanged with the contents of register D, and the contents of register L are exchanged with the contents of register E.

**Example: XCHG**

**Copy H and L registers to the stack pointer**

SPHL none

The instruction loads the contents of the H and L registers into the stack pointer register, the contents of the H register provide the high-order address and the contents of the L register provide the low-order address. The contents of the H and L registers are not altered.

**Example: SPHL**

**Exchange H and L with top of stack**

XTHL none

The contents of the L register are exchanged with the stack location pointed out by the contents of the stack pointer register. The contents of the H register are exchanged with the next stack location (SP+1); however, the contents of the stack pointer register are not altered.

**Example: XTHL**

**Push register pair onto stack**

PUSH Reg. pair

The contents of the register pair designated in the operand are copied onto the stack in the following sequence. The stack pointer register is decremented and the contents of the high- order register (B, D, H, A) are copied into that location. The stack pointer register is decremented again and the contents of the low-order register (C, E, L, flags) are copied to that location.

**Example: PUSH B or PUSH A**

**Pop off stack to register pair**

POP Reg. pair

The contents of the memory location pointed out by the stack pointer register are copied to the low-order register (C, E, L, status flags) of the operand. The stack pointer is incremented by 1 and the contents of that memory location are copied to the high-order register (B, D, H, A) of the operand. The stack pointer register is again incremented by 1.

**Example: POP H or POP A**

**Output data from accumulator to a port with 8-bit address**

OUT 8-bit port address

The contents of the accumulator are copied into the I/O port specified by the operand.

**Example: OUT F8H**

**Input data to accumulator from a port with 8-bit address**

IN 8-bit port address

The contents of the input port designated in the operand are read and loaded into the accumulator.

**Example: IN 8CH**

## 4.3 ARITHMETIC GROUP

This group provides the 8085 with mathematical ability to manipulate 8-bit data, and, by judicious use of the condition codes, to manipulate larger values. The A register (Accumulator) can perform true adds, subtracts, and compares. The other registers can only increment or decrement by 1. Unless otherwise indicated, all the condition code flags are affected.

**A few comments on the op code groups follow**:

### *1. ADD Instructions*

The Add Register (ADD r) and Add Memory (ADD M) instructions add the byte specified, either in a register or in the address contained by the H&L registers, into the accumulator. They assume that the accumulator already has in it the other value to participate in the add. The sum will remain in the accumulator. If the answer resulted in a ninth bit, it is stored in the Carry flag of the PSW.

### 2. *Add With Carry*

The Add Register with Carry (ADC r) and Add Memory with Carry (ADC M) instructions will add the specified byte, either in a register or in the address contained by the H&L registers, AND the value of the Carry bit, into the accumulator. By including the carry bit in the operation, mathematical operations on values longer than 8 bits are possible. As above, the first value must already be loaded in the A register prior to execution of these instructions. The sum remains in the accumulator. If the answer resulted in a ninth bit, it is stored in the Carry flag.

### 3. *Add Immediates*

The Add Immediate (ADI) and Add Immediate with Carry (ACI) instructions provide a means of adding a fixed value into the accumulator. These instructions assume that an initial value has already been loaded into the accumulator. The immediate data is provided by the second byte of the instruction. The ACI instruction adds the immediate value and the value of the Carry flag, while the ADI does not take the Carry flag into account. The sum remains in the accumulator. If the answer resulted in a ninth bit, it is stored in the Carry flag.

### 4. *Subtract Instructions*

The Subtract Register (SUB r) and the Subtract Memory (SUB M) instructions subtract the specified byte, in a register or in the address contained by the H&L registers, from the contents of the accumulator. The accumulator must have the first value already loaded, prior to the execution of the instructions. The subtract is accomplished by the complement-and-add technique, in which the two's complement of the specified value is computed first, and then added to the contents of the A register. The Carry flag will be set to a 1 if a borrow was required during the subtraction.

### 5. *Subtract With Borrow*

The Subtract Register with Borrow (SBB r) and the Subtract Memory with Borrow (SBB M) instructions will subtract the specified byte, either in a register or in the address contained in the H&L registers, and the value of the Carry flag, from the contents of the A register. The first value must be loaded into the A register prior to the execution of the instructions. The subtract is accomplished by the complement-and-add technique. The Carry flag will be set to a 1 if a borrow was required during the subtraction.

### 6. *Subtract Immediates*

The Subtract Immediate (SUI data) and Subtract Immediate with Borrow (SBI data) instructions provide a means of subtracting a fixed value from the contents of the accumulator. The immediate value is provided by the second byte of the instruction. The first value must be loaded into the accumulator prior to the execution of the instructions. The subtract is accomplished by the complement-and-add technique. The SBI instruction will subtract both the immediate value and the contents of the Carry flag from the A register, while the SUI does not take the Carry flag into account. The Carry flag will be set at the end of the instruction if , a borrow was required during execution.

### 7. *Increment Instructions*

The Increment Register (INR r) and Increment Memory (INR M) instructions provide a quick means of adding one to the contents of a register or memory location. These instructions allow the programmer to create counting routines and reiterations. Note that the Carry flag is not affected by these instructions.

### 8. *Decrement Instructions*

The Decrement Register (DCR r) and Decrement Memory (DCR M) instructions provide a quick means of subtracting one from the contents of a register or a memory location. These instructions allow the programmer to create counting routines and reiterations. Note that the Carry flag is not affected by these instructions.

### 9. *Register Pair Instructions*

The Increment Register Pair (INX rp) and Decrement Register Pair (DCX rp) instructions provide a means of adding to, or subtracting from, a 16-bit value contained in a register pair. In the INX instruction, this means that the carry from the sum of the low order byte of the pair and the one will be added into the upper byte automatically. In the DCX instruction, this means that a borrow from the high-order byte, if required, will be allowed into the low-order byte, if the subtraction of one from the low-order byte demands it. Note that none of the flags are affected.

### 10. *Double Add*

The Add Register Pair to H&L (DAD rp) instruction adds a 16 bit value already existing in the BC or DE register pair into the 16-bit value contained in the H&L registers. The sum remains in the H&L registers. The Carry flag will be set if a carry occurred out of the high order byte; a carry from low- to high-order bytes within the add is taken into account automatically. This instruction allows a fixed index-like value to be added to the H&L registers for Register Indirect mode.

### 11. *Decimal Adjust*

The Decimal Adjust Accumulator (DAA) instruction converts the 8-bit value in the A register, which normally is assumed to be two 4bit hexadecimal values, into two 4-bit BCD values. This allows the programmer to accept input data as BCD, process it in the accumulator using essentially hexadecimal arithmetic, and then convert the result back into BCD. This may be done by virtue of the fact that the ten numbers of BCD (0 to 9) are coded in binary exactly as are the first ten of the sixteen numbers of binary coded hexadecimal. i.e., adding 3810 and 3816 are exactly the same. The conversion be may accomplished by the use of the Auxiliary Carry flag. If the contents of the low-order four bits of the A register is >9, or if the AC flag is set, a value of 6 is added to these bits.

Then the high-order four bits of the A register are examined; again, if they contain a value >9, or if the Carry flag is on, a 6 is added to them. The Carry flag, of course, indicates that the hexadecimal value of the byte before the instruction, when translated to BCD, is too large to fit in one byte.

## Arithmetic Instructions

| **Opcode** | **Operand** | **Description** |
|---|---|---|
| **Add register or memory to accumulator** | | |
| ADD | R<br>M | The contents of the operand (register or memory) are added to the contents of the accumulator and the result is stored in the accumulator. If the operand is a memory location, its location is specified by the contents of the HL registers. All flags are modified to reflect the result of the addition.<br>**Example: ADD B or ADD M** |
| **Add register to accumulator with carry** | | |
| ADC | R<br>M | The contents of the operand (register or memory) and the Carry flag are added to the contents of the accumulator and the result is stored in the accumulator. If the operand is a memory location, its location is specified by the contents of the HL registers. All flags are modified to reflect the result of the addition.<br>**Example: ADC B or ADC M** |
| **Add immediate to accumulator** | | |
| ADI | 8-bit data | The 8-bit data (operand) is added to the contents of the accumulator and the result is stored in the accumulator. All flags are modified to reflect the result of the addition.<br>**Example: ADI 45H** |
| **Add immediate to accumulator with carry** | | |
| ACI | 8-bit data | The 8-bit data (operand) and the Carry flag are added to the contents of the accumulator and the result is stored in the accumulator. All flags are modified to reflect the result of the addition.<br>**Example: ACI 45H** |
| **Add register pair to H and L registers** | | |
| DAD | Reg. pair | The 16-bit contents of the specified register pair are added to the contents of the HL register and the sum is stored in the HL register. The contents of the source |

register pair are not altered. If the result is larger than 16 bits, the CY flag is set. No other flags are affected.

**Example: DAD H**

**Subtract register or memory from accumulator**

| | | |
|---|---|---|
| SUB | R<br>M | The contents of the operand (register or memory) are subtracted from the contents of the accumulator, and the result is stored in the accumulator. If the operand is a memory location, its location is specified by the contents of the HL registers. All flags are modified to reflect the result of the subtraction.<br>**Example: SUB B or SUB M** |

**Subtract source and borrow from accumulator**

| | | |
|---|---|---|
| SBB | R<br>M | The contents of the operand (register or memory ) and the Borrow flag are subtracted from the contents of the accumulator and the result is placed in the accumulator. If the operand is a memory location, its location is specified by the contents of the HL registers. All flags are modified to reflect the result of the subtraction.<br>**Example: SBB B or SBB M** |

**Subtract immediate from accumulator**

| | | |
|---|---|---|
| SUI | 8-bit data | The 8-bit data (operand) is subtracted from the contents of the accumulator and the result is stored in the accumulator. All flags are modified to reflect the result of the subtraction.<br>**Example: SUI 45H** |

**Subtract immediate from accumulator with borrow**

| | | |
|---|---|---|
| SBI | 8-bit data | The 8-bit data (operand) and the Borrow flag are subtracted from the contents of the accumulator and the result is stored in the accumulator. All flags are modified to reflect the result of the subtraction.<br>**Example: SBI 45H** |

**Increment register or memory by 1**

| | | |
|---|---|---|
| INR | R<br>M | The contents of the designated register or memory) are incremented by 1 and the result is stored in the same place. If the operand is a memory location, its location is specified by the contents of the HL registers.<br>**Example: INR B or INR M** |

**Increment register pair by 1**

INX R

The contents of the designated register pair are incremented by 1 and the result is stored in the same place.

**Example: INX H**

**Decrement register or memory by 1**

DCR R
M

The contents of the designated register or memory are decremented by 1 and the result is stored in the same place. If the operand is a memory location, its location is specified by the contents of the HL registers.

**Example: DCR B or DCR M**

**Decrement register pair by 1**

DCX R

The contents of the designated register pair are decremented by 1 and the result is stored in the same place.

**Example: DCX H**

**Decimal adjust accumulator**

DAA none

The contents of the accumulator are changed from a binary value to two 4-bit binary coded decimal (BCD) digits. This is the only instruction that uses the auxiliary flag to perform the binary to BCD conversion, and the conversion procedure is described below. S, Z, AC, P, CY flags are altered to reflect the results of the operation. If the value of the low-order 4-bits in the accumulator is greater than 9 or if AC flag is set, the instruction adds 6 to the low-order four bits.

If the value of the high-order 4-bits in the accumulator is greater than 9 or if the Carry flag is set, the instruction adds 6 to the high-order four bits.

*Example: DAA*

## 4.4 LOGICAL GROUP

This group of instructions provides the decision-making ability of the 8085, and includes some logically oriented utility instructions as well. By using these instructions, the condition flags may be set so that they can be tested by Jump-on-condition instructions. Unless otherwise noted, all the condition codes are affected.

**A few comments on the op code groups follow:**

### 1. *And Instructions*

The And Register (ANA r) and And Memory (ANA M) instructions perform a logical And function between the specified byte, either in a register or in the address contained in the H&L registers, and the contents of the accumulator. The accumulator must first be loaded with an initial value. The And function occurs on a bit-by-bit basis. The low order bit of the specified byte is Anded with the low order bit of the A register; if both the bit from the outside byte AND the bit from the A register are a 1, the bit in the A register is left as a 1. If either the bit position of the outside byte or the bit position in the A register, or both, contained 0's, that bit position in the A register is reset to 0.

Identical actions occur on the other seven bit positions at the same time. The result, left in the accumulator, is a bit pattern which indicates, with 1's left on, in which positions of the bytes both the A register and the outside byte contained 1's. This is valuable for testing the conditions of specific bits within a byte, and reacting accordingly. All condition flags are involved, but the Carry flag is always cleared by an And.

### 2. *And Immediate*

The And Immediate (ANI data) instruction allows the programmer to match the byte in the accumulator with a fixed mask byte, contained in the second byte of the instruction. The A register must first be loaded with the byte to be tested. The Anding function occurs exactly as shown above. All condition flags are involved, but the Carry flag is cleared.

### 3. *Or Instructions*

The Or Register (ORA r) and Or Memory (ORA M) instructions perform inclusive Or's between the specified byte, either in a register or in the address contained in the H&L registers, and the contents of the accumulator. The A register must be loaded with the first , value prior to the execution of these instructions.

The Or function occurs on a bit-by-bit basis. The low order bit of the outside byte is Ored with the low order bit of the A register; if either the bit from the outside byte OR the bit from the A register is a 1, or both, that bit position is set to a 1 in the A register. If neither bit position from the outside bit or the A register is a 1, a 0 is loaded into that bit position of the A register. (Note that this satisfies the "one, the other, or both" requirements of an inclusive Or.) Identical operations occur on the other seven bit positions of the bytes.

The result, left in the accumulator, is a bit pattern in which a 1 exists in any bit position in which either of the bytes also had a bit set. All condition flags are affected, but the Carry and Auxiliary Carry flags are always cleared.

### 4. *Or Immediate*

The Or Immediate (ORI data) instruction allows the programmer to match the contents of the accumulator against a fixed mask byte which is contained in the second byte of the instruction. The

Or function occurs on a bit-by-bit basis, exactly as shown above. The first byte must be loaded into the A register prior to execution of the instruction. All condition flags are affected, but the Carry and Auxiliary Carry flags are always cleared.

### 5. Exclusive or Instructions

The Exclusive Or Register (XRA r) and D Exclusive Or Memory (XRA M) instructions perform exclusive Or functions between a specified byte, either in a register or in a byte contained in the address in the H&L register, and the contents of the accumulator. The A register must be loaded with the first byte prior to the execution of the instruction. The Exclusive Or occurs on a bit-by-bit basis. The low order bit of the outside byte is XORed with the low bit of the accumulator; if the bit in the outside byte is a 1 and the position in the A register is a 0, or if the bit in the outside byte is a 0 and the position in the A register is a 1, a 1 is set into that bit in the A register. If the bits are either both 1's or both 0's, the bit position is reset in the A register. (Note that this agrees with the "one, the other, but not both, and not neither" parameters of an Exclusive Or.) Identical operations occur on the other bit positions of the bytes at the same time. The results are left in the accumulator, which contains a bit pattern with 1's set where there was a 1 in either the A register or the outside byte, and 0's set where there was either 0's in both bytes or 1's in both bytes, in the same bit position. All condition flags are affected, but the Carry and Auxiliary Carry flags are always cleared.

### 6. Exclusive or Immediate

The Exclusive Or Immediate (XRI data) instruction allows the programmer to perform an Exclusive Or between a mask byte stored as the second byte of the instruction and the contents of the accumulator. The first byte must be loaded into the A register prior to the execution of the instruction. The Exclusive Or function occurs on a bit-by-bit basis exactly as outlined above. All the condition flags are affected, but the Carry and Auxiliary Carry flags are cleared.

### 7. Compare Instructions

The Compare Register (CMP r) and Compare Memory (CMP M) instructions compare the contents of the specified byte, either in a register or in the address contained in the H&L registers, to the contents of the accumulator. This is accomplished by subtracting the outside byte from the contents of the accumulator. The contents of the accumulator remain unchanged, and the actual answer of subtraction is lost. The condition flags are all affected, and are set to indicate the conditions of the lost answer. Particularly, the Zero flag, if set on, will indicate that the two values compared are equal, since the result of subtracting one from the other is zero. Also, the Carry flag will be set if the value in the A reg is smaller than the outside byte. If neither the Z nor the C flags are left on, the value in the A register is larger than the outside byte.

### 8. Compare Immediate

The Compare Immediate (CPI data) instruction compares the contents of the accumulator to a fixed value provided by the second byte of the instruction. The first value must be loaded into the A register prior to the execution of the instruction. The function occurs by a subtraction with lost answer, as described above. The contents of the A register are left unchanged. The condition codes act as above.

### 9. *Rotate Instructions*

The Rotate Left (RLC) and Rotate Right (RRC) instructions rotate the accumulator's contents one bit position left or right, respectively. In the RLC, all the bits move one position to the left; the high order bit which is shifted out of the A register is moved around to the low order bit position. It is also moved to the Carry flag. In the RRC, all the bits move one position to the right; the bit shifted out of the low order position of the A register is moved around to the high order position. It is also moved to the Carry flag. Thus, the Carry flag in either case indicates whether a bit was shifted out of the accumulator. Only the Carry flag is affected by these instructions.

### 10. *Rotate through Carrys*

The Rotate Left through Carry (RAL) and the Rotate Right through Carry (RAR) instructions rotate the accumulator's contents one bit position left or right, respectively. Unlike the rotates above, however, these instructions use the Carry flag as a ninth bit in the circle. In the RAL, the bits in the A register are shifted left one position; the high order bit moved to the Carry flag; the Carry flag is moved to the low order position of the A register. In the RAR, the bits in the A register are shifted right one position; the low order bit is moved to the Carry flag; the Carry flag is moved to the high order position of the A register. Only the Carry flag is affected.

### 11. *Complement Accumulator*

The Complement Accumulator (CMA) instruction provides a 1's complement of the 8 bits in the A register, i.e., the 1's are set to 0's, and the 0's are set to 1's. A two's complement may be effected by following the CMA with an INR A instruction. No condition flags are affected.

### 12. *Carry Instructions*

The Complement Carry (CMC) and Set Carry (STC) instructions allow direct control of the Carry flag by the programmer. The CMC will change the flag from 1 to 0, or 0 to 1, depending upon its initial condition. The STC forces the flag to a 1, regardless of its previous state. No other flags are affected.

## Logical Instructins

| **Opcode** | **Operand** | **Description** |
|---|---|---|
| **Compare register or memory with accumulator** | | |
| CMP | R<br>M | The contents of the operand (register or memory) are compared with the contents of the accumulator. Both contents are preserved . The result of the comparison is shown by setting the flags of the PSW as follows:<br>if (A) < (reg/mem): carry flag is set |

if (A) = (reg/mem): zero flag is set

if (A) > (reg/mem): carry and zero flags are reset

**Example: CMP B or CMP M**

**Compare immediate with accumulator**

CPI 8-bit data

The second byte (8-bit data) is compared with the contents of the accumulator. The values being compared remain unchanged. The result of the comparison is shown by setting the flags of the PSW as follows:

if (A) < data: carry flag is set

if (A) = data: zero flag is set

if (A) > data: carry and zero flags are reset

**Example: CPI 89H**

**Logical AND register or memory with accumulator**

ANA R<br>M

The contents of the accumulator are logically ANDed with the contents of the operand (register or memory), and the result is placed in the accumulator. If the operand is a memory location, its address is specified by the contents of HL registers. S, Z, P are modified to reflect the result of the operation. CY is reset. AC is set.

**Example: ANA B or ANA M**

**Logical AND immediate with accumulator**

ANI 8-bit data

The contents of the accumulator are logically ANDed with the 8-bit data (operand) and the result is placed in the accumulator. S, Z, P are modified to reflect the result of the operation. CY is reset. AC is set.

**Example: ANI 86H**

**Exclusive OR register or memory with accumulator**

XRA R<br>M

The contents of the accumulator are Exclusive ORed with the contents of the operand (register or memory), and the result is placed in the accumulator. If the operand is a memory location, its address is specified by the contents of HL registers. S, Z, P are modified to reflect the result of the operation. CY and AC are reset.

**Example: XRA B or XRA M**

**Exclusive OR immediate with accumulator**

| | | |
|---|---|---|
| XRI | 8-bit data | The contents of the accumulator are Exclusive ORed with the 8-bit data (operand) and the result is placed in the accumulator. S, Z, P are modified to reflect the result of the operation. CY and AC are reset. |

**Example: XRI 86H**

**Logical OR register or memory with accumulator**

| | | |
|---|---|---|
| ORA | R<br>M | The contents of the accumulator are logically ORed with the contents of the operand (register or memory), and the result is placed in the accumulator. If the operand is a memory location, its address is specified by the contents of HL registers. S, Z, P are modified to reflect the result of the operation. CY and AC are reset. |

**Example: ORA B or ORA M**

**Logical OR immediate with accumulator**

| | | |
|---|---|---|
| ORI | 8-bit data | The contents of the accumulator are logically ORed with the 8-bit data (operand) and the result is placed in the accumulator. S, Z, P are modified to reflect the result of the operation. CY and AC are reset. |

**Example: ORI 86H**

**Rotate accumulator left**

| | | |
|---|---|---|
| RLC | none | Each binary bit of the accumulator is rotated left by one position. Bit D7 is placed in the position of D0 as well as in the Carry flag. CY is modified according to bit D7. S, Z, P, AC are not affected. |

**Example: RLC**

**Rotate accumulator right**

| | | |
|---|---|---|
| RRC | none | Each binary bit of the accumulator is rotated right by one position. Bit D0 is placed in the position of D7 as well as in the Carry flag. CY is modified according to bit D0. S, Z, P, AC are not affected. |

**Example: RRC**

**Rotate accumulator left through carry**

| | | |
|---|---|---|
| RAL | none | Each binary bit of the accumulator is rotated left by one position through the Carry flag. Bit D7 is placed in the Carry flag, and the Carry flag is placed in the |

least significant position D0. CY is modified according to bit D7. S, Z, P, AC are not affected.

**Example: RAL**

**Rotate accumulator right through carry**

RAR none

Each binary bit of the accumulator is rotated right by one position through the Carry flag. Bit D0 is placed in the Carry flag, and the Carry flag is placed in the most significant position D7. CY is modified according to bit D0. S, Z, P, AC are not affected.

**Example: RAR**

**Complement accumulator**

CMA none

The contents of the accumulator are complemented. No flags are affected.

**Example: CMA**

**Complement carry**

CMC none

The Carry flag is complemented. No other flags are affected.

**Example: CMC**

**Set Carry**

STC none

The Carry flag is set to 1. No other flags are affected.

**Example: STC**

## 4.5 BRANCH GROUP

This group of instructions permits the programmer to alter the flow of program execution from a normal straight line. There are two major types of these instructions in the 8085. The first type is the Jump, in which the flow is altered with no intention of returning to the place where the Jump occurred. The second type is the Call, which provides linking, via the system stack, to save the address of the next instruction following the Call, proceed to a subordinate routine, and return to the saved address when that routine is completed.

Further, both Jumps and Calls may be conditional or unconditional. An unconditional Jump or Call causes the function to be executed absolutely. The conditional Jump or Call causes the function to be executed if the conditions specified are met. In the first byte of these instructions, three bits labeled CCC will contain a code which specifies the conditions to be tested. These may be specified by the programmer in assembly language by putting together a mnemonic composed of a J, for Jump, or a C, J for Call, followed by one or two more characters which specify the conditions to be tested.

**The breakdown follows:**

| Mnemonic | Condition | CCC Bits |
|---|---|---|
| NZ | Not Zero (Z=0) | 000 |
| Z | Zero (Z=1) | 001 |
| NC | Not Carry (C=0) | 010 |
| C | Carry (C=1) | 011 |
| PO | Parity Odd (P=0) | 100 |
| PE | Parity Even (P=1) | 101 |
| P | Plus (S=0) | 110 |
| M | Minus (S=1) | 111 |

### 1. *Jump Instructions*

The Jump (JMP addr) and Jump Conditional (Jxx addr) instructions allow program flow to be altered by loading the contents of the two bytes following the instruction to be loaded into the Program Counter. The next instruction to be fetched, therefore, will the first of the new routine. The JMP instruction is unconditional; the Jump occurs absolutely. The Jxx instruction will alter program flow if the conditions specified by the "xx" bits are true; otherwise, program flow remains in a straight line. No condition codes are affected.

### 2. *Call Instructions*

The Call (CALL addr) and Call Conditional (Cxx addr) instructions allow linkage to permit a subroutine to be invoked, with the address of the next sequential instruction saved for later reference. The Call will move the high byte of the PC into the address pointed to by the Stack Pointer minus 1, and the low byte of the PC into the address below that. The SP is then decremented by two, to update it to the new stack position. The two bytes following the Call instruction will then be moved to the PC, with the second byte of the instruction containing the low order byte of the address, and the third byte of the instruction containing the high order byte of the address. Thus, the address of the instruction following the Call is saved on the system stack, and the address of the first instruction of the subroutine is fetched next. The Call Conditional executes exactly the same way, providing that the conditions specified by the CCC bits are true. None of the flags are affected.

### 3. *Return Instructions*

The Return (RET) and Return Conditional (Rxx) instructions provide a means, at the end of a subroutine, of resuming program execution at the instruction following the Call instruction which invoked the subroutine. These instructions are placed at the end of the subroutine, not in the body of the main program. When encountered, the Return will move the byte pointed to by the Stack Pointer into the lower byte of the PC, the next byte higher in RAM to the higher byte of PC, and add 2 to the contents of SP. Thus, the address of the instruction following the Call, previously saved on the stack, is now in PC, and will be fetched next. Also, the stack pointer is updated accordingly. The

Return Conditional executes exactly the same way, providing that the conditions specified by the CCC bits are true. None of the flags are affected.

### 4. Restart

The Restart (RST n) instruction provides part of the vectored interrupt system by which any one of eight different levels of interrupt may stop the execution of the program currently in progress, save the address of the next instruction onto the stack, and then jump to any one of eight different locations in low core, depending upon the contents of the bits marked NNN in the instruction. Thus, as many as eight different external events, i.e. I/O devices, etc., may ask for service; the place where the program left off is saved; and one of eight different interrupt handling routines may be entered, which correspond to the level of the interrupt. This will be fully explained in the section on interrupts.

### 5. Jump Indirect

The Jump H&L Indirect (PCHL) instruction moves the contents of the H&L registers, assumed to be a valid address, into the Program Counter. The contents of H&L must be previously built, and may be assembled by other parts of the program to the advantage of the writer. The original contents of the PC are destroyed, so this is a one-way jump.

## Branching Instructions

| Opcode | Operand | Description |
|---|---|---|
| **Jump unconditionally** | | |
| JMP | 16-bit address | The program sequence is transferred to the memory location specified by the 16-bit address given in the operand. **Example: JMP 2034H or JMP XYZ** |
| **Jump conditionally** | | |
| Operand: | 16- bit address | The program sequence is transferred to the memory location specified by the 16-bit address given in the operand based on the specified flag of the PSW as described below. **Example: JZ 2034H or JZ XYZ** |

| Opcode | Description | Flag Status |
|---|---|---|
| JC | Jump on Carry | CY = 1 |
| JNC | Jump on no Carry | CY = 0 |
| JP | Jump on positive | S = 0 |
| JM | Jump on minus | S = 1 |
| JZ | Jump on zero | Z = 1 |
| JNZ | Jump on no zero | Z = 0 |
| JPE | Jump on parity even | P = 1 |
| JPO | Jump on parity odd | P = 0 |

**Unconditional subroutine call**

CALL 16-bit address — The program sequence is transferred to the memory location specified by the 16-bit address given in the operand. Before the transfer, the address of the next instruction after CALL (the contents of the program counter) is pushed onto the stack.

**Example: CALL 2034H or CALL XYZ**

**Call conditionally**

Operand: 16-bit address — The program sequence is transferred to the memory location specified by the 16-bit address given in the operand based on the specified flag of the PSW as described below. Before the transfer, the address of the next instruction after the call (the contents of the program counter) is pushed onto the stack.

**Example: CZ 2034H or CZ XYZ**

| Opcode | Description | Flag Status |
|---|---|---|
| CC | Call on Carry | CY = 1 |
| CNC | Call on no Carry | CY = 0 |
| CP | Call on positive | S = 0 |
| CM | Call on minus | S = 1 |
| CZ | Call on zero | Z = 1 |
| CNZ | Call on no zero | Z = 0 |
| CPE | Call on parity even | P = 1 |
| CPO | Call on parity odd | P = 0 |

**Return from subroutine unconditionally**

RET none — The program sequence is transferred from the subroutine to the calling program. The two bytes from the top of the stack are copied into the program counter, and program execution begins at the new address.

**Example: RET**

**Return from subroutine conditionally**

Operand: none — The program sequence is transferred from the subroutine to the calling program based on the specified flag of the PSW as described below. The two bytes from the top of the stack are copied into the program counter, and program execution begins at the new address.

**Example: RZ**

| Opcode | Description | Flag Status |
|---|---|---|
| RC | Return on Carry | CY = 1 |
| RNC | Return on no Carry | CY = 0 |
| RP | Return on positive | S = 0 |
| RM | Return on minus | S = 1 |
| RZ | Return on zero | Z = 1 |
| RNZ | Return on no zero | Z = 0 |
| RPE | Return on parity even | P = 1 |
| RPO | Return on parity odd | P = 0 |

**Load program counter with HL contents**

PCHL none The contents of registers H and L are copied into the program counter. The contents of H are placed as the high-order byte and the contents of L as the low-order byte.

**Example: PCHL**

**Restart**

RST 0-7 The RST instruction is equivalent to a 1-byte call instruction to one of eight memory locations depending upon the number. The instructions are generally used in conjunction with interrupts and inserted using external hardware. However these can be used as software instructions in a program to transfer program execution to one of the eight locations. The addresses are:

| Instruction | Restart Address |
|---|---|
| RST 0 | 0000H |
| RST 1 | 0008H |
| RST 2 | 0010H |
| RST 3 | 0018H |
| RST 4 | 0020H |
| RST 5 | 0028H |
| RST 6 | 0030H |
| RST 7 | 0038H |

The 8085 has four additional interrupts and these interrupts generate RST instructions internally and thus do not require any external hardware. These instructions and their Restart addresses are:

| Interrupt | Restart Address |
|---|---|
| TRAP | 0024H |

| | |
|---|---|
| RST 5.5 | 002CH |
| RST 6.5 | 0034H |
| RST 7.5 | 003CH |

## 4.6 MACHINE CONTROL GROUP

This group is a collection of miscellaneous instructions which control bodily functions of the MP, or provide utilities.

**A few comments on the op code groups follow:**

### *1. Push and POP*

The Push Register Pair (PUSH rp) and Pop Register Pair (POP rp) instructions allow programmers to manipulate the system stack. The Push will place the contents of the BC, DE, or HL register pairs onto the stack, and update the SP accordingly. The Pop instruction will return the last two items on the stack to the specified register pair, and update the SP. The condition flags are not affected; the SP register pair may not be specified, for obvious reasons.

### *2. PSW Instructions*

The Push Processor Status Word (PUSH PSW) and the Pop Processor Status Word (POP PSW) instructions will allow the programmer to save the contents of the A register and of the condition flags on the stack, or to retrieve them from the stack. The Processor Status Word (PSW) of the 8085 is defined as a "Flag Byte" which contains the condition flag bits in a specific sequence:

| S | Z | 0 | AC | 0 | P | 1 | C |
|---|---|---|---|---|---|---|---|

In addition, the contents of the A register is also saved as part of the PSW. When the PUSH PSW is encountered, the contents of the A register is pushed onto the stack first, followed by the Flag byte. The SP is then updated. When the POP is executed, the Flag byte is retrieved first, and the bits are loaded into their proper flip-flops. The A register is then loaded with the next byte retrieved. This allows programmers to save conditions at the beginning of subroutines so that the execution of the instructions within the routines will not alter the conditions under which the original program was operating.

### *3. Exchange Stack Top*

The Exchange Stack Top with H&L (XTHL) instruction causes the contents of the H&L registers to be exchanged with the two bytes which are currently on the top of the system stack. These will be the last two bytes pushed. It is a two-way instruction; the stack receives the original contents of H&L, while H&L receives the two bytes from the stack. The contents of SP remain unchanged. No flags are affected.

### *4. Move H&L TO SP*

The Move H&L Register to Stack Pointer (SPHL) instruction will directly move the contents of the

H&L registers into the Stack Pointer; the original contents of SP are destroyed. This may be used to permit multiple stacks to exist at one time in the system. No flags are affected.

## 4.7 I/O INSTRUCTIONS

The Input (IN port) and Output (OUT port) instructions allow the MP to communicate with the outside world. In both cases, the address byte of the device to be used is contained in the byte following the instruction. This byte is presented at once to both the upper and lower bytes of the A0-A15 address lines. In the case of IN, the byte accepted on the D0-D7 data lines by the MP is placed in the A register. For the OUT, the byte to be sent on the data lines is placed in the A register prior to execution of the instruction. No flags are affected.

### *1. Interrupt Instructions*

The Enable Interrupts (EI) and Disable Interrupts (DI) instructions allow the MP to permit or deny interrupts under program control. For the EI, the interrupts will be enabled following the completion of the next instruction following the EI. This allows at least one more instruction, perhaps a RET or JMP, to be executed before the MP allows itself to again be interrupted. For the DI, the interrupts are disabled immediately. No flags are affected.

### *2. Halt and NO-OP*

The Halt (HLT) and No-Operation (NOP) instructions serve general utility purposes. The Halt will stop the processor from further execution; it can be restarted again only by an interrupt. A reset signal applied to the MP will abort the Halt. The MP may enter a Hold state, as the result of another device wanting the bus, from a Halt, but will return to the Halt state when the Hold is canceled. The NOP is simply a one-byte long place holder, which is passed through automatically without any data motion or action of any kind. It is used primarily as a programmer's aid in saving space within language programs for later use.

The Read Interrupt Mask (RIM) and Set Interrupt Mask (SIM) instructions are used to service both the extended interrupt system of the '85 and the Serial Input Data (SID) and Serial Output Data (SOD) pins on the device. While these items are both serviced by the same instructions, they are not electrically or logically related, and should not be confused. The Interrupt Mask is a group of bits which can be accessed by these two instructions via the accumulator. A discussion of the two instructions follows.

### *3. Read Interrupt Mask (RIM)*

This instruction permits the system to examine the interrupt mask by loading into the A register a byte which defines the condition of the mask bits for the maskable interrupts, the condition of the interrupts pending for the maskable interrupts, the condition of the Interrupt Enable flag, and the condition of the Serial Input Data (SID) pin on the MP. The format is:

| D7 | D6 | D5 | D4 | D3 | D2 | D1 | D0 |
|---|---|---|---|---|---|---|---|
| SID | I7.5 | I6.5 | I5.5 | IE | M7.5 | M6.5 | M5.5 |

From the D0 end of the byte, the purpose of these bits are as follows:

### 4. *Interrupt Mask Bits D0, D1, D2*

These bits indicate whether the interrupts for vectors 5.5, 6.5, and 7.5 are masked on or off. The bits are set to a 1 for disabled, and 0 for enabled. These bits allow the program to examine the mask and obtain the current status of it. It is a valuable tool, since one program section may not necessarily know what a second section is doing or expecting.

### 5. *Interrupt Enable Bit D3*

This bit corresponds to the Interrupts Enabled, flip-flop in the 8080. It is set or reset by the Enable Interrupts or Disable Interrupts OP codes in the same manner as in the 8080. The interrupts involved are the standard 8 vectored interrupts, and it has no effect on those special to the 8085. However, in the case of a Trap interrupt (unmaskable), the status of bit 3 may be lost, so that Intel suggests that a RIM be executed as part of the Trap service routine to preserve the condition of the Interrupts Enabled flag prior to the occurrence of Trap. The flag is not affected by the other special interrupts.

### 6. *Interrupts Pending Bits D4, D5, D6*

These bits indicate what interrupts have occurred since the last time that specific interrupt was serviced. If interrupts 5.5 or 6.5 are masked off by bits D0 or D1, bits D4 and D5 will not be set. Bit D6, which corresponds to the 7.5 interrupt, will be set on to indicate that an interrupt was requested, even if it was masked off.

### 7. *Serial Input Data Bit D7*

This bit provides the condition of the SID pin. It will be 1 if the pin is high, and 0 if it is low. The software examining this bit must have total ability to deal with whatever it finds there. The pin provides only a voltage level that exists at the time of the RIM execution.

When the RIM instruction is executed, the status of all the lines indicated are sampled, and the resulting bit pattern is placed in the A register. The instruction simply provides these conditions for display; it has no affect on the bits themselves. The bits in the A register may then be examined directly by logical instructions, or moved to a register or memory location for safekeeping.

### 8. *Set Interrupt Mask (SIM)*

This instruction is the reverse of the RIM. While the RIM simply reads the status of various lines, the SIM sets various bits to form masks or generate output data via the SOD line. The conditions that the programmer wishes to set up must be set into the A register exactly as desired first, then the SIM Instruction is executed.

The SIM will take the bit pattern it finds in the A register and loads it into the masks in the following format:

| D7 | D6 | D5 | D4 | D3 | D2 | D1 | D0 |
|---|---|---|---|---|---|---|---|
| SOD | S0E | X | R7.5 | MSE | M7.5 | M6.5 | M5.5 |

### 9. RST Masks Bits D0, D1, D2

These bits are the interrupt masks for the 5.5, 6.5, and 7.5 interrupts brought into the '85 on their own pins. The bits are 0 to enable and 1 to disable the interrupts. If bits D0 or D1 are disabled (set to 1), a signal applied to their respective pins cause no action. If D0 or D1 are set to 0 (enabled), their respective bits will be visible via the RIM instruction, and the call to the interrupt vector will occur. In the case of bit D2 for masking the 7.5 interrupt, the RIM instruction will indicate that a 7.5 interrupt is pending, but an automatic call will not occur.

### 10. Mask Set Enable Bit D3

This bit permits bits D0, D1, and D2 to be changed. If a SIM is executed with this bit low, the condition of the mask bits will not change. If a SIM is executed with this bit set high, the mask bits will take on the same arrangement as those given in the lower bits of the A register. This permits accessing of the mask byte to deal with the interrupts without affecting SOD.

### 11. RST 7.5 Reset Bit D4

This bit permits the SIM instruction to reset the interrupt pending flag indicated by bit D6 in the RIM instruction byte. Since the 7.5 interrupt is handled somewhat more importantly than the 5.5 and 6.5, it can indicate that it is pending via the RIM instruction even though it is masked off. This bit allows that pending request to be reset individually under program control.

### 12. Undefined Bit D5

This bit is unused.

### 13. SOD Enable Bit D6

This bit works in conjunction with bit D7. If it is set to 1 when the SIM is executed, the condition of bit D7, high or low, is electrically loaded into the SOD latch, and in turn appears on the SOD pin of the '85. If bit D6 is low, the SIM's execution has no affect on the bit D7. This, like bit D3, allows executing SIMs to service either interrupts or the serial I/O without affecting the other.

### 14. Serial Output Data Bit D7

This bit contains the voltage level (+5 volts = 1, 0 volts = 0) which should appear at the SOD pin of the '85. If the SIM instruction is executed and bit D6 is set to 1 (enabled), the level contained by D7 is forwarded to the SOD latch, which will in turn cause it to appear on the SOD pin. If bit D6 is low, the SIM instruction will have no affect on bit D7.

The /RESET IN line affects the flags of the interrupt masks. First, the RST 7.5 latch which catches the rising edge of the 7.5 flip-flop and holds it for a RIM instruction is reset. Next, all three mask bits are set to a 1, which will disable all interrupts. Lastly, the SOD latch is reset to a 0. These will effectively allow the new instructions following the Reset to take full control of the MP, without interrupts causing immediate problems.

The use of these instructions will be explained under Interrupts in the "Additional Information" section of the notes.

## Machine Control and I/O Instructions

| Opcode | Operand | Description |
|---|---|---|
| **No operation** | | |
| NOP | none | No operation is performed. The instruction is fetched and decoded. However no operation is executed.<br>**Example: NOP** |
| **Halt and enter wait state** | | |
| HLT | none | The CPU finishes executing the current instruction and halts any further execution. An interrupt or reset is necessary to exit from the halt state.<br>**Example: HLT** |
| **Disable interrupts** | | |
| DI | none | The interrupt enable flip-flop is reset and all the interrupts except the TRAP are disabled. No flags are affected.<br>**Example: DI** |
| **Enable interrupts** | | |
| EI | non | The interrupt enable flip-flop is set and all interrupts are enabled. No flags are affected. After a system reset or the acknowledgement of an interrupt, the interrupt enable flip-flop is reset, thus disabling the interrupts. This instruction is necessary to reenable the interrupts (except TRAP).<br>**Example: EI** |
| **Read interrupt mask** | | |
| RIM | none | This is a multipurpose instruction used to read the status of interrupts 7.5, 6.5, 5.5 and read serial data input bit. The instruction loads eight bits in the accumulator with the following interpretations.<br>**Example: RIM** |

Accumulator set up for the RIM instruction

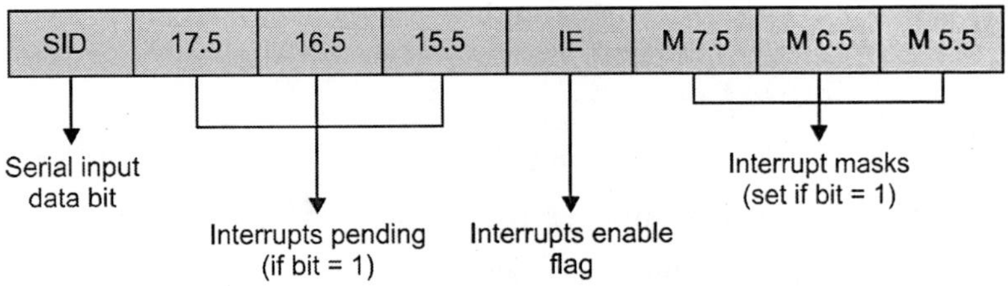

**Set interrupt mask**

SIM none This is a multipurpose instruction and used to implement the 8085 interrupts 7.5, 6.5, 5.5, and serial data output. The instruction interprets the accumulator contents as follows.

**Example: SIM**

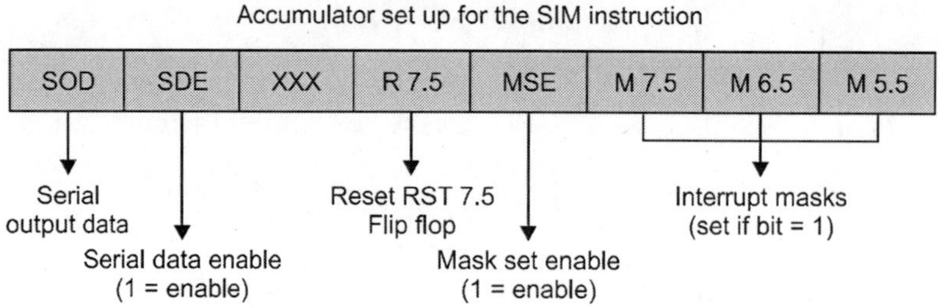


- SOD—Serial Output Data: Bit $D_7$ of the accumulator is latched into the SOD output line and made available to a serial peripheral if bit $D_6 = 1$.
- SDE —Serial Data Enable: If this bit = 1, it enables the serial output. To implement serial output, this bit needs to be enabled.
- XXX—Don't Care
- R7.5 —Reset RST 7.5: If this bit = 1, RST 7.5 flip-flop is reset. This is an additional control to reset RST 7.5.
- MSE—Mask Set Enable: If this bit is high, it enables the functions of bits $D_2$ , $D_1$, $D_0$. This is a master control over all the interrupt masking bits. If this bit is low, bits $D_2$, $D_1$, and $D_0$ do not have any effect on the masks.
- M7.5 —$D_2$ = 0, RST 7.5 is enabled.
  = 1, RST 7.5 is masked or disabled.
- M6.5 —$D_1$ = 0, RST 6.5 is enabled.
  = 1, RST 6.5 is masked or disabled.
- M5.5 —$D_0$ = 0, RST 5.5 is enabled.
  = 1, RST 5.5 is masked or disabled.

## QUESTION BANK

Q1. Give brief description of different types of instructions available in 8085 microprocessor.

Q2. Explain different types of data transfer instructions.

Q3. Explain different types of arithmetic instructions.

Q4. Explain different types of logical instructions.

Q5. Explain different types of branching instructions.

Q6. Explain different types of machine control and I/O instructions.

CHAPTER 5

# 8086 Microprocessor

## 5.1 INTRODUCTION TO 8086

16-Bit Inter microprocessors are: Intel 8086, 80186 and 80286. The Intel 8088 and 80188 are 8/16-bit microprocessors. Their internal architecture is of 16-bit but they have only 8 data lines. The 8086, 80186, 80286, 8088 and 80188 microprocessors contain the same basic set of registers, instructions and addressing modes. The 80186 is a highly integrated 16-bit microprocessor. It contains 15-20 computer components in addition to the CPU on a single chip. The 80286 is a high performance 16-bit microprocessor with memory management and protection.

The 8088 CPU became very popular and was used in cheap personal computers, PC/XTs (personal computer, extended technology). The 8088 had the attributes of both 8 and 16-bit microprocessors. As its data bus contains only 8 lines, it employed cheaper 8-bit peripherals. It was directly compatible with 8086 software.

The 80286 was also very popular and was employed in personal computers, AC/ATs (personal computer, advanced technology). PC/AT was faster, more powerful and costlier than PC/XT.

The Intel 80C86A and 80C86AL are the CHMOS versions of the industrial standard HMOS 8086 microprocessor. The Intel 80C88A and 80C88Al are the CHMOS versions of 8088 microprocessor. The Intel 80C286 is the CHMOS version of 80286. The Intel 8088 and 80286 microprocessors based personal computers are no longer manufactured.

## 5.2 INTERNAL ARCHITECTURE OF 8086

Before we can talk about how to write programs for the 8086, we need to discuss its specific internal features, such as its ALU, flags, register, instruction byte queue, and segment registers.

As shown in the block diagram in Figure 5.1, the 8086 CPU is divided into two independent functional parts, the bus interface unit or BIU, and the execution unit or EU, Dividing the work between these two units speeds up processing.

The BIU sends out addresses, fetches instructions from memory, reads data from ports and memory, and writes data to ports and memory. In other words, the BIU handles all transfers of data and addresses on the buses for the execution unit.

The EU for the 8086 tells the BIU where to fetch instructions or data from, decodes instructions and executes instructions. Let's take a look at some of the parts of the execution unit.

### 5.2.1 Execution Unit (EU)

#### *1. Control Circuitry, Instruction Decoder, and ALU*

In Figure 5.1, the EU contains control circuitry which directs internal operations. A decoder in the EU translates instructions fetched from memory into a series of actions which the EU carries out. The EU has a 16-bit arithmetic logic unit which can add, subtract AND, OR, XOR, increment decrement, complement, or shift binary numbers.

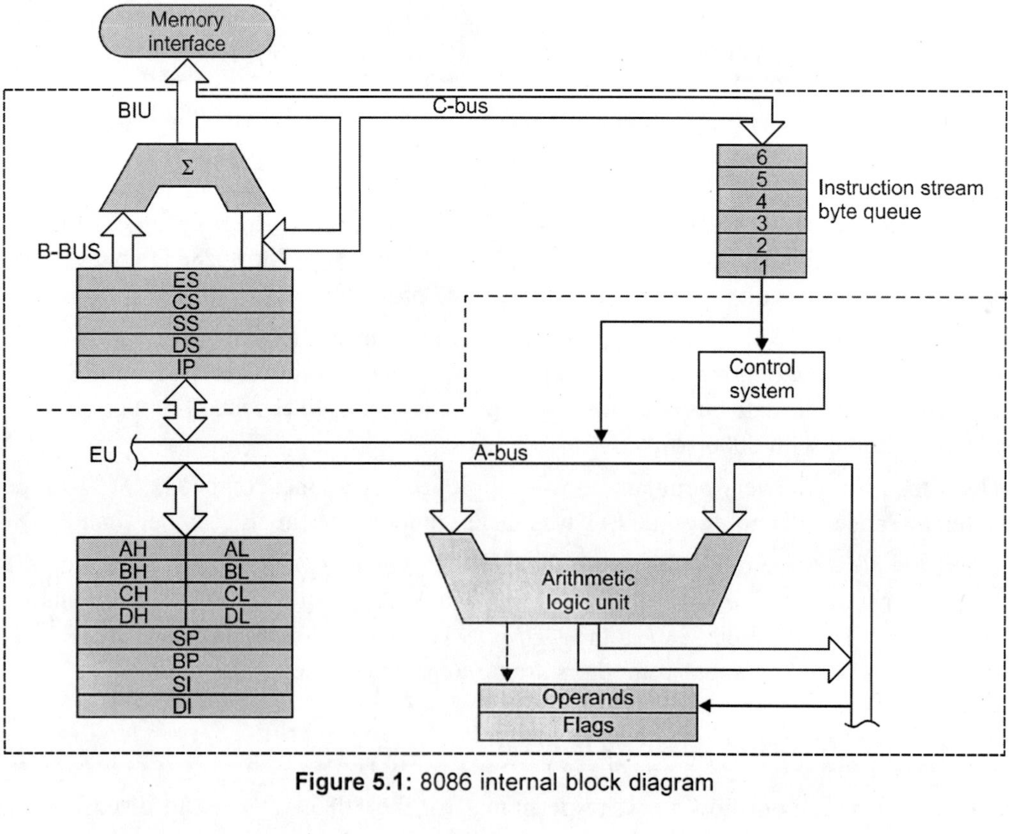


**Figure 5.1:** 8086 internal block diagram

#### *2. Flag Register*

A flag is a flip-flop which have some condition procedure by the execution of an instruction or controls certain operations of the EU. A 16-bit flag register in the EU contains nine active flags. Figure 5.2 shows the location of the nine flags in the flag register. Six of the nine flags are used to indicate some condition produced by an instruction. For example, a flip-flop called the carry flag will be set to a I if the addition of two 16-bit binary numbers produces a carry of out of the most significant bit position. If no carry out of the MSB is produced by the addition, then the carry flag will be a 0. The EU thus effectively runs up a "flag" to tell you that a carry was produced.

The six conditional flags in this group are the carry flag (CF). The parity flag (PF), the auxiliary carry flag (AF), the zero flag (ZF) the sign flag (SF), and the overflow flag (OF). The names of these flags should give you hints as to what conditions affect them. Certain 8086 instruction check these flags to determine which of two alternative actions should be done in executing the instruction.

The three remaining flags in the flag register are used to control certain operations of the processor. These flags are different from the six conditional flags describes above in the way they are set or reset. The six conditional flags are set of reset by the EU on the basis of the results of some arithmetic or logic operation. The control flags are deliberately set or reset with specific instructions you put in your program. The three control flags are the trap flag (TF), which is used for single stepping through a program; the interrupt flag (IF), which is used to allow or prohibit the interruption of a program; and the direction flag (DF), which is used with string instructions.

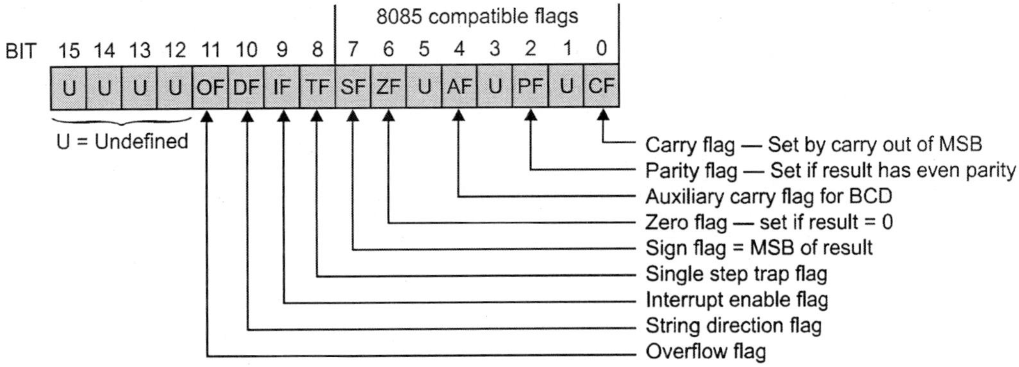


**Figure 5.2:** 8086 flag register format.

Later we will discuss in detail the operation and use of the nine flags.

### 3. *General-purpose Registers*

In Figure 5.1 that the EU has eight general-purpose registers, labeled AH,AL,BH,CH,CL,DH, and DL. These register can be used individually for temporary storage of 8-bit data. The AL register is also called the accumulator. It has some features that the other general-purpose registers do not have.

Certain pairs of these general-purpose register can be used together to store 16-bit data words. The acceptable register paris are AH and AL, BH and BL, CH and CL, AX register the BH-BL pair is referred to as the BX register, the CH-CL pair is referred to as the CX register, and the DH-DL pair is referred to as the DX register.

The 8086 general-purpose register set is very similar to those of the earlier-generation 8086 and 8085 microprocessor. It was designed this way so that the many programs written for the 8086 and 8085 could easily be translated to run on the 8086 or the 8088. The advantage of using internal registers for the temporary storage of data is that, since the data is already in the EU, it can be accessed much more quickly than it could be accessed in external memory. Now let's look at the features of the BIU.

### 5.2.2 Bus Interface Unit (BIU)

#### *1. The Queue*

While the EU is decoding an instruction or executing an instruction which does not require use of the buses, the BIU fetches up to six instruction bytes for the following instructions. The BIU stores these perfected bytes in the first-in-first-out register set called a queue. When the EU is ready for its next instruction, it simple reads the instruction byte(s) for the instruction from the queue in the BIU. This is much faster than sending out an address to the system memory and waiting for memory to send back the next instruction byte the bytes.

The process is analogous to the way a bricklayer's assistant hatches bricks ahead of time a keeps a queue of bricks lined up so that the bricklayer can just reach out and grab a brick when necessary. Except in the cases of JMP and CALL instructions, where the queue must be dumped and then reloaded starting from a new address, this prefetch-and-queue scheme greatly speeds up processing. Fetching the next instructions while the current instruction executes is called pipelining.

#### *2. Segment Register*

The 8086 BIU sends out 20-bit addresses, so it can address any of $2^{20}$ or 1,048,576 bytes in memory. However, at any given time the 8086 works with only four 65,536-byte (64-kbyte) segment within this 1,048,576 byte (1-Mbyte) segment register in the BIU address of four memory segments that the 8086 is working with at a particular time. The four segment registers are the code segment (CS) register, the stack segment (SS) register, the extra segment (ES) register, and the data segment (DS) register.

Figure 5.3 shows how these four segments might be positioned in memory at a given time. The four segments can be separated as shown, or, for small programs which do not need all 64 kbytes in each segment. They can overlap.

To repeat, than a segment register is used to hold the upper 16 bits of the starting address for each of the segments. The code segment register, for example, holds the upper 16 bits of the starting address for the segment from which the BIU is currently fetching instruction code bytes. The BIU always inserts zeros for the lowest 4 bits (nibble) of the 20-bit starting address for a segment. If the segment register contains 348AH for example, then the code segment will start at address 348A0H. In other words, a 64-kbyte segment can be located anywhere within the 1-mbyte address space, but the segment will always start at an address with zeroes in the lowest 4 bits. This constraint was put on the location of segments so that it is only necessary to store. And manipulate 16-bit numbers when working with the starting address of a segment. The part of a segment starting address stored in a segment register is often called the segment base.

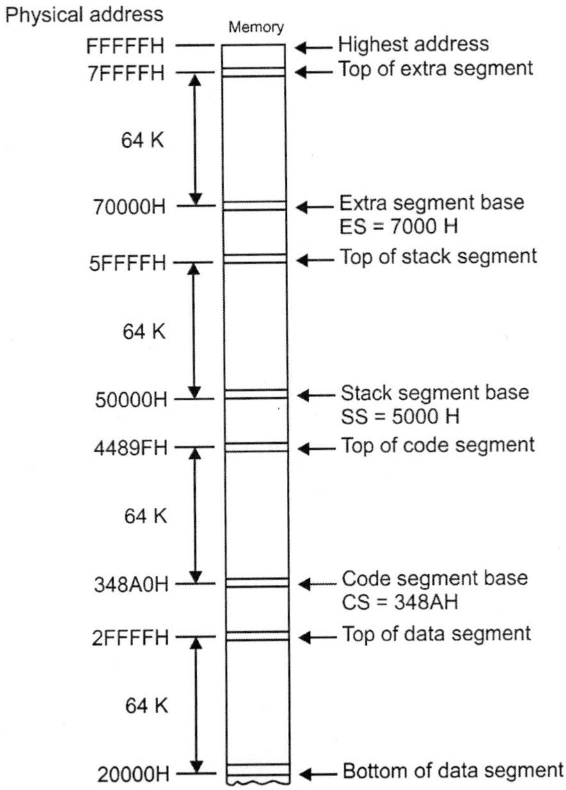


**Figure 5.3:** One way four 64-kbyte segments might be positioned within the 1-mbyte address space of an 8086.

A stack is a section of memory set aside to store addresses and data while a subprogram executes. The stack segment register is used to hold the upper 16-bits of the starting address for the program stack; we will discuss the use and operation of a stack in detail later.

The extra segment register and the data segment register are used to hold the upper 16-bits of the starting addresses of two memory segments that are used for data.

### 3. *Instruction Pointer*

The next feature to look at in the BIU is the instruction pointer (IP) register. As discussed previously, the code segment register holds the upper 16 bits of the starting address of the segment from which the BIU is currently fetching instruction code bytes. The instruction pointer register holds the 16-bit address, or offset, of the next code byte within this code segment. The value contained in the IP is referred to as an offset because this value must be offset from (added to) the segment base address in CS to produce the required 20-bit physical address sent out by the BIU. Figure 5.4a shows in diagram from how this works. The CS register points to the base or start of the current code segment. The IP contains the distance or offset from this base address to the next instructions byte to be fetched. Figure 5.4b shows how the 16-bit offset in IP is added to the 16-bit segment base address in CS to produce the 20-bit physical address. Notice that the two 16-bit numbers are not added directly in line, because the CS register contains only the upper 16 bit of the base address for the code segment. As we said before, the BIU automatically inserts zeros for the lowest 4 bits of the segment base address.

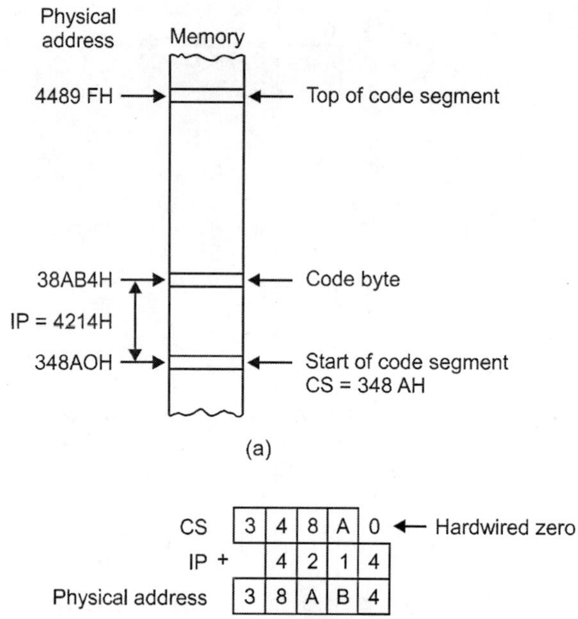


(b)

**Figure 5.4:** Addition of CS to produce the physical address of the code byte.
(a) Diagram. (b) computation.

If the CS register, for example, contains 348AH, you know that the starting address for the code segment is 348A0H. When the NIU adds the offset of 4214H in the IP to this segment base address, the result is a 20-bit physical address of 38AB4H.

An alternative way of representing a 20-bit physical address is the segment base: offset from. For the address of a code byte, the format for this alternative from will be CS:IP. As an example of this the address constructed in the preceding paragraph, 38AB4H, can also be represented as 348A:4214. To summarize, then, the CS register contain the upper 16bits of the starting address of the code segment in the 1-Mbyte address range of the 8086. The instruction pointer register contains a 16-bit offset which tells where in that 64-kbyte code segment the next instruction byte is to be fetched from. The actual physical address sent to memory is produced by adding the offset contained in the IP register to the segment base represented by the upper 16bits in the CS register.

### 4. Stack Segment Register and Stack Pointer Register

A stack, remember, is a section of memory set aside to store addresses and data while a subprogram is executing. The 8086 allows you to set aside an entire 64-kbyte segment as a stack. The upper 16 bits of the starting address for this segment are kept in the stack segment register. The stack pointer (SP) register in the execution unit holds the 16-bit offset from the start of the segment in the memory location where a word was most recently stored on the stack. The memory location where a word was most recently stored is called the top of stack. Figure 5.5a shows this in diagram form.

The physical address for a stack read on a stack write is produced by adding the contents of the stack pointer register to the segment base address represented by the upper 16bits of the base address in SS. Figure 5.5b shows an example. The 5000H in SS represents a segment base address of 50000H. When the FFE0H in the SP is address to this, the resultant physical address for the top of the stack will be 5FFE0H. The physical address can be represented either as a single number, 5FFE0H, or in SS:SP form as 5000:FFE0H.

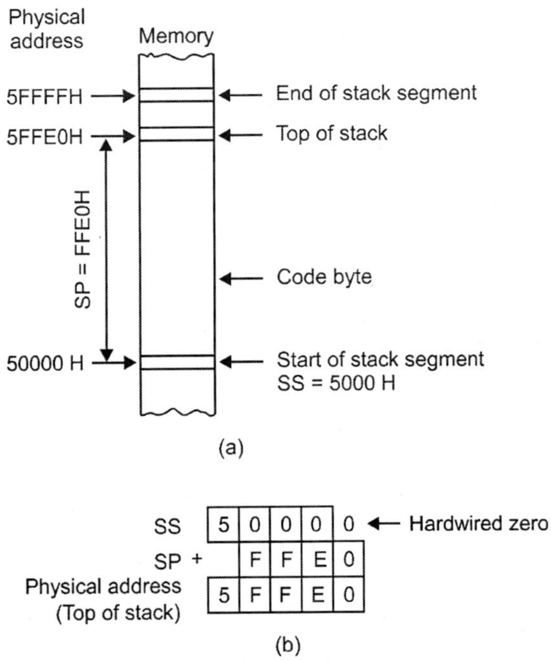


**Figure 5.5:** Addition of SS and SP to produce the physical address of the top of the stack. (a) Diagram. (b) Computation.

### 5. *Pointer and Index Registers in the Execution Unit*

In addition to the stack pointer register (SP), the EU contains a 16-bit base pointer (BP) register. It also contains a 16-bit source index (SI) register and a 16-bit destination index (DI) register. These three registers can be used for temporary storage of data just as the general-purpose registers described above. However, their main use is to hold the 16-bit offset of a data word in one of the segments. SI, for example, can be used to hold the offset of a data word in the data segment. The physical address of the data in memory will be generated in this case by adding the contents of SI to the segment base address represented by the 16-bit number in the DS register. After we give you an overview of the different levels of languages used to program a microcomputer, we will show you some examples of how we tell the 8086 to read data from or write data to a desired memory location.

## 5.3 INTEL 8086

The 8086 is a 16-bit HMOS microprocessor. It is available in a 40-pin IC, and operates at 5 volts dc supply. Its electronic circuitry consists of 29000 transistors. It is implemented in N-channel, silicon gate technology (HMOS). It is available in 3 versions: the 8086, 8086-2 and 8086-1. The clock rates are: 5 MHz for 8086, 8MHz for 8086-2 and 10MHz for 8086-1. The 8086 microprocessor is no longer used. But the concept of its principles and structure is very useful for the under-standing of other advanced Intel microprocessors.

The 8086 has 20 address lines. Using 20 address lines it can directly address up to $2^{20}$=1 megabytes (1MB) of memory. The address lines operate in time-multiplexed mode. The 16 low-order address lines carry 16 low-order address bits at certain times, and thereafter they carry 16-bit data at other times. The 4 high-order address lines are also time multiplexed lines. They carry 4 high-order address bits and thereafter certain status signals.

### 5.3.1 Functional Description of Intel 8086

The internal functions of the 8086 processor are partitioned logically into two processing units. The first is the Bus Interface Unit (BIU) and the second is the Execution Unit (EU) as shown in the block diagram of Figure 5.6.

These units can interact directly but for the most part perform as separate asynchronous operational processors. The bus interface unit provides the functions related to instruction fetching and queuing, operand fetch and store, and address relocation. This unit also provides the basic bus control. The overlap of instruction pre-fetching provided by this unit serves to increase processor performance through improved bus bandwidth utilization. Up to 6 bytes of the instruction stream can be queued while waiting for decoding and execution. The instruction stream queuing mechanism allows the BIU to keep the memory utilized very efficiently. Whenever there is space for at least 2 bytes in the queue, the BIU will attempt a word fetch memory cycle. This greatly reduces “dead time” on the memory bus. The queue acts as a First-In-First-Out (FIFO) buffer, from which the EU extracts instruction bytes as required. If the queue is empty (following a branch instruction, for example), the first byte into the queue immediately becomes available to the EU.

The execution unit receives pre-fetched instructions from the BIU queue and provides un-relocated operand addresses to the BIU. Memory operands are passed through the BIU for processing by the EU, which passes results to the BIU for storage. See the Instruction Set description for further register set and architectural descriptions.

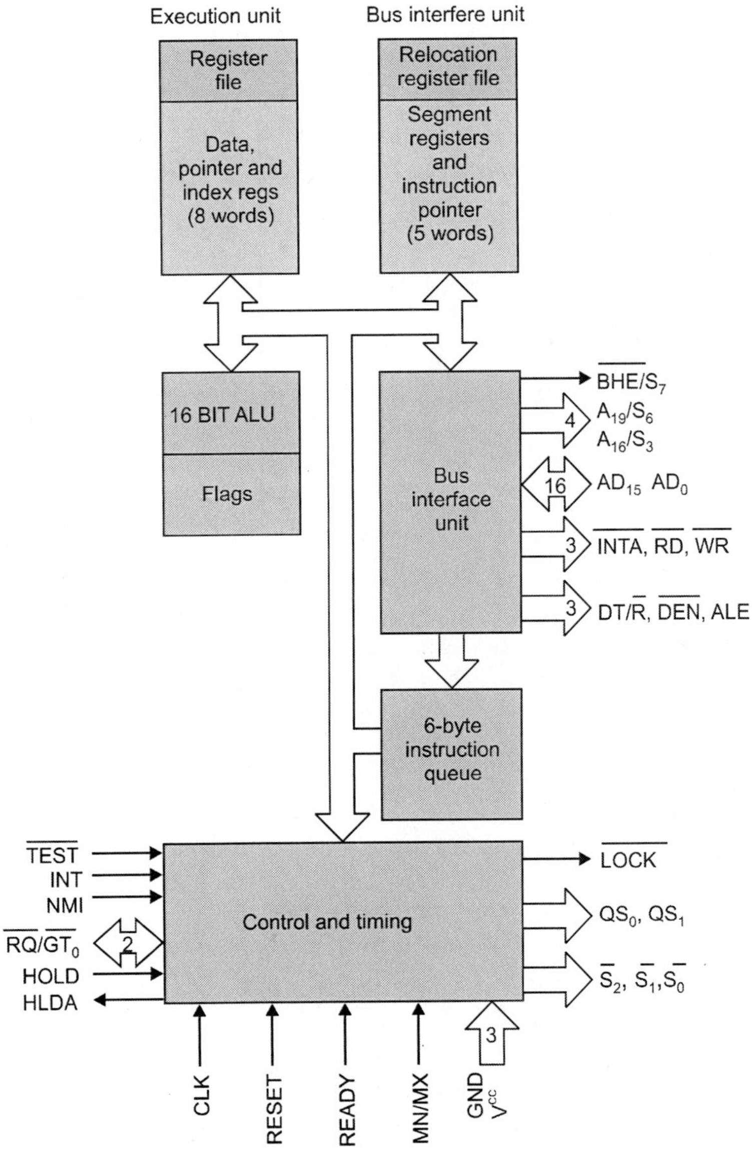


**Figure 5.6:** 8086 CPU Block Diagram

## 5.3.2 Pin Description of Intel 8086

Figure 5.7 (a) and (b) show the pin diagrams of Intel 8086. Figure 5.7(a) shows the signals of 8086 for minimum mode of operation. Figure 5.7(b) shows the signals for maximum mode of operation. These modes have been discussed in the subsequent sections. Its pins and signals are:

### Combination of 8086 maximum mode and minimum mode

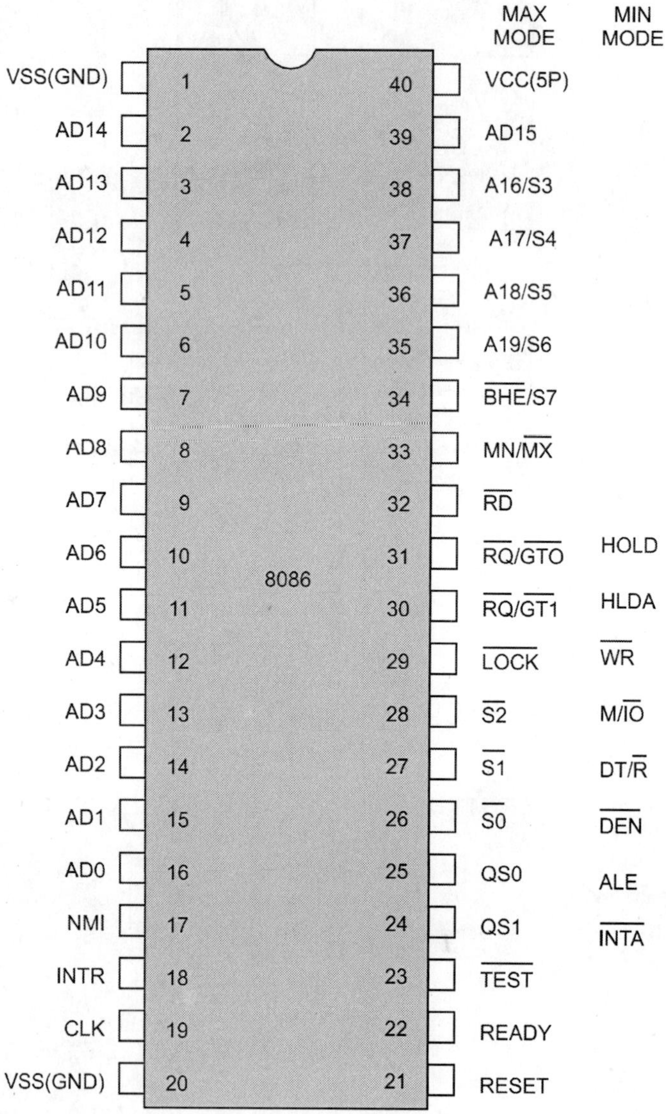


1. **$AD_0$-$AD_{15}$ (Bidirectional):** These lines constitute time-multiplexed address/data bus. During $T_1$ clock cycle of the bus cycle they carry low-order 16-bits of address During $T_2$, $T_3$ and $T_4$, they carry 16-bit data. The Intel 8284 clock generator/driver is employed to generate clock signals required for 8086, $AD_0$-$AD_7$ carry low-order byte of data and $AD_8$-$AD_{15}$ carry high-order byte of data. When AD lines carry address they are called A lines instead of AD lines, for example $A_0$-$A_{15}$.

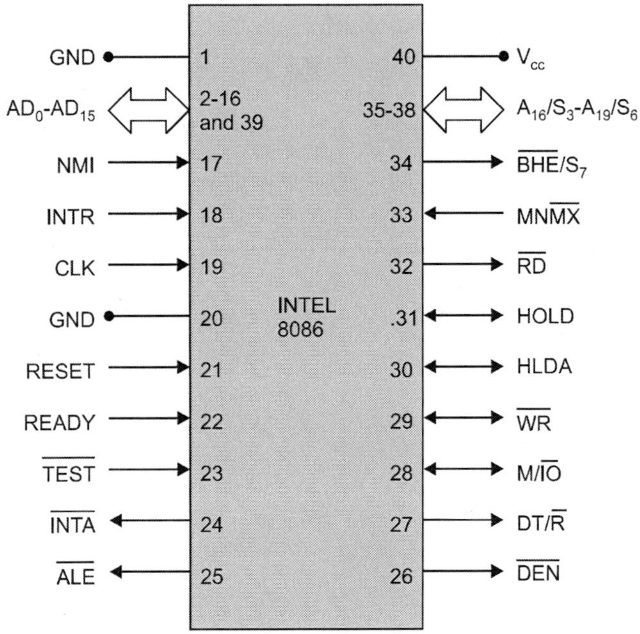


**Figure 5.7(a):** Signals of Intel 8086 for Minimum Mode of Operation

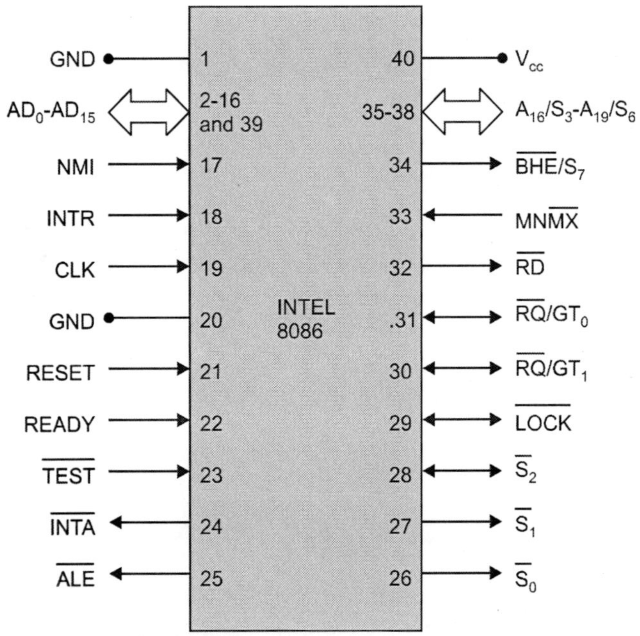


**Figure 5.7(b):** Signals of Intel 8086 for Maximum Mode of operation

When AD lines carry data they are called D lines, for example $D_0$-$D_7$, $D_8$-$D_{15}$ or $D_0$-$D_{15}$.

2. **$A_{16}$–$A_{19}$ (Output):** These are time-multiplexed lines. During $T_1$ they carry high-order 4 bits of memory address. During I/O operation, these lines remain low. During $T_2$, $T_3$ and $T_4$ these lines carry status signal.

3. **$A_{16}/S_3$, $A_{17}/S_4$ (Output):** These lines carry the address bits $A_{16}$ and $A_{17}$ during $T_1$ clock cycle of the bus cycle. During remaining bus cycle, these lines carry status signals $S_3$ and $S_4$. These status signals are identify memory segments as shown in **table 5.1.**

**Table 5.1 Encoding of $S_3$ and $S_4$**

| $S_4$ | $S_3$ | Segment Register Employed |
|---|---|---|
| 0 | 0 | ES |
| 0 | 1 | SS |
| 1 | 0 | CS or none |
| 1 | 1 | DS |

4. **$A_{18}/S_5$(Output):** The status signal $S_5$ is multiplexed with the address line $A_{18}$. During $T_1$ clock cycle of the bus cycle, eighteenth bit of the address is transmitted over this line. During the remaining bus cycle (i.e. during $T_2$ to $T_4$) the status signal $S_5$ is transmitted over this line. $S_5$ is an interrupt enable status. The status of the interrupt enable flag bit, $S_5$, is updated at the beginning of each clock cycle.

5. **$A_{19}/S_6$ (Output):** The status signal $S_6$ is multiplexed with the address line $A_{19}$. During $T_1$ clock cycle of the bus cycle, nineteenth address is transmitted over this line. During the remaining bus cycle (i.e. during $T_2$ to $T_4$) the status signal $S_6$ is available on this line. It remains low always during $T_2$ to $T_4$.

6. **$\overline{BHE}/S_7$ (Bus High Enable/Status):** The bus high enable signal, $\overline{BHE}$, remains low during $T_1$, and it is used to enable data of high-order data lines $D_8$-$D_{15}$. The 8-bit devices connected to the upper-half of the data bus use $\overline{BHE}$ signal for chip select function. $\overline{BHE}$ in conjunction with $A_0$ determines whether a byte or word will be transferred from/to memory locations.

Table 5.2 shows the status of $\overline{BHE}$ and $A_0$ for word and byte access.

**Table 5.2 $\overline{BHE}$ and $A_0$ Status**

| BHE | $A_0$ | Word/Byte Access |
|---|---|---|
| 0 | 0 | Whole word |
| 0 | 1 | Upper byte from/to odd address |
| 1 | 0 | Lower byte from/to even address |
| 1 | 1 | None |

The $\overline{BHE}/S_7$ is a time-multiplexed line. During $T_2$ to $T_4$ the status signal $S_7$ is transmitted on this line. It remains always high.

7. **$\overline{RD}$ (Output):** It is a read signal issued by the processor. It is an active low signal. It indicates that the processor is performing read operation with memory or I/O depending on the state for the M/ signal. This signal is used to read devices which are connected to the 8086 local but.
8. **READY (Input):** It is an acknowledgement received from the addressed memory or I/O device. It is an active high signal. When high, it indicates that the peripheral device is ready to transfer data.
9. **RESET (Input):** It is a system reset and an active high signal. When it goes high the processor goes into reset state. To be effective, this signal must remain high for at least four clock cycles.
10. **INTR (Input):** Interrupt request. INTERRUPT REQUEST: is a level triggered input which is sampled during the last clock cycle of each instruction to determine if the processor should enter into an interrupt acknowledge operation. A subroutine is vectored to via an interrupt vector lookup table located in system memory. It can be internally masked by software resetting the interrupt enable bit. INTR is internally synchronized. This signal is active HIGH.
11. **NMI (Input):** Non-maskable interrupt request. an edge triggered input which causes an interrupt request to the MP. A subroutine is vectored to via an interrupt vector lookup table located in system memory. NMI is not maskable internally by software.
12. **$\overline{TEST}$ (Input):** This signal is used to test the status of math coprocessor 8087. The BUSY pin of 8087 is connected to the $\overline{TEST}$ pin of 8086. Sometimes the 8086 needs the result of some computation that 8087 is doing, before it can go to the next instruction of the program. In such a situation WAIT instruction is used in the program. If $\overline{TEST}$ is high, 'WAIT' instruction causes 8086 to wait in an idle state. When 8087 completes computation, it makes BUSY signal low, thereby making $\overline{TEST}$ low. When $\overline{TEST}$ becomes low, the 8086 goes to the next instruction and continues execution of the program.
13. **MN/ $\overline{MX}$ (Input):** Indicates operating mode of 8086. When high, the processor operates in the minimum mode. When low, it operates in the maximum mode.
14. **CLK (Input):** Clock Input.
15. **$V_{CC}$ :** +5V power supply pin.
16. **GND:** Ground

The above pin description holds good for both the minimum and maximum mode of operation. The pin functions which are unique for the minimum and maximum mode of operation will be discussed in subsequent sections.

## 5.4 OPERATING MODES OF 8086

The 8086 has two modes of operation: the minimum mode and maximum mode. In a microcomputer system where only one 8086 microprocessor works as a CPU, it operates in the minimum mode of operation. In a multiprocessor and coprocessor configuration, the 8086 operates in the maximum mode of operation. The minimum mode is used in a few 8086 (or 8088) based systems with a few peripheral devices. Personal computers employing coprocessor configuration run in maximum mode. When there are large numbers of CPUs in a system, they operate in parallel. The status of the pin MN/ $\overline{\text{MX}}$ decides the operating mode of 8086. When this pin is high, the 8086 operates in minimum mode. When it is low, the 8086 operates in maximum mode. The 24 to 31 pins of 8086 have alternate function. They issue two sets of signals, one when the 8086 operates in the minimum mode of operating and the other in the maximum mode of operation.

### 5.4.1 Pin Description for Minimum Mode

The pin functions which are unique for the minimum mode of operation are discussed in this section. The other pin functions are same as already discussed in the Section 2.2. For minimum mode of operation pin MN/$\overline{\text{MX}}$ is kept high. The pins 24-31 have unique pin functions for the minimum mode of operation as shown in Figure 5.7(a) which are as follows:

1. **$\overline{\textbf{INTA}}$ (Output):** Pin 24 Interrupt Acknowledge. It is active low. On receiving interrupt signal the processor issues an interrupt acknowledge signal through this line.
2. **ALE (Output):** Pin 25 Address Latch Enable. It is a high pulse issued by the processor during $T_1$ state of a bus cycle. The processor sends this signal to latch the address into the 8282/8283 address latch.
3. **$\overline{\textbf{DEN}}$ (Output):** Pin 26 Data Enable. In the minimum mode of operation this signal is issued by the processor to enable Intel 8286/8287 bus transceiver. It is an active low signal.
4. **DT/$\overline{\textbf{R}}$ (Output):** Pin 27 Data Transmit/receive. When the minimum mode system incorporates Intel 8286/8287 octal bus transceiver, this signal is required for the data flow control. When this signal is high, data are sent out and when it is low, data are received.
5. **M/$\overline{\textbf{IO}}$ (Output):** Pin 28 Status Signal. It is issued by the processor to distinguish a memory access from an I/O access. When this signal is high memory is accessed and when it is low, an I/O device is accesses.
6. **$\overline{\textbf{WR}}$ (Output):** Pin 29 Write. When this signal is low, the processor performs memory write or I/O write operation depending on the state of M/IO signal.
7. **HLDA (Output):** Pin 30 HOLD Acknowledge. On receiving HOLD signal the processor issues a HOLD acknowledge signal through this pin. It is an active high signal.
8. **HOLD (Input):** Pin 31 When another device in the system wants to us the address and data bus, it sends a HOLD request to the processor through this line. It is an active high signal.

## 5.4.2 Pin Description for Maximum Mode

For the maximum mode of operation, the MN/MX is kept low and is grounded. The pins 24-31 have unique functions for the maximum mode of operation as shown in **Figure 5.7(b)** which are as follows:

1. **$QS_1$, $QS_0$ (Output):** Pin 24, 25 Queue Status. The queue status is valid during the clock cycle after which the queue operation is performed. These signals provide status to allow external tracking of the internal 8086 instruction queue. Table 5.3 shows the status of $QS_0$ and $QS_1$.

**Table 5.3 Status of $QS_0$ and $QS_1$**

| $QS_1$ | $QS_0$ | Characteristics |
|---|---|---|
| 0 | 0 | No Operation |
| 0 | 1 | 1st byte of opcode from queue |
| 1 | 0 | Empty queue |
| 1 | 1 | Subsequent byte from queue |

2. **$\overline{S_0}, \overline{S_1}, \overline{S_2}$ (Output):** Pin 26, 27, 28 Status Signals. These signals are required by Intel 8288 bus controller to generates all memory and I/O access control signals. Table 5.4 shows the encoding of these signals.

**Table 5.4 Encoding of $\overline{S_0}, \overline{S_1}$ and $\overline{S_1}$**

| $\overline{S_2}$ | $\overline{S_1}$ | $\overline{S_0}$ | Characteristics |
|---|---|---|---|
| 0 | 0 | 0 | Interrupt acknowledge |
| 0 | 0 | 1 | I/O read |
| 0 | 1 | 0 | I/O Write |
| 0 | 1 | 1 | Halt |
| 1 | 0 | 0 | Opcode fetch |
| 1 | 0 | 1 | Memory read |
| 1 | 1 | 1 | Memory write |
| 1 | 1 | 1 | Passive |

3. **$\overline{LOCK}$ (Output):** Pin 29 The $\overline{LOCK}$, a prefix instruction, activates $\overline{LOCK}$ signal. It is an active low signal. It remains active low signal. It remains active until the completion of the next instruction. When it goes low, all interrupts are masked and HOLD request is not granted. Consequently other devices do not get control over the system bus while $\overline{LOCK}$ is low. In a multiprocessor system, the other devices are informed through this signal that they should not issue HOLD request.

4. **$\overline{RQ}/\overline{GT_0}, \overline{RQ}/\overline{GT_1}$ (Bidirectional):** Pin 30, 31 Local Bus Priority Control. An external device sends a hold request to the CPU through these lines to release the local bus at the end

of CPU's current bus cycle. $\overline{RQ}/\overline{GT_0}$ has higher priority over $\overline{RQ}/\overline{GT_1}$. After receiving hold request, the CPU sends hold acknowledge signal through these lines. In the maximum mode of operation $\overline{WR}$, ALE, $\overline{DEN}$, DT/$\overline{R}$, etc. are not directly available from the processor. Rather they are available from the bus controller 8288.

## 5.5 REGISTER ORGANIZATION OF 8086

The 8086 contains fourteen 16-bit registers. Figure 5.8 shows register organization of 8086. They are grouped into the following categories.

(i) General purpose registers
(ii) Pointers
(iii) Index registers
(iv) Segment registers
(v) Instruction pointer and status register

### 5.5.1 General Purpose Registers

There are four 16-bit general purpose registers: AX, BX, CX and DX. Each of these general purpose registers can be used either as a 16-bit register or as two 8-bit registers as shown in Table 5.5. They are used to hold arithmetic and logical operands and results.

**Table 5.5 General Purpose registers of 8086**

| 16-Bit Registers | 8-Bit High-order Registers | 8-BitLow-order Registers |
|---|---|---|
| AX | AH | AL |
| BX | BH | BL |
| CX | CH | CL |
| DX | DH | DL |

The register AX servers as an accumulator. BX, CX and DX are primarily meant to hold data, but also serve as special purpose registers. For example, BX is used as a base register for memory address calculation. CX is used as a counter by certain instructions. It is used to hold a count value to control the number of times a loop is repeated. It is also used to hold a count for multiple bit shifts. DX is used to hold I/O address during certain I/O instructions. If the result of a multiplication is more than 16 bits, the low-order 16 bits are stored in AX, the high-order 16 bits are stored in DX register. If dividend is of 32 bits, high-order 16 bits are contained in DX register. After division, the 16-bit quotient is placed in AX and the remainder in DX.

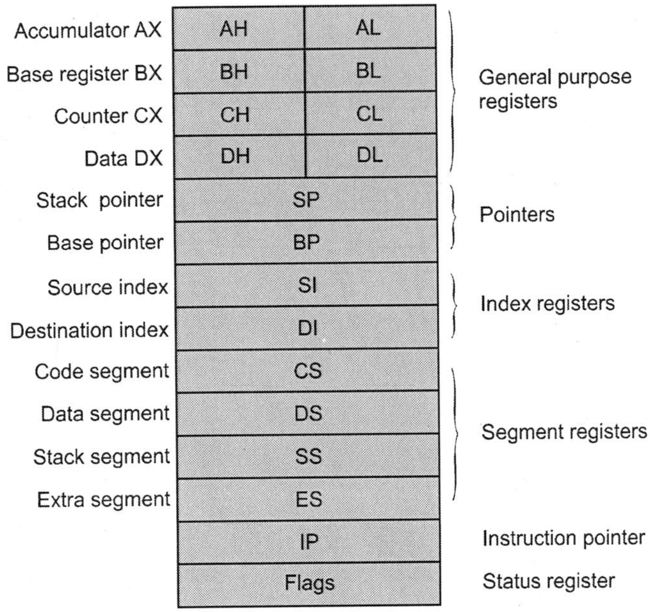


**Figure 5.8:** Schematic Diagrams of Intel 8086 Registers

Different registers and their operations are listed below:

| Register | Operation |
|---|---|
| AX | Word multiply, word divide, word I/O |
| AL | Byte multiply, byte divide, byte I/O, translate, decimal arithmetic |
| AH | Byte multiply, byte divide |
| BX | Translate |
| CX | String operations, loops |
| CL | Variable shift and rotate |
| DX | Word multiply, word divide, indirect I/O |

### 5.5.2 Pointer Registers

There are two pointers, SP and BP. The register SP is stack pointer. It points to the stack top address. It contains offset address. In PUSH, POP, CALL, INT and RETURN instructions, stack address is computed by adding the contents of SP and SS.

The register BP is the base register for accessing stack. Within stack, several data areas may exist. BP is used to hold the offset of the base of a data area in stack segment. BP can also be used as a general purpose register.

### 5.5.3 Index Registers

SI and DI are index registers. These are used in memory or stack address computation. For memory address computation in DS segment, either these are used alone or in conjunction with BX and/or displacement. For stack address computation in SS segment index registers are used with BP register and displacement.

In certain string instructions, SI is used to indicate source index and DI, destination index. In such cases the contents of SI are added to the contents of DS to get the actual source address of the data. The contents of DI are added to the contents of ES to get the actual destination address of the data. Registers SI and DI can also be used as general purpose registers.

### 5.5.4 Segment Registers

In this section, memory segments and segment registers will be discussed. The memory used with an 8086 based system is divided into the following four types of segments.

1. Code segment
2. Data segment
3. Stack segment
4. Extra segment

Each memory segment can have a capacity up to 64K bytes. The code segment contains instruction codes of a program. Program's defined data, variables and constants are contained in a data segment. A section of memory, called stack is set aside to store addresses and data of a subroutine. The PUSH instructions also uses stack to save the contents of a specified register or memory location into the stack. In 8086, a block of memory which may be up to 64K bytes long, called stack segment is set aside for such purposes. The extra segment contains the destination of some data of certain string instructions.

The maximum size of a segment is 64K bytes. Thus segmentation allows 8086 to work only with 16-bit effective address to access any memory location and therefore the 8086 manipulates and stores only 16-bit address components. In time-shared system, to change from one user's program to another user's program to another user's program, the CPU can simply reload the four segment registers with the segment starting addresses assigned to the next user's program. Thus the segmentation makes it easy to change from one user's program to another user's program. It also makes easy to keep user's programs and data separate from each other.

**The 8086 contains the following four segment registers**

1. Code Segment Register, CS
2. Data Segment Register, DS
3. Stack Segment Register, SS
4. Extra Segment Register, ES

The segment registers contain the high-order 16 bits of the starting addresses of four memory segments that the 8086 CPU works with at a particular time. The 8086 uses 20 lines for memory addressing. With 20 address lines it can directly address up to $2^{20}$ = 1,048,576 bytes (1M bytes) of memory. At a particular time, the 8086 CPU works with only four segments (each 64K bytes in size) within 1M bytes of memory. Figure 5.9 shows an example of memory segmentation. In 1MB memory space there may be several such sets of four segments. The four memory segments that the CPU is working with at any given time are called currently active segments. By changing the contents of the segment registers, the CPU can work with another set of four memory segments.

The 8086 sends 20-bit address on the address bus to access a memory location. The 20-bit address sent out by the CPU is called the physical address. A segment register contains the upper 16-bits of the starting address of a memory segment. The CPU inserts four zeros for the lowest four bits of the 20-bit starting address. For example, if the contents of the CS are 2,000 hex, the 20-bit starting address of the code segment will be 20,000 hex. A 64KB memory segment can reside anywhere within the 1MB memory space, but its starting address is always divisible by 16. It is necessary to store and manipulate 16-bit numbers.

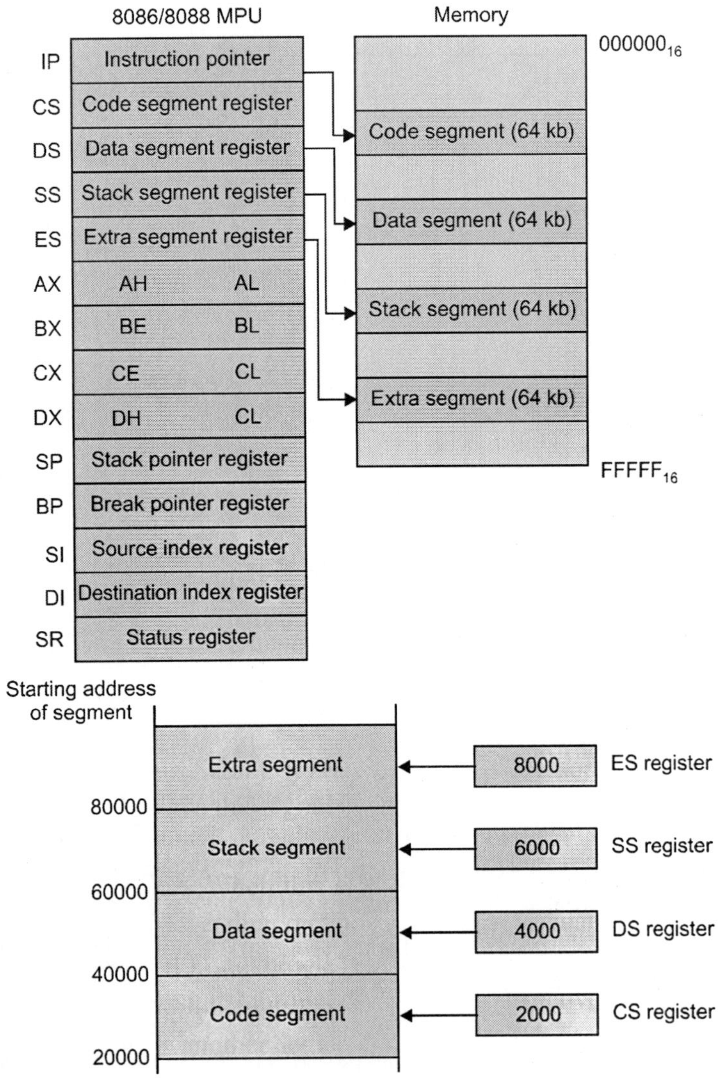


**Figure 5.9:** Memory Segmentation in Intel 8086 Based System

How far a memory location is within a memory segment from the starting address is called offset of effective address, as shown in Figure 5.10. An offset is determined by the addressing mode which is used to get an operand. To address a memory location within a memory segment the 8086 produces a 20-bit physical address. The contents of a segment register are shifted to left by 4-bits and then an offset are added to it to compute the required 20-bit physical address. For example, the contents of CS are 2000 hex and an offset is 400. The 20-bit physical address is computed as follows:

The contents of CS, 2000, are first shifted left by 4 bits. The resulting quantity is 20,000 hex.
The physical address = 20,000 + 400

= 20,400 hex.

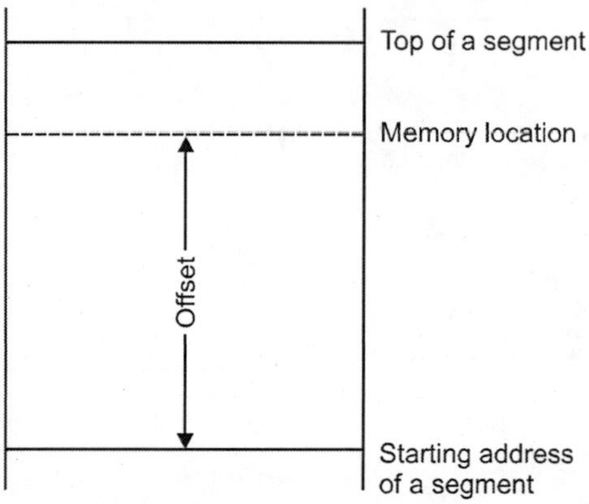


**Figure 5.10:** Effective Address or Offset

### 5.5.5 Unsegmented (or Flat) Scheme for Memory

In unsegmented or flat scheme, memory space is not divide into segments. The entire memory space appears as a single addressable memory space. In 8086 it can be achieved by initializing all the segment registers with the same base address. It will cause all memory operations to refer to the same memory space.

### 5.5.6 Instruction Pointer

The instruction pointer, IP, acts as a program counter. It points to the next instruction to be executed within the currently executing code segment. Its content is automatically incremented as the execution of a program proceeds. IP holds a 16-bit offset pointing to next instruction code within the current 64KB code segment. The actual 20-bit physical address of the next instruction code is obtained by shifting the contents of CS register 4 bits left and then adding to it the content of IP (i.e. offset).

### 5.5.7 Status Register

The 8086 has one 16-bit status register. It is also called flag register or program status word (PSW). It has nine flag. Seven bit positions of the status register remain unused. A flag is indicated by set or reset condition of a flip-flop. Out of nine flags six are condition flags. They are set or reset by the CPU depending on the results of some arithmetic or logic operation. The remaining three flags of the status register are control flags. These flags are set or reset by the programmer as required by certain instructions in a program. Figure 5.11 shows the status flags of 8086.

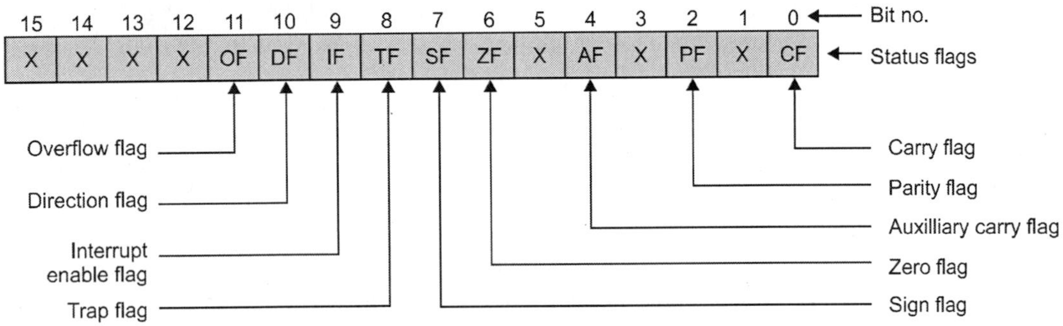


**Figure 5.11:** Status Flags of Intel 8086

The six condition flags are:

**Carry Flag (CF):** It is set to 1 if there is carry out of the most significant bit position resulting from an addition or if a borrow is needed at MSB during subtraction. If no carry out of the MSB is produced or borrow is needed, the carry flag is reset, i.e., it is zero.

**Parity Flag (PF):** If the low-order 8 bits of the result of an operation contains even number of 1 bits, the parity flag is set to 1. If the low-order 8 bits of the result contains odd number of 1s, the parity flag is zero.

**Auxiliary Carry Flag (AF):** It is set to 1 if there is a carry out of bit 3 resulting from an addition or a borrow is required from bit 4 into 3 for subtraction.

**Zero Flag (ZF):** It is set to 1 if the result of an arithmetic or logic operation is zero. In case of nonzero result it is zero.

**Sign Flag (SF):** It is set to 1 if the MSB of the result of an arithmetic operation is 1. It is set to zero, if MSB of the result is zero. This flag is used with signed numbers. In 2's complement, the MSB of a negative number is 1, and the MSB of a positive number is zero. In arithmetic operations with signed numbers, the MSB is kept reserved to represent the sign of a number. After the execution of an arithmetic instruction the result is negative if the MSB is 1; and positive if the MSB is 0.

**Overflow Flag (OF):** In case of a signed arithmetic operation, the overflow flag is set to 1, if the result is too large to fit in the number of bits available to accommodate it. The overflow flag has no significance in unsigned arithmetic operation.

Suppose two 16-bit signed numbers 2235H and 6F24H are added. These are positive numbers with most significant bit 0, which is the sign bit. The result is 9159H. Its sign bit is 1, which indicates that the result is negative. This is obviously a wrong answer. But the overflow flag is set to 1 to indicate that the result has overflowed into the sign bit. The three control flags are:

- **Trap (or Trace) Flag (TF):** It is used for single-step control. It allows user to execute one instruction of a program at a time for debugging. When TF is set to 1, a program can be run in single-step mode. When an interrupt is recognized, the TF flag is automatically cleared. When the CPU returns back to the main program after executing interrupt service subroutine, the TF flag is restored to the state in which it was before the interrupt occurred. This is achieved by executing IRET instruction at the end of the interrupt service subroutine. There is no instruction to set or reset TF flag directly. So its setting or resetting can be done by pushing the flag register on the stack, taking necessary steps to change the TF bit as desired and then popping the flag register back off the stack.

- **Interrupt Flag (IF):** It is an interrupt enable/disable flag. If it is set to 1, the maskable interrupt INTR of 8086 is enabled and if it is 0, the interrupt is disable. IF can be set to 1 by executing STI instruction. It can be cleared by executing CLI instruction. When an interrupt is recognized the IF flag is automatically cleared resulting in disabling of the INTR interrupt. IRET used at the end of the interrupt service subroutine restores IF flag to its earlier state, i.e. before the interrupt. When 8086 is reset, IF is automatically cleared.
- **Directional Flag (DF):** It is used in string operations. If it is set to 1, string bytes are accessed from high memory address to low memory address. When it is zero, the string bytes are accessed from low memory address to high memory address. For MOVS instruction if DF is set to 1, the contents of the index registers SI and DI are automatically decremented to access the string from the highest memory address down to the lowest memory address. If DF is set to 0, SI and DI are automatically incremented to access the string beginning with the lowest address. The DF can be set by executing STD instruction and it can be cleared by executing CLD instruction.

## 5.6 BUS INTERFCE AND EXECUTION UNIT (BIU AND EU)

The 8086 contains two functional units: a bus interface unit (BIU) and an execution unit (EU), as shown in Figure 5.12. The BIU handles all interfaces with the external bus and generates external memory and I/O address. It fetches instruction codes form the memory and keeps them into a 6 byte instruction queue. It fills up the queue whenever the bus is idle and 2 bytes of the queue are vacant. In case of JUMP and CALL instructions, the BIU dumps the queue and then starts reloading it from a new address. The BIU reads data from the memory and ports, and writes data to the memory and ports.

The execution unit (EU) receives prefetched instructions from the queue, decodes them and then executes them. The queue acts as the first-in-first-out for EU. If the queue is empty, the EU waits till the queue gets at least one byte. Such situation arises after the execution of a jump instruction. The EU also informs BIU where to fetch instructions or read data from. While the EU executes instructions, the BIU fetches instruction codes from the memory.

Thus, there is an overlap between instruction execution and instruction fetching. The design feature of a processor which provides overlapping among the various operations to be performed by the processor is called pipelining. To achieve pipelining, several functional units can be employed in a microprocessor. They work simultaneously in parallel. Each functional unit performs one type of operation. In 8086, BIU and EU are two functional units and they work independently in parallel.

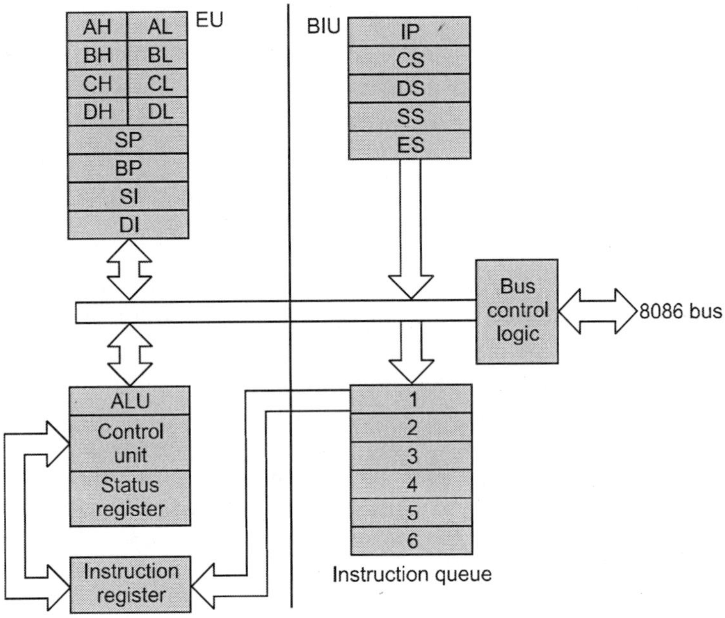


**Figure 5.12:** Functional Units of 8086

## 5.7 INTERRUPTS

The normal program execution of a microprocessor can be interrupted by some external device or by inserting a special instruction in a program. An interrupt caused by an external device is called hardware interrupt. A microprocessor has interrupt input lines to receive interrupt signals. When an external device wants to interrupt the microprocessor, it sends an interrupt signal to the microprocessor through the interrupt line to which it is connected. Slow I/O devices are connected to a microprocessor through an interrupt line. When data are ready, an input device interrupts CPU. After completing the current instruction at hand the CPU responds to the input device. The CPU executes a subroutine to transfer data from the interrupting device. This subroutine is called interrupt service subroutine (ISS). When data transfer is completed the CPU returns beck to the main program which it was executing before the interrupt occurred.

When the normal program execution of a microprocessor is interrupted by a special instruction, it is called software interrupt.

1. **Enabling, Disabling and Masking of Interrupts:** When interrupts are to be used they are enabled (made effective) by software. This is known as enabling of an interrupt. When a microprocessor is executing certain important program or subroutine, it may be desired that interrupt should not occur during that period. Then can be done by disabling (made infective) interrupts. An interrupt can be disabled by the software.

   In certain situations it may be desired that some of the several interrupts should not be entertained, and hence they are masked, I.e. made ineffective by software. On this basis the interrupts are classified into two categories, maskable interrupts and nonmaskable interrupts.

2. **Exceptions:** The interruption in normal program execution of a microprocessor by abnormal internal conditions is called an exception. On the other hand, an interruption of microprocessor caused by external events is known as an interrupt. Software interrupts caused by INT N instruction are also treated as exceptions by the microprocessor. Exceptions include faults, traps and aborts.
3. **Faults:** Program faults come under this category of exception. They are detected and corrected before the execution of the faulty instruction. Examples of faults are: attempts to executes illegal or undefined instructions, access nonexistent or protected memory, etc.
4. **Traps:** The exceptions detected immediately after the execution of an instruction that caused the problem are called traps. Examples of traps are: exceptional conditions such as division by zero, overflow, user defined interrupts, etc.
5. **Aborts:** Abnormal conditions caused by hardware faults are called aborts.

Definitions of interrupts and exceptions differ from book to book. Sometimes all sorts of internal abnormal conditions and external events are dealt under the single heading of interrupts or sometimes under exceptions.

### 5.7.1 Interrupts of 8086

1. **Hardware interrupts:** Hardware interrupts are NMI and INTR. The NMI is a nonmaskable interrupt whereas INTR is a maskable interrupt. The NMI has higher priority then INTR. The NMI is edge-triggered on a low-to-high transition. It is required to remain in high state of greater then two clock cycles. For this purpose the rising edge of NMI input is latched on-chip and is serviced at the end of the current instruction.

   The INTR is a level-triggered input which is sampled during the last clock cycle for each instruction. The single is active high.
2. **Software interrupt:** Software interrupt is generated by inserting INT instruction in a program. Any of the 256 interrupt types can generated specifying the interrupt type after INT instruction. The instruction INTO generates only overflow interrupt. It is used immediately after an arithmetic instruction.
3. The single-step interrupt.
4. Divide-by-zero interrupt.
5. Overflow interrupt.

### 5.7.2 Interrupt Pointer (or Vector)

1K bytes of memory from 00000 to 003FF (hex) is set aside to store the starting addresses of interrupt service subroutines in an 8086 based system. To store the starting address of each ISS, four bytes of memory space is required: two bytes of a store the values for CS and two bytes for IP values. Thus 1K bytes of memory space can store the starting addresses of 156 ISS. The starting address of an ISS stored in the 1KB memory space is called interrupt pointer or interrupt vector. The 1KB memory space acts as a table to store the starting addresses of ISS. This table is called the interrupt pointer table.

The 256 interrupt pointer have been numbered from 0 to 255. The number assigned to an interrupt pointer is called the type of the corresponding interrupt. For example, type 0 interrupt, type 1 interrupt, type 2 interrupt,..., and so on. The starting address of the ISS for type 0 interrupt is 00000 hex, for type 1 interrupt is 00004 hex, for type 2 interrupt is 00008 and so on Figure 5.13 shows the interrupt pointer table for 8086. The first five pointers are dedicated interrupt pointer for specific interrupts. They are for divide-by-zero, single-step control, nonmaskable interrupt NMI, breakpoint, and overflow interrupt. The Intel has kept reserved the next 27 interrupts, from type 5 to type 31; for other advanced microprocessors or they can be used for some special interrupt in an 8086 based system. The microprocessor will respond accordingly. The upper 224 interrupts from type 32 to type 255, are available to users for hardware software interrupts.

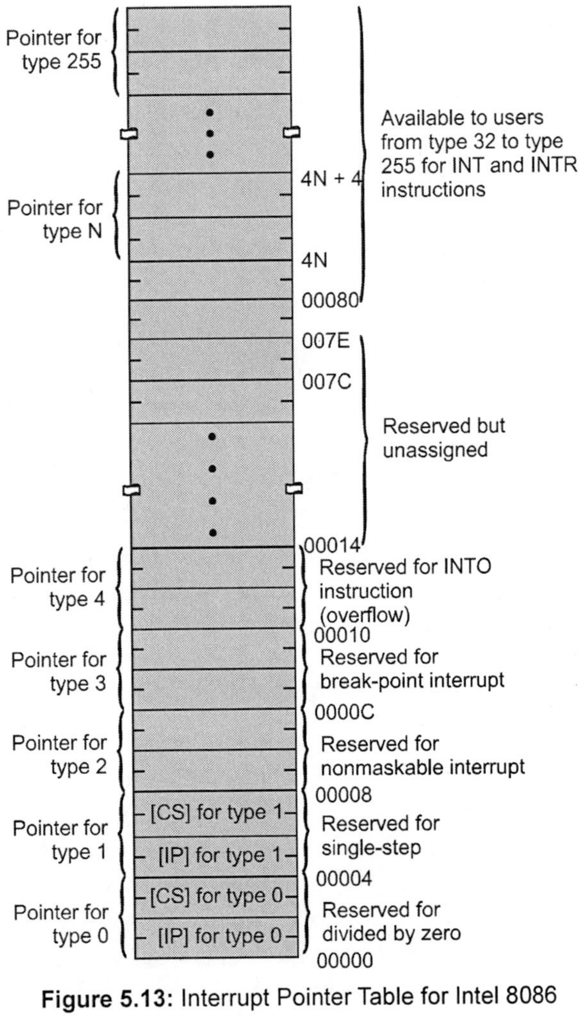


**Figure 5.13:** Interrupt Pointer Table for Intel 8086

When an interrupt occurs, the 8086 pushes the flag register on the stack, TF and IF are cleared which disable the single-step and INTR interrupt, and the contents of CS and IP are pushed on the stack. Thereafter, the program jumps automatically to the starting address of the ISS. In case of a hardware interrupt, the interrupting device or interrupt controller gives the type of the interrupt in response to the $\overline{\text{INTA}}$ signals sent to the I/O device or interrupt controller. In case of software interrupt, the type depends on the instruction used for the purpose.

At the end of an ISS, IRET instruction is used. When IRET is executed, it pops the contents of IP and then CS. Finally it pops flags into flag register for the stack. When flags are restored, IF and TF are restored to the previous values (i.e. before the interrupt occurred.). Suppose IF and TF were enabled before interrupt occurred, they will be enabled again after return to the main program from ISS. During the execution of ISS, IF and TF are cleared resulting in disabling of INTR and single step interrupt.

### 5.7.3 Type 0 Interrupt (Divide-by-zero Interrupt)

Type 0 interrupt is divide-by-zero interrupt. When the result (quotient) of division operation (execution of DIV or IDIV instruction) is too large to be stored in the destination register, the 8086 performs type 0 interrupt. The division by zero is the worst case of such division.

### 5.7.4 Type 1 Interrupt (Single-step Interrupt)

Type 1 interrupt is single-step interrupt. Single-step control is used for debugging assembly language program. In the single-step control, the processor executes one instruction and then stops. Thereafter, the contents of registers and memory locations can be examined to see the result. If the results are correct the user can give a command to execute the next instruction. In 8086 microprocessor, the trap flag TF is set to 1 to implement single-step control. When TF flag is set to 1, the 8086 automatically performs type 1 interrupt after the execution of each instruction. The ISS for type 1 interrupt is written to save the contents of all registers on stack so that they can be examined or displayed on CRT.

In 8086, there is no instruction to set the trap flag TF directly. Therefore, the flags are pushed on the stack, changes are made and then they are popped.

### 5.7.5 Type 2 Interrupt (Nonmaskable Nmi Interrupt)

Type 2 interrupt is nonmaskable NMI interrupt. It can be used for some emergency conditions. For example, it can be used to save program and data when power supply fails. An external circuit is employed to detect power failure and send an interrupt signal to 8086 through NMI interrupt line. If ac power fails, the dc voltage supplied to computer may remain for 50 milliseconds due to capacitors provided in the filter circuit. This time is sufficient to save program and data in some RAM having battery backup supply. ISS for type 2 interrupt is written to save program which the computer was executing. The program along with data is restores when power returns, and can be executed again from the point where it was interrupted.

### 5.7.6 Type 3 Interrupt (Breakpoint Interrupt)

Type 3 interrupt is a breakpoint interrupt. A breakpoint is inserted I a program for debugging assembly language program. The program is executed up to the breakpoint. An interrupt occurs at this point of the program. The ISS saves all the registers on the stack. The register contents may be displayed on CRT or in a simple system, the control is returned to the user. To implement this type of interrupt, INT 3 instruction is used. It is inserted temporarily in the program at a point up to which the user wants to execute the program and examine the result.

### 5.7.7 Type 4 Interrupt (Overflow Interrupt)

Type 4 interrupt is an overflow interrupt. When the signed result of an arithmetic operation on two signed numbers is too large to be stored in the destination register or memory location, an overflow errors occurs and the 8086 sets OF to 1. To implement type 4 interrupt INTO instruction is inserted in the program immediately after the arithmetic instruction. If overflow flag is not set, the INTO instruction simply performs no operation (functions as NOP). If OF is set to 1, the INTO instruction performs type 4 interrupt. The ISS is executed to indicate overflow error in the way user wants.

### 5.7.8 Software Interrupts Type 0-255

The software interrupt is implemented using INT n instruction. It may have any value from 0 to 255. For example, INT 33 will cause type 33 interrupt. The IP and CS values to start ISS are obtained from the interrupt pointer table. The IP value for any interrupt type is always available at the memory address having value equal to four times the interrupt type. It is because each interrupt pointer is of 4 bytes and the interrupt pointer for type zero starts from the memory address 0000. The CS value is at the memory location two addresses higher than that of IP. The starting address fro INT 33 will be 33 x 4 = 132 decimal. It is 84 hex. The IP values for ISS are at memory location 84H and CS values at 86H. INT 3 is also a software interrupt. It is used to provide a breakpoint in a program as already discussed.

In an 8086 system, many subroutines are written for input and output operations. They are called BIOS (Basic Input Output System). They are stored in a ROM. Some examples are: subroutines for reading a character from a keyboard, reading information from a disk, writing some characters to CRT, etc. Such a BIOS can be called using a software interrupt. INT n interrupt is also used to test ISS. For example, if INT 0 is executed, it will test the divide-by-zero interrupt subroutine without running an actual division program. Similarly, INT 2 will test the ISS for NMI interrupt without applying external interrupt signal to NMI.

### 5.7.9 Intr Interrupts: Type 0-255

The INTR is a maskable hardware interrupts. It is enabled if the IF flag is set to 1 and disabled (masked) if the IF flag is set to 0. The IF can be set to 1, executing STI instruction. It can be cleared executing CLI instruction.

The IF flag is automatically cleared when the 8086 is reset. When an interrupt is recognized, the IF flag is cleared. Before executing an ISS all flags are saved on the stack. At the end of an ISS,

IRET is executed. This restores the flag to values which were before the interrupt occurred. Therefore, if the IF flag was set to 1 before an interrupt occurred, it will be enabled again when the CPU returns to the main program from the ISS.

In case INTR, interrupt type is set by the external device to the CPU. After receiving INTR interrupt from an external device, the 8086 executes two consecutive interrupt acknowledge bus cycles, to get interrupt type from the external device. In the beginning of the first interrupt acknowledge cycle the 8086 floats data bus lines $AD_0$–$AD_{15}$. It then sends $\overline{INTA}$ signal to inform the external device to get ready. In he second interrupt acknowledge cycle, the 8086 sends another $\overline{INTA}$ pulse. In response to the second $\overline{INTA}$ pulse, the external device sends interrupt type to 8086 through low-order eight lines of the data bus. On receiving interrupt type from the external device, the 8086 executes corresponding ISS. During interrupt acknowledge cycles the 8086 makes $\overline{LOCK}$ signal low from $T_2$ of the first interrupt acknowledge cycle, till $T_2$ of the second interrupt acknowledge cycle so that a hold request will not be honoured during these two acknowledge cycles.

### 5.7.10 Priority of 8086 Interrupts

When two or more interrupts occur simultaneously, interrupts are taken up one by one. For this purpose, priority is assigned to interrupts. The interrupt of the highest priority is serviced first and then the interrupt of next higher priority is taken up. The priorities of interrupts of 8086 are shown in Table 5.6.

**Table 5.6 Priorities of Interrupts of 8086**

| Interrupt | Priority |
|---|---|
| Divide error, INT n, INTO | Highest |
| NMI | |
| INTR | |
| Single step | Lowest |

## 5.8 8086 BASED COMPUTER SYSTEM

Figure 5.14 shows an example of an 8086-based computer system in minimum mode configuration. This configuration is used for a vary small 8086 (or 8088) based system with a few peripheral devices. The 8282 is an octal latch. It is used to latch the address sent out by the CPU. The STB is the strobe pin of 8282. It is active high to admit the input signals. The input is latched at the high to low transition of STB. The $\overline{OE}$ is an output enable pin. It is made low to enable the output of 8282.

The 8286 is an octal bus transceiver. It is a data bus buffer driver. is an output enable control signal. When it goes low the output of 8286 is enabled. T is a control input signal to control the direction of data flow. When it goes low the 8086 receives data through the transceiver. When it goes high, the 8086 sends data through the 8286. The 8286 may not be used in a very small system. Its need arises when a number of devices are connected to the system. Figure 5.15 shows an example of an 8086 based computer system in the maximum mode configuration. In addition to 8282 and 8286, a bus controller 8288 is also used. The $\overline{S_0}$, $\overline{S_1}$ and $\overline{S_2}$ are input to 8288.

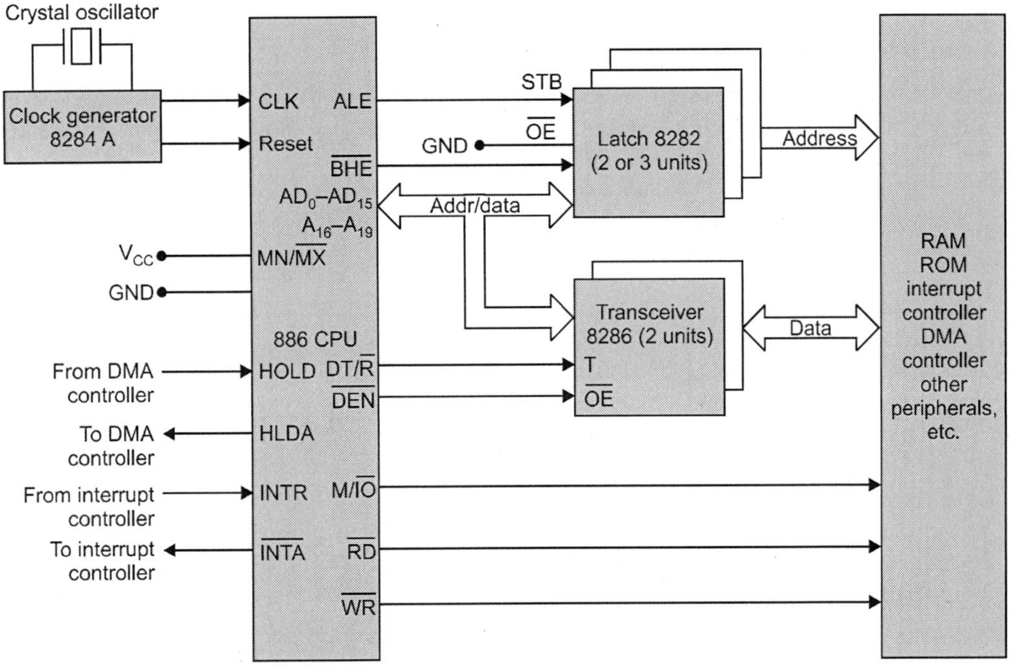


**Figure 5.14:** 8086 Based Computer in Minimum Mode Configuration

A number of control signals, as shown in Figure 5.15, are generated by 8288. The important control signals are

$\overline{MRDC}$ — Memory read command. It is same as $\overline{MEMR}$.

$\overline{MWTC}$ — Memory write command. It is same as $\overline{MEMW}$.

$\overline{IORC}$ — I/O read command. It is same as $\overline{IOR}$.

$\overline{IOWC}$ — I/O write command. It is same as $\overline{IOW}$.

$\overline{AMWC}$ — Advanced memory write command. It gives memory an early indication of a write instruction.

$\overline{ALOWC}$ — Advanced I/O write command. It gives I/O devices an early indication of write instruction.

$\overline{INTA}$ — Interrupt acknowledge.

DEN — Data enable. It is to be applied to $\overline{OE}$ of 8286. As $\overline{OE}$ is low active, DEN is inverted and then applied to  of 8286.

$DT/\overline{R}$ Data transmit/receive.

ALE Address latch enable.

The 8288 generates control signals compatible with Intel's multibus architecture, which is suitable for a multiprocessor system.

An 8086 based computer system uses 8284 clock generator, 8087 math processor, 8259 interrupt controller, 8237 DMA controller, 8255 programmable peripheral interface, etc.

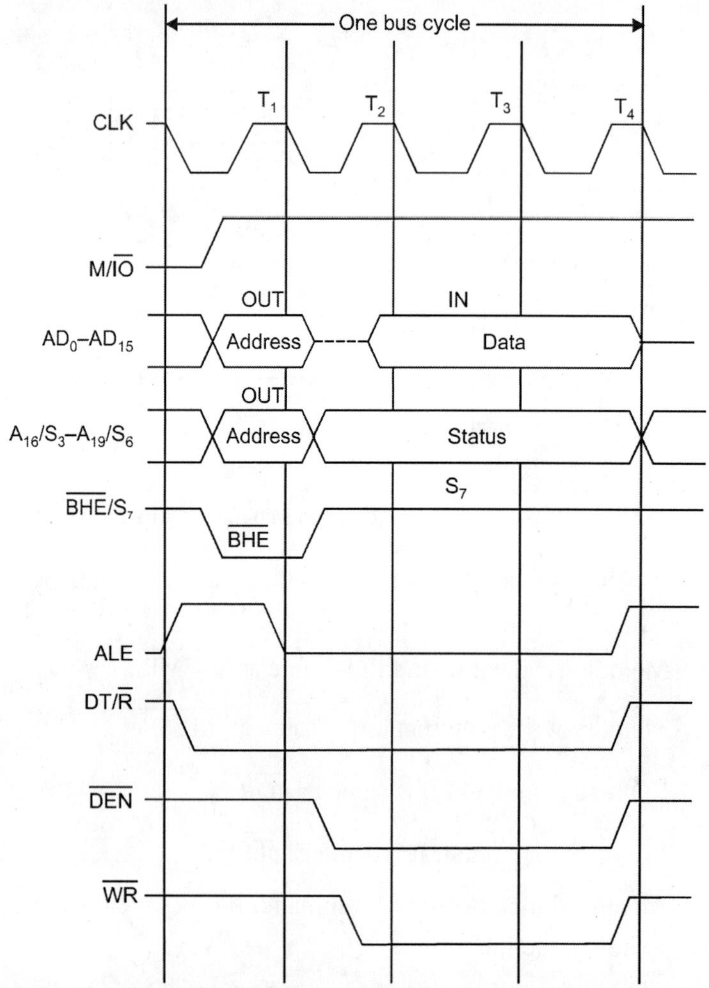


**Figure 5.15:** 8086 Based Computer in Maximum Mode Configuration

## 5.9 8086 READ AND WRITE BUS CYCLES

A bus cycle is a bus interface unit (BIU) phenomenon. The EU executes instructions in a number of clock periods which are not grouped in any way to form a machine cycle. Hence, the EU does not use machine cycles. The BIU fetches instruction codes from memory, reads data from memory or I/O devices or writes data to memory or I/O devices. The clock periods are grouped only when the BIU wishes to access memory or I/O devices. Any external access to memory or an I/O devices requires four clock periods to perform either a read or write operation. These four clock periods are grouped, and this group is called a bus cycle. Since this is the only time the 8086 groups clock periods, it is more appropriate to call it a bus cycle.

### 5.9.1 8086 Bus Cycle for Memory Read for the Minimum Mode

Figure 5.16 shows the timing diagram for 8086 memory read bus cycle for minimum mode of operation. During $T_1$ clock cycle of bus cycle, the 8086 sends out the 20-bit memory address on the address bus.

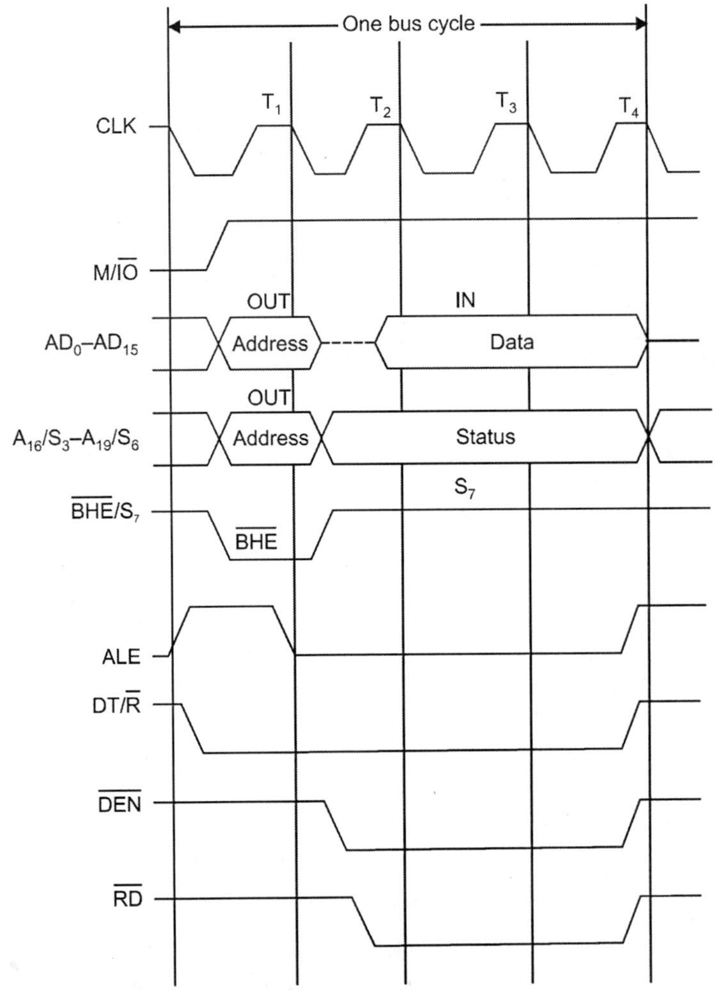


**Figure 5.16:** Timing diagram for 8086 Memory Read Bus Cycle for Minimum Mode of Operation

The 16 least significant bits of the address are sent out on $AD_0$-$AD_{15}$ lines, and 4 most significant bits of the address on $A_{16}$-$A_{19}$ lines. As the address lines operate in time multiplexed mode, the address remains on these lines only for the clock period $T_1$. For the minimum mode of operation, $MN/\overline{MX}$ is kept high. The 8086 sends our a high signal on $M/\overline{IO}$ line to indicate that data transfer will take place with memory.

During $T_1$, an address latch enable signal, ALE, is sent out so that the address is latched in the external devices.

During $T_2$ state the $AD_0$-$AD_{15}$ go into high impedance state. This is shown by dotted line in Fig. 5.16, as their drivers are disabled.

During $T_3$ and $T_4$ data are transmitted on $AD_0$-$AD_{15}$-. During $T_2$, $T_3$ and $T_4$ the address lines $A_{16}/S_3$-$A_{19}/S_6$ carry status signals $S_3$, $S_4$, $S_5$ and $S_6$.

$\overline{BHE}$ goes low during $T_1$. The $\overline{BHE}$ in conjunction with $A_0$ decides whether a byte or word is to be transmitted from/to the addressed memory location, Table 5.2. During $T_2$, $T_3$ and $T_4$ the status signal $S_7$ is sent out on $\overline{BHE}/S_7$ line.

The $\overline{RD}$ goes low during $T_2$. This signal can be used by the selected memory IC to enable its output buffer. Now, the addressed memory puts data on the data line. $\overline{RD}$ becomes high in $T_4$.

If the READY signal is high during $T_2$, it indicates that the memory device is ready and the 8086 operates normally according to the bus cycle shown in the Figure 5.16. The READY signal is used by slow memory devices. If the memory is not ready to transfer data, it does not make READY signal high, i.e. READY is low. If READY is low during $T_2$, the 8086 inserts a wait state between $T_3$ and $T_4$. The 8086 takes necessary steps to read data from the memory only when READY becomes high.

The $\overline{DEN}$ signal is an output enable signal for the 8286/8287 octal bus transceivers. The $DT/\overline{R}$ signal controls the direction of data flow through the bus transceivers. For read operation, $DT/\overline{R}$ is made low (refer to Figure 5.16). For a very small system, these transceivers may not be used but they become necessary when large number of devices are connected to the system. The outputs of the transceivers (buffers) should be enabled only when required and this is achieved employing $\overline{DEN}$ signal. To avoid interference these should not be enabled when AD bus is carrying 16 LSBs of the address. For a read or $\overline{INTA}$ cycle $\overline{DEN}$ is active from the middle of $T_2$ till the middle of $T_4$, while for a write cycle, it is active from the beginning of $T_2$ till the middle of $T_{-4}$.

$S_3$ and $S_4$ identify the memory segment which is to be accessed refer to Table 5.1. $S_5$ is interrupt enable status. The status of the interrupt enable flag bit, $S_5$ is updated at the beginning of each clock cycle.

### 5.9.2 8086 Bus Cycle for Memory Write for the Minimum Mode

The memory write bus cycle as shown in Figure 5.17 is similar to the read bus cycle. In case of write bus cycle, goes low in $T_2$ to enable write operation and again becomes high in $T_4$.

The signal $DT/\overline{R}$ is made high for transmitting data to the memory through transceivers. All other signals remain similar to those for read cycle. $AD_0$–$AD_{15}$ is not disabled in $T_2$ and the data to be sent out are put on these lines.

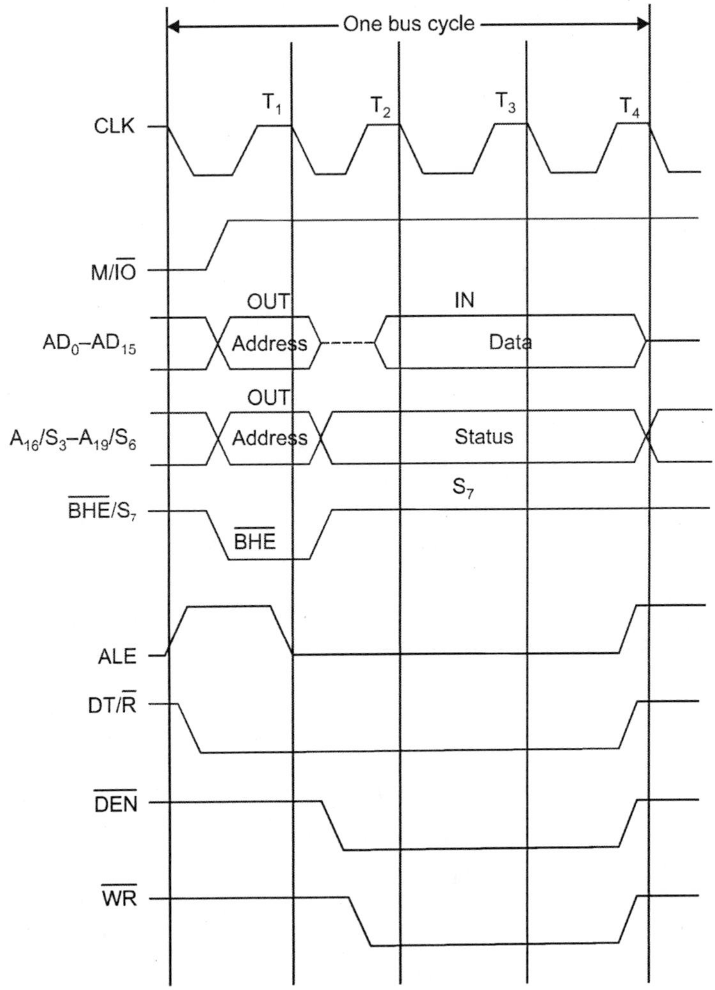


**Figure 5.17:** Timing Diagram for 8086 Memory Write Bus Cycle for Minimum Mode of Operation

### 5.9.3 I/O READ/WRITE BUS CYCLE

The I/O read bus cycle is very similar to memory read cycle. For I/O read operation, M/IO single is made low. All other signals are similar to those for memory read bus cycle as shown in Fig. 5.16.

The I/O write bus cycle is very similar to memory write bus cycle. The M/IO signal is made low for I/O operation. All other signals are similar to those for memory write operation as shown in Fig. 5.17.

### 5.9.4 Read/Write Bus Cycle For Maximum Mode Of 8086

When the 8086 works in the maximum mode of operation, the bus controller 8288 is also employed. The three status signals $\overline{S_0}, \overline{S_1}$ and $\overline{S_2}$ generated by 8086 are applied to the bus controller 8288. The 8288 gives a number of output signals as shown in Fig. 5.14 $\overline{MRDC}$ is memory read and $\overline{MWTC}$ is memory write. These signals can be used for memory read and memory write operation respectively.

Similarly, $\overline{IORC}$ is for I/O read and $\overline{IOWC}$ is for I/O write operation. These signals can be used for I/O read and write operation respectively. DEN, ALE ad DT/$\overline{R}$ is also applied to 8286. In the timing diagram for the maximum mode of operation the clock, $AD_0$-$AD_{15}$, $A_{16}/S_3 - A_{19}/S_6$ and $\overline{BHE}/S_7$ are drawn as they have been shown in the timing diagram for the minimum mode of operation. The other signals for the maximum mode of operation are the status signals $\overline{S_0}, \overline{S_1}$ and $\overline{S_2}$

Their encoding for various types of bus operations is as shown in Table 5.4. These remain active during $T_1$, $T_2$ and $T_4$ and become passive during $T_3$ and $T_w$, $T_w$ denotes wait state. $QS_0$ and $QS_1$ signals should also be shown in the timing diagram for the maximum mode of operation. These indicate the status of the instruction queue as shown in the Table 5.3.

## 5.10 LOCK

The LOCK is a prefix instruction. An instruction of 8086 can be made non-interruptible using LOCK prefix before it. In a multiprocessor system, each processor uses the system bus to access common resources. When a processor is executing certain instruction and making use of the system bus, it should not be interrupted by any other processor.

During this period the hold request of any other processor is prevented. This is achieved using LOCK prefix before the instruction being executed. However, any hold request by the other processor is recorded and honored after completing the execution of the instruction. The LOCK prefix makes $\overline{LOCK}$ pin of 8086 low. It remains low until the instruction before which LOCK prefix has been used is executed. The $\overline{LOCK}$ signal is applied to the bus controller. As long as the $\overline{LOCK}$ remains low, the bus controller does not allow any other processor to take control of the system bus. When $\overline{LOCK}$ is low, all interrupts are masked.

## QUESTION BANK

Q1. Explain the internal architecture of 8086 microprocessor.

Q2. Explain the Execution unit and Bus Interface unit in detail.

Q3. Explain the functional description of intel 8086.

Q4. Explain the pin description of intel 8086.

Q5. Discuss the different operating modes of 8086 microprocessor.

Q6. Explain the register organization of 8086.

Q7. What are interrupts? Explain different types of interrupts available in 8086 microprocessor.

Q8. Explain the 8086 based computer system.

Q9. Explain 8086 read and write bus cycles.

Q10. What is lock? Explain it.

CHAPTER 6

# 8086 Instruction Set

## 6.1 INSTRUCTION FORMAT

Instruction formats vary between microprocessors and minicomputers and mainframe computers. As the machine instructions are generally longer in larger computers with their larger memory words, the instruction format or how the instruction is translated differs. Each instruction is composed of fields. The lengths of instructions and the lengths and positions of the fields differ depending on the instruction and the computer. An operation (function) code is part of all instructions. How the remainder of the instruction is translated and the names assigned to the parts vary.

## 6.2 GENERAL INSTRUCTION FORMAT

The general instruction format is:

**[label:] instruction [operand(s)] [; comments]**

The square brackets indicate optional items. Depending on the specific instruction there may be 0, 1, or 2 operands.

### 6.2.1 Instructions With No Operands

These are instructions that do not use data or ones for which the operand is implied.

```
E.g.    NOP      ;no operation
        CLC      ;Clear Carry Flag
        RET      ;return from subroutine
```

### 6.2.2 Instructions With One Operand

```
E.g.    PUSH     BX         ;put contents of BX on stack
        CALL     ADDLIST    ;call subroutine ADDLIST
```

| | | |
|---|---|---|
| INC | memory | ;increment |
| JNE | SAME | ;jump if not equal |

### 6.2.3 Instructions With Two Operands

The order of the operands is always: destination, source. If the two operands are used by the operation the result is placed in the first operand (the destination). The 8086 CPU does not support two memory address operations therefore only one of the operands can be a memory address the other a register or immediate data. The only exception is with the special string instructions.

| | | | |
|---|---|---|---|
| E.g., | ADD | AX, BX | ; AX = (AX)+(BX) |
| | MOV | BP, SP | ; BP = (SP) |

## 6.3 MACHINE INSTRUCTIONS

Machine instructions are the codes that tell the CPU what to do. The major types of instructions in the 8086 instruction set are:

- Data Movement instructions
- Arithmetic instructions
- Logical instructions
- Program Control and loop instructions
- CPU control and flag instructions
- String instructions

**The following is a brief summary of the 8086 instruction set:**

### *Data Transfer Instructions*

| | |
|---|---|
| MOV | Move byte or word to register or memory |
| IN, OUT | Input byte or word from port, output word to port |
| LEA | Load effective address |
| LDS, LES | Load pointer using data segment, extra segment |
| PUSH, POP | Push word onto stack, pop word off stack |
| XCHG | Exchange byte or word |
| XLAT | Translate byte using look-up table |

### *Logical Instructions*

| | |
|---|---|
| NOT | Logical NOT of byte or word (one's complement) |
| AND | Logical AND of byte or word |
| OR | Logical OR of byte or word |
| XOR | Logical exclusive-OR of byte or word |

| | |
|---|---|
| TEST | Test byte or word (AND without storing) |

### *Shift and Rotate Instructions*

| | |
|---|---|
| SHL, SHR | Logical shift left, right byte or word by 1 or CL |
| SAL, SAR | Arithmetic shift left, right byte or word by 1 or CL |
| ROL, ROR | Rotate left, right byte or word by 1 or CL |
| RCL, RCR | Rotate left, right through carry byte or word by 1 or CL |

### *Arithmetic Instructions*

| | |
|---|---|
| ADD, SUB | Add, subtract byte or word |
| ADC, SBB | Add, subtract byte or word and carry (borrow) |
| INC, DEC | Increment, decrement byte or word |
| NEG | Negate byte or word (two's complement) |
| CMP | Compare byte or word (subtract without storing) |
| MUL, DIV | Multiply, divide byte or word (unsigned) |
| IMUL, IDIV | Integer multiply, divide byte or word (signed) |
| CBW, CWD | Convert byte to word, word to double word (useful before multiply/divide) |
| AAA, AAS, AAM, | adjust for addition, subtraction, multiplication, |
| AAD ASCII | division (ASCII codes 30-39) |
| DAA, DAS | Decimal adjust for addition, subtraction (binary coded decimal numbers) |

### *Transfer Instructions*

| | |
|---|---|
| JMP | Unconditional jump |
| JA (JNBE) | Jump if above (not below or equal) |
| JAE (JNB) | Jump if above or equal (not below) |
| JB (JNAE) | Jump if below (not above or equal) |
| JBE (JNA) | Jump if below or equal (not above) |
| JE (JZ) | Jump if equal (zero) |
| JG (JNLE) | Jump if greater (not less or equal) |
| JGE (JNL) | Jump if greater or equal (not less) |
| JL (JNGE) | Jump if less (not greater nor equal) |
| JLE (JNG) | Jump if less or equal (not greater) |
| JC, JNC | Jump if carry set, carry not set |
| JO, JNO | Jump if overflow, no overflow |
| JS, JNS | Jump if sign, no sign |

| | |
|---|---|
| JNP (JPO) | Jump if no parity (parity odd) |
| JP (JPE) | Jump if parity (parity even) |
| LOOP | Loop unconditional, count in CX |
| LOOPE (LOOPZ) | Loop if equal (zero), count in CX |
| LOOPNE (LOOPNZ) | Loop if not equal (not zero), count in CX |
| JCXZ | Jump if CX equals zero |

### *Subroutine and Interrupt Instructions*

| | |
|---|---|
| CALL, RET Call, | Return from procedure |
| INT, INTO | Software interrupt, interrupt if overflow |
| IRET | Return from interrupt |

### *String Instructions*

| | |
|---|---|
| MOVS | Move byte or word string |
| MOVSB, MOVSW | Move byte, word string |
| CMPS | Compare byte or word string |
| SCAS | Scan byte or word string |
| LODS, STOS | Load, store byte or word string |
| REP | Repeat |
| REPE, REPZ | Repeat while equal, zero |
| REPNE, REPNZ | Repeat while not equal (zero) |

### *Processor Control Instructions*

| | |
|---|---|
| STC, CLC, CMC | Set, clear, complement carry flag |
| STD, CLD | Set, clear direction flag |
| STI, CLI | Set, clear interrupt enable flag |
| LAHF, SAHF | Load AH from flags, store AH into flags |
| PUSHF, POPF | Push flags onto stack, pop flags off stack |
| ESC | Escape to external processor interface |
| LOCK | Lock bus during next instruction |
| NOP | No operation (do nothing) |
| WAIT | Wait for signal on TEST input |
| HLT | Halt processor |

## 6.4 ADDRESSING MODES OF 8086

Each instruction performs an operation on the specified data (operand). Hence, an operand must be specified for an instruction to be executed. The operand may reside in the accumulator, in a general purpose register or in a memory location. The way by which an operand is specified for an instruction is called addressing mode.

There are in total eight addressing modes for 8086 instructions to specify operands. Out of eight addressing modes mere two addressing modes are provided for instruction which operate on the contents of a register or immediate operands. These two addressing modes are.

1. **Register Addressing:** In this mode of addressing an 8-bit or 16-bit general purpose register contains an operand. Some examples are:
   (i) MOV BX, CX ; move the content of CX register to BX resister.
   (ii) ADD AL, CH ; add the content of CH to the content of AL.
   (iii) ADD CX, DX ; add the content of DX to the content of CX.
2. **Immediate Addressing:** In this immediate addressing mode, the operand is contained in the instruction itself. The operand forms a part of the instruction. Some examples are:
   (i) MOV AL, 6=58H ; move 58h TO AL register.
   (ii) MOV BX, 0354H ; move 0354H to BX register.
   (iii) MOV [offset address], data; such as MOV [0400], 5238H; This instruction will move 38H to the offset address 0400H and 52 to the offset address 0401 H.
   (iv) ADD AX, 0395H; Add 0395H to the content of AX register.

   The remaining six addressing modes are for specifying an operand stored in the memory. The memory address of an operand consists of two 16-bit components: the starting address of a memory segment and resides in the corresponding segment register. The operand is placed at an offset within the segment with reference to the starting address of the segment. An offset is also called an effective address. The offset is determined by adding any combination of the following three offset address elements: displacement, base and index. The combination depends on the addressing mode of the instruction to be executed. The actual memory address is given by:

   Memory address = Starting address of the memory segment + Offset.

   The 8086 uses 20-bit memory address. The segment register gives 16 MSBs of the starting address of the memory segment. The BIU generates 20-bit starting address of the memory segment by shifting the content of the segment register left by 4 bits. In other words, it puts a memory segment to get the memory address.

   **Displacement:** It is an 8-bit or 16-bit immediate value given in the instruction.

   **Base:** It is content of the base register, BX or BP.

   **Index:** It is the content of the index register, SI or DI.

   The combination of these three address elements result in the following six addressing modes.

3. **Direct Addressing:** In this mode of addressing an effective address for offset) is given in the instruction itself. Some examples are:

   (i) MOV AL, [0300H]: This instruction will move the content of the offset address 0300H (relative to data segment register DS which contains the starting address of the data segment) to AL.

   (ii) MOV [0401H], AX: This instruction will move the content of AL to the offset address 0401H, and content of AH to 0402H relative to DS register.

4. **Register Indirect Addressing:** The operand's offset is in the base register, BX or base pointer BP or in an index register (SI or DI) specified in the instruction. Some examples are:

   (i) ADD, CX, [BX]: This will add the contents of memory locations addressed by register BX to the register CX. For example BX contains 0301H. The content of 0301H is 64H and the content of next memory location 0302H is 50H. Now 5064H will be added to the content of CX register and the result will be placed in CX.

   (ii) MOV DX, [SI]: The content of the memory location addressed by SI will move to DL and the content of next memory location will move to DH.

5. **Based Addressing:** The operand's offset is the sum of the contents of the base register, BX or BP and an 8-bit or 16-bit displacement.

   Offset (effective address) = [BX or BP + 8-bit or 16-bit displacement]

   Some examples are:

   (i) ADD, AL, [BX + 04]; case of 8-bit displacement

   Suppose, BX contains 0301H. So 0301 + 4 = 0305H. The content of [BX + 4] will be added to the content of AL, and the result will be placed in AL.

   (ii) ADD, AL [BX + 1243H]; case of 16-bit displacement.

6. **Indexed Addressing**: The operand's offset is computed by adding an 8-bit or 16-bit displacement to the contents of an index register SI or DI.

   Offset = [SI or DI + 8-bit or 16-bit displacement]

   Some examples are:

   (i) ADD AX, [SI + 08]

   (ii) MOV CX, [SI + 1523H]

7. **Based Indexed Addressing:** The operand's offset is computed by adding the contents of a base register to the contents of an index register.

   Offset = [BX or BP] + [SI or DI]

   BX is used as a base register for data segment. BP is used as a base register for stack segment.

   Some examples are:

   (i) MOV AX, [BX + SI]

   (ii) ADD CX, [BX + SI]

8. **Based Indexed with Displacement:** The operand's offset is computed by adding a base register's contents, an index register's contents and an 8-bit or 16-bit displacement.

   Offset = [BX or BP] + [SI or DI] + Displacement

   BX is used with data segment, whereas BP is used stack segment.

   Some examples are:

   (i) MOV AX, [BX + SI + 04]

   (ii) ADD CX, [BX + SI + 1523H]

**8086 Address Modes**

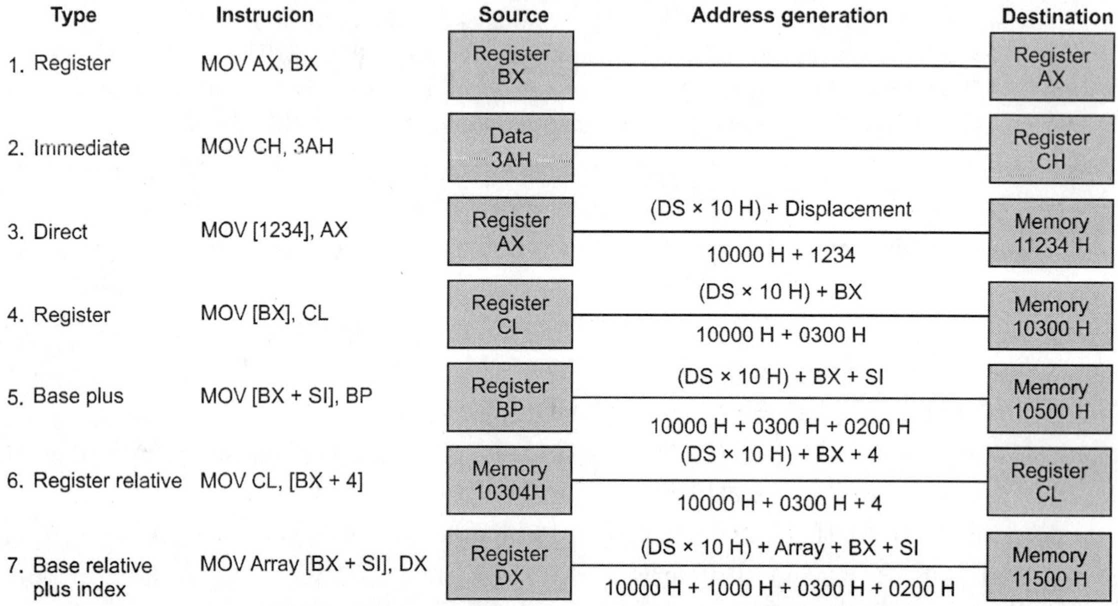


Assume: BX = 0300 H; SI = 0200 H; Array = 1000 H; DS = 1000H

## Getting Effective Addresses

There are two ways of having a program load the effective address of an operand into a register - using the OFFSET operation or the LEA (Load Effective Address) instruction

E.g. MOV SI, OFFSET message

LEA SI, message

The first instruction uses the assembler OFFSET operation which will result in the offset of the label being made into immediate data. The LEA instruction has the advantage being able to get the effective address of operands specified using all of the addressing modes. E.g.

**LEA SI, [BX+6]**

## Creating A Program

1. Define the problem and write the algorithm in pseudo-code

   This may have to be repeated a number of times with each version of the pseudo-code having more detail.

   You could use a flow chart to represent the program.

2. Define data structure
   - Number, type and size of data items - unsigned, signed, BCD or a more complicated structure such as an array, etc.
   - Will they be stored in memory or in registers? How will you use the registers?

3. Initialization
   - Initialize registers e.g. DS
   - Initialize variables

4. Choose the instructions and code the program.

5. Test and repeat steps 1 to 5 as needed.

### Example Programs

**Example 1:** Add two numbers (bytes).

**Pseudo-code**

Get first number

Add second number

Save result

**Data Definition**

The two numbers are bytes in memory and the result will be a byte in memory.

**Initialization**

The DS will have to point to the segment the data is in. This is done in the template (TPEXE or TPCOM) files. The number variables should be initialized with test data.

**Coding**

```
NUM1    DB    5             ;test data
NUM2    DB    4             ; "
RESULT  DB    ?

START:
        MOV   AL, NUM1      ;Get first number
        ADD   AL, NUM2      ;Add second number
        MOV   RESULT, AL    ;Save result
DONE:
```

Comments: This program will add two bytes and save the result in a byte location. The result will be correct as long as the result is not larger than one byte.

**Example 2:** Add two bytes together allowing the result to be greater than a byte.

**Pseudo-code**

Get first number

Add second number keeping carry generated

Save result

**Data Definition:** The two numbers are bytes in memory and the result will be a word in memory. The AX register will be used to do the calculation.

**Initialization**

Do it the same as in example 1.

**Coding**

```
NUM1     DB     5           ;test data
NUM2     DB     4           ; "
RESULT   DW     ?
START:
         MOV    AX, 0       ;Clear AX
         MOV    AL, NUM1        ;Get first number
         ADD    AL, NUM2        ;Add second number
         ADC    AH, 0       ;Add carry to upper byte of answer
         MOV    RESULT, AX      ;Save result
DONE:
```

**Example 3:** Add a list of numbers. The list is a series of bytes and the first item in the list will be the count of numbers to be added followed by the numbers themselves.

**Pseudo-code**

```
Point to start of list
Get count
Sum=0
For i=count to 0 by -1
          sum=sum+1
result = sum
```

**Data Definition:** The list consists of bytes. The result could be larger than a byte so we will use a word. AX will be used to keep a running total. CX will be used as the counter and SI as a pointer to the list.

**Coding**

```
        LIST      DB     5,1,2,3,4,5        ;5 items and test data
        RESULT    DW     ?
        START:
                  MOV    SI, OFFSET LIST    ;point to list
                  MOV    CL, [SI]           ;Get count
                  INC    SI
                  XOR    AX, AX             ;Clear sum
        NEXT:
                  ADD    AL, [SI]           ;Sum=sum+1
                  ADC    AH, 0
                  INC    SI
                  DEC    CL
                  JNZ    NEXT               ;Repeat if count not zero
                  MOV    RESULT, AX         ;result = sum
```

## 6.5 8086 INSTRUCTION SET

### AAA - Ascii Adjust for Addition

**Usage:** AAA

**Modifies flags:** AF CF (OF,PF,SF,ZF undefined)

Changes contents of AL to valid unpacked decimal. The high order nibble is zeroed.

| | Clocks | Size |
|---|---|---|
| Operands | 8086 | Bytes |
| none | 8 | 1 |

### AAD - Ascii Adjust for Division

**Usage:** AAD

**Modifies flags:** SF ZF PF (AF,CF,OF undefined)

Used before dividing unpacked decimal numbers. Multiplies AH by 10 and the adds result into AL. Sets AH to zero. This instruction is also known to have an undocumented behavior.

AL := 10*AH+AL

AH := 0

| | Clocks | Size |
|---|---|---|
| Operands | 8086 | Bytes |
| none | 60 | 2 |

**AAM - Ascii Adjust for Multiplication**

**Usage:** AAM

**Modifies flags:** PF SF ZF (AF,CF,OF undefined)

AH = AL / 10

AL = AL mod 10

Used after multiplication of two unpacked decimal numbers, this instruction adjusts an unpacked decimal number. The high order nibble of each byte must be zeroed before using this instruction. This instruction is also known to have an undocumented behavior.

| | Clocks | Size |
|---|---|---|
| Operands | 8086 | Bytes |
| none | 83 | 2 |

**AAS - Ascii Adjust for Subtraction**

**Usage:** AAS

**Modifies flags:** AF CF (OF,PF,SF,ZF undefined)

Corrects result of a previous unpacked decimal subtraction in AL. High order nibble is zeroed.

| | Clocks | Size |
|---|---|---|
| Operands | 8086 | Bytes |
| none | 8 | 1 |

ADC - Add With Carry

**Usage:** ADC dest, src

**Modifies flags:** AF CF OF SF PF ZF

Sums two binary operands placing the result in the destination.

If CF is set, a 1 is added to the destination.

| | Clocks | Size |
|---|---|---|
| Operands | 8086 | Bytes |
| reg,reg | 3 | 2 |
| mem,reg | 16+EA | 2-4 (W88=24+EA) |
| reg,mem | 9+EA | 2-4 (W88=13+EA) |
| reg,immed | 4 | 3-4 |
| mem,immed | 17+EA | 3-6 (W88=23+EA) |
| accum,immed | 4 | 2-3 |

**ADD - Arithmetic Addition**

**Usage:** ADD dest,src

**Modifies flags:** AF CF OF PF SF ZF

Adds "src" to "dest" and replacing the original contents of "dest".

Both operands are binary.

| | Clocks | Size |
|---|---|---|
| Operands | 8086 | Bytes |
| reg,reg | 3 | 2 |
| mem,reg | 16+EA | 2-4 (W88=24+EA) |
| reg,mem | 9+EA | 2-4 (W88=13+EA) |
| reg,immed | 4 | 3-4 |
| mem,immed | 17+EA | 3-6 (W88=23+EA) |
| accum,immed | 4 | 2-3 |

**AND - Logical And**

**Usage:** AND dest,src

**Modifies flags:** CF OF PF SF ZF (AF undefined)

Performs a logical AND of the two operands replacing the destination with the result.

| | Clocks | Size |
|---|---|---|
| Operands | 8086 | Bytes |
| reg,reg | 3 | 2 |
| mem,reg | 16+EA | 2-4 (W88=24+EA) |
| reg,mem | 9+EA | 2-4 (W88=13+EA) |
| reg,immed | 4 | 3-4 |
| mem,immed | 17+EA | 3-6 (W88=23+EA) |
| accum,immed | 4 | 2-3 |

**CALL - Procedure Call**

Usage: CALL destination

Modifies flags: None

Pushes Instruction Pointer (and Code Segment for far calls) onto stack and loads Instruction Pointer with the address of proc-name. Code continues with execution at CS:IP.

**CBW - Convert Byte to Word**

Usage: CBW

Modifies flags: None

Converts byte in AL to word Value in AX by extending sign of AL throughout register AH.

| Operands | Clocks 8086 | Size Bytes |
|---|---|---|
| none | 2 | 1 |

**CLC - Clear Carry**

Usage: CLC

Modifies flags: CF

Clears the Carry Flag.

| Operands | Clocks 8086 | Size Bytes |
|---|---|---|
| none | 2 | 1 |

**CLD - Clear Direction Flag**

Usage: CLD

Modifies flags: DF

Clears the Direction Flag causing string instructions to increment the SI and DI index registers.

| Operands | Clocks 8086 | Size Bytes |
|---|---|---|
| none | 2 | 1 |

**CLI - Clear Interrupt Flag (disable)**

Usage: CLI

Modifies flags: IF

Disables the maskable hardware interrupts by clearing the Interrupt flag. NMI's and software interrupts are not inhibited.

| Operands | Clocks 8086 | Size Bytes |
|---|---|---|
| none | 2 | 1 |

**CMC - Complement Carry Flag**

Usage: CMC

Modifies flags: CF

Toggles (inverts) the Carry Flag

| Operands | Clocks 8086 | Size Bytes |
|---|---|---|
| none | 2 | 1 |

### CMP - Compare

Usage: CMP dest,src

Modifies flags: AF CF OF PF SF ZF

Subtracts source from destination and updates the flags but does not save result. Flags can subsequently be checked for conditions.

| | Clocks | Size |
|---|---|---|
| Operands | 8086 | Bytes |
| reg,reg | 3 | 2 |
| mem,reg | 9+EA | 2-4 (W88=13+EA) |
| reg,mem | 9+EA | 2-4 (W88=13+EA) |
| reg,immed | 4 | 3-4 |
| mem,immed | 10+EA | 3-6 (W88=14+EA) |
| accum,immed | 4 | 2-3 |

### CMPS - Compare String (Byte, Word or Doubleword)

Usage: CMPS dest,src

CMPSB

CMPSW

Modifies flags: AF CF OF PF SF ZF

Subtracts destination value from source without saving results. Updates flags based on the subtraction and the index registers (E)SI and (E)DI are incremented or decremented depending on the state of the Direction Flag. CMPSB inc/decrements the index registers by 1, CMPSW inc/ decrements by 2, while CMPSD increments or decrements by 4. The REP prefixes can be used to process entire data items.

| | Clocks | Size |
|---|---|---|
| Operands | 8086 | Bytes |
| dest,src | 22 | 1 (W88=30) |

### CWD - Convert Word to Doubleword

Usage: CWD

Modifies flags: None

Extends sign of word in register AX throughout register DX forming a doubleword quantity in DX:AX.

| | Clocks | Size |
|---|---|---|
| Operands | 8086 | Bytes |
| none | 5 | 1 |

**DAA - Decimal Adjust for Addition**

Usage: DAA

Modifies flags: AF CF PF SF ZF (OF undefined)

Corrects result (in AL) of a previous BCD addition operation.

Contents of AL are changed to a pair of packed decimal digits.

| | Clocks | Size |
|---|---|---|
| Operands | 8086 | Bytes |
| none | 4 | 1 |

**DAS - Decimal Adjust for Subtraction**

Usage: DAS

Modifies flags: AF CF PF SF ZF (OF undefined)

Corrects result (in AL) of a previous BCD subtraction operation.

Contents of AL are changed to a pair of packed decimal digits.

| | Clocks | Size |
|---|---|---|
| Operands | 8086 | Bytes |
| none | 4 | 1 |

**DEC - Decrement**

Usage: DEC dest

Modifies flags: AF OF PF SF ZF

Unsigned binary subtraction of one from the destination.

| | Clocks | Size |
|---|---|---|
| Operands | 8086 | Bytes |
| reg8 | 3 | 2 |
| mem | 15+EA | 2-4 |
| reg16/32 | 3 | 1 |

**DIV - Divide**

Usage: DIV src

Modifies flags: (AF,CF,OF,PF,SF,ZF undefined)

Unsigned binary division of accumulator by source. If the source divisor is a byte value then AX is divided by "src" and the quotient is placed in AL and the remainder in AH. If source operand is a word value, then DX:AX is divided by "src" and the quotient is stored in AX and the remainder in DX.

| | Clocks | Size |
|---|---|---|
| Operands | 8086 | Bytes |

| | | |
|---|---|---|
| reg8 | 80-90 | 2 |
| reg16 | 144-162 | 2 |
| mem8 | (86-96)+EA | 2-4 |
| mem16 | (150-168)+EA | 2-4 (W88=158-176+EA) |
| mem32 | - | 2-4 |

**ESC - Escape**

Usage: ESC immed,src

Modifies flags: None

Provides access to the data bus for other resident processors.

The CPU treats it as a NOP but places memory operand on bus.

| | Clocks | Size |
|---|---|---|
| Operands | 8086 | Bytes |
| immed,reg | 2 | 2 |
| immed,mem | 2 | 2-4 |

**HLT - Halt CPU**

Usage: HLT

Modifies flags: None

Halts CPU until RESET line is activated, NMI or maskable interrupt received. The CPU becomes dormant but retains the current CS:IP for later restart.

| | Clocks | Size |
|---|---|---|
| Operands | 8086 | Bytes |
| none | 2 | 1 |

**IDIV - Signed Integer Division**

Usage: IDIV src

Modifies flags: (AF,CF,OF,PF,SF,ZF undefined)

Signed binary division of accumulator by source. If source is a byte value, AX is divided by "src" and the quotient is stored in AL and the remainder in AH. If source is a word value, DX:AX is divided by "src", and the quotient is stored in AL and the remainder in DX.

| | Clocks | Size |
|---|---|---|
| Operands | 8086 | Bytes |
| reg8 | 101-112 | 2 |
| reg16 | 165-184 | 2 |
| mem8 | (107-118)+EA | 2-4 |
| mem16 | (171-190)+EA | 2-4 (W88=175-194) |

IMUL - Signed Multiply

Usage: IMUL src

Modifies flags: CF OF (AF,PF,SF,ZF undefined)

Signed multiplication of accumulator by "src" with result placed in the accumulator. If the source operand is a byte value, it is multiplied by AL and the result stored in AX. If the source operand is a word value it is multiplied by AX and the result is stored in DX:AX. Other variations of this instruction allow specification of source and destination registers as well as a third immediate factor.

| | Clocks | Size |
|---|---|---|
| Operands | 8086 | Bytes |
| reg8 | 80-98 | 2 |
| reg16 | 128-154 | 2 |
| mcm8 | 86-104 | 2-4 |
| mem16 | 134-160 | 2-4 |

**IN - Input Byte or Word From Port**

Usage: IN accum,port

Modifies flags: None

A byte, word or dword is read from "port" and placed in AL, AX or EAX respectively. If the port number is in the range of 0-255 it can be specified as an immediate, otherwise the port number must be specified in DX. Valid port ranges on the PC are 0-1024, though values through 65535 may be specified and recognized by third party vendors and PS/2's.

| | Clocks | Size |
|---|---|---|
| Operands | 8086 | Bytes |
| accum,immed8 | 10/14 | 2 |
| accum,immed8 | (PM) | 2 |
| accum,DX | 8/12 | 1 |
| accum,DX | (PM) | 1 |

**INC - Increment**

Usage: INC dest

Modifies flags: AF OF PF SF ZF

Adds one to destination unsigned binary operand.

| | Clocks | Size |
|---|---|---|
| Operands | 8086 | Bytes |
| reg8 | 3 | 2 |
| reg16 | 3 | 1 |

| | | |
|---|---|---|
| reg32 | 3 | 1 |
| mem | 15+EA | 2-4 (W88=23+EA) |

**INT - Interrupt**

Usage: INT num

Modifies flags: TF IF

Initiates a software interrupt by pushing the flags, clearing the Trap and Interrupt Flags, pushing CS followed by IP and loading CS:IP with the value found in the interrupt vector table. Execution then begins at the location addressed by the new CS:IP

| | Clocks | Size |
|---|---|---|
| Operands | 8086 | Bytes |
| 3 (constant) | 52/72 | 2 |
| immed8 | 51/71 | 1 |

**INTO - Interrupt on Overflow**

Usage: INTO

Modifies flags: IF TF

If the Overflow Flag is set this instruction generates an INT 4 which causes the code addressed by 0000:0010 to be executed.

| | Clocks | Size |
|---|---|---|
| Operands | 8086 | Bytes |
| none: jump | 53/73 | 1 |

**IRET/IRETD - Interrupt Return**

Usage: IRET

Modifies flags: AF CF DF IF PF SF TF ZF

Returns control to point of interruption by popping IP, CS and then the Flags from the stack and continues execution at this location. CPU exception interrupts will return to the instruction that cause the exception because the CS:IP placed on the stack during the interrupt is the address of the offending instruction.

| | Clocks | Size |
|---|---|---|
| Operands | 8086 | Bytes |

iret 32/44 1

**Jxx - Jump Instructions Table**

| **Mnemonic** | **Meaning** | **Jump Condition** |
|---|---|---|
| JA | Jump if Above | CF=0 and ZF=0 |
| JAE | Jump if Above or Equal | CF=0 |

| | | |
|---|---|---|
| JB | Jump if Below | CF=1 |
| JBE | Jump if Below or Equal | CF=1 or ZF=1 |
| JC | Jump if Carry | CF=1 |
| JCXZ | Jump if CX Zero | CX=0 |
| JE | Jump if Equal | ZF=1 |
| JG | Jump if Greater (signed) | ZF=0 and SF=OF |
| JGE | Jump if Greater or Equal (signed) | SF=OF |
| JL | Jump if Less (signed) | SF != OF |
| JLE | Jump if Less or Equal (signed) | ZF=1 or SF != OF |
| JMP | Unconditional Jump | unconditional |
| JNA | Jump if Not Above | CF=1 or ZF=1 |
| JNAE | Jump if Not Above or Equal | CF=1 |
| JNB | Jump if Not Below | CF=0 |
| JNBE | Jump if Not Below or Equal | CF=0 and ZF=0 |
| JNC | Jump if Not Carry | CF=0 |
| JNE | Jump if Not Equal | ZF=0 |
| JNG | Jump if Not Greater (signed) | ZF=1 or SF != OF |
| JNGE | Jump if Not Greater or Equal (signed) | SF != OF |
| JNL | Jump if Not Less (signed) | SF=OF |
| JNLE | Jump if Not Less or Equal (signed) | ZF=0 and SF=OF |
| JNO | Jump if Not Overflow (signed) | OF=0 |
| JNP | Jump if No Parity | PF=0 |
| JNS | Jump if Not Signed (signed) | SF=0 |
| JNZ | Jump if Not Zero | ZF=0 |
| JO | Jump if Overflow (signed) | OF=1 |
| JP | Jump if Parity | PF=1 |
| JPE | Jump if Parity Even | PF=1 |
| JPO | Jump if Parity Odd | PF=0 |
| JS | Jump if Signed (signed) | SF=1 |
| JZ | Jump if Zero | ZF=1 |

| | Clocks | Size |
|---|---|---|
| Operands | 8086 | Bytes |
| Jx: jump | 16 | 2 |
| no jump | 4 | |

- It's a good programming practice to organize code so the expected case is executed without a jump since the actual jump takes longer to execute than falling through the test.
- see JCXZ and JMP for their respective timings

**JCXZ/JECXZ - Jump if Register (E)CX is Zero**

Usage: JCXZ label

Modifies flags: None

Causes execution to branch to "label" if register CX is zero. Uses unsigned comparision.

| | Clocks | Size |
|---|---|---|
| Operands | 8086 | Bytes |
| label: jump | 18 | 2 |
| no jump | 6 | |

JMP - Unconditional Jump

Usage: JMP target

Modifies flags: None

Unconditionally transfers control to "label". Jumps by default are within -32768 to 32767 bytes from the instruction following the jump. NEAR and SHORT jumps cause the IP to be updated while FAR jumps cause CS and IP to be updated.

| | Clocks |
|---|---|
| Operands | 8086 |
| rel8 (relative) | 15 |
| rel16 (relative) | 15 |
| reg16 (near, register indirect) | 11 |
| mem16 (near, mem indirect) | 18+EA |
| mem32 (near, mem indirect) | 24+EA |

**LAHF - Load Register AH From Flags**

Usage: LAHF

Modifies flags: None

Copies bits 0-7 of the flags register into AH. This includes flags AF, CF, PF, SF and ZF other bits are undefined.

AH := SF ZF xx AF xx PF xx CF

| | Clocks | Size |
|---|---|---|
| Operands | 8086 | Bytes |
| none | 4 | 1 |

**LDS - Load Pointer Using DS**

Usage: LDS dest,src

Modifies flags: None

Loads 32-bit pointer from memory source to destination register and DS. The offset is placed in the destination register and the segment is placed in DS. To use this instruction the word at the lower memory address must contain the offset and the word at the higher address must contain the segment. This simplifies the loading of far pointers from the stack and the interrupt vector table.

| | Clocks | Size |
|---|---|---|
| Operands | 8086 | Bytes |
| reg16,mem32 | 16+EA | 2-4 |

**LEA - Load Effective Address**

Usage: LEA dest,src

Modifies flags: None

Transfers offset address of "src" to the destination register.

| | Clocks | Size |
|---|---|---|
| Operands | 8086 | Bytes |
| reg,mem | 2+EA 2-4 | |

- the MOV instruction can often save clock cycles when used in place of LEA on 8088 processors

**LES - Load Pointer Using ES**

Usage: LES dest,src

Modifies flags: None

Loads 32-bit pointer from memory source to destination register and ES. The offset is placed in the destination register and the segment is placed in ES. To use this instruction the word at the lower memory address must contain the offset and the word at the higher address must contain the segment. This simplifies the loading of far pointers from the stack and the interrupt vector table.

| | Clocks | Size |
|---|---|---|
| Operands | 8086 | Bytes |
| reg,mem | 16+EA | 2-4 (W88=24+EA) |

**LOCK - Lock Bus**

Usage: LOCK

Modifies flags: None

This instruction is a prefix that causes the CPU assert bus lock signal during the execution of the next instruction. Used to avoid two processors from updating the same data location. The 286 always asserts lock during an XCHG with memory operands. This should only be used to lock the bus prior to XCHG, MOV, IN and OUT instructions.

| Operands | Clocks 8086 | Size Bytes |
|---|---|---|
| none | 2 | 1 |

### LODS - Load String (Byte, Word or Double)

Usage: LODS src

LODSB

LODSW

Modifies flags: None

Transfers string element addressed by DS:SI (even if an operand is supplied) to the accumulator. SI is incremented based on the size of the operand or based on the instruction used. If the Direction Flag is set SI is decremented, if the Direction Flag is clear SI is incremented. Use with REP prefixes.

| Operands | Clocks 8086 | Size Bytes |
|---|---|---|
| src | 12/16 | 1 |

### LOOP - Decrement CX and Loop if CX Not Zero

Usage: LOOP label

Modifies flags: None

Decrements CX by 1 and transfers control to "label" if CX is not Zero. The "label" operand must be within -128 or 127 bytes of the instruction following the loop instruction

| Operands | Clocks 8086 | Size Bytes |
|---|---|---|
| label: jump | 18 2 | |
| no jump | 5 | |

### LOOPE/LOOPZ - Loop While Equal / Loop While Zero

Usage: LOOPE label

LOOPZ label

Modifies flags: None

Decrements CX by 1 (without modifying the flags) and transfers control to "label" if CX != 0 and the Zero Flag is set. The "label" operand must be within -128 or 127 bytes of the instruction following the loop instruction.

| Operands | Clocks 8086 | Size Bytes |
|---|---|---|
| label: jump | 18 | 2 |
| no jump | 5 | |

### LOOPNZ/LOOPNE - Loop While Not Zero / Loop While Not Equal

Usage: LOOPNZ label

LOOPNE label

Modifies flags: None

Decrements CX by 1 (without modifying the flags) and transfers control to "label" if CX != 0 and the Zero Flag is clear. The "label" operand must be within -128 or 127 bytes of the instruction following the loop instruction.

| | Clocks | Size |
|---|---|---|
| Operands | 8086 | Bytes |
| label: jump | 19 | 2 |
| no jump | 5 | |

### MOV - Move Byte or Word

Usage: MOV dest,src

Modifies flags: None

Copies byte or word from the source operand to the destination operand. If the destination is SS interrupts are disabled except on early buggy 808x CPUs. Some CPUs disable interrupts if the destination is any of the segment registers

| | Clocks | Size |
|---|---|---|
| Operands | 8086 | Bytes |
| reg,reg | 2 | 2 |
| mem,reg | 9+EA | 2-4 (W88=13+EA) |
| reg,mem | 8+EA | 2-4 (W88=12+EA) |
| mem,immed | 10+EA | 3-6 (W88=14+EA) |
| reg,immed | 4 | 2-3 |
| mem,accum | 10 | 3 (W88=14) |
| accum,mem | 10 | 3 (W88=14) |
| segreg,reg16 | 2 | 2 |
| segreg,mem16 | 8+EA | 2-4 (W88=12+EA) |
| reg16,segreg | 2 | 2 |
| mem16,segreg | 9+EA | 2-4 (W88=13+EA) |

### MOVS - Move String (Byte or Word)

Usage: MOVS dest,src

MOVSB

MOVSW

Modifies flags: None

Copies data from addressed by DS:SI (even if operands are given) to the location ES:DI destination and updates SI and DI based on the size of the operand or instruction used. SI and DI are incremented when the Direction Flag is cleared and decremented when the Direction Flag is Set. Use with REP prefixes.

| | Clocks | Size |
|---|---|---|
| Operands | 8086 | Bytes |
| dest,src | 18 | 1 (W88=26) |

**MUL - Unsigned Multiply**

Usage: MUL src

Modifies flags: CF OF (AF,PF,SF,ZF undefined)

Unsigned multiply of the accumulator by the source. If “src” is a byte value, then AL is used as the other multiplicand and the result is placed in AX. If “src” is a word value, then AX is multiplied by “src” and DX:AX receives the result. If “src” is a double word value, then EAX is multiplied by “src” and EDX:EAX receives the result. The 386+ uses an early out algorithm which makes multiplying any size value in EAX as fast as in the 8 or 16 bit registers.

| | Clocks | Size |
|---|---|---|
| Operands | 8086 | Bytes |
| reg8 | 70-77 | 2 |
| reg16 | 118-113 | 2 |
| mem8 | (76-83)+EA | 2-4 |
| mem16 | (124-139)+EA | 2-4 |

**NEG - Two’s Complement Negation**

Usage: NEG dest

Modifies flags: AF CF OF PF SF ZF

Subtracts the destination from 0 and saves the 2s complement of “dest” back into “dest”.

| | Clocks | Size |
|---|---|---|
| Operands | 8086 | Bytes |
| reg | 3 | 2 |
| mem | 16+EA | 2-4 (W88=24+EA) |

**NOP - No Operation (90h)**

Usage: NOP

Modifies flags: None

This is a do nothing instruction. It results in occupation of both space and time and is most useful for patching code segments. (This is the original XCHG AL,AL instruction)

| Operands | Clocks 8086 | Size Bytes |
|---|---|---|
| none | 3 | 1 |

**NOT - One's Compliment Negation (Logical NOT)**

Usage: NOT dest

Modifies flags: None

Inverts the bits of the "dest" operand forming the 1s complement.

| Operands | Clocks 8086 | Size Bytes |
|---|---|---|
| reg | 3 | 2 |
| mem | 16+EA | 2-4 (W88=24+EA) |

**OR - Inclusive Logical OR**

Usage: OR dest,src

Modifies flags: CF OF PF SF ZF (AF undefined)

Logical inclusive OR of the two operands returning the result in the destination. Any bit set in either operand will be set in the destination.

| Operands | Clocks 8086 | Size Bytes |
|---|---|---|
| reg,reg | 3 | 2 |
| mem,reg | 16+EA | 2-4 (W88=24+EA) |
| reg,mem | 9+EA | 2-4 (W88=13+EA) |
| reg,immed | 4 | 3-4 |
| mem8,immed8 | 17+EA | 3-6 |
| mem16,immed16 | 25+EA | 3-6 |
| accum,immed | 4 | 2-3 |

**POP - Pop Word off Stack**

Usage: POP dest

Modifies flags: None

Transfers word at the current stack top (SS:SP) to the destination then increments SP by two to point to the new stack top. CS is not a valid destination.

| Operands | Clocks 8086 | Size Bytes |
|---|---|---|
| reg16 | 8 | 1 |
| reg32 | 4 | 1 |

| | | |
|---|---|---|
| segreg | 8 | 1 |
| mem16 | 17+EA | 2-4 |
| mem32 | 5 | 2-4 |

**PUSH - Push Word onto Stack**

Usage: PUSH src

PUSH immed

Modifies flags: None

Decrements SP by the size of the operand (two or four, byte values are sign extended) and transfers one word from source to the stack top (SS:SP).

| | Clocks | Size |
|---|---|---|
| Operands | 8086 | Bytes |
| reg16 | 11/15 | 1 |
| mem16 | 16+EA | 2-4 (W88=24+EA) |
| segreg | 10/14 | 1 |

**PUSHF- Push Flags onto Stack**

Usage: PUSHF

Modifies flags: None

Transfers the Flags Register onto the stack. PUSHF saves a 16 bit value while PUSHFD saves a 32 bit value.

| | Clocks | Size |
|---|---|---|
| Operands | 8086 | Bytes |
| none | 10/14 | 1 |

**RCL - Rotate Through Carry Left**

Usage: RCL dest,count

Modifies flags: CF OF

Rotates the bits in the destination to the left "count" times with all data pushed out the left side re-entering on the right. The Carry Flag holds the last bit rotated out.

| | Clocks | Size |
|---|---|---|
| Operands | 8086 | Bytes |
| reg,1 | 2 | 2 |
| mem,1 | 15+EA | 2-4 (W88=23+EA) |
| reg,CL | 8+4n | 2 |
| mem,CL | 20+EA+4n | 2-4 (W88=28+EA+4n) |

### RCR - Rotate Through Carry Right

Usage: RCR dest,count

Modifies flags: CF OF

Rotates the bits in the destination to the right "count" times with all data pushed out the right side re-entering on the left. The Carry Flag holds the last bit rotated out.

| Operands | Clocks 8086 | Size Bytes |
|---|---|---|
| reg,1 2 | 2 | |
| mem,1 | 15+EA | 2-4 (W88=23+EA) |
| reg,CL | 8+4n | 2 |
| mem,CL | 20+EA+4n | 2-4 (W88=28+EA+4n) |

### REP - Repeat String Operation

Usage: REP

Modifies flags: None

Repeats execution of string instructions while CX != 0. After each string operation, CX is decremented and the Zero Flag is tested. The combination of a repeat prefix and a segment override on CPU's before the 386 may result in errors if an interrupt occurs before CX=0. The following code shows code that is susceptible to this and how to avoid it:

```
again:  rep movs byte ptr ES:[DI],ES:[SI] ; vulnerable instr.
        jcxz next            ; continue if REP successful
        loop again           ; interrupt goofed count
next:
```

| Operands | Clocks 8086 | Size Bytes |
|---|---|---|
| none | 2 | 1 |

### REPE/REPZ - Repeat Equal / Repeat Zero

Usage: REPE

REPZ

Modifies flags: None

Repeats execution of string instructions while CX != 0 and the Zero Flag is set. CX is decremented and the Zero Flag tested after each string operation. The combination of a repeat prefix and a segment override on processors other than the 386 may result in errors if an interrupt occurs before CX=0.

| Operands | Clocks 8086 | Size Bytes |
|---|---|---|
| none | 2 | 1 |

**REPNE/REPNZ - Repeat Not Equal / Repeat Not Zero**

Usage: REPNE

REPNZ

Modifies flags: None

Repeats execution of string instructions while CX != 0 and the Zero Flag is clear. CX is decremented and the Zero Flag tested after each string operation. The combination of a repeat prefix and a segment override on processors other than the 386 may result in errors if an interrupt occurs before CX=0.

| | Clocks | Size |
|---|---|---|
| Operands | 8086 | Bytes |
| none | 2 | 1 |

**RET/RETF - Return From Procedure**

Usage: RET nBytes

RETF nBytes

RETN nBytes

Modifies flags: None

Transfers control from a procedure back to the instruction address saved on the stack. "n bytes" is an optional number of bytes to release. Far returns pop the IP followed by the CS, while near returns pop only the IP register.

| | Clocks | Size |
|---|---|---|
| Operands | 8086 | Bytes |
| retn | 16/20 | 1 |
| retn immed | 20/24 | 3 |
| retf | 26/34 | 1 |
| retf (PM, same priv.) | | 1 |
| retf (PM, lesser priv.) | | 1 |
| retf immed | 25/33 | 3 |
| retf immed (PM, same priv.) | | 1 |
| retf immed (PM, lesser priv.) | | 1 |

**ROL - Rotate Left**

Usage: ROL dest,count

Modifies flags: CF OF

Rotates the bits in the destination to the left "count" times with all data pushed out the left side re-entering on the right. The Carry Flag will contain the value of the last bit rotated out.

| Operands | Clocks 8086 | Size Bytes |
|---|---|---|
| reg,1 | 2 | 2 |
| mem,1 | 15+EA | 2-4 (W88=23+EA) |
| reg,CL | 8+4n | 2 |
| mem,CL | 20+EA+4n | 2-4 (W88=28+EA+4n) |

**ROR - Rotate Right**

Usage: ROR dest,count

Modifies flags: CF OF

Rotates the bits in the destination to the right "count" times with all data pushed out the right side re-entering on the left. The Carry Flag will contain the value of the last bit rotated out.

| Operands | Clocks 8086 | Size Bytes |
|---|---|---|
| reg,1 | 2 | 2 |
| mem,1 | 15+EA | 2-4 (W88=23+EA) |
| reg,CL | 8+4n | 2 |
| mem,CL | 20+EA+4n | 2-4 (W88=28+EA+4n) |

**SAHF - Store AH Register into FLAGS**

Usage: SAHF

Modifies flags: AF CF PF SF ZF

Transfers bits 0-7 of AH into the Flags Register. This includes AF, CF, PF, SF and ZF.

| Operands | Clocks 8086 | Size Bytes |
|---|---|---|
| none | 4 | 1 |

**SAL/SHL - Shift Arithmetic Left / Shift Logical Left**

Usage: SAL dest,count

SHL dest,count

Modifies flags: CF OF PF SF ZF (AF undefined)

Shifts the destination left by "count" bits with zeroes shifted in on right. The Carry Flag contains the last bit shifted out.

| Operands | Clocks 8086 | Size Bytes |
|---|---|---|
| reg,1 | 2 | 2 |
| mem,1 | 15+EA | 2-4 (W88=23+EA) |

| | | |
|---|---|---|
| reg,CL | 8+4n | 2 |
| mem,CL | 20+EA+4n | 2-4 (W88=28+EA+4n) |

**SAR - Shift Arithmetic Right**

Usage: SAR dest,count

Modifies flags: CF OF PF SF ZF (AF undefined)

Shifts the destination right by "count" bits with the current sign bit replicated in the leftmost bit. The Carry Flag contains the last bit shifted out.

| | Clocks | Size |
|---|---|---|
| Operands | 8086 | Bytes |
| reg,1 | 2 | 2 |
| mem,1 | 15+EA | 2-4 (W88=23+EA) |
| reg,CL | 8+4n | 2 |
| mem,CL | 20+EA+4n | 2-4 (W88=28+EA+4n) |

**SBB - Subtract with Borrow/Carry**

Usage: SBB dest,src

Modifies flags: AF CF OF PF SF ZF

Subtracts the source from the destination, and subtracts 1 extra if the Carry Flag is set. Results are returned in "dest".

| | Clocks | Size |
|---|---|---|
| Operands | 8086 | Bytes |
| reg,reg | 3 | 2 |
| mem,reg | 16+EA | 2-4 (W88=24+EA) |
| reg,mem | 9+EA | 2-4 (W88=13+EA) |
| reg,immed | 4 | 3-4 |
| mem,immed | 17+EA | 3-6 (W88=25+EA) |
| accum,immed | 4 | 2-3 |

**SCAS - Scan String (Byte, Word or Doubleword)**

Usage: SCAS string

SCASB

SCASW

Modifies flags: AF CF OF PF SF ZF

Compares value at ES:DI (even if operand is specified) from the accumulator and sets the flags similar to a subtraction. DI is incremented/decremented based on the instruction format (or operand size) and the state of the Direction Flag. Use with REP prefixes.

| Operands | Clocks 8086 | Size Bytes |
|---|---|---|
| string | 15 | 1 (W88=19) |

**SHL - Shift Logical Left**

See: SAL

**SHR - Shift Logical Right**

Usage: SHR dest,count

Modifies flags: CF OF PF SF ZF (AF undefined)

Shifts the destination right by "count" bits with zeroes shifted in on the left. The Carry Flag contains the last bit shifted out.

| Operands | Clocks 8086 | Size Bytes |
|---|---|---|
| reg,1 | 2 | 2 |
| mem,1 | 15+EA | 2-4 (W88=23+EA) |
| reg,CL | 8+4n | 2 |
| mem,CL | 20+EA+4n | 2-4 (W88=28+EA+4n) |

**STC - Set Carry**

Usage: STC

Modifies flags: CF

Sets the Carry Flag to 1.

| Operands | Clocks 8086 | Size Bytes |
|---|---|---|
| none | 2 | 1 |

**STD - Set Direction Flag**

Usage: STD

Modifies flags: DF

Sets the Direction Flag to 1 causing string instructions to auto-decrement SI and DI instead of auto-increment.

| Operands | Clocks 8086 | Size Bytes |
|---|---|---|
| none | 2 | 1 |

**STI - Set Interrupt Flag (Enable Interrupts)**

Usage: STI

Modifies flags: IF

Sets the Interrupt Flag to 1, which enables recognition of all hardware interrupts. If an interrupt is generated by a hardware device, an End of Interrupt (EOI) must also be issued to enable other hardware interrupts of the same or lower priority.

| Operands | Clocks 8086 | Size Bytes |
|---|---|---|
| none | 2 | 1 |

**STOS - Store String (Byte, Word or Doubleword)**

Usage: STOS dest
STOSB
STOSW
STOSD

Modifies flags: None

Stores value in accumulator to location at ES:(E)DI (even if operand is given). (E)DI is incremented/decremented based on the size of the operand (or instruction format) and the state of the Direction Flag. Use with REP prefixes.

| Operands | Clocks 8086 | Size Bytes |
|---|---|---|
| dest | 11 | 1 (W88=15) |

**SUB - Subtract**

Usage: SUB dest,src

Modifies flags: AF CF OF PF SF ZF

The source is subtracted from the destination and the result is stored in the destination.

| Operands | Clocks 8086 | Size Bytes |
|---|---|---|
| reg,reg | 3 | 2 |
| mem,reg | 16+EA | 2-4 (W88=24+EA) |
| reg,mem | 9+EA | 2-4 (W88=13+EA) |
| reg,immed | 4 | 3-4 |
| mem,immed | 17+EA | 3-6 (W88=25+EA) |
| accum,immed | 4 | 2-3 |

**TEST - Test For Bit Pattern**

Usage: TEST dest,src

Modifies flags: CF OF PF SF ZF (AF undefined)

Performs a logical AND of the two operands updating the flags register without saving the result.

| Operands | Clocks 8086 | Size Bytes |
|---|---|---|
| reg,reg | 3 | 2 |
| reg,mem | 9+EA | 2-4 (W88=13+EA) |
| mem,reg | 9+EA | 2-4 (W88=13+EA) |
| reg,immed | 5 | 3-4 |
| mem,immed | 11+EA | 3-6 |
| accum,immed | 4 | 2-3 |

**WAIT/FWAIT - Event Wait**

Usage: WAIT
FWAIT

Modifies flags: None

CPU enters wait state until the coprocessor signals it has finished its operation. This instruction is used to prevent the CPU from accessing memory that may be temporarily in use by the coprocessor. WAIT and FWAIT are identical.

| Operands | Clocks 8086 | Size Bytes |
|---|---|---|
| none | 4 | 1 |

**XCHG - Exchange**

Usage: XCHG dest,src

Modifies flags: None

Exchanges contents of source and destination.

| Operands | Clocks 8086 | Size Bytes |
|---|---|---|
| reg,reg | 4 | 2 |
| mem,reg | 17+EA | 2-4 (W88=25+EA) |
| reg,mem | 17+EA | 2-4 (W88=25+EA) |
| accum,reg | 3 | 1 |
| reg,accum | 3 | 1 |

**XLAT/XLATB - Translate**

Usage: XLAT translation-table
XLATB (masm 5.x)

Modifies flags: None

Replaces the byte in AL with byte from a user table addressed by BX. The original value of AL is the index into the translate table. The best way to discripe this is MOV AL,[BX+AL]

| | Clocks | Size |
|---|---|---|
| Operands | 8086 | Bytes |
| table offset | 11 | 1 |

**XOR - Exclusive OR**

Usage: XOR dest,src

Modifies flags: CF OF PF SF ZF (AF undefined)

Performs a bitwise exclusive OR of the operands and returns the result in the destination.

| | Clocks | Size |
|---|---|---|
| Operands | 8086 | Bytes |
| reg,reg | 3 | 2 |
| mem,reg | 16+EA | 2-4 (W88=24+EA) |
| reg,mem | 9+EA | 2-4 (W88=13+EA) |
| reg,immed | 4 | 3-4 |
| mem,immed | 17+EA | 3-6 (W88=25+EA) |
| accum,immed | 4 | 2-3 |

## QUESTION BANK

Q1. What is instruction format? Explain different types of instruction formats in detail.

Q2. Give brief description of different types of instructions available in 8086 microprocessor.

Q3. Explain different types of addressing modes available in 8086 microprocessor.

Q4. Explain different types of data transfer instructions.

Q5. Explain different types of arithmetic instructions.

Q6. Explain different types of logical instructions.

Q7. Explain different types of branching instructions.

Q8. Explain different types of looping instructions.

CHAPTER 7

# Assembly Language–I

## 7.1 INTRODUCTION

Assembly languages are a family of low-level languages for programming computers, microprocessors, microcontrollers, and other (usually) integrated circuits. They implement a symbolic representation of the numeric machine codes and other constants needed to program a particular CPU architecture. This representation is usually defined by the hardware manufacturer, and is based on abbreviations (called mnemonics) that help the programmer remember individual instructions, registers, etc. An assembly language is thus specific to certain physical or virtual computer architecture (as opposed to most high-level languages, which are usually portable).

This is a brief introduction to assembly language. Assembly language is the most basic programming language available for any processor. With assembly language, a programmer works only with operations implemented directly on the physical CPU. Assembly language lacks high-level conveniences such as variables and functions, and it is not portable between various families of processors. Nevertheless, assembly language is the most powerful computer programming language available, and it gives programmers the insight required to write effective code in high-level languages. Learning assembly language is well worth the time and effort of every serious programmer.

## 7.2 BASICS OF ASSEMBLY LANGUAGE

Before we can explore the process of writing computer programs, we have to go back to the basics and learn exactly what a computer is and how it works. Every computer, no matter how simple or complex, has at its heart exactly two things: a CPU and some memory. Together, these two things are what make it possible for your computer to run programs.

On the most basic level, a computer program is nothing more than a collection of numbers stored in memory. Different numbers tell the CPU to do different things. The CPU reads the numbers one at a time, decodes them, and does what the numbers say. For example, if the CPU reads the number 64 as part of a program, it will add 1 to the number stored in a special location called AX. If the CPU reads the number 146, it will swap the number stored in AX with the number stored in another location called BX. By combining many simple operations such these into a program, a programmer can make the computer perform many incredible things.

As an example, here are the numbers of a simple computer program: 184, 0, 184, 142, 216, 198, 6, 158, 15, 36, 205, 32. If you were to enter these numbers into your computer's memory and run them under MS-DOS, you would see a dollar sign placed in the lower right hand corner of your screen, since that is what these numbers tell the computer to do.

### 7.2.1 Basic Elements

Any Assembly language consists of 3 types of instruction statements which are used to define the program operations:

1. Opcode mnemonics
2. Data sections
3. Assembly directives

#### *1. Opcode Mnemonics*

Instructions (statements) in assembly language are generally very simple, unlike those in high-level languages. Generally, an opcode is a symbolic name for a single executable machine language instruction, and there is at least one opcode mnemonic defined for each machine language instruction. Each instruction typically consists of an *operation* or *opcode* plus zero or more *operands*. Most instructions refer to a single value, or a pair of values. Operands can be either immediate (typically one byte values, coded in the instruction itself) or the addresses of data located elsewhere in storage. This is determined by the underlying processor architecture: the assembler merely reflects how this architecture works.

#### *2. Data Sections*

There are instructions used to define data elements to hold data and variables. They define what type of data, length and alignment of data. These instructions can also define whether the data is available to outside programs (programs assembled separately) or only to the program in which the data section is defined.

#### *3. Assembly Directives/Pseudo-ops*

Assembly Directives are instructions that are executed by the Assembler at assembly time, not by the CPU at run time. They can make the assembly of the program dependent on parameters input by the programmer, so that one program can be assembled different ways, perhaps for different applications. They also can be used to manipulate presentation of the program to make it easier for the programmer to read and maintain.

(For example, pseudo-ops would be used to reserve storage areas and optionally their initial contents.) The names of pseudo-ops often start with a dot to distinguish them from machine instructions.

Some assemblers also support *pseudo-instructions*, which generate two or more machine instructions.

Symbolic assemblers allow programmers to associate arbitrary names (*labels* or *symbols*) with memory locations. Usually, every constant and variable is given a name so instructions can reference those locations by name, thus promoting self-documenting code. In executable code, the name of each subroutine is associated with its entry point, so any calls to a subroutine can use its name. Inside subroutines, GOTO destinations are given labels. Some assemblers support *local symbols* which are lexically distinct from normal symbols (e.g., the use of "10$" as a GOTO destination).

Most assemblers provide flexible symbol management, allowing programmers to manage different namespaces, automatically calculate offsets within data structures, and assign labels that refer to literal values or the result of simple computations performed by the assembler. Labels can also be used to initialize constants and variables with relocatable addresses.

Assembly languages, like most other computer languages, allow comments to be added to assembly source code that are ignored by the assembler. Good use of comments is even more important with assembly code than with higher-level languages, as the meaning and purpose of a sequence of instructions is harder to decipher from the code itself.

Wise use of these facilities can greatly simplify the problems of coding and maintaining low-level code. *Raw* assembly source code as generated by compilers or disassemblers — code without any comments, meaningful symbols, or data definitions — is quite difficult to read when changes must be made.

### 7.2.2 Macros

Many assemblers support *macros*, programmer-defined symbols that stand for some sequence of text lines. This sequence of text lines may include a sequence of instructions, or a sequence of data storage pseudo-ops. Once a macro has been defined using the appropriate pseudo-op, its name may be used in place of a mnemonic. When the assembler processes such a statement, it replaces the statement with the text lines associated with that macro, then processes them just as though they had appeared in the source code file all along (including, in better assemblers, expansion of any macros appearing in the replacement text).

Since macros can have 'short' names but expand to several or indeed many lines of code, they can be used to make assembly language programs appear to be much shorter (require less lines of source code from the application programmer - as with a higher level language). They can also be used to add higher levels of structure to assembly programs, optionally introduce embedded de-bugging code via parameters and other similar features. Many assemblers have built-in macros for system calls and other special code sequences.

Macro assemblers often allow macros to take parameters. Some assemblers include quite sophisticated macro languages, incorporating such high-level language elements as optional parameters, symbolic variables, conditionals, string manipulation, and arithmetic operations, all usable during the execution of a given macros, and allowing macros to save context or exchange information. Thus a macro might generate a large number of assembly language instructions or data definitions, based on the macro arguments. This could be used to generate record-style data structures or "unrolled" loops, for example, or could generate entire algorithms based on complex parameters. An organization using assembly language that has been heavily extended using such a macro suite can be considered to be working in a higher-level language, since such programmers are not working with a computer's lowest-level conceptual elements.

Macros were used to customize large scale software systems for specific customers in the mainframe era and were also used by customer personnel to satisfy their employers' needs by making specific versions of manufacturer operating systems; this was done, for example, by systems programmers working with IBM's Conversational Monitor System/Virtual Machine (CMS/VM) and with IBM's "real time transaction processing" add-ons, CICS, Customer Information Control System, and ACP/TPF, the airline/financial system that began in the 1970s and still runs many large Global Distribution Systems (GDS) and credit card systems today.

It was also possible to use solely the macro processing capabilities of an assembler to generate code written in completely different languages, for example, to generate a version of a program in Cobol using a pure macro assembler program containing lines of Cobol code inside assembly time operators instructing the assembler to generate arbitrary code.

This was because, as was realized in the 1970s, the concept of "macro processing" is independent of the concept of "assembly", the former being in modern terms more word processing, text processing, than generating object code. The concept of macro processing in fact appeared in and appears in the C programming language, which supports "preprocessor instructions" to set variables, and make conditional tests on their values. Note that unlike certain previous macro processors inside assemblers, the C preprocessor was not Turing-complete because it lacked the ability to either loop or "go to", the latter allowing the programmer to loop.

Despite the power of macro processing, it fell into disuse in high level languages while remaining a perennial for assemblers.

This was because many programmers were rather confused by macro parameter substitution and did not disambiguate macro processing from assembly and execution.

Macro parameter substitution is strictly by name: at macro processing time, the value of a parameter is textually substituted for its name. The most famous class of bugs resulting was the use of a parameter that itself was an expression and not a simple name when the macro writer expected a name. In the macro: foo: macro a load a*b the intention was that the caller would provide the name of a variable, and the "global" variable or constant b would be used to multiply "a". If foo is called with the parameter a-c, an unexpected macro expansion occurs.

To avoid this, users of macro processors learned to religiously parenthesize formal parameters inside macro definitions, and callers had to do the same to their "actual" parameters.

PL/I and C feature macros, but this facility was underused or dangerous when used because they can only manipulate text. On the other hand, homoiconic languages, such as Lisp, Prolog, and

Forth, retain the power of assembly language macros because they are able to manipulate their own code as data

### 7.2.3 Assembler

Typically a modern **assembler** creates object code by translating assembly instruction mnemonics into opcodes, and by resolving symbolic names for memory locations and other entities.[1] The use of symbolic references is a key feature of assemblers, saving tedious calculations and manual address updates after program modifications. Most assemblers also include macro facilities for performing textual substitution—e.g., to generate common short sequences of instructions to run inline, instead of in a subroutine.

Assemblers are generally simpler to write than compilers for high-level languages, and have been available since the 1950s. Modern assemblers, especially for RISC based architectures, such as MIPS, Sun SPARC, and HP PA-RISC, as well as x86(-64), optimize instruction scheduling to exploit the CPU pipeline efficiently.

There are two types of assemblers based on how many passes through the source are needed to produce the executable program. One-pass assemblers go through the source code once and assumes that all symbols will be defined before any instruction that references them. Two-pass assemblers (and multi-pass assemblers) create a table with all unresolved symbols in the first pass, then use the 2nd pass to resolve these addresses. The advantage in one-pass assemblers is speed - which is not as important as it once was with advances in computer speed and capabilities. The advantage of the two-pass assembler is that symbols can be defined anywhere in the program source. As a result, the program can be defined in a more logical and meaningful way. This makes two-pass assembler programs easier to read and maintain.

More sophisticated high-level assemblers provide language abstractions such as:

- Advanced control structures
- High-level procedure/function declarations and invocations
- High-level abstract data types, including structures/records, unions, classes, and sets
- Sophisticated macro processing
- Object-Oriented features such as encapsulation, polymorphism, inheritance, interfaces

## 7.3 WHY ASSEMBLY LANGUAGE

Although the numbers of the above program make perfect sense to a computer, they are about as clear as mud to a human. Who would have guessed that they put a dollar sign on the screen? Clearly, entering numbers by hand is a lousy way to write a program.

It doesn't have to be this way, though. A long time ago, someone came up with the idea that computer programs could be written using words instead of numbers. A special program called an assembler would then take the programmer's words and convert them to numbers that the computer could understand. This new method, called writing a program in assembly language, saved programmers thousands of hours, since they no longer had to look up hard-to-remember numbers in the backs of programming books. The program above, written in assembly language, looks like this:

```
MOV AX, 47104
MOV DS, AX
MOV [3998], 36
INT 32
```

When an assembler reads this sample program, it converts each line of code into one CPU-level instruction. This program uses two types of instructions, MOV and INT. On Intel processors, the MOV instruction moves data around, while the INT instruction transfers processor control to the device drivers or operating system.

The program still isn't quite clear, but it is much easier to understand than it was before. The first instruction, MOV AX, 47104, tells the computer to copy the number 47104 into the location AX. The next instruction, MOV DS, AX, tells the computer to copy the number in AX into the location DS. The next instruction, MOV [3998], 36 tells the computer to put the number 36 into memory location 3998. Finally, INT 32 exits the program by returning to the operating system.

Before we go on, I would like to explain just how this program works. Inside the CPU are a number of locations, called registers, which can store a number. Some registers, such as AX, are general purpose, and don't do anything special. Other registers, such as DS, control the way the CPU works. DS just happens to be a segment register, and is used to pick which area of memory the CPU can write to. In our program, we put the number 47104 into DS, which tells the CPU to access the memory on the video card. The next thing our program does is to put the number 36 into location 3998 of the video card's memory. Since 36 is the code for the dollar sign, and 3998 is the memory location of the bottom right hand corner of the screen, a dollar sign shows up on the screen a few microseconds later. Finally, our program tells the CPU to perform what is called an interrupt. An interrupt is used to stop one program and execute another in its place. In our case, we want interrupt 32, which ends our program and goes back to MS-DOS, or whatever other program was used to start our program.

## 7.4 ASSEMBLY FOR MICROPROCESSORS

First of all, we'll talk about the registers and then about the instructions to manipulate (change) them. The **8086** has 14 16-bit registers, all with different usage (see below). You might not understand some of the registers purposes, but be patient, I'll explain everything later.

### 7.4.1 Segment Registers

| | | |
|---|---|---|
| CS | Code Segment | 16-bit number that points to the active code-segment |
| DS | Data Segment | 16-bit number that points to the active data-segment |
| SS | Stack Segment | 16-bit number that points to the active stack-segment |
| ES | Extra Segment | 16-bit number that points to the active extra-segment |

### 7.4.2 Pointer Registers

| | | |
|---|---|---|
| IP | Instruction Pointer | 16-bit number that points to the offset of the next instruction |
| SP | Stack Pointer | 16-bit number that points to the offset that the stack is using |
| BP | Base Pointer | used to pass data to and from the stack |

### 7.4.3 General-purpose Registers

| | | |
|---|---|---|
| AX | Accumulator Register | mostly used for calculations and for input/output |
| BX | Base Register | Only register that can be used as an index |
| CX | Count Register | register used for the loop instruction |
| DX | Data Register | input/output and used by multiply and divide |

### 7.4.4 Index Registers

| | | |
|---|---|---|
| SI | Source Index | used by string operations as source |
| DI | Destination Index | used by string operations as destination |

*(The general purpose registers can be "split". You have the AH and the AL register for example. AH contains the high byte of AX and AL contains the lowbyte. You also have: BH, BL, CH, CL, DL, DH So if eg. DX contains the value 1234h DH would be 12h and DL would be 34h).*

And a 16-bit FLAG Register. All "flags" (see below) are stored here. The FLAGS Register consists of 9 status bits. These bits are also called flags, because they can either be **SET (1)** or **NOT SET (0)**. All these flags have a name and purpose.

***Flags Register***

| Abr. | Name | bit nє | Description |
|---|---|---|---|
| **OF** | Overflow Flag | 11 | indicates an overflow when *set* |
| **DF** | Direction Flag | 10 | used for string operations to check direction |
| **IF** | Interrupt Flag | 9 | if *set*, interrupt are enabled, else disabled |
| **TF** | Trap Flag | 8 | if *set*, CPU can work in single step mode |
| **SF** | Sign Flag | 7 | if *set*, resulting number of calculation is negative |
| **ZF** | Zero Flag | 6 | if *set*, resulting number of calculation is zero |
| **AF** | Auxiliary Carry | 4 | some sort of second carry flag |
| **PF** | Parity Flag | 2 | indicates even or odd parity |
| **CF** | Carry Flag | 0 | contains the left-most bit after calculations |

### *Test It!*

If you want to see all these register and flags, you can go to DOS and then start "debug" (just type debug) When you're in debug, just type "r" and you'll see all the registers and some abbreviations for the flags. Type "q" to quit again. We won't use debug to program in this tutorial, we'll use a real assembler. I use TASM 3.2, but MASM or any other assembler works just fine too.

## 7.5 MEMORY SEGMENTATION

Now I've to explain something about the way the **8086** uses memory (actually about how DOS uses memory). Since the databus of the **8086** is 16-bits, it can move and store 16-bits (1 word=2 bytes) at a time. If the processor stores a "word" (16-bits) it stores the bytes in reverse order in the memory. It looks like this:

**1234h** ***(word)*** **—> memory 34h** ***(byte)*** **12h** ***(byte)***

So if the memory looks like this: 78h 56h and you get a word from memory you'll get the value 5678h. (Note, I use the "h" after a number to indicate it's hexadecimal) However, if you just get a byte from memory it goes this way: memory 78h 56h ——> first byte you get 78h.

Now let's talk about segments. The **8086** divides it's memory into segments. Segments are (standard in DOS) 64 KB big and have a number. These numbers are stored in the segment registers (see above). Three main segments are the code, data and stack segment. Segments overlap each other almost completely. If you start debug again and type "d" you can see some addresses at the left of the screen. The format is like this: 4576:0100. that's a memory address. The first number is the segment number and the second number is the offset within the segment. So FFFF:FFF0 means: Segment FFFFh and FFF0h bytes from the beginning of the segment.

As I said before, segments overlap. The address 0000:0010 is EXACTLY the same address as 0001:0000. That means that segment begin at paragraph boundaries. (a paragraph=16 bytes, so the segment starts at an address divisible by 16) Now you can start calculating REAL addresses in memory. An example: 0000:0010 means: segment 0000h offset 10h Now we multiply the segment number with 16 and add the offset.

*Note that the offset 10h means the value 16 in decimal*: $(0 * 16 = 0 + 16 = 16)$ this is the linear address.

Next, the other addresses 0001:0000: $(16 * 1 = 16 + 0 = 16)$. Same linear address! Like I told you. These are some basic things you need to know when you want to program in Assembly. Learn the registers and flags by heart and try to understand the segmentation of memory.

## 7.6 DATA MOVEMENT INSTRUCTIONS

Data movement instructions move data from one location to another. The source and destination locations are determined by the addressing modes, and can be registers or memory. Some processors have different instructions for loading registers and storing to memory, while other processors have a single instruction with flexible addressing modes. Data movement instructions generally have the greatest options for addressing modes. Data movement instructions typically come in a variety of sizes. Data movement instructions destroy the previous contents of the destination. Data movement instructions typically set and clear processor flags. When the destination is a register and the data is smaller than the full register size, the data might be placed only in the low order bits (leaving high order bits unchanged), or might be zero- or sign-extended to fill the entire register (some processors only use one choice, others permit the programmer to choose how this is handled). Register to register operations can usually have the same source and destination register.

Earlier processors had different instructions and different names for different kinds of data movement, while most modern processors group data movement into a single symbolic name, with different kinds of data movement being indicated by address mode and size designation. A **load** instruction loads a register from memory. A **store** instruction stores the contents of a register into memory. A **transfer** instruction loads a register from another register. In processors that have separate names for different kinds of data moves, a memory to memory data move might be specially designated as a "move" instruction.

An **exchange** instruction exchanges the contents of two registers, two memory locations, or a register and a memory location (although some processors only have register-register exchanges or other limitations).

Some processors include versions of data movement instructions that can perform simple operations during the data move (such as compliment, negate, or absolute value).

Some processors include instructions that can **save** (to memory) or **restore** (from memory) a block of registers at one time (useful for implementing subroutines).

Some processors include instructions that can move a block of memory from one location to another at one time. If a processor includes string instructions, then there will usually be a string instruction that moves a string from one location in memory to another.

- **MOVE** Move Data; Motorola 680x0, Motorola 68300; move a byte (MOVE.B 8 bits), word (MOVE.W 16 bits), or longword (MOVE.L 32 bits) of data; memory to memory, memory to register, register to memory, or register to register; moves of byte and word data to registers leaves high order bits unchanged; sets or clears flags
- **MOV** Move Data; Intel 80x86; move a byte (8 bits), word (16 bits), or doubleword (32 bits) of data; memory to register, register to memory, or register to register (cannot move data from memory to memory or from segment register to segment register); does not affect flags
- **MOV** Move Data; DEC VAX; move a byte (MOVB 8 bits), word (MOVW 16 bits), longword (MOVL 32 bits), quadword (MOVQ 64 bits), octaword (MOVQ 128 bits), single precision floating (MOVF 32 bits), double precision floating (MOVD 64 bits), G floating (MOVG 64 bits), or H floating (MOVH 128 bits) of data; memory to memory, memory to register, register to memory, or register to register; moves of byte and word data to registers leaves high order bits unchanged; quadword, D float, and G float moves to or from registers are consecutive register pairs, octaword, and H float moves to or from registers are four consecutive registers; and sets or clears flags

- **PUSH** Push; Intel 80x86; decrement stack pointer and move a word (16 bits) or doubleword (32 bits) of data from memory or register (or byte of immediate data) onto stack; does not affect flags
- **PUSHL** Push Long; DEC VAX; decrement stack pointer (R14) and move a longword (32 bits) of data from memory or register onto stack; equivalent to MOVL src, -(SP), but shorter and executes faster; sets or clears flags
- **POP** Pop; Intel 80x86; move a word (16 bits) or doubleword (32 bits) of data from top of stack to register or memory and increment stack pointer; does not affect flags
- **LR** Load from Register; IBM 360/370; RR format; move a full word (32 bits) of data; register to register only; does not affect condition code
- **L** Load (from main storage); IBM 360/370; RX format; move a full word (32 bits) of data; main storage to register only; does not affect condition code
- **LH** Load Half-word; IBM 360/370; RX format; move a half-word (16 bits) of data; main storage to register only; does not affect condition code
- **LDA** Load A-register; MIX; move word or partial word field of data; main storage to accumulator only
- **LDX** Load X-register; MIX; move word or partial word field of data; main storage to extension register only
- **LD*i*** Load index-register; MIX; move word or partial word field of data; main storage to one of five index registers only
- **ST** Store (into main storage); IBM 360/370; RX format; move a full word (32 bits) of data; register to main storage only; does not affect condition code
- **STH** Store Half-word; IBM 360/370; RX format; move a half-word (16 bits) of data; register to main storage only; does not affect condition code
- **STA** Store A-register; MIX; move word or partial word field of data; accumulator to main storage only
- **STX** Store X-register; MIX; move word or partial word field of data; extension register to main storage only
- **ST*i*** Store index-register; MIX; move word or partial word field of data; one of five index registers to main storage only
- **MVI** MoVe Immediate; IBM 360/370; SI format; move a character (8 bits) of data; immediate data to register only; does not affect condition code
- **MOVEQ** Move Quick; Motorola 680x0, Motorola 68300; moves byte (8 bits) of sign-extended data (32 bits) to a data register; sets or clears flags
- **CLR** Clear; Motorola 680x0, Motorola 68300; clears a register or contents of a memory location (.B 8, .W 16, or .L 32 bits) to zero; clears flags for memory and data registers, does not modify flags for address register

- **CLR** Clear; DEC VAX; clears a scalar quantity in register or memory to zero (CLRB 8 bits, CLRW 16 bits, CLRL 32 bits, CLRQ 64 bits, CLRO 128 bits, CLRF 32 bit float, or CLRD 64 bit float), an integer CLR will clear the same size floating point quantity because VAX floating point zero is represented as all zero bits; quadword and D float clears of registers are consecutive register pairs, octaword clears to registers are four consecutive registers; equivalent to MOVx #0, dst, but shorter and executes faster; sets or clears flags
- **STZ** Store Zero; MIX; move word or partial word field of data, store zero into designated word or field of word of memory
- **EXG** Exchange; Motorola 680x0, Motorola 68300; exchanges the data (32 bits) in two data registers; does not affect flags
- **XCHG** Exchange; Intel 80x86; exchanges the data (16 bits or 32 bits) in a register with the AX or EAX register or exchanges the data (8 bits, 16 bits, or 32 bits) in a register with the contents of an effective address (register or memory); LOCK prefix and LOCK# signal asserted in XCGHs involving memory; does not affect flags
- **MOVSX** Move with Sign Extension; Intel 80x86; moves data from a register or memory to a register, with a sign extension (conversion to larger binary integer: byte to word, byte to doubleword, or word to doubleword); does not affect flags
- **MOVZX** Move with Zero Extension; Intel 80x86; moves data from a register or memory to a register, with a zero extension (conversion to larger binary integer: byte to word, byte to doubleword, or word to doubleword); does not affect flags
- **MOVZ** Move Zero Extended; DEC VAX; moves an unsigned integer to a larger unsigned integer with zero extend, source and destination in register or memory (MOVZBW Byte to Word, MOVZBL Byte to Long, MOVZWL Word to Long); sets or clears flags
- **MCOM** Move Complemented; DEC VAX; moves the logical complement (one's complement) of an integer to register or memory (MCOMB 8 bits, MCOMW 16 bits, or MCOML 32 bits); sets or clears flags
- **LCR** Load Complement from Register; IBM 360/370; RR format; fetches a full word (32 bits) of data from one of 16 general purpose registers, complements the data, and stores a full word (32 bits) of data in one of 16 general purpose registers; register to register only; sets or clears flags
- **LPR** Load Positive from Register (absolute value); IBM 360/370; RR format; fetches a full word (32 bits) of data from one of 16 general purpose registers, creates the absolute value (positive) the data, and stores a full word (32 bits) of data in one of 16 general purpose registers; register to register only; sets or clears flags
- **MNEG** Move Negated; DEC VAX; moves the arithmetic negative of a scalar quantity to register or memory (MNEGB 8 bits, MNEGW 16 bits, MNEGL 32 bits, MNEGQ 64 bits, MNEGF 32 bit float, or MNEGD 64 bit float); if source is positive zero, result is also positive zero; sets or clears flags

- **LNR** Load Negative from Register (negative of absolute value); IBM 360/370; RR format; fetches a full word (32 bits) of data from one of 16 general purpose registers, creates the absolute value the data, complements (negative) the absolute value of the data, and stores a full word (32 bits) of data in one of 16 general purpose registers; register to register only; sets or clears flags
- **LDAN** Load A-register Negative; MIX; move word or partial word field of data, load sign field with opposite sign; main storage to accumulator only
- **LDXN** Load X-register Negative; MIX; move word or partial word field of data, load sign field with opposite sign; main storage to extension register only
- **LD*i*N** Load index-register Negative; MIX; move word or partial word field of data, load sign field with opposite sign; main storage to one of five index registers only
- **STZ** Store Zero; MIX; move word or partial word field of data, store zero into designated word or field of word of memory
- **MVC** MoVe Character; IBM 360/370; SS format; moves one to 256 characters (8 bits each) of data; main storage to main storage only; does not affect condition code
- **MOVE** Move (block); MIX; move the number of words specified by the F field from location M to the location specified by the contents of index register 1, incrementing the index register on each word moved
- **MOVEM** Move Multiple; Motorola 680x0, Motorola 68300; move contents of a list of registers to memory or restore from memory to a list of registers; does not affect condition code
- **LM** Load Multiple; IBM 360/370; RS format; moves a series of full words (32 bits) of data from memory to a series of general purpose registers; main storage to register only; does not affect condition code
- **STM** STore Multiple; IBM 360/370; RS format; moves contents of a series of general purpose registers to a series of full words (32 bits) in memory; register to main storage only; does not affect condition code
- **PUSHA** Push All Registers; Intel 80x86; move contents all 16-bit general purpose registers to memory pointed to by stack pointer (in the order AX, CX, DX, BX, original SP, BP, SI, and DI ); does not affect flags
- **PUSHAD** Push All Registers; Intel 80386; move contents all 32-bit general purpose registers to memory pointed to by stack pointer (in the order EAX, ECX, EDX, EBX, original ESP, EBP, ESI, and EDI ); does not affect flags
- **POPA** Pop All Registers; Intel 80x86; move memory pointed to by stack pointer to all 16-bit general purpose registers (except for SP); does not affect flags
- **POPAD** Pop All Registers; Intel 80386; move memory pointed to by stack pointer to all 32-bit general purpose registers (except for ESP); does not affect flags
- **STJ** Store jump-register; MIX; move word or partial word field of data; jump register to main storage only

- **MOVEP** Move Peripheral Data; Motorola 680x0, Motorola 68300; moves data (16 bits or 32 bits) from a data register to memory mapped peripherals or moves data from memory mapped peripherals to a data register, skipping every other byte

## 7.7 ARITHMETIC INSTRUCTIONS

The 80x86 provides many arithmetic operations: addition, subtraction, negation, multiplication, division/modulo (remainder), and comparing two values. The instructions that handle these operations are add, adc, sub, sbb, mul, imul, div, idiv, cmp, neg, inc, dec, xadd, cmpxchg, and some miscellaneous conversion instructions: aaa, aad, aam, aas, daa, and das. The following sections describe these instructions in detail.

### 7.7.1 Integer Arithmetic

For most processors, integer arithmetic is faster than floating point arithmetic. This can be reversed in special cases such digital signal processors.

The basic four integer arithmetic operations are **addition**, **subtraction**, **multiplication**, and **division**. Arithmetic operations can be **signed** or **unsigned** (unsigned is useful for effective address computations). Some older processors don't include hardware multiplication and division. Some processors don't include actual multiplication or division hardware, instead looking up the answer in a massive table of results embedded in the processor.

A specialized, but common, form of addition is an **increment** instruction, which adds one to the contents of a register or memory location. For address computations, "increment" may mean the addition of a constant other than one. Some processors have "short" or "quick" addition instructions that extend increment to include a small range of positive values.

A specialized, but common, form of subtraction is an **decrement** instruction, which subtracts one from the contents of a register or memory location. For address computations, "decrement" may mean the subtraction of a constant other than one. Some processors have "short" or "quick" subtraction instructions that extend decrement to include a small range of values.

**Compare** instructions are used to examine one or more integers non-destructively. These are usually implemented by performing a subtraction in some shadow register or accumulator and then setting flags accordingly. Compare instructions can compare two integers, or can compare a single integer to zero. Triadic compare instructions compare a test value to an upper and lower limit, which can be useful for bounds and range checking.

Some processors have specific hardware support for large multi-byte integer arithmetic. Even if there is no specific support, generally carry and borrow flags can be used to implement software multi-byte arithmetic routines.

Some processors have other special integer arithmetic operations. A **clear** instruction sets a register or memory location to zero. Some processors have special instructions for setting a register to a special value (such as pi) with additional guard bits also being set appropriately. A sign **extend** operation takes a small value and sign extends it to a larger storage format (such as byte to word).

An arithmetic **complement** gives the arithmetic complement of a number (one's complement). An arithmetic **negate** gives the arithmetic inverse of a number (subtract from zero; two's complement).

- **ADD** Arithmetic Addition; DEC VAX; signed addition of scalar quantities (8, 16, or 32 bit integer or 32 or 64 bit floating point) in general purpose registers or memory, available in two operand (first operand added to second operand with result replacing second operand) and three operand (first operand added to second operand with result placed in third operand) (ADDB2 add byte 2 operand, ADDB3 add byte 3 operand, ADDW2 add word 2 operand, ADDW3 add word 3 operand, ADDL2 add long 2 operand, ADDL3 add long 3 operand); clears or sets flags
- **ADD** Add Integers; Intel 80x86; integer add of the contents of a register or memory (8, 16, or 32 bits) to a memory location or a register; sets or clear flags
- **ADD** Add; Motorola 680x0, Motorola 68300; signed add of the contents of a data register (8, 16, or 32 bits) to a memory location or adds the contents of a memory location (8, 16, or 32 bits) to a data register; sets or clear flags
- **ADD** Add; MIX; add word or partial word field contents of memory to A-register (accumulator), overflow toggle set if result is too large for A-register
- **AR** Add Register; IBM 360/370; RR format; signed add of the contents of a general purpose register (32 bits) to a general purpose register (32 bits); register to register only; sets or clears flags
- **A** Add; IBM 360/370; RX format; signed add of the contents of a memory location (32 bits) to a general purpose register (32 bits); main memory to register only; sets or clears flags
- **AH** Add Half-word; IBM 360/370; RX format; signed add of the contents of a memory location (16 bits) to a general purpose register (low order 16 bits); main memory to register only; sets or clears flags
- **ADDA** Add Address; Motorola 680x0, Motorola 68300; unsigned add of the contents of a memory location or register (16 or 32 bits) to an address register; does not modify flags
- **ADDI** Add Immediate; Motorola 680x0, Motorola 68300; signed add of immediate data (8, 16, or 32 bits) to a register or memory location; sets or clears flags
- **ADDQ** Add Quick; Motorola 680x0, Motorola 68300; signed add of an immediate value of 1 to 8 inclusive to a register or memory lcoation; sets or clears flags for data registers and memory locations, does not modify flags for an address register
- **INC** Increment; DEC VAX; increments the integer contents of a general purpose register or contents of memory (INCB byte, INCW word, INCL longword); equivalent to ADDx2 #1, sum, but shorter and executes faster; clears or sets flags
- **INC** Increment by 1; Intel 80x86; increments the contents of a register or memory (8, 16, or 32 bits); sets or clear flags (does not modify carry flag)
- **ADWC** Add With Carry; DEC VAX; integer addition (32 bit) in general purpose registers or memory, first operand added to second operand and the C (carry) flag with result replacing second operand; used for multiprecision arithmetic; clears or sets flags

- **ADC** Add Integers with Carry; Intel 80x86; integer add of the contents of a register or memory (8, 16, or 32 bits) and the carry flag to a memory location or a register, used to implement multi-precision integer arithmetic; sets or clear flags
- **ADDX** Add Extended; Motorola 680x0, Motorola 68300; (signed add of a data register [8, 16, or 32 bits] and the extend bit to a data register) or (signed add of the contents of memory location [8, 16, or 32 bits] and the extend bit to the contents of another memory location while predecrementing both the source and destination address pointer registers), used to implement multi-precision integer arithmetic; sets or clears flags
- **ADAWI** Add Aligned Word Interlocked; DEC VAX; adds (16 bit integer) a source operand from a register or memory to a memory location that is word aligned while interlocking the memory location so that no other processor or device can read or write to the interlocked memory location, used to maintain operating system resource usage counts; and sets or clears flags
- **SUB** Subtract; DEC VAX; signed subtraction of scalar quantities (8, 16, or 32 bit integer) in general purpose registers or memory, available in two operand (first operand subtracted from second operand with result replacing second operand) and three operand (first operand subtracted from second operand with result placed in third operand) (SUBB2 subtract byte 2 operand, SUBB3 subtract byte 3 operand, SUBW2 subtract word 2 operand, SUBW3 subtract word 3 operand, SUBL2 subtract long 2 operand, SUBL3 subtract long 3 operand); clears or sets flags
- **SUB** Subtract Integers; Intel 80x86; integer subtraction of the contents of a register or memory (8, 16, or 32 bits) from a memory location or a register; sets or clear flags
- **SUB** Subtract; Motorola 680x0, Motorola 68300; signed subtract of the contents of a data register (8, 16, or 32 bits) from a memory location or subtracts the contents of a memory location (8, 16, or 32 bits) from a data register; sets or clear flags
- **SUB** Subtract; MIX; subtract word or partial word field contents of memory from A-register (accumulator), overflow toggle possibly set
- **SR** Subtract Register; IBM 360/370; RR format; signed subtract of the contents of a general purpose register (32 bits) from a general purpose register (32 bits); register to register only; sets or clears flags
- **S** Subtract; IBM 360/370; RX format; signed subtract of the contents of a memory location (32 bits) from a general purpose register (32 bits); main memory to register only; sets or clears flags
- **SH** Subtract Half-word; IBM 360/370; RX format; signed subtract of the contents of a memory location (16 bits) from a general purpose register (low order 16 bits); main memory to register only; sets or clears flags
- **SUBA** Subtract Address; Motorola 680x0, Motorola 68300; unsigned subtract of the contents of a memory location or register (16 or 32 bits) from an address register; does not modify flags

- **SUBI** Subtract Immediate; Motorola 680x0, Motorola 68300; signed subtract of immediate data (8, 16, or 32 bits) from a register or memory location; sets or clears flags
- **SUBQ** Subtract Quick; Motorola 680x0, Motorola 68300; signed subtract of an immediate value of 1 to 8 inclusive from a register or memory lcoation; sets or clears flags for data registers and memory locations, does not modify flags for an address register
- **DEC** Decrement; DEC VAX; decrements the integer contents of a general purpose register or contents of memory (DECB byte, DECW word, DECL longword); equivalent to SUBx2 #1, sum, but shorter and executes faster; clears or sets flags
- **DEC** Decrement by 1; Intel 80x86; decrements the contents of a register or memory (8, 16, or 32 bits); sets or clear flags (does not modify carry flag)
- **SBWC** Subtract With Carry; DEC VAX; integer subtraction (32 bit) in general purpose registers or memory, first operand and the C (carry) flag subtracted from second operand with result replacing second operand; used for extended precision subtraction; clears or sets flags
- **SBB** Subtract Integers with Borrow; Intel 80x86; integer subtraction of the contents of a register or memory (8, 16, or 32 bits) and carry flag from a memory location or a register; sets or clear flags
- **SUBX** Subtract Extended; Motorola 680x0, Motorola 68300; (signed subtract of a data register [8, 16, or 32 bits] and the extend bit from a data register) or (signed subtract of the contents of memory location [8, 16, or 32 bits] and the extend bit from the contents of another memory location while predecrementing both the source and destination address pointer registers), used to implement multi-precision integer arithmetic; sets or clears flags
- **MUL** Multiply; DEC VAX; signed multiplication of scalar quantities (8, 16, or 32 bit integer) in general purpose registers or memory, available in two operand (first operand multiplied by second operand with result replacing second operand) and three operand (first operand multiplied by second operand with result placed in third operand) (MULB2 multiply byte 2 operand, MULB3 multiply byte 3 operand, MULW2 multiply word 2 operand, MULW3 multiply word 3 operand, MULL2 multiply long 2 operand, MULL3 multiply long 3 operand); clears or sets flags
- **MULS.W** Signed Multiply; Motorola 680x0, Motorola 68300; signed multiplication of a word (16 bits) from memory or a register by a word (16 bits) in a data register with a longword (32 bit) result stored in the entire data register; sets or clears flags
- **MULS.L** Signed Multiply; Motorola 680x0, Motorola 68300; signed multiplication of a longword (32 bits) from memory or a register by a longword (32 bits) in a data register with a longword (32 bit) result stored in the data register (high order 32 bits of product are discarded); sets or clears flags
- **MULS.L <ea>,Dh:Dl** Signed Multiply; Motorola 680x0, Motorola 68300; signed multiplication of a longword (32 bits) from a data register by a longword (32 bits) in a data register with a quadword (64 bit) result stored in the data registers (high order 32 bits of

product in first register, low order 32 bits of product in second data register); sets or clears flags

- **MULU.W** Unsigned Multiply; Motorola 680x0, Motorola 68300; unsigned multiplication of a word (16 bits) from memory or a register by a word (16 bits) in a data register with a longword (32 bit) result stored in the entire data register; sets or clears flags
- **MULU.L** Unsigned Multiply; Motorola 680x0, Motorola 68300; unsigned multiplication of a longword (32 bits) from memory or a register by a longword (32 bits) in a data register with a longword (32 bit) result stored in the data register (high order 32 bits of product are discarded); sets or clears flags
- **MULU.L <ea>,Dh:Dl** Unsigned Multiply; Motorola 680x0, Motorola 68300; unsigned multiplication of a longword (32 bits) from a data register by a longword (32 bits) in a data register with a quadword (64 bit) result stored in the data registers (high order 32 bits of product in first register, low order 32 bits of product in second data register); sets or clears flags
- **MUL** Unsigned Multiplication of AL or AX; Intel 80x86; unsigned multiplication of a byte (8 bits) from register or memory by the contents of the AL register with a word (16-bit) result in AX register, or unsigned multiplication of a word (16 bits) from register or memory by the contents of the AX register with a doubleword (32-bit) result in DX:AX register pair, or unsigned multiplication of a doubleword (32 bits) from register or memory by the contents of the EAX register with a quadword (64-bit) result in EDX:EAX register pair; uses an early out algorithm to speed up computations when possible; sets or clears flags
- **IMUL** Signed Integer Multiply; Intel 80x86; signed multiplication of a byte (8 bits), word (16 bits), or doubleword (32 bits) from register or memory by the contents of the EAX and EDX registers with result stored in the EAX and EDX registers, signed multiplication of a byte (8 bits), word (16 bits), or doubleword (32 bits) from register or memory by the contents of a register with truncated results (to size as operands) stored in the register, or signed multiplication of a byte (8 bits), word (16 bits), or doubleword (32 bits) from register or memory by the contents of an immediate value with truncated results (to size as operands) stored in any general register; uses an early out algorithm to speed up computations when possible; sets or clears flags
- **MUL** Multiply; MIX; multiply word or partial word field contents of memory to A-register (accumulator) with results stored in X-register and A-register pair, overflow toggle possibly set
- **MR** Multiply Register; IBM 360/370; RR format; signed multiply of the contents of an even numbered general purpose register (32 bits) by the contents of the immediately following odd numbered general purpose register (32 bits) with a 64-bit product in the register pair; register to register only; does not affect condition code
- **M** Multiply; IBM 360/370; RX format; signed multiply of the contents of a memory location (32 bits) by the contents of an odd numbered general purpose register (32 bits) with a 64-bit product in the register pair; main storage to register only; does not affect condition code

- **MH** Multiply Half-word; IBM 360/370; RX format; signed multiply of the contents of a memory location (16 bits) by the contents of a general purpose register (32 bits) with a 32-bit product (the high 8 bits of the true 48-bit product are discarded) in the register; main storage to register only; does not affect condition code
- **EMUL** Extended Multiply; DEC VAX; extended precision multiplication on operands in registers or memory, the first (longword) operand (multiplicand) is multiplied by the second (longword) operand (multiplier) giving a (doubleword) intermediary result which is stored in the fourth (doubleword) operand (product); clears or sets flags
- **DIVS.W** Signed Divide; Motorola 680x0, Motorola 68300; signed division of a longword (32 bits) in memory or a register by a word (16 bits) in a data register with a result of the quotient (16 bits) in the lower word and the remainder (16 bits) in the upper word of the data register; clears or sets flags
- **DIVS.L <ea>,Dq** Signed Divide; Motorola 680x0, Motorola 68300; signed division of a longword (32 bits) in memory or a register by a longword (32 bits) in a data register with a result of a longword (32 bit) quotient in the data register and the remainder being discarded; clears or sets flags
- **DIVS.L <ea>,Dr:Dq** Signed Divide; Motorola 680x0, Motorola 68300; signed division of a quadword (64 bits) in any two data registers by a longword (32 bits) in a data register with a result of the quotient (32 bits) in the second data register and the remainder (32 bits) in the third data register; clears or sets flags
- **DIVSL.L <ea>,Dr:Dq** Signed Divide; Motorola 680x0, Motorola 68300; signed division of a longword (32 bits) in a data register by a longword (32 bits) in a second data register with a result of the quotient (32 bits) in the first data register and the remainder (32 bits) in the second data register; clears or sets flags
- **DIVU.W** Unsigned Divide; Motorola 680x0, Motorola 68300; unsigned division of a longword (32 bits) in memory or a register by a word (16 bits) in a data register with a result of the quotient (16 bits) in the lower word and the remainder (16 bits) in the upper word of the data register; clears or sets flags
- **DIVU.L <ea>,Dq** Unsigned Divide; Motorola 680x0, Motorola 68300; unsigned division of a longword (32 bits) in memory or a register by a longword (32 bits) in a data register with a result of a longword (32 bit) quotient in the data register and the remainder being discarded; clears or sets flags
- **DIVU.L <ea>,Dr:Dq** Unsigned Divide; Motorola 680x0, Motorola 68300; unsigned division of a quadword (64 bits) in any two data registers by a longword (32 bits) in a data register with a result of the quotient (32 bits) in the second data register and the remainder (32 bits) in the third data register; clears or sets flags
- **DIVUL.L <ea>,Dr:Dq** Unsigned Divide; Motorola 680x0, Motorola 68300; unsigned division of a longword (32 bits) in a data register by a longword (32 bits) in a second data register with a result of the quotient (32 bits) in the first data register and the remainder (32 bits) in the second data register; clears or sets flags

- **DR** Divide Register; IBM 360/370; RR format; signed divide of the contents of a general purpose register pair (64 bits) by the contents of a general purpose register (32 bits) with a 32-bit quotient in the odd numbered register and a 32-bit remainder in the even numbered register of the register pair; register to register only; does not affect condition code
- **D** Divide; IBM 360/370; RX format; signed divide of the contents of a general purpose register pair (64 bits) by the contents of a memory location (32 bits) with a 32-bit quotient in the odd numbered register and a 32-bit remainder in the even numbered register of the register pair; main storage to register only; does not affect condition code
- **DIV** Divide; DEC VAX; arithmetic division of scalar quantities (8, 16, or 32 bit integer) in general purpose registers or memory, available in two operand (first operand [divisor] divided from second operand [dividend] with result [quotient] replacing second operand) and three operand (first operand [divisor] divided from second operand [dividend] with result placed in third operand [quotient]) (DIVB2 divide byte 2 operand, DIVB3 divide byte 3 operand, DIVW2 divide word 2 operand, DIVW3 divide word 3 operand, DIVL2 divide long 2 operand, DIVL3 divide long 3 operand); clears or sets flags
- **EDIV** Extended Divide; DEC VAX; extended precision multiplication on operands in registers or memory, the second (longword) operand (dividend) is divided from the first (longword) operand (divisor) giving the third (longword) operand (quotient) and the the fourth (longword) operand (remainder); clears or sets flags
- **DIV** Unsigned Divide; Intel 80x86; unsigned division of the accumulator by a byte (8 bits), word (16 bits), or doubleword (32 bits) divisor of half the size of the dividend in the accumulator, with the results stored in the accumulator (byte divisor: dividend is in the AX register, quotient in the AL register, and remainder in the AH register; word divisor: dividend is in the DX:AX register pair, quotient in the AX register, and remainder in the DX register; doubleword divisor: dividend is in the EDX:AEX register pair, quotient in the EAX register, and remainder in the EDX register); non-integral quotients are truncated to integers toward 0; sets or clears flags
- **IDIV** Signed Integer Division; Intel 80x86; Intel 80x86; signed division of the accumulator by a byte (8 bits), word (16 bits), or doubleword (32 bits) divisor of half the size of the dividend in the accumulator, with the results stored in the accumulator (byte divisor: dividend is in the AX register, quotient in the AL register, and remainder in the AH register; word divisor: dividend is in the DX:AX register pair, quotient in the AX register, and remainder in the DX register; doubleword divisor: dividend is in the EDX:AEX register pair, quotient in the EAX register, and remainder in the EDX register); non-integral quotients are truncated to integers toward 0; sets or clears flags
- **DIV** Divide; MIX; divide word or partial word field contents of memory from A-register (accumulator) and X-register (extension) pair with quotient stored in A-register and remainder stored in X-register, overflow toggle possibly set

- **CMP** Compare; DEC VAX; arithmetic comparison between two scalar quantities (8, 16, or 32 bit integer or 32 or 64 bit floating point) in general purpose registers or memory (CMPB Byte, CMPW Word, CMPL Longword); clears or sets flags
- **TST** Test; DEC VAX; arithmetic comparison of a scalar quantities (8, 16, or 32 bit integer or 32 or 64 bit floating point) in general purpose registers or memory (TSTB Byte, TSTW Word, TSTL Longword) to zero; equivalent to CMPs src, #0, but shorter and executes faster; clears or sets flags
- **CMP** Compare; Motorola 680x0, Motorola 68300; compares a register or contents of a memory location (8, 16, or 32 bits) to contents of a data register (data register minus effective address contents); clears or sets flags
- **CMP** Compare Two Operands; Intel 80x86; compares a register or contents of a memory location (8, 16, or 32 bits) to contents of a register or memory location (subtract of second operand from first operand with no storage of results, but setting or clearing of flags); clears or sets flags
- **CMPA** Compare A-register; MIX; compare word or partial word field contents of memory with same word or partial field of A-register (accumulator), set comparison indicator
- **CMPX** Compare X-register; MIX; compare word or partial word field contents of memory with same word or partial field of X-register (extension), set comparison indicator
- **CMP*i*** Compare I-register; MIX; compare word or partial word field contents of memory with same word or partial field of designated index register, set comparison indicator
- **CMPA** Compare Address; Motorola 680x0, Motorola 68300; compares a register or contents of a memory location (16 or 32 bits) to contents of an address register (adress register minus effective address contents); clears or sets flags
- **CMPI** Compare Immediate; Motorola 680x0, Motorola 68300; compares immediate data (8, 16, or 32 bits) to contents of a register or memory (effective address contents minus immediate data); clears or sets flags
- **CMPM** Compare Memory; Motorola 680x0, Motorola 68300; compares the contents of two memory locations (8, 16, or 32 bits) with a post increment of both address pointer registers (second location minus first location); clears of sets flags
- **CMP2** Compare Register Against Bounds; Motorola 680x0, Motorola 68300; compares the contents of register (8, 16, or 32 bits) to a bounds pair (lower bound followed by upper bound), if both bounds are equal then this operation tests for a specific value; sets or clears flags
- **CLR** Clear; Motorola 680x0, Motorola 68300; clears a register or contents of a memory location (.B 8, .W 16, or .L 32 bits) to zero; clears flags for memory and data registers, does not modify flags for address register
- **CLR** Clear; DEC VAX; clears a scalar quantity in register or memory to zero (CLRB 8 bits, CLRW 16 bits, CLRL 32 bits, CLRQ 64 bits, CLRO 128 bits, CLRF 32 bit float, or CLRD 64 bit float), an integer CLR will clear the same size floating point quantity because VAX

floating point zero is represented as all zero bits; quadword and D float clears of registers are consecutive register pairs, octaword clears to registers are four consecutive registers; equivalent to MOVx #0, dst, but shorter and executes faster; sets or clears flags

- **STZ** Store Zero; MIX; move word or partial word field of data, store zero into designated word or field of word of memory
- **EXT** Sign Extend; Motorola 680x0, Motorola 68300; sign extends a byte (8 bits) in a data register to a word (16 bits) or sign extends a word (16 bits) in a data register to a longword (32 bits); sets or clears flags
- **EXTB** Sign Extend Byte; Motorola 680x0, Motorola 68300; sign extends a byte (8 bits) in a data register to a longword (32 bits); sets or clears flags
- **NEG** Two's Complement Negation; Intel 80x86; subtracts the contents of a register or memory (8, 16, or 32 bits) from zero and store the results in the original register or memory location (arithmetic negation or arithmetic inverse); sets or clears flags
- **NEG** Negate; Motorola 680x0, Motorola 68300; subtracts the contents of a register or memory (8, 16, or 32 bits) from zero and store the results in the original register or memory location (arithmetic negation or arithmetic inverse); sets or clears flags
- **NEGX** Negate with Extend; Motorola 680x0, Motorola 68300; subtracts the contents of a register or memory location (8, 16, or 32 bits) and the extend bit from zero and stores the results in the original register or memory location (multi-precision negation); sets or clears flags

### 7.7.2 Floating Point Arithmetic

For most processors, integer arithmetic is faster than floating point arithmetic. This can be reversed in special cases such digital signal processors.

On many processors, floating point arithmetic is in an optional unit or optional coprocessor rather than being included on the main processor. This allows the manufacturer to charge less for the business machines that don't need floating point arithmetic. The basic four floating point arithmetic operations are **addition**, **subtraction**, **multiplication**, and **division**. Some processors don't include actual multiplication or division hardware, instead looking up the answer in a massive table of results embedded in the processor.

**Compare** instructions are used to examine one or more floating point numbers non-destructively. These are usually implemented by performing a subtraction in some shadow register or accumulator and then setting flags accordingly. Compare instructions can compare two floating point numbers, or can compare a single floating point number to zero.

- **ADD** Arithmetic Addition; DEC VAX; signed addition of scalar quantities (32, 64, or 128 bit floating point) in general purpose registers or memory, available in two operand (first operand added to second operand with result replacing second operand) and three operand (first operand added to second operand with result placed in third operand) (ADDF2 add float 2 operand, ADDF3 add float 3 operand, ADDD2 add double float 2 operand, ADDD3 add double float 3 operand, ADDG2 add G float 2 operand, ADDG3 add G float 3 operand, ADDH2 add H float 2 operand, ADDH3 add H float 3 operand); clears or sets flags

- **SUB** Subtract; DEC VAX; signed subtraction of scalar quantities (32, 64, or 128 bit floating point) in general purpose registers or memory, available in two operand (first operand subtracted from second operand with result replacing second operand) and three operand (first operand subtracted from second operand with result placed in third operand) (SUBF2 subtract float 2 operand, SUBF3 subtract float 3 operand, SUBD2 subtract double float 2 operand, SUBD3 subtract double float 3 operand, SUBG2 subtract G float 2 operand, SUBG3 subtract G float 3 operand, SUBH2 subtract H float 2 operand, SUBH3 subtract H float 3 operand); clears or sets flags
- **MUL** Multiply; DEC VAX; signed multiplication of scalar quantities (32, 64, or 128 bit floating point) in general purpose registers or memory, available in two operand (first operand multiplied by second operand with result replacing second operand) and three operand (first operand multiplied by second operand with result placed in third operand) (MULF2 multiply float 2 operand, MULF3 multiply float 3 operand, MULD2 multiply double float 2 operand, MULD3 multiply double float 3 operand, MULG2 multiply G float 2 operand, MULG3 multiply G float 3 operand, MULH2 multiply H float 2 operand, MULH3 multiply H float 3 operand); clears or sets flags
- **EMOD** Extended Multiply and Integerize; DEC VAX; performs accurate range reduction of math function arguments, the floating point multiplier extension operand (second operand) is concatenated with the floating point multiplier (first operand) to gain eight additional low order fraction bits, the multiplicand operand (third operand) is multiplied by the extended multiplier operand, after multiplication the integer portion (fourth operand) is extracted and a 32 bit (EMODF) or 64 bit (EMODD) floating point number is formed from the fractional part of the product by truncating extra bits, the multiplication is such that the result is equivalent to the exact product truncated (before normalization) to a fraction field of 32 bits in floating or 64 bits in double (fifth operand); clears or sets flags
- **DIV** Divide; DEC VAX; arithmetic division of scalar quantities (32, 64, or 128 bit floating point) in general purpose registers or memory, available in two operand (first operand [divisor] divided from second operand [dividend] with result [quotient] replacing second operand) and three operand (first operand [divisor] divided from second operand [dividend] with result placed in third operand [quotient]) (DIVF2 divide float 2 operand, DIVF3 divide float 3 operand, DIVD2 divide double float 2 operand, DIVD3 divide double float 3 operand, DIVG2 divide G float 2 operand, DIVG3 divide G float 3 operand, DIVH2 divide H float 2 operand, DIVH3 divide H float 3 operand); clears or sets flags
- **POLY** Polynomial Evaluation; DEC VAX; performs fast calculation of math functions, for degree times (second operand) evaluate a function using Horner's method, where d=degree (second operand), x=argument (first operand), and result = C[0] + x*(C[1] + x*(C[2] + … x*C[d])), float result stored in D0 register, double float result stored in D0:D1 register pair, the table address operand (third operand) points to a table of polynomial coefficients ordered

from highest order term of the polynomial through lower order coefficients stored at increasing addresses, the data type of the coefficients must be the same as the data type of the argument operand (first operand), the unsigned word degree operand (second operand) specifies the highest numbered coefficient to participate in the evaluation (POLYF polynomial evaluation floating, POLYD polynomial evaluation double float); D0 through D4 registers modified by POLYF, D0 through D5 registers modified by POLYD; sets or clears flags

- **CMP** Compare; DEC VAX; arithmetic comparison between two scalar quantities (8, 16, or 32 bit integer or 32 or 64 bit floating point) in general purpose registers or memory (CMPF Floating, CMPD Double Float); clears or sets flags
- **TST** Test; DEC VAX; arithmetic comparison of a scalar quantities (8, 16, or 32 bit integer or 32 or 64 bit floating point) in general purpose registers or memory (TSTF Floating, TSTD Double Float) to zero; equivalent to CMPs src, #0, but shorter and executes faster; clears or sets flags

## 7.8 LOGICAL INSTRUCTIONS

Logical instructions typically work on a bit by bit basis, although some processors use the entire contents of the operands as whole flags (zero or not zero input, zero or negative one output). Typical logical operations include logical negation or logical complement (NOT), logical and (AND), logical inclusive or (OR or IOR), and logical exclusive or (XOR or EOR). Logical tests are a comparison of a value to a bit string (or operand treated as a bit string) of all zeros. Some processors have an instruction that sets or clears a bit or byte in registers or memory based on the processor condition codes.

- **NOT** Logical Complement; Motorola 680x0, Motorola 68300; calculates the one's complement (logical negation) of the contents of memory or a register (8, 16, or 32 bits); sets or clears flags
- **NOT** One's Complement Negation; Intel 80x86; calculates the one's complement (logical negation) of the contents of memory or a register (8, 16, or 32 bits); does not modify flags
- **AND** And Logical; Motorola 680x0, Motorola 68300; performs a logical AND of a source operand with a destination operand and stores the result in the destination operand (8, 16, or 32 bits), one of the two operands must be a data register, the other operand can be the contents of any register or memory location; sets or clears flags
- **ANDI** And Immediate; Motorola 680x0, Motorola 68300; performs a logical AND of the contents of a register or memory location (8, 16, or 32 bits) with an immediate; sets or clears flags
- **AND** Logical AND; Intel 80x86; performs a logical AND between two registers, a register and contents of a memory location, or an immediate operand and either the contents of a register or the contents of a memory location; byte (8 bits), word (16 bits), or doubleword (32 bits); sets or clears flags

- **OR** Inclusive Or Logical; Motorola 680x0, Motorola 68300; performs a logical inclusive OR of a source operand with a destination operand and stores the result in the destination operand (8, 16, or 32 bits), one of the two operands must be a data register, the other operand can be the contents of any register or memory location; sets or clears flags
- **ORI** Inclusive Or Immediate; Motorola 680x0, Motorola 68300; performs a logical inclusive OR of the contents of a register or memory location (8, 16, or 32 bits) with an immediate; sets or clears flags
- **OR** Logical Inclusive OR; Intel 80x86; performs a logical OR between two registers, a register and contents of a memory location, or an immediate operand and either the contents of a register or the contents of a memory location; byte (8 bits), word (16 bits), or doubleword (32 bits); sets or clears flags
- **BIS** Bit Set; DEC VAX; performs a logical inclusive OR of the bit mask (first operand, register or memory) with the source (second operand, register or memory), available in two operand (results stored in second operand) and three operand (results stored in third operand) (BISB2 bit set byte 2 operand, BISB3 bit set byte 3 operand, BISW2 bit set word 2 operand, BISW3 bit set word 3 operand, BISL2 bit set longword 2 operand, BISL3 bit set longword 3 operand); sets or clears flags
- **XOR** Exclusive OR; DEC VAX; performs a logical exclusive OR of the bit mask (first operand, register or memory) with the source (second operand, register or memory), available in two operand (results stored in second operand) and three operand (results stored in third operand) (XORB2 exclusive or byte 2 operand, XORB3 exclusive or byte 3 operand, XORW2 exclusive or word 2 operand, XORW3 exclusive or word 3 operand, XORL2 exclusive or longword 2 operand, XORL3 exclusive or longword 3 operand); sets or clears flags
- **EOR** Exclusive Or Logical; Motorola 680x0, Motorola 68300; performs a logical exclusive or (XOR) of a source operand with a destination operand and stores the result in the destination operand (8, 16, or 32 bits), one of the two operands must be a data register, the other operand can be the contents of any register or memory location; sets or clears flags
- **EORI** Exclusive Or Immediate; Motorola 680x0, Motorola 68300; performs a logical exclusive or (XOR) of the contents of a register or memory location (8, 16, or 32 bits) with an immediate; sets or clears flags
- **XOR** Logical Exclusive OR; Intel 80x86; performs a logical exclusive OR between two registers, a register and contents of a memory location, or an immediate operand and either the contents of a register or the contents of a memory location; byte (8 bits), word (16 bits), or doubleword (32 bits); sets or clears flags
- **BIC** Bit Clear; DEC VAX; performs a complimented logical AND of the bit mask (first operand, register or memory) with the source (second operand, register or memory), available in two operand (results stored in second operand) and three operand (results stored in third operand) (BICB2 bit clear byte 2 operand, BICB3 bit clear byte 3 operand, BICW2 bit clear word 2 operand, BICW3 bit clear word 3 operand, BICL2 bit clear longword 2 operand, BICL3 bit clear longword 3 operand); sets or clears flags

- **TST** Test an Operand; Motorola 680x0, Motorola 68300; compares the contents of a register or memory location (8, 16, or 32 bits) with zero; sets or clears flags
- **TEST** Logical Compare; Intel 80x86; compares the contents of a register or memory location (8, 16, or 32 bits) with an immediate value or the contents of a register; sets or clears flags
- **BIT** Bit Test; DEC VAX; performs a logical AND of the bit mask (first operand, register or memory) with the source (second operand, register or memory) without modifying either operand and tests the resulting bits for being all zero (BITB byte, BITW word, BITL longword); sets or clears flags
- **Scc** Set According to Condition; Motorola 680x0, Motorola 68300; tests a condition code, if the condition is true then sets a byte (8 bits) of a data register or memory location to TRUE (all ones), if the condition is false then sets a byte (8 bits) of a data register or memory location to FALSE (all zeros): SCC, SCS, SEQ, SF, SGE, SGT, SHI, SLE, SLS, SLT, SMI, SNE, SPL, ST, SVC, SVS
- **SETcc** Set Byte on Condition cc; Intel 80x86; tests a condition code, if the condition is true then sets a byte (8 bits) of a data register or memory location to TRUE (all ones), if the condition is false then sets a byte (8 bits) of a data register or memory location to FALSE (all zeros): SETA, SETAE, SETB, SETBE, SETC, SETE, SETG, SETGE, SETL, SETLE, SETNA, SETNAE, SETNB, SETNBE, SETNC, SETNE, SETNG, SETNGE, SETNL, SETNLE, SETNO, SETNP, SETNS, SETNZ, SETO, SETP, SETPE, SETPO, SETS, SETZ

## 7.9 SHIFT AND ROTATE INSTRUCTIONS

Shift and rotate instructions move bit strings (or operand treated as a bit string). **Shift** instructions move a bit string (or operand treated as a bit string) to the **right** or **left**, with excess bits discarded (although one or more bits might be preserved in flags). In **arithmetic shift left** or **logical shift left** zeros are shifted into the low-order bit. In **arithmetic shift right** the sign bit (most significant bit) is shifted into the high-order bit. In **logical shift right** zeros are shifted into the high-order bit.

**Rotate** instructions are similar to shift instructions, ecept that rotate instructions are circular, with the bits shifted out one end returning on the other end. Rotates can be to the left or right. Rotates can also employ an extend bit for multi-precision rotates.

A **swap** instruction swaps the high and low order portions of a register or contents of a series of memory locations.

The carry bit typically receives the last bit shifted out of the operand. Sometimes an extend bit will receive the last bit shifted out also. Somtimes an overflow bit is used to indicate a sign change has occurred.

- **ASH** Arithmetic Shift; DEC VAX; performs a bit shift on a longword or quadword, the first operand is a byte count, the second operand is the source longword or quadword in registers or memory, the third operand is the destination longword or quadword in registers or memory, positive counts causes a left shift with zeros entering in the least significant bits, negative count causes a right shift with the most significant bit being copied into the most significant bit, a zero count results in the destination being replaced by the unmodified source (ASHL arithmetic shift longword, ASHQ arithmetic shift quadword); sets or clears flags

- **ASL** Arithmetic Shift Left; Motorola 680x0, Motorola 68300; shifts the contents of a data register (8, 16, or 32 bits) or memory location (16 bits) to the left (towards most significant bit) by a specified amount (by 1 to 8 bits for an immediate operation on a data register, by the contents of a data register modulo 64 for a data register, or by 1 bit only for a memory location), with the high-order bit being shifted into the carry and extend flags, zeros shifted into the low-order bit, overflow flag indicating a change of sign; sets or clear flags
- **SAL** Shift Arithmetic Left; Intel 80x86; shifts the contents of a data register or memory location (8, 16, or 32 bits) to the left (towards most significant bit) by a specified amount (by 1, by 0 to 31 bits specified by an immediate operand, or by 0-31 bits specified by the contents of the CL register), with the high-order bit being shifted into the carry flag, zeros shifted into the low-order bit; sets or clear flags
- **ASR** Arithmetic Shift Right; Motorola 680x0, Motorola 68300; shifts the contents of a data register (8, 16, or 32 bits) or memory location (16 bits) to the right (towards the least significant bit) by a specified amount (by 1 to 8 bits for an immediate operation on a data register, by the contents of a data register modulo 64 for a data register, or by 1 bit only for a memory location), with the low-order bit being shifted into the carry and extend flags, the original high-order bit being replicated and shifted into the high-order bit; sets or clear flags
- **SAR** Shift Arithmetic Right; Intel 80x86; shifts the contents of a data register or memory location (8, 16, or 32 bits) to the right (towards least significant bit) by a specified amount (by 1, by 0 to 31 bits specified by an immediate operand, or by 0-31 bits specified by the contents of the CL register), with the low-order bit being shifted into the carry flag, the original high-order bit being replicated and shifted into the high-order bit; sets or clear flags
- **SLA** Shift Left A-register; MIX; *byte* shift the contents of the A-register (leaving sign unchanged) to the left by the designated number of bytes, with zeros shifted in to low order bytes
- **SLAX** Shift Left AX-register; MIX; *byte* shift the contents of the A-register and X-register pair (leaving signs unchanged) to the left by the designated number of bytes, with zeros shifted in to low order bytes
- **LSL** Logical Shift Left; Motorola 680x0, Motorola 68300; shifts the contents of a data register (8, 16, or 32 bits) or memory location (16 bits) to the left (towards most significant bit) by a specified amount (by 1 to 8 bits for an immediate operation on a data register, by the contents of a data register modulo 64 for a data register, or by 1 bit only for a memory location), with the high-order bit being shifted into the carry and extend flags, zeros shifted into the low-order bit; sets or clear flags
- **SHL** Shift Logical Left; Intel 80x86; shifts the contents of a data register or memory location (8, 16, or 32 bits) to the left (towards most significant bit) by a specified amount (by 1, by 0 to 31 bits specified by an immediate operand, or by 0-31 bits specified by the contents of the CL register), with the high-order bit being shifted into the carry flag, zeros shifted into the low-order bit; sets or clear flags

- **SRA** Shift Right A-register; MIX; *byte* shift the contents of the A-register (leaving sign unchanged) to the right by the designated number of bytes, with zeros shifted in to high order bytes
- **SRAX** Shift Right AX-register; MIX; *byte* shift the contents of the A-register and X-register pair (leaving signs unchanged) to the right by the designated number of bytes, with zeros shifted in to high order bytes
- **LSR** Logical Shift Right; Motorola 680x0, Motorola 68300; shifts the contents of a data register (8, 16, or 32 bits) or memory location (16 bits) to the right (towards the least significant bit) by a specified amount (by 1 to 8 bits for an immediate operation on a data register, by the contents of a data register modulo 64 for a data register, or by 1 bit only for a memory location), with the low-order bit being shifted into the carry and extend flags, and zeros shifted into the high-order bit; sets or clear flags
- **SHR** Shift Logical Right; Intel 80x86; shifts the contents of a data register or memory location (8, 16, or 32 bits) to the right (towards least significant bit) by a specified amount (by 1, by 0 to 31 bits specified by an immediate operand, or by 0-31 bits specified by the contents of the CL register), with the low-order bit being shifted into the carry flag, and zeros shifted into the high-order bit; sets or clear flags
- **SHLD** Double Precision Shift Left; Intel 80x86; shifts the contents of a general purpose register (16 or 32 bits) to the left (towards most significant bit) by a specified amount (by 0 to 31 bits specified by an immediate operand or by 0-31 bits specified by the contents of the CL register) with the high-order bits being shifted into a general purpose register or memory location (the source register is unchanged); used to implement multiprecision shifts, "bit blts" (BIT BLock Transfers), or bit string extracts and inserts; sets or clear flags
- **SHRD** Double Precision Shift Right; Intel 80x86; shifts the contents of a general purpose register (16 or 32 bits) to the right (towards least significant bit) by a specified amount (by 0 to 31 bits specified by an immediate operand or by 0-31 bits specified by the contents of the CL register) with the low-order bits being shifted into a general purpose register or memory location (the source register is unchanged); used to implement multiprecision shifts, "bit blts" (BIT BLock Transfers), or bit string extracts and inserts; sets or clear flags
- **ROTL** Rotate Long; DEC VAX; performs a bit rotate on a longword, the first operand is a byte count, the second operand is the source longword in registers or memory, the third operand is the destination longword in registers or memory, positive counts causes a left rotate, negative count causes a right rotate, a zero count results in the destination being replaced by the unmodified source, bits shifted out one end are rotated back in the other end; sets or clears flags
- **ROL** Rotate Left; Motorola 680x0, Motorola 68300; rotates the contents of a data register (8, 16, or 32 bits) or a memory location (16 bits) to the left (towards the most significant bit) by a specified amount (by 1 to 8 bits for an immediate operation on a data register, by the contents of a data register modulo 64 for a data register, or by 1 bit only for a memory location), with the high-order bit rotating into both the carry flag and the low-order bit; sets or clear flags

- **ROL** Rotate Left; Intel 80x86; rotates the contents of a general purpose register or a memory location (8, 16, or 32 bits) to the left (towards the most significant bit) by a specified amount (by 1 bit or by 0 to 31 bits specified by an immediate operand or by the contents of the CL register); sets or clear flags
- **SLC** Shift Left AX-register Circularly; MIX; *byte* shift the contents of the A-register and X-register pair (leaving signs unchanged) to the left by the designated number of bytes, with bytes shifted off low order end entering on high order end
- **ROR** Rotate Right; Motorola 680x0, Motorola 68300; rotates the contents of a data register (8, 16, or 32 bits) or a memory location (16 bits) to the right (towards the least significant bit) by a specified amount (by 1 to 8 bits for an immediate operation on a data register, by the contents of a data register modulo 64 for a data register, or by 1 bit only for a memory location), with the low-order bit rotating into both the carry flag and the high-order bit; sets or clear flags
- **ROR** Rotate Right; Intel 80x86; rotates the contents of a general purpose register or a memory location (8, 16, or 32 bits) to the right (towards the least significant bit) by a specified amount (by 1 bit or by 0 to 31 bits specified by an immediate operand or by the contents of the CL register); sets or clear flags
- **SRC** Shift Right AX-register Circularly; MIX; *byte* shift the contents of the A-register and X-register pair (leaving signs unchanged) to the right by the designated number of bytes, with bytes shifted off high order end entering on low order end
- **ROXL** Rotate Left with Extend; Motorola 680x0, Motorola 68300; rotates the contents of a data register (8, 16, or 32 bits) or a memory location (16 bits) to the left (towards the most significant bit) through the extend bit by a specified amount (by 1 to 8 bits for an immediate operation on a data register, by the contents of a data register modulo 64 for a data register, or by 1 bit only for a memory location), with the high-order bit rotating into both the carry flag and the extend bit and the extend bit rotating into the low-order bit; sets or clear flags
- **RCL** Rotate Through Carry Left; Intel 80x86; rotates the contents of a general purpose register or a memory location (8, 16, or 32 bits) to the left (towards the most significant bit) by a specified amount (by 1 bit or by 0 to 31 bits specified by an immediate operand or by the contents of the CL register) with the carry flag (CF) being treated as a high-order one-bit extension of the destination operand; sets or clear flags
- **ROXR** Rotate Right with Extend; Motorola 680x0, Motorola 68300; rotates the contents of a data register (8, 16, or 32 bits) or a memory location (16 bits) to the right (towards the least significant bit) through the extend bit by a specified amount (by 1 to 8 bits for an immediate operation on a data register, by the contents of a data register modulo 64 for a data register, or by 1 bit only for a memory location), with the low-order bit rotating into both the carry flag and the extend bit and the extend bit rotating into the high-order bit; sets or clear flags
- **RCR** Rotate Through Carry Right; Intel 80x86; rotates the contents of a general purpose register or a memory location (8, 16, or 32 bits) to the right (towards the least significant bit) by a specified amount (by 1 bit or by 0 to 31 bits specified by an immediate operand or by

the contents of the CL register) with the carry flag (CF) being treated as as a low-order one-bit extension of the destination operand; sets or clear flags

- **SWAP** Swap; Motorola 680x0, Motorola 68300; exchanges the high order word (16 bits) with the low order word (16 bits) of a data register; sets or clears flags

## QUESTION BANK

Q1. What do you mean by assembly language?

Q2. Give brief description of basic elements of assembly language.

Q3. What do you mean by macros?

Q4. What do you mean by assembler?

Q5. Why we need assembly language.

Q6. Explain assembly language for microprocessors.

Q7. Explain the concept of memory segmentation.

Q8. Explain different type's data movement instructions.

Q9. Explain different type's arithmetic instructions.

Q10. Explain different type's logical instructions.

Q11. Explain different type's shift and rotate instructions.

CHAPTER 8

# Assembly Language–II

## 8.1 DATA MOVEMENT INSTRUCTIONS

The data movement instructions copy values from one location to another. These instructions include mov, xchg, lds, lea, les,push, pushf, pop, popf, lahf, and sahf.

### 8.1.1 The mov Instruction

The mov instruction takes several different forms:

```
mov     reg, reg
mov     mem, reg
mov     reg, mem
mov     mem, immediate data
mov     reg, immediate data
mov     ax/al, mem
mov     mem, ax/al
mov     segreg, mem16
mov     segreg, reg16
mov     mem16, segreg
mov     reg16, segreg
```

First, there are variations of the mov instruction that are faster and shorter than other mov instructions that do the same job. For example, both the mov ax, mem and mov reg, mem instructions can load the ax register from a memory location. On all processors the first version is shorter. On the earlier members of the 80x86 family, it is faster as well.

There are two very important details to note about the mov instruction. First, there is no memory to memory move operation. The mod-reg-r/m addressing mode byte (see Chapter Four) allows two register operands or a single register and a single memory operand. There is no form of the mov instruction that allows you to encode two memory addresses into the same instruction. Second, you cannot move immediate data into a segment register. The only instructions that move data into or out of a segment register have mod-reg-r/m bytes associated with them; there is no format that moves an immediate value into a segment register. Two common errors beginning programmers make are attempting a memory to memory move and trying to load a segment register with a constant.

The operands to the mov instruction may be bytes, words, or double words. Both operands must be the same size or MASM will generate an error while assembling your program. This applies to memory operands and register operands. If you declare a variable, B, using byte and attempt to load this variable into the ax register, MASM will complain about a type conflict.

The CPU extends immediate data to the size of the destination operand (unless it is too big to fit in the destination operand, which is an error). Note that you can move an immediate value into a memory location. The same rules concerning size apply. However, MASM cannot determine the size of certain memory operands. For example, does the instruction mov [bx], 0 store an eight bit, sixteen bit, or thirty-two bit value? MASM cannot tell, so it reports an error. This problem does not exist when you move an immediate value into a variable you've declared in your program. For example, if you've declared B as a byte variable, MASM knows to store an eight bit zero into B for the instruction mov B, 0. Only those memory operands involving pointers with no variable operands suffer from this problem. The solution is to explicitly tell MASM whether the operand is a byte, word, or double word. You can accomplish this with the following instruction forms:

```
mov     byte ptr [bx], 0
mov     word ptr [bx], 0
```

Moves to and from segment registers are always 16 bits; the mod-reg-r/m operand must be 16 bits or MASM will generate an error. Since you cannot load a constant directly into a segment register, a common solution is to load the constant into an 80x86 general purpose register and then copy it to the segment register. For example, the following two instruction sequence loads the es register with the value 40h:

```
mov     ax, 40h
mov     es, ax
```

Note that almost any general purpose register would suffice. Here, ax was chosen arbitrarily. The mov instructions do not affect any flags. In particular, the 80x86 preserves the flag values across the execution of a mov instruction.

### 8.1.2 The Xchg Instruction

The xchg (exchange) instruction swaps two values. The general form is

```
xchg    operand1, operand2
```

There are four specific forms of this instruction on the 80x86:

```
xchg    reg, mem
```

```
xchg    reg, reg
xchg    ax, reg16
```

The first two general forms require two or more bytes for the opcode and mod-reg-r/m bytes (a displacement, if necessary, requires additional bytes). The third and fourth forms are special forms of the second that exchange data in the (e)ax register with another 16 or 32 bit register. The 16 bit form uses a single byte opcode that is shorter than the other two forms that use a one byte opcode and a mod-reg-r/m byte.

Already you should note a pattern developing: the 80x86 family often provides shorter and faster versions of instructions that use the ax register. Therefore, you should try to arrange your computations so that they use the (e)ax register as much as possible. The xchg instruction is a perfect example, the form that exchanges 16 bit registers is only one byte long.

Note that the order of the xchg's operands does not matter. That is, you could enter xchg mem, reg and get the same result as xchg reg, mem. Most modern assemblers will automatically emit the opcode for the shorter xchg ax, reg instruction if you specify xchg reg, ax.

### 8.1.3 The LDS and LES Instructions

The lds, les, lfs, lgs, and lss instructions let you load a 16 bit general purpose register and segment register pair with a single instruction. The general form is

```
LxS   dest, source
```

These instructions take the specific forms:

```
lds   reg16, mem32
les   reg16, mem32
```

Reg16 is any general purpose 16 bit register and mem32 is a double word memory location (declared with the dword statement).

These instructions will load the 32 bit double word at the address specified by mem32 into reg16 and the ds, es registers. They load the general purpose register from the L.O. word of the memory operand and the segment register from the H.O. word. The following algorithms describe the exact operation:

```
lds reg16, mem32:
        reg16 := [mem32]
        ds := [mem32 + 2]
les reg16, mem32:
        reg16 := [mem32]
        es := [mem32 + 2]
```

Since the LxS instructions load the 80x86's segment registers, you must not use these instructions for arbitrary purposes. Keep in mind that these instructions load the four bytes at a given memory location into the register pair; they do not load the address of a variable into the register pair (i.e., this instruction does not have an immediate mode). The LxS instructions do not affect any of the 80x86's flag bits.

### 8.1.4 The Lea Instruction

The lea (Load Effective Address) instruction is another instruction used to prepare pointer values. The lea instruction takes the form:

```
lea   dest, source
```

The specific forms on the 80x86 are

```
lea   reg16, mem
```

It loads the specified 16 bit general purpose register with the effective address of the specified memory location. The effective address is the final memory address obtained after all addressing mode computations. For example, lea ax, ds:[1234h] loads the ax register with the address of memory location 1234h; here it just loads the ax register with the value 1234h. If you think about it for a moment, this isn't a very exciting operation. After all, the mov ax, immediate_data instruction can do this. So why bother with the lea instruction at all? Well, there are many other forms of a memory operand besides displacement-only operands. Consider the following lea instructions:

```
lea   ax, [bx]
lea   bx, 3[bx]
lea   ax, 3[bx]
lea   bx, 4[bp+si]
lea   ax, -123[di]
```

The lea ax, [bx] instruction copies the address of the expression [bx] into the ax register. Since the effective address is the value in the bx register, this instruction copies bx's value into the ax register. Again, this instruction isn't very interesting because mov can do the same thing, even faster.

The lea bx,3[bx] instruction copies the effective address of 3[bx] into the bx register. Since this effective address is equal to the current value of bx plus three, this lea instruction effectively adds three to the bx register. There is an add instruction that will let you add three to the bx register, so again, the lea instruction is superfluous for this purpose.

The third lea instruction above shows where lea really begins to shine. lea ax, 3[bx] copies the address of the memory location 3[bx] into the ax register; i.e., it adds three with the value in the bx register and moves the sum into ax. This is an excellent example of how you can use the lea instruction to do a mov operation and an addition with a single instruction.

The final two instructions above, lea bx,4[bp+si] and lea ax,-123[di] provide additional examples of lea instructions that are more efficient than their mov/add counterparts.

The (real) purpose of lea is to load a register with a memory address. For example, lea bx, 128[bp+di] sets up bx with the address of the byte referenced by 128[BP+DI]. As it turns out, an instruction of the form mov al,[bx] runs faster than an instruction of the form mov al,128[bp+di]. If this instruction executes several times, it is probably more efficient to load the effective address of 128[bp+di] into the bx register and use the [bx] addressing mode. This is a common optimization in high performance programs.

The lea instruction does not affect any of the 80x86's flag bits.

### 8.1.5 The Push and Pop Instructions

The 80x86 push and pop instructions manipulate data on the 80x86's hardware stack. There are 19 varieties of the push and pop instructions, they are

```
push    reg16
pop     reg16
push    segreg
pop     segreg    (except CS)
push    memory
pop     memory
pushf
popf
```

The first two instructions push and pop a 16 bit general purpose register. This is a compact (one byte) version designed specifically for registers. Note that there is a second form that provides a mod-reg-r/m byte that could push registers as well; most assemblers only use that form for pushing the value of a memory location.

The second pair of push/pop instructions let you push or pop an 80x86 segment register. You can only push the cs register (popping the cs register would create some interesting program flow control problems).

The pushf and popf instructions allow you to push/pop the processor status register (the flags). Note that these two instructions provide a mechanism to modify the 80x86's trace flag. "So what do these instructions do?" you're probably asking by now. The push instructions move data onto the 80x86 hardware stack and the pop instructions move data from the stack to memory or to a register. The following is an algorithmic description of each instruction:

push instructions (16 bits):

SP := SP - 2

[SS:SP] := 16 bit operand (store result at location SS:SP.)

pop instructions (16 bits):

16-bit operand := [SS:SP]

SP := SP + 2

push instructions (32 bits):

SP := SP - 4

[SS:SP] := 32 bit operand

pop instructions (32 bits):

32 bit operand := [SS:SP]

SP := SP + 4

You can treat the pusha/pushad and popa/popad instructions as equivalent to the corresponding sequence of 16 or 32 bit push/pop operations (e.g., push ax, push cx, push dx, push bx, etc.).

Notice three things about the 80x86 hardware stack. First, it is always in the stack segment (wherever ss points). Second, the stack grows down in memory. That is, as you push values onto the stack the CPU stores them into successively lower memory locations. Finally, the 80x86 hardware stack pointer (ss:sp) always contains the address of the value on the top of the stack (the last value pushed on the stack).

You can use the 80x86 hardware stacks for temporarily saving registers and variables, passing parameters to a procedure, allocating storage for local variables, and other uses. The push and pop instructions are extremely valuable for manipulating these items on the stack. You'll get a chance to see how to use them later in this text.

Most of the push and pop instructions do not affect any of the flags in the 80x86 processor status register. The popf instructions, by their very nature, can modify all the flag bits in the 80x86 processor status register (flags register). Pushf push the flags onto the stack, but they do not change any flags while doing so.

All pushes and pops are 16 or 32 bit operations. There is no (easy) way to push a single eight bit value onto the stack. To push an eight bit value you would need to load it into the H.O. byte of a 16 bit register, push that register, and then add one to the stack pointer. On all processors except the 8088, this would slow future stack access since sp now contains an odd address, misaligning any further pushes and pops. Therefore, most programs push or pop 16 bits, even when dealing with eight bit values.

Although it is relatively safe to push an eight bit memory variable, be careful when popping the stack to an eight bit memory location. Pushing an eight bit variable with push word ptr ByteVar pushes two bytes, the byte in the variable ByteVar and the byte immediately following it. Your code can simply ignore the extra byte this instruction pushes onto the stack. Popping such values is not quite so straight forward. Generally, it doesn't hurt if you push these two bytes. However, it can be a disaster if you pop a value and wipe out the following byte in memory. There are only two solutions to this problem. First, you could pop the 16 bit value into a register like ax and then store the L.O. byte of that register into the byte variable. The second solution is to reserve an extra byte of padding after the byte variable to hold the whole word you will pop. Most programs use the former approach.

### 8.1.6 The LAHF and SAHF Instructions

The lahf (load ah from flags) and sahf (store ah into flags) instructions are archaic instructions included in the 80x86's instruction set to help improve compatibility with Intel's older 8080 mP chip. As such, these instructions have very little use in modern day 80x86 programs. The lahf instruction does not affect any of the flag bits. The sahf instruction, by its very nature, modifies the S, Z, A, P, and C bits in the processor status register. These instructions do not require any operands and you use them in the following manner:

```
sahf
lahf
```

Sahf only affects the L.O. eight bits of the flags register. Likewise, lahf only loads the L.O. eight bits of the flags register into the AH register. These instructions do not deal with the overflow, direction, interrupt disable, or trace flags. The fact that these instructions do not deal with the overflow flag is an important limitation.

## 8.2 CONVERSIONS

The 80x86 instruction set provides several conversion instructions. They include cbw, cwd and xlat. Most of these instructions sign or zero extend values, the last two convert between storage formats and translate values via a lookup table. These instructions take the general form:

```
cbw
cwd
xlat                    ;Special form allows an operand.
```

### 8.2.1 The CBW and CWD Instructions

These instructions zero and sign extend values. The cbw and cwd instructions are available on all 80x86 processors.

The cbw (convert byte to word) instruction sign extends the eight bit value in al to ax. That is, it copies bit seven of AL throughout bits 8-15 of ax. This instruction is especially important before executing an eight bit division (as you'll see in the section "Arithmetic Instructions"). This instruction requires no operands and you use it as follows:

```
cbw
```

The cwd (convert word to double word) instruction sign extends the 16 bit value in ax to 32 bits and places the result in dx:ax. It copies bit 15 of ax throughout the bits in dx. It is available on all 80x86 processors which explains why it doesn't sign extend the value into eax. Like the cbw instruction, this instruction is very important for division operations. Cwd requires no operands and you use it as follows

```
cwd
```

If you want to sign extend an eight bit value to 32 or 64 bits using these instructions, you could use sequences like the following:

```
; Sign extend al to dx:ax
        cbw
        cwd
```

```
; Sign extend al to eax
        cbw
; Sign extend al to edx:eax
        cbw
```

You can also use the movsx for sign extensions from eight to sixteen or thirty-two bits.

### 8.2.2 The Xlat Instruction

The xlat instruction translates the value in the al register based on a lookup table in memory. It does the following:

```
temp := al+bx
al := ds:[temp]
```

that is, bx points at a table in the current data segment. Xlat replaces the value in al with the byte at the offset originally in al. If al contains four, xlat replaces the value in al with the fifth item (offset four) within the table pointed at by ds:bx. The xlat instruction takes the form:

```
xlat
```

Typically it has no operand. You can specify one but the assembler virtually ignores it. The only purpose for specifying an operand is so you can provide a segment override prefix:

```
xlat        es:Table
```

This tells the assembler to emit an es: segment prefix byte before the instruction. You must still load bx with the address of Table; the form above does not provide the address of Table to the instruction. Only the segment override prefix in the operand is significant.

## 8.3 ARITHMETIC INSTRUCTIONS

The 80x86 provides many arithmetic operations: addition, subtraction, negation, multiplication, division/modulo (remainder), and comparing two values. The instructions that handle these operations are add, adc, sub, sbb, mul, imul, div, idiv, cmp, neg, inc, dec, and some miscellaneous conversion instructions: aaa, aad, aam, aas, daa, and das. The following sections describe these instructions in detail.

The generic forms for these instructions are

```
add     dest, src               dest := dest + src
adc     dest, src               dest := dest + src + C
SUB     dest, src               dest := dest - src
sbb     dest, src               dest := dest - src - C
mul     src                     acc := acc * src
imul    src                     acc := acc * src
imul    dest, src1, imm_src     dest := src1 * imm_src
```

| | | |
|---|---|---|
| imul | dest, imm_src | dest := dest * imm_src |
| imul | dest, src | dest := dest * src |
| div | src | acc := xacc /-mod src |
| idiv | src | acc := xacc /-mod src |
| cmp | dest, src | dest - src (and set flags) |
| neg | dest | dest := - dest |
| inc | dest | dest := dest + 1 |
| dec | dest | dest := dest - 1 |

### 8.3.1 The Addition Instructions: ADD, ADC, INC, AAA, AND DAA

These instructions take the forms:

```
add     reg, reg
add     reg, mem
add     mem, reg
add     reg, immediate data
add     mem, immediate data
add     eax/ax/al, immediate data
adc forms are identical to ADD.
inc     reg
inc     mem
inc     reg16
aaa
daa
```

Note that the aaa and daa instructions use the implied addressing mode and allow no operands.

#### *1. The ADD And ADC Instructions*

The syntax of add and adc (add with carry) is similar to mov. Like mov, there are special forms for the ax/eax register that are more efficient. Unlike mov, you cannot add a value to a segment register with these instructions.

The add instruction adds the contents of the source operand to the destination operand. For example, add ax, bx adds bx to ax leaving the sum in the ax register. Add computes dest :=dest+source while adc computes dest :=dest+source+C where C represents the value in the carry flag. Therefore, if the carry flag is clear before execution, adc behaves exactly like the add instruction.

Both instructions affect the flags identically. They set the flags as follows:

- The overflow flag denotes a signed arithmetic overflow.
- The carry flag denotes an unsigned arithmetic overflow.

- The sign flag denotes a negative result (i.e., the H.O. bit of the result is one).
- The zero flag is set if the result of the addition is zero.
- The auxiliary carry flag contains one if a BCD overflow out of the L.O. nibble occurs.
- The parity flag is set or cleared depending on the parity of the L.O. eight bits of the result. If there are an even number of one bits in the result, the ADD instructions will set the parity flag to one (to denote even parity). If there are an odd number of one bits in the result, the ADD instructions clear the parity flag (to denote odd parity).

The add and adc instructions do not affect any other flags.

Since there are no memory to memory additions, you must load memory operands into registers if you want to add two variables together. The following code examples demonstrate possible forms for the add instruction:

```
; J:= K + M
        mov    ax, K
        add    ax, M
        mov    J, ax
```

If you want to add several values together, you can easily compute the sum in a single register:

```
; J := K + M + N + P
        mov    ax, K
        add    ax, M
        add    ax, N
        add    ax, P
        mov    J, ax
```

One thing that beginning assembly language programmers often forget is that you can add a register to a memory location. The 80x86 is a CISC processor that allows you to use memory addressing modes with various instructions like add. It is often more efficient to take advantages of the 80x86's memory addressing capabilities

```
; J := K + J
        mov    ax, K        ;This works because addition is
        add    J, ax        ; commutative!
; Often, beginners will code the above as one of the following two sequences.
; This is unnecessary!
        mov    ax, J        ;Really BAD way to compute
        mov    bx, K        ; J := J + K.
        add    ax, bx
        mov    J, ax
```

```
mov    ax, J        ;Better, but still not a good way to
add    ax, K        ; compute J := J + K
mov    J, ax
```

Of course, if you want to add a constant to a memory location, you only need a single instruction. The 80x86 lets you directly add a constant to memory:

```
; J := J + 2
        add    J, 2
```

There are special forms of the add and adc instructions that add an immediate constant to the al, ax, or eax register. These forms are shorter than the standard add reg, immediate instruction. Other instructions also provide shorter forms when using these registers; therefore, you should try to keep computations in the accumulator registers (al, ax, and eax) as much as possible.

```
add    bl, 2        ;Three bytes long
add    al, 2        ;Two bytes long
add    bx, 2        ;Four bytes long
add    ax, 2        ;Three bytes long
etc.
```

Another optimization concerns the use of small signed constants with the add and adc instructions. If a value is in the range -128,,+127, the add and adc instructions will sign extend an eight bit immediate constant to the necessary destination size (eight, sixteen, or thirty-two bits). Therefore, you should try to use small constants, if possible, with the add and adc instructions.

### 2. *The Inc Instruction*

The inc (increment) instruction adds one to its operand. Except for the carry flag, inc sets the flags the same way as add operand, 1 would.

Note that there are two forms of inc for 16 bit registers. They are the inc reg and inc reg16 instructions. The inc reg and inc mem instructions are the same. This instruction consists of an opcode byte followed by a mod-reg-r/m byte. The inc reg16 instruction has a single byte opcode. Therefore, it is shorter and usually faster.

The inc operand may be an eight bit, sixteen bit register or memory location.

The inc instruction is more compact and often faster than the comparable add reg, 1 or add mem, 1 instruction. Indeed, the inc reg16 instruction is one byte long, so it turns out that two such instructions are shorter than the comparable add reg, 1 instruction; however, the two increment instructions will run slower on most modern members of the 80x86 family.

The inc instruction is very important because adding one to a register is a very common operation. Incrementing loop control variables or indices into an array is a very common operation, perfect for the inc instruction. The fact that inc does not affect the carry flag is very important. This allows you to increment array indices without affecting the result of a multiprecision arithmetic operation.

### 3. The AAA and DAA Instructions

The aaa (ASCII adjust after addition) and daa (decimal adjust for addition) instructions support BCD arithmetic. Beyond this chapter, this text will not cover BCD or ASCII arithmetic since it is mainly for controller applications, not general purpose programming applications. BCD values are decimal integer coded in binary form with one decimal digit (0..9) per nibble. ASCII (numeric) values contain a single decimal digit per byte, the H.O. nibble of the byte should contain zero.

The aaa and daa instructions modify the result of a binary addition to correct it for ASCII or decimal arithmetic. For example, to add two BCD values, you would add them as though they were binary numbers and then execute the daa instruction afterwards to correct the results. Likewise, you can use the aaa instruction to adjust the result of an ASCII addition after executing an add instruction. Please note that these two instructions assume that the add operands were proper decimal or ASCII values. If you add binary (non-decimal or non-ASCII) values together and try to adjust them with these instructions, you will not produce correct results.

The choice of the name "ASCII arithmetic" is unfortunate, since these values are not true ASCII characters. A name like "unpacked BCD" would be more appropriate. However, Intel uses the name ASCII, so this text will do so as well to avoid confusion. However, you will often hear the term "unpacked BCD" to describe this data type.

Aaa (which you generally execute after an add, adc, or xadd instruction) checks the value in al for BCD overflow. It works according to the following basic algorithm:

```
if ( (al and 0Fh) > 9 or (AuxC =1) ) then
    if (8088 or 8086) then
        al := al + 6
    else
        ax := ax + 6
    endif
    ah := ah + 1
    AuxC := 1          ;Set auxilliary carry
    Carry := 1         ; and carry flags.
else
    AuxC := 0          ;Clear auxilliary carry
    Carry := 0         ; and carry flags.
endif
al := al and 0Fh
```

The aaa instruction is mainly useful for adding strings of digits where there is exactly one decimal digit per byte in a string of numbers. This text will not deal with BCD or ASCII numeric strings, so you can safely ignore this instruction for now. Of course, you can use the aaa instruction any time you need to use the algorithm above, but that would probably be a rare situation. The daa instruction functions like aaa except it handles packed BCD (binary code decimal) values rather than the one digit per byte unpacked values aaa handles. As for aaa, daa's main purpose is to add strings of BCD digits (with two digits per byte). The algorithm for daa is

```
if ( (AL and 0Fh) > 9 or (AuxC = 1)) then
        al := al + 6
        AuxC := 1              ;Set Auxilliary carry.
endif
if ( (al > 9Fh) or (Carry = 1)) then
        al := al + 60h
        Carry := 1;            ;Set carry flag.
endif
```

### 8.3.2 The Subtraction Instructions: SUB, SBB, DEC, AAS, AND DAS

The sub (subtract), sbb (subtract with borrow), dec (decrement), aas (ASCII adjust for subtraction), and das (decimal adjust for subtraction) instructions work as you expect. Their syntax is very similar to that of the add instructions:

```
sub   reg, reg
sub   reg, mem
sub   mem, reg
sub   reg, immediate data
sub   mem, immediate data
sub   eax/ax/al, immediate data
sbb forms are identical to sub.
dec   reg
dec   mem
dec   reg16
aas
das
```

The sub instruction computes the value dest := dest - src. The sbb instruction computes dest := dest - src - C. Note that subtraction is not commutative. If you want to compute the result for dest := src - dest you will need to use several instructions, assuming you need to preserve the source operand).

One last subject worth discussing is how the sub instruction affects the 80x86 flags register. The sub, sbb, and dec instructions affect the flags as follows:

- They set the zero flag if the result is zero. This occurs only if the operands are equal for sub and sbb. The dec instruction sets the zero flag only when it decrements the value one.

- These instructions set the sign flag if the result is negative.
- These instructions set the overflow flag if signed overflow/underflow occurs.
- They set the auxiliary carry flag as necessary for BCD/ASCII arithmetic.
- They set the parity flag according to the number of one bits appearing in the result value.
- The sub and sbb instructions set the carry flag if an unsigned overflow occurs. Note that the dec instruction does not affect the carry flag.

The aas instruction, like its aaa counterpart, lets you operate on strings of ASCII numbers with one decimal digit (in the range 0..9) per byte. You would use this instruction after a sub or sbb instruction on the ASCII value. This instruction uses the following algorithm:

```
if ( (al and 0Fh) > 9 or AuxC = 1) then
        al := al - 6
        ah := ah - 1
        AuxC := 1       ;Set auxilliary carry
        Carry := 1      ; and carry flags.
else
        AuxC := 0       ;Clear Auxilliary carry
        Carry := 0      ; and carry flags.
endif
al := al and 0Fh
```

The das instruction handles the same operation for BCD values, it uses the following algorithm:

```
if ( (al and 0Fh) > 9 or (AuxC = 1)) then
        al := al -6
        AuxC = 1
endif
if (al > 9Fh or Carry = 1) then
        al := al - 60h
        Carry := 1          ;Set the Carry flag.
endif
```

Since subtraction is not commutative, you cannot use the sub instruction as freely as the add instruction. The following examples demonstrate some of the problems you may encounter.

```
; J := K - J
            mov   ax, K        ;This is a nice try, but it computes
            sub   J, ax        ; J := J - K, subtraction isn't
                               ; commutative!
            mov   ax, K        ;Correct solution.
```

```
        sub   ax, J
        mov   J, ax
; J := J - (K + M) — Don't forget this is equivalent to J := J - K - M
        mov   ax, K        ;Computes AX := K + M
        add   ax, M
        sub   J, ax        ;Computes J := J - (K + M)
        mov   ax, J        ;Another solution, though less
        sub   ax, K        ;Efficient
        sub   ax, M
        mov   J, ax
```

Note that the sub and sbb instructions, like add and adc, provide short forms to subtract a constant from an accumulator register (al, ax, or eax). For this reason, you should try to keep arithmetic operations in the accumulator registers as much as possible. The sub and sbb instructions also provide a shorter form when subtracting constants in the range -128..+127 from a memory location or register. The instruction will automatically sign extend an eight bit signed value to the necessary size before the subtraction occurs.

In practice, there really isn't a need for an instruction that subtracts a constant from a register or memory location - adding a negative value achieves the same result. Nevertheless, Intel provides a subtract immediate instruction.

After the execution of a sub instruction, the condition code bits (carry, sign, overflow, and zero) in the flags register contain values you can test to see if one of sub's operands is equal, not equal, less than, less than or equal, greater than, or greater than or equal to the other operand. See the cmp instruction for more details.

### 8.3.3 The CMP Instruction

The cmp (compare) instruction is identical to the sub instruction with one crucial difference - it does not store the difference back into the destination operand. The syntax for the cmp instruction is very similar to sub, the generic form is

```
        cmp   dest, src
```

The specific forms are

```
        cmp   reg, reg
        cmp   reg, mem
        cmp   mem, reg
        cmp   reg, immediate data
        cmp   mem, immediate data
        cmp   eax/ax/al, immediate data
```

The cmp instruction updates the 80x86's flags according to the result of the subtraction operation (dest - src). You can test the result of the comparison by checking the appropriate flags in the flags register. For details on how this is done, see "The "Set on Condition" Instructions" and "The Conditional Jump Instructions".

Usually you'll want to execute a conditional jump instruction after a cmp instruction. This two step process, comparing two values and setting the flag bits then testing the flag bits with the conditional jump instructions, is a very efficient mechanism for making decisions in a program.

Probably the first place to start when exploring the cmp instruction is to take a look at exactly how the cmp instruction affects the flags. Consider the following cmp instruction:

```
cmp        ax, bx
```

This instruction performs the computation ax-bx and sets the flags depending upon the result of the computation. The flags are set as follows:

Z: The zero flag is set if and only if ax = bx. This is the only time ax-bx produces a zero result. Hence, you can use the zero flag to test for equality or inequality.

S: The sign flag is set to one if the result is negative. At first glance, you might think that this flag would be set if ax is less than bx but this isn't always the case. If ax=7FFFh and bx=-1 (0FFFFh) subtracting ax from bx produces 8000h, which is negative (and so the sign flag will be set). So, for signed comparisons anyway, the sign flag doesn't contain the proper status. For unsigned operands, consider ax=0FFFFh and bx=1. Ax is greater than bx but their difference is 0FFFEh which is still negative. As it turns out, the sign flag and the overflow flag, taken together, can be used for comparing two signed values.

O: The overflow flag is set after a cmp operation if the difference of ax and bx produced an overflow or underflow. As mentioned above, the sign flag and the overflow flag are both used when performing signed comparisons.

C: The carry flag is set after a cmp operation if subtracting bx from ax requires a borrow. This occurs only when ax is less than bx where ax and bx are both unsigned values.

The cmp instruction also affects the parity and auxiliary carry flags, but you'll rarely test these two flags after a compare operation. Given that the cmp instruction sets the flags in this fashion, you can test the comparison of the two operands with the following flags:

```
cmp Oprnd1, Oprnd2
```

| **Condition Code Settings After CMP** | |
|---|---|
| Unsigned operands: | Signed operands: |
| Z: equality/inequality | Z: equality/inequality |
| C: Oprnd1 < Oprnd2 (C=1) Oprnd1 >= Oprnd2 (C=0) | C: no meaning |
| S: no meaning | S: see below |
| O: no meaning | O: see below |

For signed comparisons, the S (sign) and O (overflow) flags, taken together, have the following meaning:

If ((S=0) and (O=1)) or ((S=1) and (O=0)) then Oprnd1 < Oprnd2 when using a signed comparison.

If ((S=0) and (O=0)) or ((S=1) and (O=1)) then Oprnd1 >= Oprnd2 when using a signed comparison.

To understand why these flags are set in this manner, consider the following examples:

| Oprnd1 | minus | Oprnd2 | S | O |
|---|---|---|---|---|
| ------ | ------ | - | - | |
| 0FFFF (-1) | - | 0FFFE (-2) | 0 | 0 |
| 08000 | - | 00001 | 0 | 1 |
| 0FFFE (-2) | - | 0FFFF (-1) | 1 | 0 |
| 07FFF (32767) | - | 0FFFF (-1) | 1 | 1 |

Remember, the cmp operation is really a subtraction, therefore, the first example above computes (-1)-(-2) which is (+1). The result is positive and an overflow did not occur so both the S and O flags are zero. Since (S xor O) is zero, Oprnd1 is greater than or equal to Oprnd2.

In the second example, the cmp instruction would compute (-32768)-(+1) which is (-32769). Since a 16-bit signed integer cannot represent this value, the value wraps around to 7FFFh (+32767) and sets the overflow flag. Since the result is positive (at least within the confines of 16 bits) the sign flag is cleared. Since (S xor O) is one here, Oprnd1 is less than Oprnd2.

In the third example above, cmp computes (-2)-(-1) which produces (-1). No overflow occurred so the O flag is zero, the result is negative so the sign flag is one. Since (S xor O) is one, Oprnd1 is less than Oprnd2.

In the fourth (and final) example, cmp computes (+32767)-(-1). This produces (+32768), setting the overflow flag. Furthermore, the value wraps around to 8000h (-32768) so the sign flag is set as well. Since (S xor O) is zero, Oprnd1 is greater than or equal to Oprnd2.

### 8.3.4 The Neg Instruction

The neg (negate) instruction takes the two's complement of a byte or word. It takes a single (destination) operation and negates it. The syntax for this instruction is

```
neg        dest
```

It computes the following:

```
dest := 0 - dest
```

This effectively reverses the sign of the destination operand.

If the operand is zero, its sign does not change, although this clears the carry flag. Negating any other value sets the carry flag. Negating a byte containing -128, a word containing -32,768, or a double word containing -2,147,483,648 does not change the operand, but will set the overflow flag. Neg always updates the A, S, P, and Z flags as though you were using the sub instruction.

The allowable forms are:

```
        neg   reg
        neg   mem
```

The operands may be eight, sixteen, or (on the 80386 and later) thirty-two bit values.

Some examples:

```
; J := - J
        neg   J
; J := -K
        mov   ax, K
        neg   ax
        mov   J, ax
```

### 8.3.5 The Multiplication Instructions: MUL, IMUL, AND AAM

The multiplication instructions provide you with your first taste of irregularity in the 8086's instruction set. Instructions like add, adc, sub, sbb, and many others in the 8086 instruction set use a mod-reg-r/m byte to support two operands. Unfortunately, there aren't enough bits in the 8086's opcode byte to support all instructions, so the 8086 uses the reg bits in the mod-reg-r/m byte as an opcode extension. For example, inc, dec, and neg do not require two operands, so the 80x86 CPUs use the reg bits as an extension to the eight bit opcode. This works great for single operand instructions, allowing Intel's designers to encode several instructions (eight, in fact) with a single opcode.

Unfortunately, the multiply instructions require special treatment and Intel's designers were still short on opcodes, so they designed the multiply instructions to use a single operand. The reg field contains an opcode extension rather than a register value. Of course, multiplication is a two operand function. The 8086 always assumes the accumulator (al, ax, or eax) is the destination operand. This irregularity makes using multiplication on the 8086 a little more difficult than other instructions because one operand has to be in the accumulator. Intel adopted this unorthogonal approach because they felt that programmers would use multiplication far less often than instructions like add and sub.

There are two forms of the multiply instruction: an unsigned multiplication (mul) and a signed multiplication (imul). Unlike addition and subtraction, you need separate instructions for these two operations.

The multiply instructions take the following forms:

Unsigned Multiplication:

```
        mul   reg
        mul   mem
```

Signed (Integer) Multiplication:

```
imul   reg
imul   mem
```

BCD Multiplication Operations:

```
aam
```

Mul, available on all processors, multiplies unsigned eight, sixteen, or thirty-two bit operands. Note that when multiplying two n-bit values, the result may require as many as 2*n bits. Therefore, if the operand is an eight bit quantity, the result will require sixteen bits. Likewise, a 16 bit operand produces a 32 bit result and a 32 bit operand requires 64 bits for the result.

The mul instruction, with an eight bit operand, multiplies the al register by the operand and stores the 16 bit result in ax. So

```
        mul       operand8
or      imul      operand8
```

computes:

ax := al * operand8

"*" represents an unsigned multiplication for mul and a signed multiplication for imul.

If you specify a 16 bit operand, then mul and imul compute:

dx:ax := ax * operand16

"*" has the same meanings as above and dx:ax means that dx contains the H.O. word of the 32 bit result and ax contains the L.O. word of the 32 bit result.

If you specify a 32 bit operand, then mul and imul compute the following:

edx:eax := eax * operand32

"*" has the same meanings as above and edx:eax means that edx contains the H.O. double word of the 64 bit result and eax contains the L.O. double word of the 64 bit result.

If an 8x8, 16x16, or 32x32 bit product requires more than eight, sixteen, or thirty-two bits (respectively), the mul and imul instructions set the carry and overflow flags.

Mul and imul scramble the A, P, S, and Z flags. Especially note that the sign and zero flags do not contain meaningful values after the execution of these two instructions.

Imul (integer multiplication) operates on signed operands. There are many different forms of this instruction as Intel attempted to generalize this instruction with successive processors. Expanded to show allowable operand sizes, they are

```
imul   operand1, operand2, immediate   ;General form
imul   reg16, reg16, immediate8
imul   reg16, reg16, immediate16
imul   reg16, mem16, immediate8
imul   reg16, mem16, immediate16
```

imul reg16, immediate8

imul reg16, immediate16

The imul reg, immediate instructions are a special syntax the assembler provides. The encodings for these instructions are the same as imul reg, reg, immediate. The assembler simply supplies the same register value for both operands.

These instructions compute:

operand1 := operand2 * immediate

operand1 := operand1 * immediate

Besides the number of operands, there are several differences between these forms and the single operand mul/imul instructions:

There isn't an 8x8 bit multiplication available (the immediate8 operands simply provide a shorter form of the instruction. Internally, the CPU sign extends the operand to 16 or 32 bits as necessary).

These instructions do not produce a 2*n bit result. That is, a 16x16 multiply produces a 16 bit result. These instructions set the carry and overflow flags if the result does not fit into the destination register.

### 8.3.6 The Division Instructions: DIV, IDIV, AND AAD

The 80x86 divide instructions perform a 32/16 division or a 16/8 division. These instructions take the form:

div reg For unsigned division

div mem

idiv reg For signed division

idiv mem

aad ASCII adjust for division

The div instruction computes an unsigned division. If the operand is an eight bit operand, div divides the ax register by the operand leaving the quotient in al and the remainder (modulo) in ah. If the operand is a 16 bit quantity, then the div instruction divides the 32 bit quantity in dx:ax by the operand leaving the quotient in ax and the remainder in .

You cannot, on the 80x86, simply divide one eight bit value by another. If the denominator is an eight bit value, the numerator must be a sixteen bit value. If you need to divide one unsigned eight bit value by another, you must zero extend the numerator to sixteen bits. You can accomplish this by loading the numerator into the al register and then moving zero into the ah register. Then you can divide ax by the denominator operand to produce the correct result. Failing to zero extend al before executing div may cause the 80x86 to produce incorrect results!

When you need to divide two 16 bit unsigned values, you must zero extend the ax register (which contains the numerator) into the dx register. Just load the immediate value zero into the dx register. If you need to divide one 32 bit value by another, you must zero extend the eax register into edx (by loading a zero into edx) before the division.

When dealing with signed integer values, you will need to sign extend al to ax, ax to dx or eax into edx before executing idiv. To do so, use the cbw, cwd, cdq, or movsx instructions. If the H.O. byte or word does not already contain significant bits, then you must sign extend the value in the accumulator (al/ax/eax) before doing the idiv operation. Failure to do so may produce incorrect results.

There is one other catch to the 80x86's divide instructions: you can get a fatal error when using this instruction. First, of course, you can attempt to divide a value by zero. Furthermore, the quotient may be too large to fit into the eax, ax, or al register. For example, the 16/8 division "8000h / 2" produces the quotient 4000h with a remainder of zero. 4000h will not fit into eight bits. If this happens, or you attempt to divide by zero, the 80x86 will generate an int 0 trap. This usually means BIOS will print "division by zero" or "divide error" and abort your program. If this happens to you, chances are you didn't sign or zero extend your numerator before executing the division operation. Since this error will cause your program to crash, you should be very careful about the values you select when using division.

The auxiliary carry, carry, overflow, parity, sign, and zero flags are undefined after a division operation. If an overflow occurs (or you attempt a division by zero) then the 80x86 executes an INT 0 (interrupt zero).

The aad (ASCII Adjust before Division) instruction is another unpacked decimal operation. It splits apart unpacked binary coded decimal values before an ASCII division operation. Although this text will not cover BCD arithmetic, the aad instruction is useful for other operations. The algorithm that describes this instruction is

```
al := ah*10 + al
ah := 0
```

This instruction is quite useful for converting strings of digits into integer values.

The following examples show how to divide one sixteen bit value by another.

```
; J := K / M (unsigned)
        mov   ax, K   ;Get dividend
        mov   dx, 0   ;Zero extend unsigned value in AX to DX.
    < In practice, we should verify that M does not contain zero here >
        div   M
        mov   J, ax
; J := K / M (signed)
        mov   ax, K   ;Get dividend
        cwd           ;Sign extend signed value in AX to DX.
    < In practice, we should verify that M does not contain zero here >
        idiv  M
```

```
        mov    J, ax
; J := (K*M)/P
        mov    ax, K   ;Note that the imul instruction produces
        imul   M       ; a 32 bit result in DX:AX, so we don't
        idiv   P       ; need to sign extend AX here.
```

## 8.4 LOGICAL, SHIFT AND ROTATE INSTRUCTIONS

The 80x86 family provides five logical instructions, four rotate instructions, and three shift instructions. The logical instructions are and, or, xor, test, and not; the rotates are ror, rol, rcr, and rcl; the shift instructions are shl/sal, shr, and sar.

### 8.4.1 The Logical Instructions: AND, OR, XOR, AND NOT

The 80x86 logical instructions operate on a bit-by-bit basis. Both eight, sixteen, and thirty-two bit versions of each instruction exist. The and, not, or, and xor instructions do the following:

```
and    dest, source        ;dest := dest and source
or     dest, source        ;dest := dest or source
xor    dest, source        ;dest := dest xor source
not    dest                ;dest := not dest
```

The specific variations are

```
and    reg, reg
and    mem, reg
and    reg, mem
and    reg, immediate data
and    mem, immediate data
and    eax/ax/al, immediate data
or uses the same formats as AND
xor uses the same formats as AND
not    register
not    mem
```

Except not, these instructions affect the flags as follows:

- They clear the carry flag.
- They clear the overflow flag.
- They set the zero flag if the result is zero, they clear it otherwise.
- They copy the H.O. bit of the result into the sign flag.
- They set the parity flag according to the parity (number of one bits) in the result.

- They scramble the auxiliary carry flag.

The not instruction does not affect any flags.

Testing the zero flag after these instructions is particularly useful. The and instruction sets the zero flag if the two operands do not have any ones in corresponding bit positions (since this would produce a zero result); for example, if the source operand contained a single one bit, then the zero flag will be set if the corresponding destination bit is zero, it will be one otherwise. The or instruction will only set the zero flag if both operands contain zero. The xor instruction will set the zero flag only if both operands are equal. Notice that the xor operation will produce a zero result if and only if the two operands are equal. Many programmers commonly use this fact to clear a sixteen bit register to zero since an instruction of the form

```
xor     reg16, reg16
```

is shorter than the comparable mov reg,0 instruction.

Like the addition and subtraction instructions, the and, or, and xor instructions provide special forms involving the accumulator register and immediate data. These forms are shorter and sometimes faster than the general "register, immediate" forms. Although one does not normally think of operating on signed data with these instructions, the 80x86 does provide a special form of the "reg/mem, immediate" instructions that sign extend a value in the range -128..+127 to sixteen or thirty-two bits, as necessary.

### 8.4.2 The Shift Instructions: SHL/SAL, SHR, SAR

The 80x86 supports three different shift instructions (shl and sal are the same instruction): shl (shift left), sal (shift arithmetic left), shr (shift right), and sar (shift arithmetic right). The shift instructions move bits around in a register or memory location. The general format for a shift instruction is

```
shl   dest, count
sal   dest, count
shr   dest, count
sar   dest, count
```

Dest is the value to shift and count specifies the number of bit positions to shift. For example, the shl instruction shifts the bits in the destination operand to the left the number of bit positions specified by the count operand.

The specific forms for these instructions are

```
shl   reg, 1
shl   mem, 1
shl   reg, cl
shl   mem, cl
```

sal is a synonym for shl and uses the same formats.

shr uses the same formats as shl.

sar uses the same formats as shl.

For 8088 and 8086 CPUs, the number of bits to shift is either "1" or the value in cl. On 80286 and later processors you can use an eight bit immediate constant. Of course, the value in cl or the immediate constant should be less than or equal to the number of bits in the destination operand. It would be a waste of time to shift left al by nine bits (eight would produce the same result, as you will soon see). Algorithmically, you can think of the shift operations with a count other than one as follows:

```
for temp := 1 to count do
        shift dest, 1
```

There are minor differences in the way the shift instructions treat the overflow flag when the count is not one, but you can ignore this most of the time.

The shl, sal, shr, and sar instructions work on eight, sixteen, and thirty-two bit operands. The shld and shrd instructions work on 16 and 32 bit destination operands only.

### 1. SHL/SAL

The shl and sal mnemonics are synonyms. They represent the same instruction and use identical binary encodings. These instructions move each bit in the destination operand one bit position to the left the number of times specified by the count operand. Zeros fill vacated positions at the L.O. bit; the H.O. bit shifts into the carry flag:

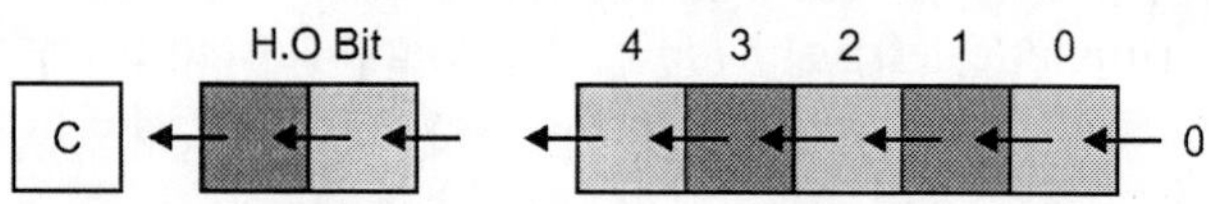


The shl/sal instruction sets the condition code bits as follows:

- If the shift count is zero, the shl instruction doesn't affect any flags.
- The carry flag contains the last bit shifted out of the H.O. bit of the operand.
- The overflow flag will contain one if the two H.O. bits were different prior to a single bit shift. The overflow flag is undefined if the shift count is not one.
- The zero flag will be one if the shift produces a zero result.
- The sign flag will contain the H.O. bit of the result.
- The parity flag will contain one if there are an even number of one bits in the L.O. byte of the result.
- The A flag is always undefined after the shl/sal instruction.

The shift left instruction is especially useful for packing data. For example, suppose you have two nibbles in al and ah that you want to combine. You could use the following code to do this:

```
or      al, ah  ;Merge in H.O. four bits.
```

Of course, al must contain a value in the range 0..F for this code to work properly (the shift left operation automatically clears the L.O. four bits of ah before the or instruction). If the H.O. four bits of al are not zero before this operation, you can easily clear them with an and instruction:

```
shl   ah, 4      ;Move L.O. bits to H.O. position.
and   al, 0Fh    ;Clear H.O. four bits.
or    al, ah     ;Merge the bits.
```

Since shifting an integer value to the left one position is equivalent to multiplying that value by two, you can also use the shift left instruction for multiplication by powers of two:

```
shl   ax, 1  ;Equivalent to AX*2
shl   ax, 2  ;Equivalent to AX*4
shl   ax, 3  ;Equivalent to AX*8
shl   ax, 4  ;Equivalent to AX*16
shl   ax, 5  ;Equivlaent to AX*32
shl   ax, 6  ;Equivalent to AX*64
shl   ax, 7  ;Equivalent to AX*128
shl   ax, 8  ;Equivalent to AX*256
etc.
```

Note that shl ax, 8 is equivalent to the following two instructions:

```
mov   ah, al
mov   al, 0
```

The shl/sal instruction multiplies both signed and unsigned values by two for each shift. This instruction sets the carry flag if the result does not fit in the destination operand (i.e., unsigned overflow occurs). Likewise, this instruction sets the overflow flag if the signed result does not fit in the destination operation. This occurs when you shift a zero into the H.O. bit of a negative number or you shift a one into the H.O. bit of a non-negative number.

### 2. SAR

The sar instruction shifts all the bits in the destination operand to the right one bit, replicating the H.O. bit:

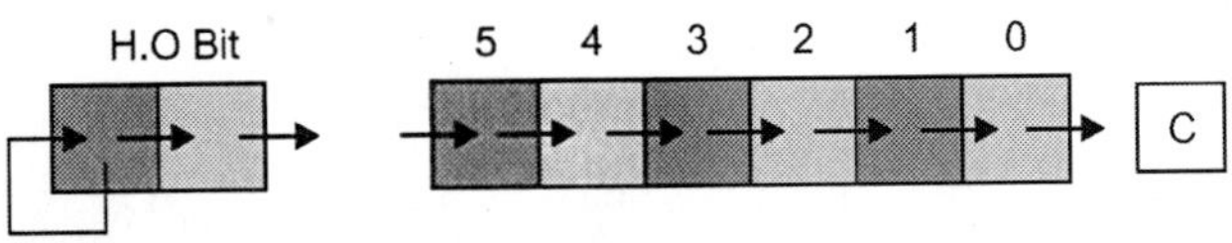


The sar instruction sets the flag bits as follows:

- If the shift count is zero, the sar instruction doesn't affect any flags.
- The carry flag contains the last bit shifted out of the L.O. bit of the operand.

- The overflow flag will contain zero if the shift count is one. Overflow can never occur with this instruction. However, if the count is not one, the value of the overflow flag is undefined.
- The zero flag will be one if the shift produces a zero result.
- The sign flag will contain the H.O. bit of the result.
- The parity flag will contain one if there are an even number of one bits in the L.O. byte of the result.
- The auxiliary carry flag is always undefined after the sar instruction.

The sar instruction's main purpose is to perform a signed division by some power of two. Each shift to the right divides the value by two. Multiple right shifts divide the previous shifted result by two, so multiple shifts produce the following results:

```
sar  ax, 1  ;Signed division by 2
sar  ax, 2  ;Signed division by 4
sar  ax, 3  ;Signed division by 8
sar  ax, 4  ;Signed division by 16
sar  ax, 5  ;Signed division by 32
sar  ax, 6  ;Signed division by 64
sar  ax, 7  ;Signed division by 128
sar  ax, 8  ;Signed division by 256
```

There is a very important difference between the sar and idiv instructions. The idiv instruction always truncates towards zero while sar truncates results toward the smaller result. For positive results, an arithmetic shift right by one position produces the same result as an integer division by two. However, if the quotient is negative, idiv truncates towards zero while sar truncates towards negative infinity. The following examples demonstrate the difference:

```
mov   ax, -15
cwd
mov   bx, 2
idiv             ;Produces -7
mov   ax, -15
sar   ax, 1      ;Produces -8
```

Keep this in mind if you use sar for integer division operations.

The sar ax, 8 instruction effectively copies ah into al and then sign extends al into ax. This is because sar ax, 8 will shift ah down into al but leave a copy of ah's H.O. bit in all the bit positions of ah. Indeed, you can use the sar instruction on 80286 and later processors to sign extend one register into another. The following code sequences provide examples of this usage:

```
; Equivalent to CBW:
    mov   ah, al
```

```
        sar    ah, 7
; Equivalent to CWD:
        mov    dx, ax
        sar    dx, 15
; Equivalent to CDQ:
        mov    edx, eax
        sar    edx, 31
```

Of course it may seem silly to use two instructions where a single instruction might suffice; however, the cbw, cwd, and cdq instructions only sign extend al into ax, ax into dx:ax, and eax into edx:eax. Likewise, the movsx instruction copies its sign extended operand into a destination operand twice the size of the source operand. The sar instruction lets you sign extend one register into another register of the same size, with the second register containing the sign extension bits:

```
; Sign extend bx into cx:bx
        mov    cx, bx
        sar    cx, 15
```

### 3. SHR

The shr instruction shifts all the bits in the destination operand to the right one bit shifting a zero into the H.O. bit:

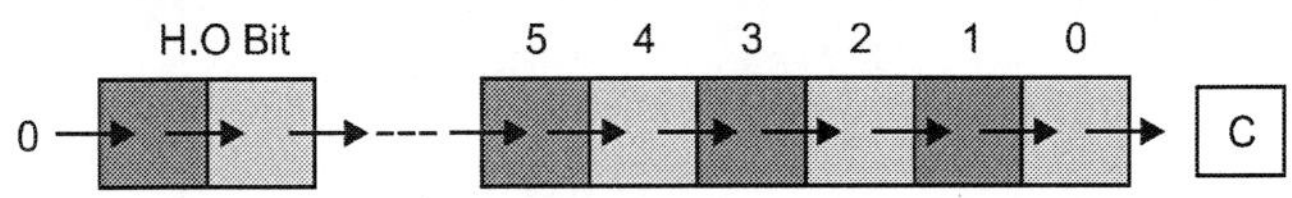


The shr instruction sets the flag bits as follows:

- If the shift count is zero, the shr instruction doesn't affect any flags.
- The carry flag contains the last bit shifted out of the L.O. bit of the operand.
- If the shift count is one, the overflow flag will contain the value of the H.O. bit of the operand prior to the shift (i.e., this instruction sets the overflow flag if the sign changes). However, if the count is not one, the value of the overflow flag is undefined.
- The zero flag will be one if the shift produces a zero result.
- The sign flag will contain the H.O. bit of the result, which is always zero.
- The parity flag will contain one if there are an even number of one bits in the L.O. byte of the result.
- The auxiliary carry flag is always undefined after the shr instruction.

The shift right instruction is especially useful for unpacking data. For example, suppose you want to extract the two nibbles in the al register, leaving the H.O. nibble in ah and the L.O. nibble in al. You could use the following code to do this:

```
mov   ah, al ;Get a copy of the H.O. nibble
shr   ah, 4  ;Move H.O. to L.O. and clear H.O. nibble
and   al, 0Fh ;Remove H.O. nibble from al
```

Since shifting an unsigned integer value to the right one position is equivalent to dividing that value by two, you can also use the shift right instruction for division by powers of two:

```
shr   ax, 1      ;Equivalent to AX/2
shr   ax, 2      ;Equivalent to AX/4
shr   ax, 3      ;Equivalent to AX/8
shr   ax, 4      ;Equivalent to AX/16
shr   ax, 5      ;Equivlaent to AX/32
shr   ax, 6      ;Equivalent to AX/64
shr   ax, 7      ;Equivalent to AX/128
shr   ax, 8      ;Equivalent to AX/256
etc.
```

Note that shr ax, 8 is equivalent to the following two instructions:

```
mov          al, ah
mov          ah, 0
```

Remember that division by two using shr only works for unsigned operands. If ax contains -1 and you execute shr ax, 1 the result in ax will be 32767 (7FFFh), not -1 or zero as you would expect. Use the sar instruction if you need to divide a signed integer by some power of two.

### 8.4.3 The Rotate Instructions: RCL, RCR, ROL, AND ROR

The rotate instructions shift the bits around, just like the shift instructions, except the bits shifted out of the operand by the rotate instructions recirculate through the operand. They include rcl (rotate through carry left), rcr (rotate through carry right), rol (rotate left), and ror (rotate right). These instructions all take the forms:

```
rcl   dest, count
rol   dest, count
rcr   dest, count
ror   dest, count
```

The specific forms are

```
rcl   reg, 1
rcl   mem, 1
rcl   reg, cl
rcl   mem, cl
```

rol uses the same formats as rcl.

rcr uses the same formats as rcl.

ror uses the same formats as rcl.

### 1. RCL

The rcl (rotate through carry left), as its name implies, rotates bits to the left, through the carry flag, and back into bit zero on the right:

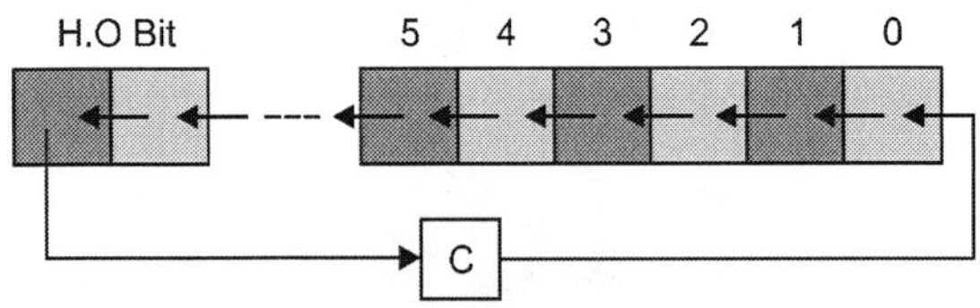


Note that if you rotate through carry an object n+1 times, where n is the number of bits in the object, you wind up with your original value. Keep in mind, however, that some flags may contain different values after n+1 rcl operations.

The rcl instruction sets the flag bits as follows:

- The carry flag contains the last bit shifted out of the H.O. bit of the operand.
- If the shift count is one, rcl sets the overflow flag if the sign changes as a result of the rotate. If the count is not one, the overflow flag is undefined.
- The rcl instruction does not modify the zero, sign, parity, or auxiliary carry flags.

Important warning: unlike the shift instructions, the rotate instructions do not affect the sign, zero, parity, or auxiliary carry flags. This lack of orthogonality can cause you lots of grief if you forget it and attempt to test these flags after an rcl operation. If you need to test one of these flags after an rcl operation, test the carry and overflow flags first (if necessary) then compare the result to zero to set the other flags.

### 2. RCR

The rcr (rotate through carry right) instruction is the complement to the rcl instruction. It shifts its bits right through the carry flag and back into the H.O. bit:

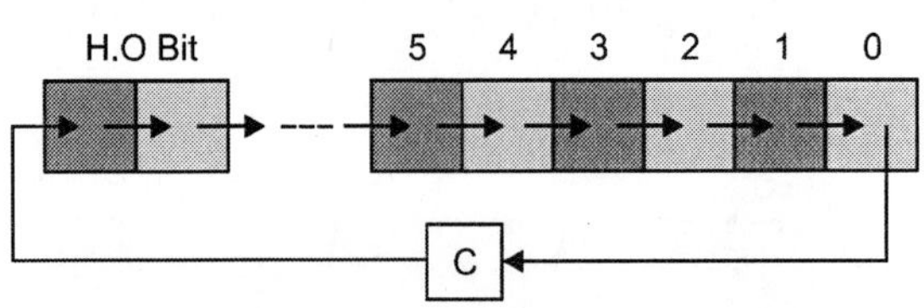


This instruction sets the flags in a manner analogous to rcl:

- The carry flag contains the last bit shifted out of the L.O. bit of the operand.

- If the shift count is one, then rcr sets the overflow flag if the sign changes (meaning the values of the H.O. bit and carry flag were not the same before the execution of the instruction). However, if the count is not one, the value of the overflow flag is undefined.
- The rcr instruction does not affect the zero, sign, parity, or auxiliary carry flags.

Keep in mind the warning given for rcl above.

### 3. ROL

The rol instruction is similar to the rcl instruction in that it rotates its operand to the left the specified number of bits. The major difference is that rol shifts its operand's H.O. bit, rather than the carry, into bit zero. Rol also copies the output of the H.O. bit into the carry flag:

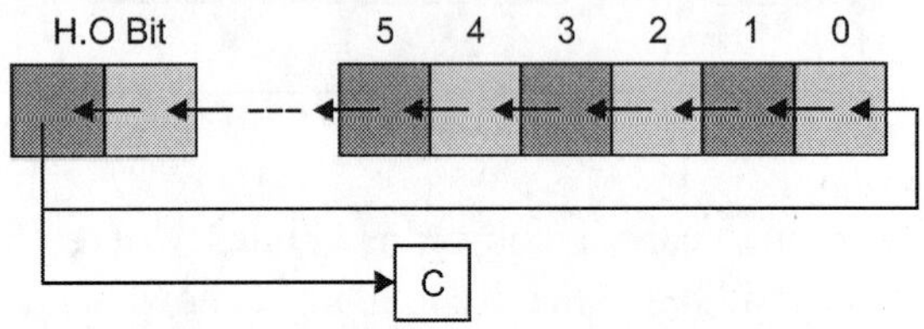


The rol instruction sets the flags identically to rcl. Other than the source of the value shifted into bit zero, this instruction behaves exactly like the rcl instruction. Don't forget the warning about the flags!

Like shl, the rol instruction is often useful for packing and unpacking data. For example, suppose you want to extract bits 10..14 in ax and leave these bits in bits 0..4. The following code sequences will both accomplish this:

```
shr  ax, 10
and  ax, 1Fh
rol  ax, 6
and  ax, 1Fh
```

### 4. ROR

The ror instruction relates to the rcr instruction in much the same way that the rol instruction relates to rcl. That is, it is almost the same operation other than the source of the input bit to the operand. Rather than shifting the previous carry flag into the H.O. bit of the destination operation, ror shifts bit zero into the H.O. bit:

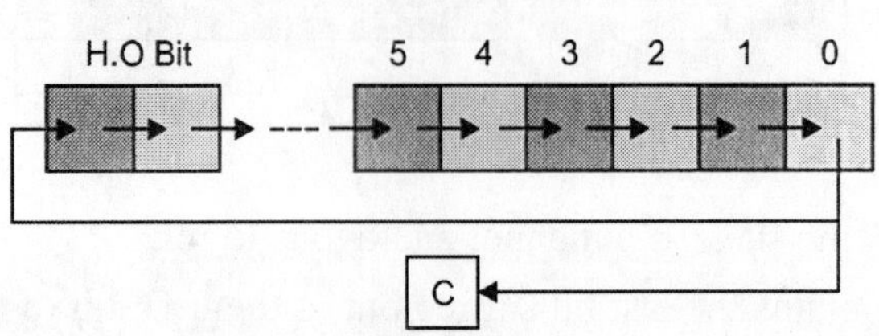

The ror instruction sets the flags identically to rcr. Other than the source of the bit shifted into the H.O. bit, this instruction behaves exactly like the rcr instruction. Don't forget the warning about the flags!

## 8.5 I/O INSTRUCTIONS

The 80x86 supports two I/O instructions: in and out. They take the forms:

```
in    eax/ax/al, port
in    eax/ax/al, dx
out   port, eax/ax/al
out   dx, eax/ax/al
```

port is a value between 0 and 255.

The 80x86 supports up to 65,536 different I/O ports (requiring a 16 bit I/O address). The port value above, however, is a single byte value. Therefore, you can only directly address the first 256 I/O ports in the 80x86's I/O address space. To address all 65,536 different I/O ports, you must load the address of the desired port (assuming it's above 255) into the dx register and access the port indirectly. The in instruction reads the data at the specified I/O port and copies it into the accumulator. The out instruction writes the value in the accumulator to the specified I/O port.

Please realize that there is nothing magical about the 80x86's in and out instructions. They're simply another form of the mov instruction that accesses a different memory space (the I/O address space) rather than the 80x86's normal 1 Mbyte memory address space.

The in and out instructions do not affect any 80x86 flags.

Examples of the 80x86 I/O instructions:

```
in    al, 60h     ;Read keyboard port
mov   dx, 378h    ;Point at LPT1: data port
in    al, dx      ;Read data from printer port.
inc   ax          ;Bump the ASCII code by one.
out   dx, al      ;Write data in AL to printer port.
```

## 8.6 STRING INSTRUCTIONS

The 80x86 supports twelve string instructions:

- movs (move string)
- lods (load string element into the accumulator)
- stos (store accumulator into string element)
- scas (Scan string and check for match against the value in the accumulator)
- cmps (compare two strings).

- ins (input a string from an I/O port)
- outs (output a string to an I/O port
- rep (repeat a string operation)
- repz (repeat while zero)
- repe (repeat while equal)
- repnz (repeat while not zero)
- repne (repeat while not equal)

You can use the movs, stos, scas, cmps, ins and outs instructions to manipulate a single element (byte, word, or double word) in a string, or to process an entire string. Generally, you would only use the lods instruction to manipulate a single item at a time.

These instructions can operate on strings of bytes, words, or double words. To specify the object size, simply append a b, w, or d to the end of the instruction's mnemonic, i.e., lodsb, movsw, cmpsd, etc. Of course, the double word forms are only available on 80386 and later processors.

The movs and cmps instructions assume that ds:si contains the segmented address of a source string and that es:di contains the segmented address of a destination string. The lods instruction assumes that ds:si points at a source string, the accumulator (al/ax/eax) is the destination location. The scas and stos instructions assume that es:di points at a destination string and the accumulator contains the source value.

The movs instruction moves one string element (byte, word, or dword) from memory location ds:si to es:di. After moving the data, the instruction increments or decrements si and di by one, two, or four if processing bytes, words, or dwords, respectively. The CPU increments these registers if the direction flag is clear, the CPU decrements them if the direction flag is set.

The movs instruction can move blocks of data around in memory. You can use it to move strings, arrays, and other multi-byte data structures.

```
movs{b,w,d}:  es:[di] := ds:[si]
          if direction_flag = 0 then
              si := si + size;
              di := di + size;
          else
              si := si - size;
              di := di - size;
          endif;
```

Note: size is one for bytes, two for words, and four for dwords.

The cmps instruction compares the byte, word, or dword at location ds:si to es:di and sets the processor flags accordingly. After the comparison, cmps increments or decrements si and di by one, two, or four depending on the size of the instruction and the status of the direction flag in the flags register.

```
cmps{b,w,d}:   cmp ds:[si], es:[di]
          if direction_flag = 0 then
               si := si + size;
               di := di + size;
          else
               si := si - size;
               di := di - size;
          endif;
```

The lods instruction moves the byte, word, or dword at ds:si into the al, ax, or eax register. It then increments or decrements the si register by one, two, or four depending on the instruction size and the value of the direction flag. The lods instruction is useful for fetching a sequence of bytes, words, or double words from an array, performing some operation(s) on those values and then processing the next element from the string.

```
lods{b,w,d}:   eax/ax/al := ds:[si]
          if direction_flag = 0 then
               si := si + size;
          else
               si := si - size;
          endif;
```

The stos instruction stores al, ax, or eax at the address specified by es:di. Again, di is incremented or decremented according to the size of the instruction and the value of the direction flag. The stos instruction has several uses. Paired with the lods instruction above, you can load (via lods), manipulate, and store string elements. By itself, the stos instruction can quickly store a single value throughout a multi-byte data structure.

```
stos{b,w,d}:   es:[di] := eax/ax/al
          if direction_flag = 0 then
               di := di + size;
          else
               di := di - size;
          endif;
```

The scas instruction compares al, ax or eax against the value at location es:di and then adjusts di accordingly. This instruction sets the flags in the processor status register just like the cmp and cmps instructions. The scas instruction is great for searching for a particular value throughout some multi-byte data structure.

```
scas{b,w,d}:   cmp eax/ax/al, es:[di]
          if direction_flag = 0 then
```

```
        di := di + size;
    else
        di := di - size;
    endif;
```

The ins instruction inputs a byte, word, or double word from the I/O port specified in the dx register. It then stores the input value at memory location es:di and increments or decrements di appropriately.

The outs instruction fetches the byte, word, or double word at address ds:si, increments or decrements si accordingly, and then outputs the value to the port specified in the dx register.

```
outs{b,w,d}:  port(dx) := ds:[si]
    if direction_flag = 0 then
        si := si + size;
    else
        si := si - size;
    endif;
```

As explained here, the string instructions are useful, but it gets even better! When combined with the rep, repz, repe, repnz, and repne prefixes, a single string instruction can process an entire string. For more information on these prefixes see the chapter on strings.

## 8.7 PROGRAM FLOW CONTROL INSTRUCTIONS

The instructions discussed thus far execute sequentially; that is, the CPU executes each instruction in the sequence it appears in your program. To write real programs requires several control structures, not just the sequence. Examples include the if statement, loops, and subroutine invocation (a call). Since compilers reduce all other languages to assembly language, it should come as no surprise that assembly language supports the instructions necessary to implement these control structures. 80x86 program control instructions belong to three groups: unconditional transfers, conditional transfers, and subroutine call and return instructions. The following sections describe these instructions:

### 8.7.1 Unconditional Jumps

The jmp (jump) instruction unconditionally transfers control to another point in the program. There are six forms of this instruction: an intersegment/direct jump, two intrasegment/direct jumps, an intersegment/indirect jump, and two intrasegment/indirect jumps. Intrasegment jumps are always between statements in the same code segment. Intersegment jumps can transfer control to a statement in a different code segment.

These instructions generally use the same syntax, it is

```
jmp        target
```

The assembler differentiates them by their operands:

```
jmp   disp8   ;direct intrasegment, 8 bit displacement.
jmp   disp16  ;direct intrasegment, 16 bit displacement.
jmp   adrs32  ;direct intersegment, 32 bit segmented address.
jmp   mem16   ;indirect intrasegment, 16 bit memory operand.
jmp   reg16   ;register indirect intrasegment.
jmp   mem32   ;indirect intersegment, 32 bit memory operand.
```

Intersegment is a synonym for far, intrasegment is a synonym for near.

The two direct intrasegment jumps differ only in their length. The first form consists of an opcode and a single byte displacement. The CPU sign extends this displacement to 16 bits and adds it to the ip register. This instruction can branch to a location -128..+127 from the beginning of the next instruction following it (i.e., -126..+129 bytes around the current instruction).

The second form of the intrasegment jump is three bytes long with a two byte displacement. This instruction allows an effective range of -32,768..+32,767 bytes and can transfer control to anywhere in the current code segment. The CPU simply adds the two byte displacement to the ip register.

These first two jumps use a relative addressing scheme. The offset encoded as part of the opcode byte is not the target address in the current code segment, but the distance to the target address. Fortunately, MASM will compute the distance for you automatically, so you do not have to compute this displacement value yourself. In many respects, these instructions are really nothing more than add ip, disp instructions.

The direct intersegment jump is five bytes long, the last four bytes containing a segmented address (the offset in the second and third bytes, the segment in the fourth and fifth bytes). This instruction copies the offset into the ip register and the segment into the cs register. Execution of the next instruction continues at the new address in cs:ip. Unlike the previous two jumps, the address following the opcode is the absolute memory address of the target instruction; this version does not use relative addressing. This instruction loads cs:ip with a 32 bit immediate value.

For the three direct jumps described above, you normally specify the target address using a statement label. A statement label is usually an identifier followed by a colon, usually on the same line as an executable machine instruction. The assembler determines the offset of the statement after the label and automatically computes the distance from the jump instruction to the statement label. Therefore, you do not have to worry about computing displacements manually. For example, the following short little loop continuously reads the parallel printer data port and inverts the L.O. bit. This produces a square wave electrical signal on one of the printer port output lines:

```
              mov   dx, 378h     ;Parallel printer port address.
LoopForever:  in    al, dx       ;Read character from input port.
              xor   al, 1        ;Invert the L.O. bit.
              out   dx, al       ;Output data back to port.
```

```
jmp    LoopForever    ;Repeat forever.
```

The fourth form of the unconditional jump instruction is the indirect intrasegment jump instruction. It requires a 16 bit memory operand. This form transfers control to the address within the offset given by the two bytes of the memory operand. For example,

```
WordVar       word    TargetAddress
        :
        jmp    WordVar
```

transfers control to the address specified by the value in the 16 bit memory location WordVar. This does not jump to the statement at address WordVar, it jumps to the statement at the address held in the WordVar variable. Note that this form of the jmp instruction is roughly equivalent to:

```
mov            ip, WordVar
```

Although the example above uses a single word variable containing the indirect address, you can use any valid memory address mode, not just the displacement only addressing mode. You can use memory indirect addressing modes like the following:

```
jmp    DispOnly       ;Word variable
jmp    Disp[bx]       ;Disp is an array of words
jmp    Disp[bx][si]
jmp    [bx]
etc.
```

Consider the indexed addressing mode above for a moment (disp[bx]). This addressing mode fetches the word from location disp+bx and copies this value to the ip register; this lets you create an array of pointers and jump to a specified pointer using an array index. Consider the following example:

```
AdrsArray     word    stmt1, stmt2, stmt3, stmt4
        :
        mov    bx, I          ;I is in the range 0..3
        add    bx, bx         ;Index into an array of words.
        jmp    AdrsArray[bx]  ;Jump to stmt1, stmt2, etc., depending
                              ; on the value of I.
```

The important thing to remember is that the near indirect jump fetches a word from memory and copies it into the ip register; it does not jump to the memory location specified, it jumps indirectly through the 16 bit pointer at the specified memory location.

The fifth jmp instruction transfers control to the offset given in a 16 bit general purpose register. Note that you can use any general purpose register, not just bx, si, di, or bp. An instruction of the form

```
jmp            ax
```

is roughly equivalent to

```
mov     ip, ax
```

Note that the previous two forms (register or memory indirect) are really the same instruction. The mod and r/m fields of a mod-reg-r/m byte specify a register or memory indirect address. See Appendix D for the details.

The sixth form of the jmp instruction, the indirect intersegment jump, has a memory operand that contains a double word pointer. The CPU copies the double word at that address into the cs:ip register pair. For example,

```
FarPointer    dword   TargetAddress
        ⋮
        jmp   FarPointer
```

transfers control to the segmented address specified by the four bytes at address FarPointer. This instruction is semantically identical to the (mythical) instruction

```
lcs ip, FarPointer                  ;load cs, ip from FarPointer
```

As for the near indirect jump described earlier, this far indirect jump lets you specify any arbitrary (valid) memory addressing mode. You are not limited to the displacement only addressing mode the example above uses.

MASM uses a near indirect or far indirect addressing mode depending upon the type of the memory location you specify. If the variable you specify is a word variable, MASM will automatically generate a near indirect jump; if the variable is a dword, MASM emits the opcode for a far indirect jump. Some forms of memory addressing, unfortunately, do not intrinsically specify a size. For example, [bx] is definitely a memory operand, but does bx point at a word variable or a double word variable? It could point at either. Therefore, MASM will reject a statement of the form:

```
jmp            [bx]
```

MASM cannot tell whether this should be a near indirect or far indirect jump. To resolve the ambiguity, you will need to use a type coercion operator. Chapter Eight will fully describe type coercion operators, for now, just use one of the following two instructions for a near or far jump, respectively:

```
jmp            word ptr [bx]
jmp            dword ptr [bx]
```

The register indirect addressing modes are not the only ones that could be type ambiguous. You could also run into this problem with indexed and base plus index addressing modes:

```
jmp            word ptr 5[bx]
jmp            dword ptr 9[bx][si]
```

For more information on the type coercion operators.

In theory, you could use the indirect jump instructions and the setcc instructions to conditionally transfer control to some given location. For example, the following code transfers control to iftrue if word variable X is equal to word variable Y. It transfers control to iffalse, otherwise.

```
JmpTbl      word   iffalse, iftrue
            :
            mov    ax, X
            cmp    ax, Y
            sete   bl
            movzx  ebx, bl
            jmp    JmpTbl[ebx*2]
```

As you will soon see, there is a much better way to do this using the conditional jump instructions.

### 8.7.2 The Call and Ret Instructions

The call and ret instructions handle subroutine calls and returns. There are five different call instructions and six different forms of the return instruction:

```
call  disp16  ;direct intrasegment, 16 bit relative.
call  adrs32  ;direct intersegment, 32 bit segmented address.
call  mem16   ;indirect intrasegment, 16 bit memory pointer.
call  reg16   ;indirect intrasegment, 16 bit register pointer.
call  mem32   ;indirect intersegment, 32 bit memory pointer.
ret           ;near or far return
retn          ;near return
retf          ;far return
ret   disp    ;near or far return and pop
retn  disp    ;near return and pop
retf  disp    ;far return and pop
```

The call instructions take the same forms as the jmp instructions except there is no short (two byte) intrasegment call.

The far call instruction does the following:

- It pushes the cs register onto the stack.
- It pushes the 16 bit offset of the next instruction following the call onto the stack.
- It copies the 32 bit effective address into the cs: ip registers. Since the call instruction allows the same addressing modes as jmp, call can obtain the target address using a relative, memory, or register addressing mode.
- Execution continues at the first instruction of the subroutine. This first instruction is the opcode at the target address computed in the previous step.

The near call instruction does the following:

- It pushes the 16 bit offset of the next instruction following the call onto the stack.
- It copies the 16 bit effective address into the ip register. Since the call instruction allows the same addressing modes as jmp, call can obtain the target address using a relative, memory, or register addressing mode.
- Execution continues at the first instruction of the subroutine. This first instruction is the opcode at the target address computed in the previous step.

The call disp16 instruction uses relative addressing. You can compute the effective address of the target by adding this 16 bit displacement with the return address (like the relative jmp instructions, the displacement is the distance from the instruction following the call to the target address).

The call adrs32 instruction uses the direct addressing mode. A 32 bit segmented address immediately follows the call opcode. This form of the call instruction copies that value directly into the cs:ip register pair. In many respects, this is equivalent to the immediate addressing mode since the value this instruction copies into the cs:ip register pair immediately follows the instruction.

Call mem16 uses the memory indirect addressing mode. Like the jmp instruction, this form of the call instruction fetches the word at the specified memory location and uses that word's value as the target address. Remember, you can use any memory addressing mode with this instruction. The displacement-only addressing mode is the most common form, but the others are just as valid:

```
call    CallTbl[bx]        ;Index into an array of pointers.
call    word ptr [bx]      ;BX points at word to use.
call    WordTbl[bx][si]    ; etc.
```

Note that the selection of addressing mode only affects the effective address computation for the target subroutine. These call instructions still push the offset of the next instruction following the call onto the stack. Since these are near calls (they obtain their target address from a 16 bit memory location), they all push a 16 bit return address onto the stack.

Call reg16 works just like the memory indirect call above, except it uses the 16 bit value in a register for the target address. This instruction is really the same instruction as the call mem16 instruction. Both forms specify their effective address using a mod-reg-r/m byte. For the call reg16 form, the mod bits contain 11b so the r/m field specifies a register rather than a memory addressing mode. Of course, this instruction also pushes the 16 bit offset of the next instruction onto the stack as the return address.

The call mem32 instruction is a far indirect call. The memory address specified by this instruction must be a double word value. This form of the call instruction fetches the 32 bit segmented address at the computed effective address and copies this double word value into the cs:ip register pair. This instruction also copies the 32 bit segmented address of the next instruction onto the stack (it pushes the segment value first and the offset portion second). Like the call mem16 instruction, you can use any valid memory addressing mode with this instruction:

```
call  DWordVar
call  DwordTbl[bx]
```

```
call dword ptr [bx]
etc.
```

It is relatively easy to synthesize the call instruction using two or three other 80x86 instructions. You could create the equivalent of a near call using a push and a jmp instruction:

```
push    <offset of instruction after jmp>
jmp     subroutine
```

A far call would be similar, you'd need to add a push cs instruction before the two instructions above to push a far return address on the stack.

The ret (return) instruction returns control to the caller of a subroutine. It does so by popping the return address off the stack and transferring control to the instruction at this return address. Intrasegment (near) returns pop a 16 bit return address off the stack into the ip register. An intersegment (far) return pops a 16 bit offset into the ip register and then a 16 bit segment value into the cs register. These instructions are effectively equal to the following:

```
retn:    pop     ip
retf:    popd    cs:ip
```

Clearly, you must match a near subroutine call with a near return and a far subroutine call with a corresponding far return. If you mix near calls with far returns or vice versa, you will leave the stack in an inconsistent state and you probably will not return to the proper instruction after the call. Of course, another important issue when using the call and ret instructions is that you must make sure your subroutine doesn't push something onto the stack and then fail to pop it off before trying to return to the caller. Stack problems are a major cause of errors in assembly language subroutines. Consider the following code:

```
Subroutine: push  ax
            push  bx
              ⋮
            pop   bx
            ret
              ⋮
            call  Subroutine
```

The call instruction pushes the return address onto the stack and then transfers control to the first instruction of subroutine. The first two push instructions push the ax and bx registers onto the stack, presumably in order to preserve their value because subroutine modifies them. Unfortunately, a programming error exists in the code above, subroutine only pops bx from the stack, it fails to pop ax as well. This means that when subroutine tries to return to the caller, the value of ax rather than the return address is sitting on the top of the stack. Therefore, this subroutine returns control to the address specified by the initial value of the ax register rather than to the true return address. Since there are 65,536 different values ax can have, there is a 1/65,536th of a chance that your code will return to the real return address. The odds are not in your favor! Most likely, code like this will hang up the machine. Moral of the story - always make sure the return address is sitting on the stack before executing the return instruction.

Like the call instruction, it is very easy to simulate the ret instruction using two 80x86 instructions. All you need to do is pop the return address off the stack and then copy it into the ip register. For near returns, this is a very simple operation, just pop the near return address off the stack and then jump indirectly through that register:

```
pop     ax
jmp     ax
```

Simulating a far return is a little more difficult because you must load cs:ip in a single operation. The only instruction that does this (other than a far return) is the jmp mem32 instruction. See the exercises at the end of this chapter for more details.

There are two other forms of the ret instruction. They are identical to those above except a 16 bit displacement follows their opcodes. The CPU adds this value to the stack pointer immediately after popping the return address from the stack. This mechanism removes parameters pushed onto the stack before returning to the caller. See Chapter Eleven for more details.

The assembler allows you to type ret without the "f" or "n" suffix. If you do so, the assembler will figure out whether it should generate a near return or a far return. See the chapter on procedures and functions for details on this.

### 8.7.3 The Int, Into, Bound, And Iret Instructions

The int (for software interrupt) instruction is a very special form of a call instruction. Whereas the call instruction calls subroutines within your program, the int instruction calls system routines and other special subroutines. The major difference between interrupt service routines and standard procedures is that you can have any number of different procedures in an assembly language program, while the system supports a maximum of 256 different interrupt service routines. A program calls a subroutine by specifying the address of that subroutine; it calls an interrupt service routine by specifying the interrupt number for that particular interrupt service routine. This chapter will only describe how to call an interrupt service routine using the int, into, and bound instructions, and how to return from an interrupt service routine using the iret instruction.

There are four different forms of the int instruction. The first form is

```
int     nn
```

(where "nn" is a value between 0 and 255). It allows you to call one of 256 different interrupt routines. This form of the int instruction is two bytes long. The first byte is the int opcode. The second byte is immediate data containing the interrupt number.

Although you can use the int instruction to call procedures (interrupt service routines) you've written, the primary purpose of this instruction is to make a system call. A system call is a subroutine call to a procedure provided by the system, such as a DOS , PC-BIOS, mouse, or some other piece of software resident in the machine before your program began execution. Since you always refer to a specific system call by its interrupt number, rather than its address, your program does not need to know the actual address of the subroutine in memory. The int instruction provides dynamic linking to your program. The CPU determines the actual address of an interrupt service routine at run time by looking up the address in an interrupt vector table. This allows the authors of such system routines to change their code (including the entry point) without fear of breaking any older programs that call their interrupt service routines. As long as the system call uses the same interrupt number, the CPU will automatically call the interrupt service routine at its new address.

The only problem with the int instruction is that it supports only 256 different interrupt service routines. MS-DOS alone supports well over 100 different calls. BIOS and other system utilities provide thousands more. This is above and beyond all the interrupts reserved by Intel for hardware interrupts and traps. The common solution most of the system calls use is to employ a single interrupt number for a given class of calls and then pass a function number in one of the 80x86 registers (typically the ah register). For example, MS-DOS uses only a single interrupt number, 21h. To choose a particular DOS function, you load a DOS function code into the ah register before executing the int 21h instruction. For example, to terminate a program and return control to MS-DOS, you would normally load ah with 4Ch and call DOS with the int 21h instruction:

```
mov  ah, 4ch  ;DOS terminate opcode.
int  21h      ;DOS call
```

The BIOS keyboard interrupt is another good example. Interrupt 16h is responsible for testing the keyboard and reading data from the keyboard. This BIOS routine provides several calls to read a character and scan code from the keyboard, see if any keys are available in the system type ahead buffer, check the status of the keyboard modifier flags, and so on. To choose a particular operation, you load the function number into the ah register before executing int 16h. The following table lists the possible functions:

BIOS Keyboard Support Functions

| Function # (AH) | Input Parameters | Output Parameters | Description |
|---|---|---|---|
| 0 | - | al- ASCII character ah- scan code | Read character. Reads next available character from the system's type ahead buffer. Wait for a keystroke if the buffer is empty. |
| 1 | - | ZF- Set if no key. ZF- Clear if key available. al- ASCII code ah- scan code | Checks to see if a character is available in the type ahead buffer. Sets the zero flag if not key is available, clears the zero flag if a key is available. If there is an available key, this function returns the ASCII and scan code value in ax. The value in ax is undefined if no key is available. |
| 2 | - | al- shift flags | Returns the current status of the shift flags in al. The shift flags are defined as follows: bit |

| | | | |
|---|---|---|---|
| | | | 7: Insert toggle bit 6: Capslock toggle bit 5: Numlock toggle bit 4: Scroll lock toggle bit 3: Alt key is down bit 2: Ctrl key is down bit 1: Left shift key is down bit 0: Right shift key is down |
| 3 | al = 5 bh = 0, 1, 2, 3 for 1/4, 1/2, 3/4, or 1 second delay bl= 0..1Fh for 30/sec to 2/sec. | - | Set auto repeat rate. The bh register contains the amount of time to wait before starting the autorepeat operation, the bl register contains the autorepeat rate. |
| 5 | ch = scan code cl = ASCII code | - | Store keycode in buffer. This function stores the value in the cx register at the end of the type ahead buffer. Note that the scan code in ch doesn't have to correspond to the ASCII code appearing in cl. This routine will simply insert the data you provide into the system type ahead buffer. |
| 10h | - | al- ASCII character ah- scan code | Read extended character. Like ah=0 call, except this one passes all key codes, the ah=0 call throws away codes that are not PC/XT compatible. |
| 11h | - | ZF- Set if no key. ZF- Clear if key available. al- ASCII code ah- scan code | Like the ah=01h call except this one does not throw away keycodes that are not PC/XT compatible (i.e., the extra keys found on the 101 key keyboard). |
| 12h | - | al- shift flags ah- extended shift flags | Returns the current status of the shift flags in ax. The shift flags are defined as follows: bit 15: SysReq key pressed bit 14: Capslock key currently down bit 13: Numlock key currently down bit 12: Scroll lock key currently down bit 11: Right alt key is down bit 10:Right ctrl key is down bit 9: Left alt key is down bit 8: Left ctrl key is down bit 7: Insert toggle bit 6: Capslock toggle bit 5: Numlock toggle bit 4: Scroll lock toggle bit 3: Either alt key is down (some machines, left only) bit 2: Either ctrl key is down bit 1: Left shift key is down bit 0: Right shift key is down |

For example, to read a character from the system type ahead buffer, leaving the ASCII code in al, you could use the following code:

```
mov   ah, 0          ;Wait for key available, and then
int   16h            ; read that key.
mov   character, al  ;Save character read.
```

Likewise, if you wanted to test the type ahead buffer to see if a key is available, without reading that keystroke, you could use the following code:

```
mov   ah, 1  ;Test to see if key is available.
int   16h    ;Sets the zero flag if a key is not
             ; available.
```

The second form of the int instruction is a special case:

```
int        3
```

Int 3 is a special form of the interrupt instruction that is only one byte long. CodeView and other debuggers use it as a software breakpoint instruction. Whenever you set a breakpoint on an instruction in your program, the debugger will typically replace the first byte of the instruction's opcode with an int 3 instruction. When your program executes the int 3 instruction, this makes a "system call" to the debugger so the debugger can regain control of the CPU. When this happens, the debugger will replace the int 3 instruction with the original opcode.

While operating inside a debugger, you can explicitly use the int 3 instruction to stop program executing and return control to the debugger. This is not, however, the normal way to terminate a program. If you attempt to execute an int 3 instruction while running under DOS, rather than under the control of a debugger program, you will likely crash the system.

The third form of the int instruction is into. Into will cause a software breakpoint if the 80x86 overflow flag is set. You can use this instruction to quickly test for arithmetic overflow after executing an arithmetic instruction. Semantically, this instruction is equivalent to

```
if overflow = 1 then int 4
```

You should not use this instruction unless you've supplied a corresponding trap handler (interrupt service routine). Doing so would probably crash the system.

### 8.7.4 The Conditional Jump Instructions

Although the jmp, call, and ret instructions provide transfer of control, they do not allow you to make any serious decisions. The 80x86's conditional jump instructions handle this task. The conditional jump instructions are the basic tool for creating loops and other conditionally executable statements like the if..then statement.

The conditional jumps test one or more flags in the flags register to see if they match some particular pattern (just like the setcc instructions). If the pattern matches, control transfers to the target location. If the match fails, the CPU ignores the conditional jump and execution continues with the next instruction. Some instructions, for example, test the conditions of the sign, carry, overflow, and zero flags. For example, after the execution of a shift left instruction, you could test the carry flag to determine if it shifted a one out of the H.O. bit of its operand. Likewise, you could test the condition of the zero flag after a test instruction to see if any specified bits were one. Most of the time, however, you will probably execute a conditional jump after a cmp instruction. The cmp instruction sets the flags so that you can test for less than, greater than, equality, etc.

Note: Intel's documentation defines various synonyms or instruction aliases for many conditional jump instructions. The following tables list all the aliases for a particular instruction. These tables also list out the opposite branches. You'll soon see the purpose of the opposite branches.

| Jcc Instructions That Test Flags | | | | |
|---|---|---|---|---|
| **Instruction** | **Description** | **Condition** | **Aliases** | **Opposite** |
| JC | Jump if carry | Carry = 1 | JB, JNAE | JNC |
| JNC | Jump if no carry | Carry = 0 | JNB, JAE | JC |
| JZ | Jump if zero | Zero = 1 | JE | JNZ |
| JNZ | Jump if not zero | Zero = 0 | JNE | JZ |
| JS | Jump if sign | Sign = 1 | - | JNS |
| JNS | Jump if no sign | Sign = 0 | - | JS |
| JO | Jump if overflow | Ovrflw=1 | - | JNO |
| JNO | Jump if no Ovrflw | Ovrflw=0 | - | JO |
| JP | Jump if parity | Parity = 1 | JPE | JNP |
| JPE | Jump if parity even | Parity = 1 | JP | JPO |
| JNP | Jump if no parity | Parity = 0 | JPO | JP |
| JPO | Jump if parity odd | Parity = 0 | JNP | JPE |

| Jcc Instructions for Unsigned Comparisons | | | | |
|---|---|---|---|---|
| **Instruction** | **Description** | **Condition** | **Aliases** | **Opposite** |
| JA | Jump if above (>) | Carry=0, Zero=0 | JNBE | JNA |
| JNBE | Jump if not below or equal (not <=) | Carry=0, Zero=0 | JA | JBE |
| JAE | Jump if above or equal (>=) | Carry = 0 | JNC, JNB | JNAE |
| JNB | Jump if not below (not <) | Carry = 0 | JNC, JAE | JB |
| JB | Jump if below (<) | Carry = 1 | JC, JNAE | JNB |
| JNAE | Jump if not above or equal (not >=) | Carry = 1 | JC, JB | JAE |
| JBE | Jump if below or equal (<=) | Carry = 1 or Zero = 1 | | JNA JNBE |
| JNA | Jump if not above (not >) | Carry = 1 or Zero = 1 | | JBE JA |
| JE | Jump if equal (=) | Zero = 1 | JZ | JNE |
| JNE | Jump if not equal () | Zero = 0 | JNZ | JE |

| Jcc Instructions for Signed Comparisons | | | | |
|---|---|---|---|---|
| **Instruction** | **Description** | **Condition** | **Aliases** | **Opposite** |
| JG | Jump if greater (>) | Sign = Ovrflw or Zero=0 | JNLE | JNG |
| JNLE | Jump if not less than or equal (not <=) | Sign = Ovrflw or Zero=0 | JG | JLE |
| JGE | Jump if greater than or equal (>=) | Sign = Ovrflw | JNL | JGE |
| JNL | Jump if not less than (not <) | Sign = Ovrflw | JGE | JL |
| JL | Jump if less than (<) | Sign Ovrflw | JNGE | JNL |
| JNGE | Jump if not greater or equal (not >=) | Sign Ovrflw | JL | JGE |
| JLE | Jump if less than or equal (<=) | Sign Ovrflw or Zero = 1 | JNG | JNLE |
| JNG | Jump if not greater than (not >) | Sign Ovrflw or Zero = 1 | | JLE JG |
| JE | Jump if equal (=) | Zero = 1 | JZ | JNE |
| JNE | Jump if not equal () | Zero = 0 | JNZ | JE |

These instructions are all two bytes long. The first byte is a one byte opcode followed by a one byte displacement. Although this leads to very compact instructions, a single byte displacement only allows a range of ?128 bytes. There is a simple trick you can use to overcome this limitation on these earlier processors:

- Whatever jump you're using, switch to its opposite form. (given in the tables above).
- Once you've selected the opposite branch, use it to jump over a jmp instruction whose target address is the original target address.

For example, to convert:

```
        jc          Target
```

to the long form, use the following sequence of instructions:

```
        jnc         SkipJmp
        jmp         Target
SkipJmp:
```

If the carry flag is clear (NC=no carry), then control transfers to label SkipJmp, at the same point you'd be if you were using the jc instruction above. If the carry flag is set when encountering this sequence, control will fall through the jnc instruction to the jmp instruction that will transfer control to Target. Since the jmp instruction allows 16 bit displacement and far operands, you can jump anywhere in the memory using this trick.

One brief comment about the "opposites" column is in order. As mentioned above, when you need to manually extend a branch from ?128 you should choose the opposite branch to branch around a jump to the target location. As you can see in the "aliases" column above, many conditional jump instructions have aliases. This means that there will be aliases for the opposite jumps as well. Do not use any aliases when extending branches that are out of range. With only two exceptions, a very simple rule completely describes how to generate an opposite branch:

- If the second letter of the jcc instruction is not an "n", insert an "n" after the "j". E.g., je becomes jne and jl becomes jnl.
- If the second letter of the jcc instruction is an "n", then remove that "n" from the instruction. E.g., jng becomes jg, jne becomes je.

The two exceptions to this rule are jpe (jump parity even) and jpo (jump parity odd). These exceptions cause few problems because (a) you'll hardly ever need to test the parity flag, and (b) you can use the aliases jp and jnp synonyms for jpe and jpo. The "N/No N" rule applies to jp and jnp.

Though you know that jge is the opposite of jl, get in the habit of using jnl rather than jge. It's too easy in an important situation to start thinking "greater is the opposite of less" and substitute jg instead. You can avoid this confusion by always using the "N/No N" rule.

MASM 6.x and many other modern 80x86 assemblers will automatically convert out of range branches to this sequence for you. There is an option that will allow you to disable this feature. For performance critical code that runs on 80286 and earlier processors, you may want to disable this feature so you can fix the branches yourself. The reason is quite simple, this simple fix always wipes out the pipeline no matter which condition is true since the CPU jumps in either case. One thing nice about conditional jumps is that you do not flush the pipeline or the prefetch queue if you do not take the branch. If one condition is true far more often than the other, you might want to use the conditional jump to transfer control to a jmp nearby, so you can continue to fall through as before. For example, if you have a je target instruction and target is out of range, you could convert it to the following code:

```
            je      GotoTarget
            :
GotoTarget: jmp     Target
```

Although a branch to target now requires executing two jumps, this is much more efficient than the standard conversion if the zero flag is normally clear when executing the je instruction.

The 80x86 conditional jump instruction give you the ability to split program flow into one of two paths depending upon some logical condition. Suppose you want to increment the ax register if bx is or equal to cx. You can accomplish this with the following code:

```
            cmp     bx, cx
```

```
        jne     SkipStmts
        inc     ax
SkipStmts:
```

The trick is to use the opposite branch to skip over the instructions you want to execute if the condition is true. Always use the "opposite branch (N/no N)" rule given earlier to select the opposite branch. You can make the same mistake choosing an opposite branch here as you could when extending out of range jumps. You can also use the conditional jump instructions to synthesize loops. For example, the following code sequence reads a sequence of characters from the user and stores each character in successive elements of an array until the user presses the Enter key (carriage return):

```
            mov     di, 0
ReadLnLoop: mov     ah, 0           ;INT 16h read key opcode.
            int     16h
            mov     Input[di], al
            inc     di
            cmp     al, 0dh         ;Carriage return ASCII code.
            jne     ReadLnLoop
            mov     Input[di-1],0   ;Replace carriage return with zero.
```

For more information concerning the use of the conditional jumps to synthesize IF statements, loops, and other control structures.

Like the setcc instructions, the conditional jump instructions come in two basic categories - those that test specific process flag values (e.g., jz, jc, jno) and those that test some condition ( less than, greater than, etc.). When testing a condition, the conditional jump instructions almost always follow a cmp instruction. The cmp instruction sets the flags so you can use a ja, jae, jb, jbe, je, or jne instruction to test for unsigned less than, less than or equal, equality, inequality, greater than, or greater than or equal. Simultaneously, the cmp instruction sets the flags so you can also do a signed comparison using the jl, jle, je, jne, jg, and jge instructions.

The conditional jump instructions only test flags, they do not affect any of the 80x86 flags.

### 8.7.5 The JCXZ Instructions

The jcxz (jump if cx is zero) instruction branches to the target address if cx contains zero. Although you can use it anytime you need to see if cx contains zero, you would normally use it before a loop you've constructed with the loop instructions. The loop instruction can repeat a sequence of operations cx times. If cx equals zero, loop will repeat the operation 65,536 times. You can use jcxz to skip over such a loop when cx is zero.

There are no "opposite" jcxz or jecxz instructions. Therefore, you cannot use "N/No N" rule to extend the jcxz and jecxz instructions. The easiest way to solve this problem is to break the instruction up into two instructions that accomplish the same task:

```
        jcxz        Target
becomes
        test        cx, cx              ;Sets the zero flag if cx=0
        je          Target
```

Now you can easily extend the je instruction using the techniques from the previous section.

The test instruction above will set the zero flag if and only if cx contains zero. After all, if there are any non-zero bits in cx, logically anding them with themselves will produce a non-zero result. This is an efficient way to see if a 16 or 32 bit register contains zero. In fact, this two instruction sequence is faster than the jcxz instruction on the 80486 and later processors. Indeed, Intel recommends the use of this sequence rather than the jcxz instruction if you are concerned with speed. Of course, the jcxz instruction is shorter than the two instruction sequence, but it is not faster. This is a good example of an exception to the rule "shorter is usually faster."

The jcxz instruction does not affect any flags.

## 8.8 THE LOOP INSTRUCTION

This instruction decrements the cx register and then branches to the target location if the cx register does not contain zero. Since this instruction decrements cx then checks for zero, if cx originally contained zero, any loop you create using the loop instruction will repeat 65,536 times. If you do not want to execute the loop when cx contains zero, use jcxz to skip over the loop.

There is no "opposite" form of the loop instruction, and like the jcxz/jecxz instructions the range is limited to ?128 bytes on all processors. If you want to extend the range of this instruction, you will need to break it down into discrete components:

```
; "loop lbl" becomes:
        dec         cx
        jne         lbl
```

You can easily extend this jne to any distance.

There is no eloop instruction that decrements ecx and branches if not zero (there is a loope instruction, but it does something else entirely). The reason is quite simple.

Although the loop instruction's name suggests that you would normally create loops with it, keep in mind that all it is really doing is decrementing cx and branching to the target address if cx does not contain zero after the decrement. You can use this instruction anywhere you want to decrement cx and then check for a zero result, not just when creating loops. Nonetheless, it is a very convenient instruction to use if you simply want to repeat a sequence of instructions some number of times. For example, the following loop initializes a 256 element array of bytes to the values 1, 2, 3, ...

```
        mov     ecx, 255
ArrayLp:    mov     Array[ecx], cl
```

```
        loop   ArrayLp
        mov    Array[0], 0
```

The last instruction is necessary because the loop does not repeat when cx is zero. Therefore, the last element of the array that this loop processes is Array[1], hence the last instruction.

The loop instruction does not affect any flags.

## 8.9 THE LOOPE/LOOPZ INSTRUCTION

Loope/loopz (loop while equal/zero, they are synonyms for one another) will branch to the target address if cx is not zero and the zero flag is set. This instruction is quite useful after cmp or cmps instruction, and is marginally faster than the comparable 80386/486 instructions if you use all the features of this instruction. However, this instruction plays havoc with the pipeline and superscalar operation of the Pentium so you're probably better off sticking with discrete instructions rather than using this instruction. This instruction does the following:

```
        cx := cx - 1
        if ZeroFlag = 1 and cx  0, goto target
```

The loope instruction falls through on one of two conditions. Either the zero flag is clear or the instruction decremented cx to zero. By testing the zero flag after the loop instruction (with a je or jne instruction, for example), you can determine the cause of termination.

This instruction is useful if you need to repeat a loop while some value is equal to another, but there is a maximum number of iterations you want to allow. For example, the following loop scans through an array looking for the first non-zero byte, but it does not scan beyond the end of the array:

```
           mov    cx, 16           ;Max 16 array elements.
           mov    bx, -1           ;Index into the array (note next inc).
SearchLp:  inc    bx               ;Move on to next array element.
           cmp    Array[bx], 0     ;See if this element is zero.
           loope  SearchLp         ;Repeat if it is.
           je     AllZero          ;Jump if all elements were zero.
```

Note that this instruction is not the opposite of loopnz/loopne. If you need to extend this jump beyond ?128 bytes, you will need to synthesize this instruction using discrete instructions. For example, if loope target is out of range, you would need to use an instruction sequence like the following:

```
           jne    quit
           dec    cx
           je     Quit2
           jmp    Target
quit:      dec    cx       ;loope decrements cx, even if ZF=0.
```

```
quit2:
```

The loope/loopz instruction does not affect any flags.

## 8.10 THE LOOPNE/LOOPNZ INSTRUCTION

This instruction is just like the loope/loopz instruction in the previous section except loopne/loopnz (loop while not equal/not zero) repeats while cx is not zero and the zero flag is clear. The algorithm is

```
cx := cx - 1
if ZeroFlag = 0 and cx  0, goto target
```

You can determine if the loopne instruction terminated because cx was zero or if the zero flag was set by testing the zero flag immediately after the loopne instruction. If the zero flag is clear at that point, the loopne instruction fell through because it decremented cx to zero. Otherwise it fell through because the zero flag was set.

This instruction is not the opposite of loope/loopz. If the target address is out of range, you will need to use an instruction sequence like the following:

```
        je    quit
        dec   cx
        je    Quit2
        jmp   Target
quit:   dec   cx    ;loopne decrements cx, even if ZF=1.
quit2:
```

You can use the loopne instruction to repeat some maximum number of times while waiting for some other condition to be true. For example, you could scan through an array until you exhaust the number of array elements or until you find a certain byte using a loop like the following:

```
              mov    cx, 16        ;Maximum # of array elements.
              mov    bx, -1        ;Index into array.
LoopWhlNot0:  inc    bx            ;Move on to next array element.
              cmp    Array[bx],0   ;Does this element contain zero?
              loopne LoopWhlNot0   ;Quit if it does, or more than 16 bytes.
```

Although the loope/loopz and loopne/loopnz instructions are slower than the individual instruction from which they could be synthesized, there is one main use for these instruction forms where speed is rarely important; indeed, being faster would make them less useful - timeout loops during I/O operations. Suppose bit #7 of input port 379h contains a one if the device is busy and contains a zero if the device is not busy. If you want to output data to the port, you could use code like the following:

```
        mov   dx, 379h
```

```
WaitNotBusy:  in     al, dx          ;Get port
              test   al, 80h         ;See if bit #7 is one
              jne    WaitNotBusy     ;Wait for "not busy"
```

The only problem with this loop is that it is conceivable that it would loop forever. In a real system, a cable could come unplugged, someone could shut off the peripheral device, and any number of other things could go wrong that would hang up the system. Robust programs usually apply a timeout to a loop like this. If the device fails to become busy within some specified amount of time, then the loop exits and raises an error condition. The following code will accomplish this:

```
              mov    dx, 379h        ;Input port address
              mov    cx, 0           ;Loop 65,536 times and then quit.
WaitNotBusy:  in     al, dx          ;Get data at port.
              test   al, 80h         ;See if busy
              loopne WaitNotBusy     ;Repeat if busy and no time out.
              jne    TimedOut        ;Branch if CX=0 because we timed out.
```

You could use the loope/loopz instruction if the bit were zero rather than one.

The loopne/loopnz instruction does not affect any flags.

## 8.11 MISCELLANEOUS INSTRUCTIONS

There are various miscellaneous instructions on the 80x86 that don't fall into any category above. Generally these are instructions that manipulate individual flags, provide special processor services, or handle privileged mode operations.

There are several instructions that directly manipulate flags in the 80x86 flags register. They are

- clc Clears the carry flag
- stc Sets the carry flag
- cmc Complements the carry flag
- cld Clears the direction flag
- std Sets the direction flag
- cli Clears the interrupt enable/disable flag
- sti Sets the interrupt enable/disable flag

Note: you should be careful when using the cli instruction in your programs. Improper use could lock up your machine until you cycle the power.

The nop instruction doesn't do anything except waste a few processor cycles and take up a byte of memory. Programmers often use it as a place holder or a debugging aid. As it turns out, this isn't a unique instruction, it's just a synonym for the xchg ax, ax instruction.

The hlt instruction halts the processor until a reset, non-maskable interrupt, or other interrupt (assuming interrupts are enabled) comes along. Generally, you shouldn't use this instruction on the IBM PC unless you really know what you are doing. This instruction is not equivalent to the x86 halt instruction. Do not use it to stop your programs.

The 80x86 provides another prefix instruction, lock, that, like the rep instruction, affects the following instruction. However, this instruction has little meaning on most PC systems. Its purpose is to coordinate systems that have multiple CPUs. As systems become available with multiple processors, this prefix may finally become valuable. You need not be too concerned about this here.

The Pentium provides two additional instructions of interest to real-mode DOS programmers. These instructions are cpuid and rdtsc. If you load eax with zero and execute the cpuid instruction, the Pentium (and later processors) will return the maximum value cpuid allows as a parameter in eax. For the Pentium, this value is one. If you load the eax register with one and execute the cpuid instruction, the Pentium will return CPU identification information in eax. Since this instruction is of little value until Intel produces several additional chips in the family, there is no need to consider it further, here.

The second Pentium instruction of interest is the rdtsc (read time stamp counter) instruction. The Pentium maintains a 64 bit counter that counts clock cycles starting at reset. The rdtsc instruction copies the current counter value into the edx:eax register pair. You can use this instruction to accurately time sequences of code.

Besides the instructions presented thus far, the 80286 and later processors provide a set of protected mode instructions. This text will not consider those protected most instructions that are useful only to those who are writing operating systems. You would not even use these instructions in your applications when running under a protected mode operating system like Windows, UNIX, or OS/2. These instructions are reserved for the individuals who write such operating systems and drivers for them.

## 8.12 OUR FIRST PROGRAM

Our first program will be a real simple one. I'll first give you the code and then I'll explain it. Here's the code, cut it out and put it in a file called FIRST.

```
.model small
.stack
.data
message  db "Hello world, I'm learning Assembly !!!", "$"
.code
main  proc
   mov  ax,seg message
   mov  ds,ax
   mov  ah,09
```

```
    lea  dx,message
    int  21h
    mov  ax,4c00h
    int  21h
main  endp
end main
```

You can assemble this by typing: "tasm first [enter] tlink first [enter]" or something like: "masm first [enter] link first [enter] You must have an assembler and the link/tlink program. I'll explain the code now.

***.model small*** : Lines that start with a "." are used to provide the assembler with infomation. The word(s) behind it say what kind of info. In this case it just tells the assembler the program is small and doesn't need a lot of memory. I'll get back on this later.

***.stack*** : Another line with info. This one tells the assembler that the "stack" segment starts here. The stack is used to store temporary data. It isn't used in the program, but it must be there, because we make an .EXE file and these files MUST have a stack.

***.data*** : indicates that the data segment starts here and that the stack segment ends there.

***.code*** : indicates that the code segment starts there and the data segment ends there.

***main proc*** : Code must be in procedures, just like in C or any other language. This indicates a procedure called main starts here. ***main endp*** states that the procedure is finished. Procedures MUST have a start and end. ***end main*** : tells the assembler that the program is finished. It also tells the assembler were to start the program. At the procedure called main in this case.

***message db "xxxx"*** : DB means Define Byte and so it does. In the data-segment it defines a couple of bytes. These bytes contain the information between the brackets. "Message" is a name to indentify this byte-string. It's called an "indentifier".

***mov ax, seg message*** : AX is a register. You use registers all the time, so that's why you had to know about them before I could explain this. MOV is an instruction that moves data. It can have a few "operands" (don't worry, I'll explain these names later) Here the operands are AX and seg message. Seg message can be seen as a number. It's the number of the segment "message" is in (The data-segment) We have to know this number, so we can load the DS register with it. Else we can't get to the bit-string in memory. We need to know WHERE the bit-string is located in memory. The number is loaded in the AX register. MOV always moves data to the operand left of the comma and from the operand right of the comma.

***mov ds,ax*** : The MOV instruction again. Here it moves the number in the AX register (the number of the data segment) into the DS register. We have to load this DS register this way (with two instructions) Just typing: "mov ds,segment message" isn't possible.

***mov ah, 09*** : MOV again. This time it load the AH register with the constant value nine.

***lea dx, message*** : LEA Load Efective Address. This intructions stores the offset within the datasegment of the bit-string message into the DX register. This offset is the second thing we need to know, when we want to know where "message" is in the memory. So now we have DS:DX. See the segment explanation above.

***int 21h*** : This instruction causes an Interrupt. The processor calls a routine somewhere in memory. 21h tells the processor what kind of routine, in this case a DOS routine. INT's are very important and I'll explain more of them later, since they're also very, very complex. However, for now assume that it just calls a procedure from DOS. The procedure looks at the AH register to find out out what it has to do. In this example the value 9 in the AH register indicates that the procedure should write a bit-string to the screen.

***mov ax, 4c00h*** : Load the Ax register with the constant value 4c00h

***int 21h*** : The same INT again. But this time the AH register contains the value 4ch (AX=4c00h) and to the DOS procedure that means "exit program". The value of AL is used as an "exit-code" 00h means "No error"

That's it!!! You now fully understand this program

## QUESTION BANK

Q1. What do you mean by data movement instructions? Give description of each instruction with example.

Q2. Give brief description of conversions instructions.

Q3. What do you mean by arithmetic instructions? Give description of each instruction with example.

Q4. What do you mean by **logical instructions**? Give description of each instruction with example.

Q5. What do you mean by **shift and rotate** instructions? Give description of each instruction with example.

Q6. What do you mean by I/O instructions? Give description of each instruction with example.

Q7. What do you mean by string instructions?

Q8. What do you mean by program flow control instructions? Give description of each instruction with example.

Q9. Explain the loop instruction and loopne/loopnz instruction with the help of example.

Q10. Explain the various miscellaneous instructions like ctc, stc, cmc etc.

CHAPTER 9

# 8237 DMA Controller

## 9.1 DMA DATA TRANSFER

Direct Memory Access (DMA) is a useful and powerful mechanism for transferring data in data acquisition and control applications. A personal computer equipped with a DMA controller can transfer large amounts of data at high speeds between an I/O device and system memory. Programming a DMA operation generally consists of two tasks: programming the I/O device to use DMA, and programming the DMA controller. Programming an I/O device for DMA transfers is relatively simple. For example, a data acquisition board is programmed to collect data as usual. Typically, the only additional programming required is to set a configuration bit which enables the generation of DMA requests when data is available to be transferred. On the other hand, programming a DMA controller is more involved and most available documentation for programming DMA controllers is difficult to decipher. This application note explains how DMA works in PC/XT/AT computers and discusses how to program the 8237A DMA controller that is used in these systems.

DMA is a method of transferring data at high speeds between an I/O device and memory. A specialized processor, the DMA controller, is used to transfer the data, bypassing the CPU. The DMA controller requests control of the system bus from the CPU, transfers the data, and relinquishes the bus to the CPU at the completion of the data transfer.

The DMA controller is a special-purpose device designed explicitly for this data transfer function. It can perform all the operations required for data trsansfer (increment the memory address, decrement the count, input, write, and test for operation complete) in *one* bus cycle. Thus the bus is only tied up for one bus cycle per byte transferred and the CPU can continue to execute programs while the I/O is in progress.

**DMA controller can operates in two modes:** Cycle Stealing and Burst Modes. DMA controllers can operate in a ***cycle stealing*** mode in which they take over the bus for each byte of data to be transferred and then return control to the CPU. They can also operate in ***burst mode*** in which a block of data is transferred before returning bus control to the CPU. The choice depends on the speed at which data is arriving relative to the bus bandwidth and whether a particular application will allow the CPU to be locked off the bus for the duration of one transfer. DMA controllers can transfer data in a two-step process by reading a value from one port or address in one bus cycle and writing that value to another port or address in a second bus cycle. It is also possible for the DMA controller to carry out read and write operations simultaneously. In this case the data is transferred directly between the I/O device and memory in the same bus cycle. This is the mode of operation used in the IBM PC.

## 9.2 DMA CONTROLLER FOR 8085

Direct Memory Access is an I/O technique commonly used for high-speed data transfer; for example, data transfer between system memory and a floppy disk. In status check I/O and interrupt I/O, data transfer is relatively slow because each instruction needs to be fetched and executed. In DMA, the MPU releases the control of the buses to a device called a DMA controller. The controller manages data transfer between memory and a peripheral under its control, thus bypassing the MPU. Conceptually, this is an important I/O technique; it introduces two new signals available on the 8085-HOLD and HLDA (Hold Acknowledge).

- **HOLD** – This is an active high input signal to the 8085 from another master requesting the use of the address and data buses. After receiving the Hold request, the MPU relinquishes the buses in the following machine cycle. All buses are tri-stated and the Hold Acknowledge (HLDA) signal is sent out. The MPU regains the control of the buses after HOLD goes low.
- **HLDA (Hold Acknowledge)** – This is an active high output signal indicating that the MPU is relinquishing control of the buses.

A DMA controller uses these signals as if it were a peripheral requesting the MPU for the control of the buses. The MPU communicates with the controller by using the Chip Select line, buses, and control signals. However, once the controller has gained control, it plays the role of a processor for data transfer. To perform this function the DMA controller should have

1. A data bus,
2. An address bus,
3. Read/Write control signals, and
4. Control signals to disable its role as a peripheral and to enable its role as a processor.

This process is called switching from the slave mode to the master mode. For all practical purpose, the DMA controller is a processor capable only of copying data at high speed from one location to another location. As an illustration, a programmable DMA controller, the Intel 8237, is described below.

### 9.2.1 The 8237 DMA Controller

The 82C37A is an enhanced version of the industry standard 8237A Direct Memory Access (DMA) controller, fabricated using advanced 2 micron CMOS process. Pin compatible with NMOS designs, the 82C37A offers increased functionality, improved performance, and dramatically reduced power consumption. The fully static design permits gated clock operation for even further reduction of power.

The 82C37A controller can improve system performance by allowing external devices to transfer data directly to or from system memory. Memory-to-memory transfer capability is also provided, along with a memory block initialization feature. DMA requests may be generated by either hardware or software, and each channel is independently programmable with a variety of features for flexible operation. The 82C37A is designed to be used with an external address latch, such as the 82C82, to demultiplex the most significant 8-bits of address. The 82C37A can be used with industry standard microprocessors such as 80C286, 80286, 80C86, 80C88, 8086, 8088, 8085, Z80, NSC800, 80186 and many more.

Multimode programmability allows the user to select from three basic types of DMA services, and reconfiguration under program control is possible even with the clock to the controller stopped. Each channel has a full 64K address and word count range, and may be programmed to autoinitialize these registers following DMA termination (end of process).

The 8237 is a programmable Direct Memory Access controller (DMA) houses in a 40-pin package. It has four independent channels with each channel capable of transferring 64K bytes. It must interface with two types of devices: the MPU and peripherals such as floppy disks. As mentioned earlier, the DMA plays two roles in a given system: It is an I/O to the microprocessor (slave mode) and it is a data transfer processor to peripherals such a floppy disks (master mode). Many of its signals that are input in the I/O mode become outputs in the processor mode. It also needs additional signal lines to communicate with the addresses of 64K data bytes, and these signals must be generated externally by the using latches and buffers. The 8237 is a complex device. To maintain clarity, the following discussion is divided into five segments: DMA channels and interfacing, DMA signals, system interface, programming DMA execution.

### 9.2.2 DMA Channels And Interfacing

Figure 9.1 shows a logical pin out and internal registers of the 8237. It also shows the interface with the 8085 using a 3-to-8 decoder.

The 8237 has four independent channels, CH0 to CH3. Internally, two 16-bit registers are associated with each channel: One is used to load a starting address of the byte to be copied and the second is used to load a count of the number of bytes to be copied. Figure 9.1 shows eight such registers that can be accessed by the MPU. The addresses of these registers are determined by four address lines, $A_3$ to $A_0$, and the Chip Select (CS) signal. Address 0000 on lines $A_3$-$A_0$ selects CH0 Memory Address Register (MAR) and address 0001 selects the next register, CH0 Count.

Similarly, all the remaining registers are selected in sequential order. The last eight registers are used to write commands or read status as shown. In Figure 9.1, the MPU accesses the DMA controller by asserting the signal $Y_0$ of the decoder. Therefore, the addresses of these internal registers range from 00 to 0FH as follows:

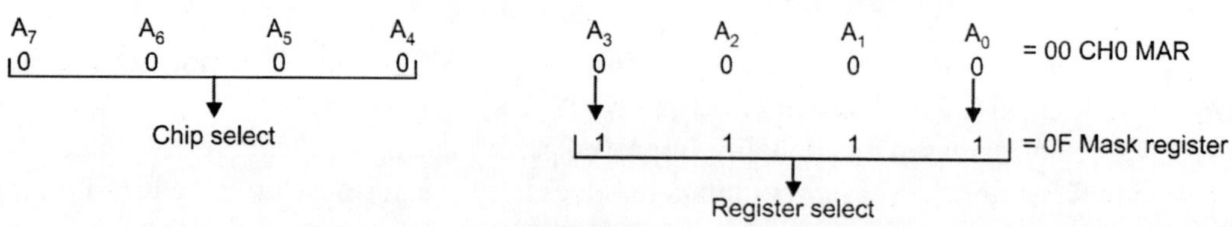

### 9.2.3 DMA Signals

In Figure 9.1, signals are divided into two groups: (1) one group of signals shown on the left of the 8237 is used for interfacing with the MPU; (2) the second group shown on the right-hand side of the 8237 is for communicating with peripherals. Some of these signals are bidirectional and their functions are determined by the DMA mode of operation (I/O or processor). The signals that are necessary to understand the DMA operation are explained as follows; the remaining signals are listed in Appendix D.

- **DREQ0-DREQ3-DMA Request:** These are four independent, asynchronous input signals to the DMA channels from peripheral such as floppy disks and the hard disk. To obtain DMA service, a request is generated by activating the DREQ line of the channel.

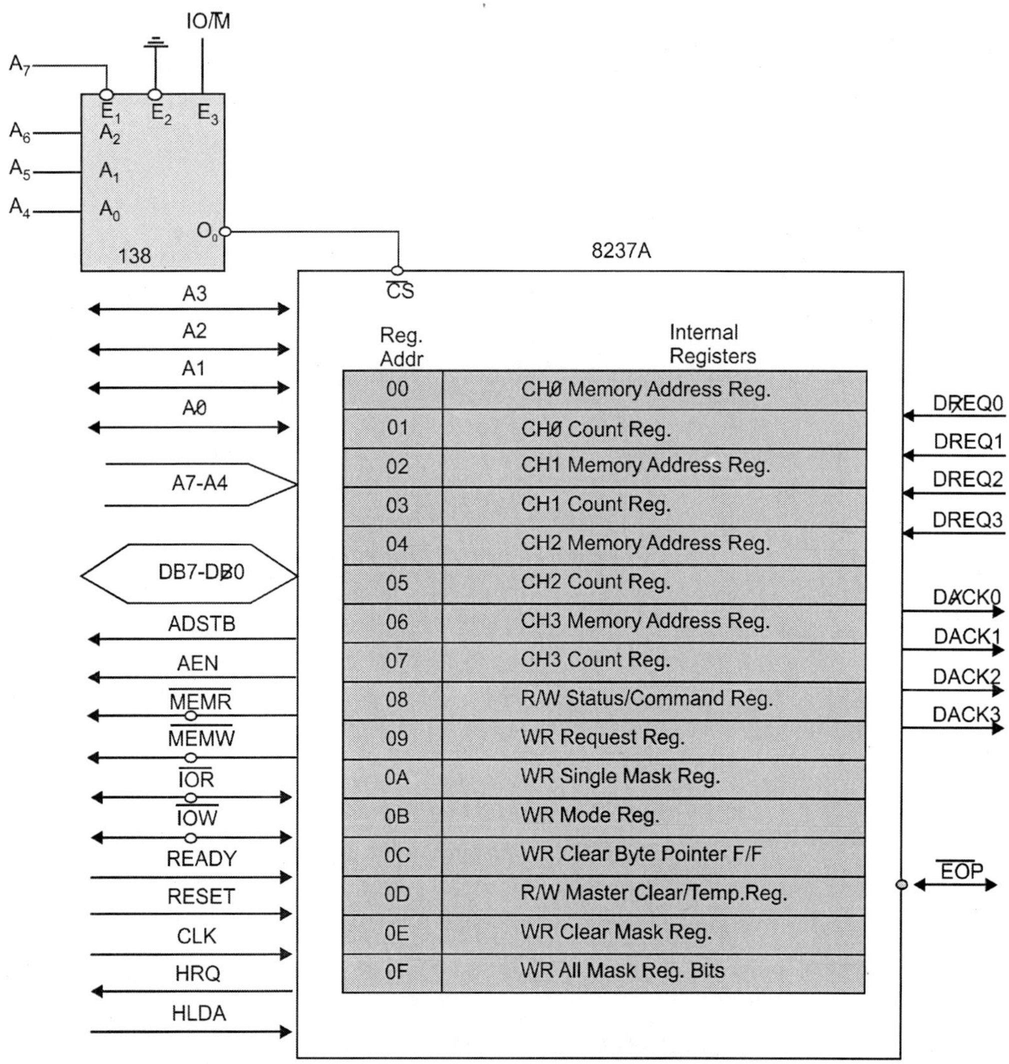


**Figure 9.1:** 8237A-DMA Controller with Internal registers

- **DCK0-DACK3-DMA Acknowledge:** These are output lines to inform the individual peripherals that a DMA is granted DREQ and DACK are equivalent to handshake signals in I/O devices.
- **AEN and ADSTB-Address Enable and Address Strobe:** These are active high output signals that that are used to latch a high-order address byte to generate a 16-bit address.
- $\overline{\text{MEMR}}$ **and** $\overline{\text{MEMW}}$ **– Memory Read and Memory Write:** These are output signals used during the DMA cycle to write and read from memory.
- **$A_3$-$A_0$ and $A_7$-$A_4$-Address:** $A_3$-$A_0$ are bidirectional address lines. They are used as inputs to access control registers as shown in the previous section. During the DMA cycle, these lines are used as output lines to generate a low-order address that is combined with the remaining address lines $A_7$-$A_4$.
- **HRQ and HLDA – Hold request and Hold Acknowledge:** HRQ is an output signal used to request the MPU control of the system bus. After receiving the HRQ, the MPU completes the bus cycle in process and issues the HLDA signal.

### 9.2.4 System Interface

The DMA is used to transfer data bytes between I/O (such as a floppy disk) and system memory (or from memory to memory) at high speed. It includes eight data lines, four control signals ($\overline{\text{IOR}}$, $\overline{\text{IOW}}$, $\overline{\text{MEMR}}$, and $\overline{\text{MEMW}}$), and eight address lines ($A_7$-$A_0$). However, it needs 16 address lines to access 64K bytes. Therefore, an additional eight lines must be generated as shown in Figure 9.2.

When a transfer begins, the DMA places the low-order byte on the address bus and the high-order byte on the data bus and asserts AEN (Address Enable) and ADSTB (Address Strobe). These two signals are used to latch the high-order byte from the data bus; thus, it places the 16-bit address on the system bus. After the transfer of the first byte, the latch is updated when the lower bytes generates a carry (or borrow). Figure 9.2 shows two latches: one latch (373 # 1) to latch a high-order address from the data bus by using the AEN and ADSTB signals, and the second latch (373 # 2) to demultiplex the 8085 bus and generate the low-order address bus by using the ALE (Address Latch Enable from the 8085) signal. The AEN signal is connected to the OE signal of the second latch to disable the low-order address bus from the 8085 when the first latch is enabled to latch the high-order byte of the address.

### 9.2.5 DMA Execution

The process of data transfer from the peripheral to the system memory under the DMA controller can be classified under two modes: the slave mode and the master mode.

**SLAVE MODE:** In the slave mode, the DMA controller is treated as a peripheral, using the following steps:

1. The MPU selects the DMA controller through Chip Select.

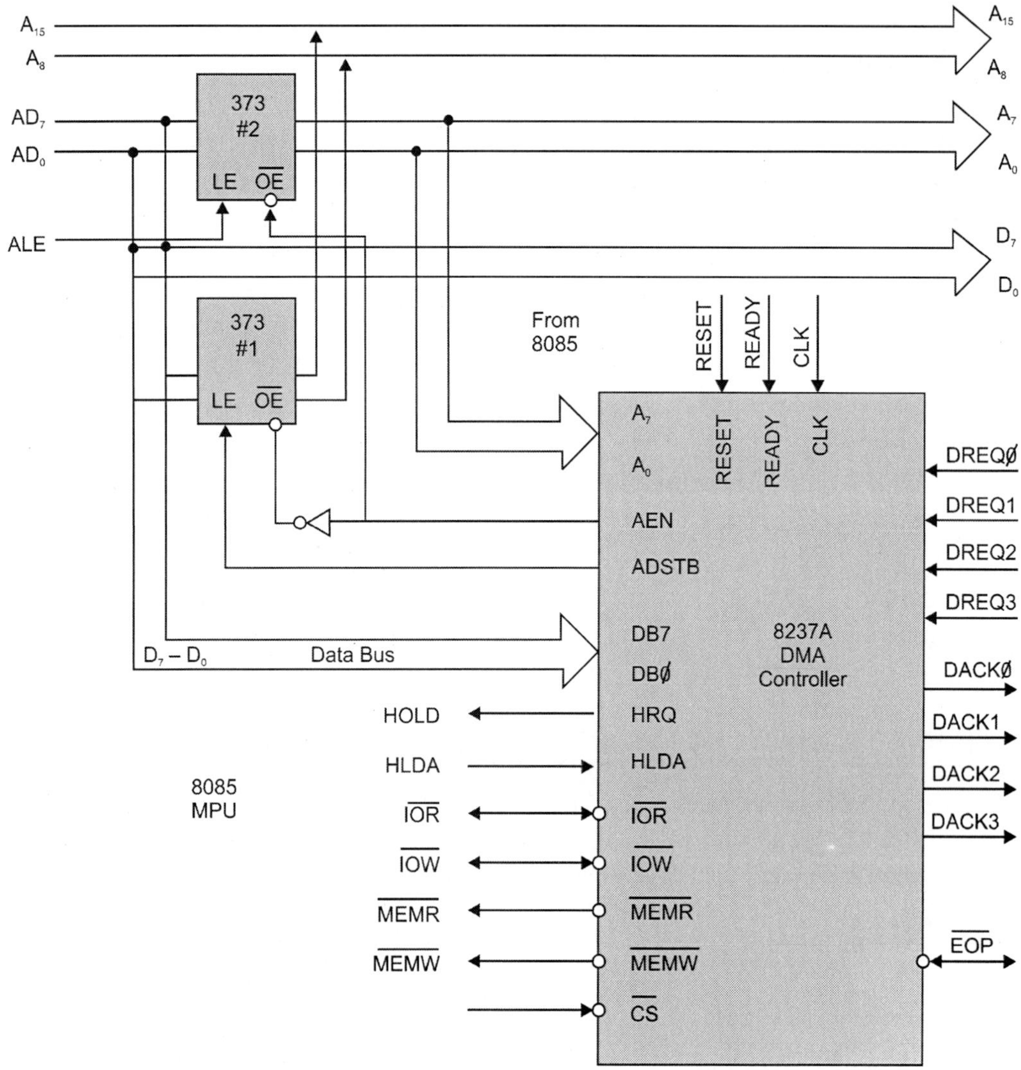


**Figure 9.2:** Interfacing 8237A-DMA Controller with the 8085

2. The MPU writes the control words as illustrated in channel registers and command/status registers by using control signals $\overline{IOW}$ and $\overline{IOR}$.

In this mode, the output signals of the 8237, such as $A_7$-$A_4$, $\overline{MEMW}$, and $\overline{MEMR}$, are in tri-state.

**MASTER MODE:** After the initialization, the 8237 in master mode keeps checking for a DMA request, and the steps in data transfer can be listed as follows:

1. When the peripheral is ready for data transfer, it sends a high signal to DRQ.

2. When the DRQ has been received and the channel enabled, the control logic sets HRQ (Hold Request) high. (HRQ is connected to the HOLD signal of the 8085.)
3. In the next cycle, the MPU relinquishes the buses and sends the HLDA (Hold Acknowledge) signal to the 8237.
4. After receiving the HLDA signal, the DMA asserts AEN (Address Enable) signal high. The high AEN signal disables 373 Latch #2, thus disconnecting the demultiplexed bus $A_7$-$A_0$ of the MPU and enables 373 Latch #1 through an inverter. Next, the DMA asserts ADSTB (Address Strobe) high that is connected to Latch Enable (LE) of 373 Latch #1 and places the contents of the data bus, which is a high-order byte of the starting address, on $A_{15}$-$A_8$. At the same time, the DMA also outputs the low order address $A_7$-$A_0$ on the low-order address bus.
5. When the entire address $A_{15}$-$A_0$ is available on the address bus, the DMA sends DACK to the peripheral.
6. The DMA controller continues the data transfer by asserting the necessary control signals ($\overline{IOR}$, $\overline{IOW}$, $\overline{MEMR}$, or $\overline{MEMW}$) until DACK remains high.
7. At the end of the data transfer, the DMA asserts $\overline{EOP}$ (End of Process) signal low that can be used to inform the peripheral that the data transfer is complete. The DMA data transfer can also be terminated by sending a low signal to EOP from outside.

## 9.3 BLOCK DIAGRAM OF 82C37A DMA CONTROLLER

The 82C37A direct memory access controller is designed to improve the data transfer rate in systems which must transfer data from an I/O device to memory, or move a block of memory to an I/O device. It will also perform memory-to-memory block moves, or fill a block of memory with data from a single location. Operating modes are provided to handle single byte transfers as well as discontinuous data streams, which allows the 82C37A to control data movement with software transparency.

The DMA controller is a state-driven address and control signal generator, which permits data to be transferred directly from an I/O device to memory or vice versa without ever being stored in a temporary register. This can greatly increase the data transfer rate for sequential operations, compared with processor move or repeated string instructions. Memory-to-memory operations require temporary internal storage of the data byte between generation of the source and destination addresses, so memory-to-memory transfers take place at less than half the rate of I/O operations, but still much faster than with central processor techniques.

The block diagram of the 82C37A is shown in Figure 9.3. The timing and control block, priority block, and internal registers are the main components. Pin diagram of 82C37 is also shown in Figure 9.4 with pin descriptions.

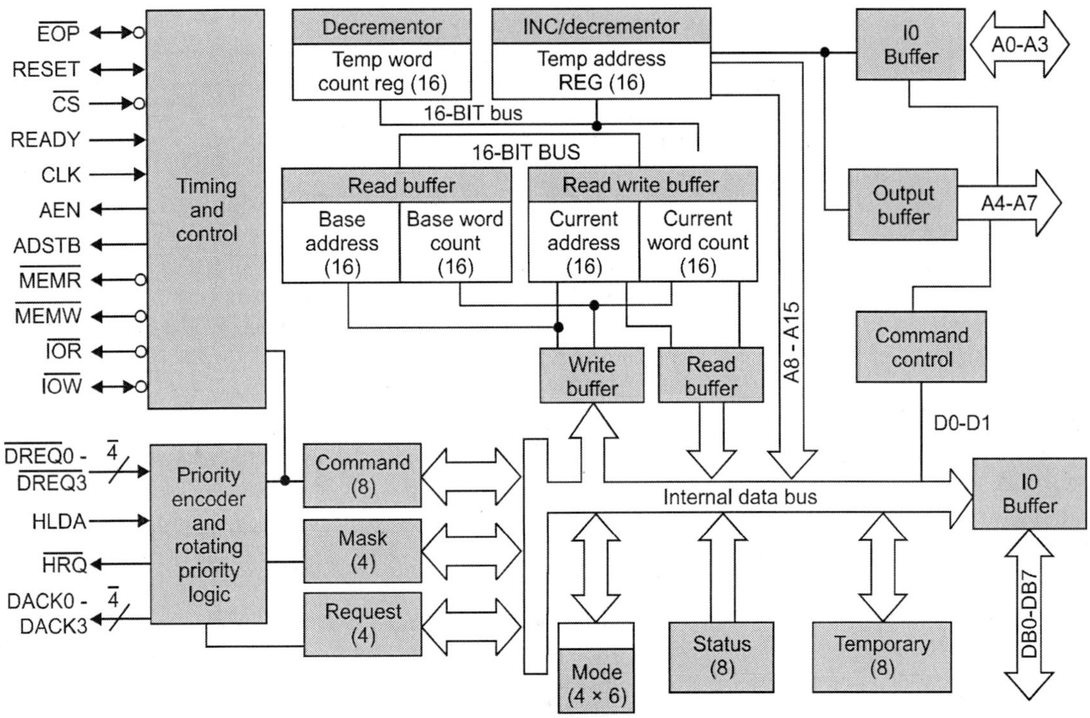


**Figure 9.3:** Block Diagram of 8237

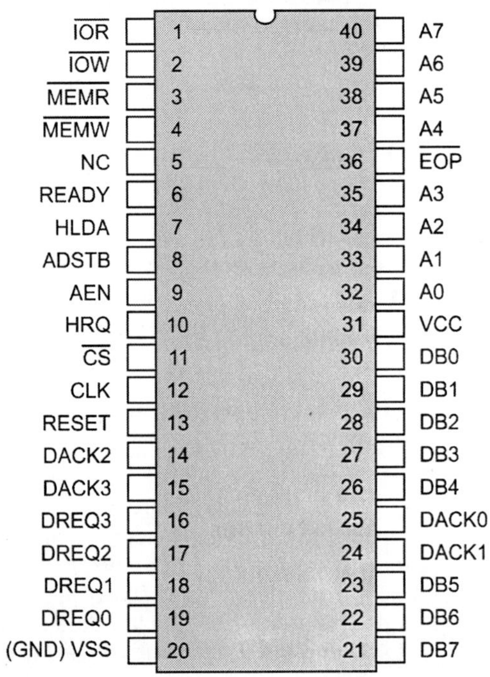


**Figure 9.4:** Pin Diagram of 8237 DMA Controller

## 9.4 PIN DESCRIPTION OF 8237 DMA CONTROLLER

| Symbol | Pin Number | Type | Description |
|---|---|---|---|
| $V_{CC}$ | 31 | | $V_{CC}$. is the + 5v power supply pin. A 0.1 μF capacitor between pins 31 and 20 is recommended fro decoupling. |
| GND | 20 | | Ground |
| CLK | 12 | I | **CLOCK INPUT:** The Clock input is used to generate the timing signals which control 82C37A operations. This input may be driven from DC to 12.5MHz for the 82C37A-12, from DC to 8MHz for the 82C37A, or from DC to 5MHz for the 82C37A-5. The clock may be stopped in either state for standby operation. |
| $\overline{CS}$ | 11 | I | **CHIP SELECT:** Chip Select is an active low input used to enable the controller onto the data bus for CPU communications. |
| RESET | 13 | I | **RESET:** This is an active high input which clears the command, stauts, Request, and Temporary registers, the First/Last Flip--Flop, and the mode register counter. The Mask register is set to ignore requests. Following a Reset, the controller is in an idle cycle. |
| READY | 6 | I | **READY:** This signal can be used to extend the memory read and write pulses from the 82C37A to accommodate slow memories or 1/O devices. READY must not make transitions during its specified set-up and hold times. See Figure 12 for timing. READY is ignored in verify transfer mode. |
| HLDA | 7 | I | **HOLD ACKNOWLEDGE:** The active high Hold Acknowledge from the CPU indicates that it has relinquished control of the system busses. HLDA is a synchronous input and must not transition during its specified set-up time. There is an implied hold time (HLDA inactive) of TCH from the rising edge of CLK, during which time HLDA must not transition. |
| DREQ0-dREQ3 | 16–19 | I | **DMA REQUEST:** The DMA Request (DREQ) lines are individual asynchronous channel request inputs used by peripheral circuits to obtain DMA service. In Fixed priority, DREQ0 has the highest priority and DREQ3 has the lowest priority. A request is generated by activating the DREQ line of a channel, DACK will acknowledge the recognition of a DREQ signal. Polarity of DREQ is programmable. RESET initializes these lines to active high, DREQ must be maintained until the corresponding DACK goes active. DREQ will not be recognized while the clock is stopped. Unused DREQ inputs should be pulled High or Low (inactive) and the corresponding mask bit set. |
| DBO-DB7 | 21–23<br>26–30 | I/O | **DATA BUS:** The Data Bus lines are bidirectional three-state signals connected to the system data bus. The outputs are enabled in the program condition during the I/O Read to output the contents of a register to the CPU. The outputs are disabled and the inputs are read during an I/O Write cycle when the CPU is programming the 82C37A control registers. During DMA cycles, the most significant 8-bits of the address are output onto the data bus to be strobed into an external latch by ADSTB. In memory-to-memory operations, data from the |

| Symbol | Pin Number | Type | Description |
|---|---|---|---|
| | | | memory enters the 82C37A on the data bus during the read-from-memory transfer then during the write-to memory transfer, the data bus outputs writ the data into the new memory location. |
| $\overline{IOR}$ | 1 | I/O | **I/O READ:** I/O Read is a bidirectional active low three-state line. In the idle cycle, it is an input control signal used by the CPU to read the control registers. IN the Active cycle, it is an output control signal used by the 82C37A to access data from the peripheral during a DMA Write transfer. |
| $\overline{IOW}$ | 2 | I/O | **I/O WRITE:** I/O Write is a bidirectional active low three-state line. In the Idle cycle, it is an input control signal used by the CPU to load information into the 82C37A, In the active cycle, it is an output control signal used by the 82C37A to load data to the peripheral during a DMA Read transfer. |
| $\overline{EOP}$ | 36 | I/O | **END OF PROCESS:** End of Process $\overline{\text{EOP}}$ is an active low bidirectional signal. Infromation concering the completion of DMA services is available at the bidirectional $\overline{\text{EOP}}$ pin.<br>The 82C37A allows an external signal to terminate an active DMA service by pulling the $\overline{\text{EOP}}$ pin low. A pulse is generated by the 82C37A when terminal count (TC) for any channel is reached, except for channel 0 in memory-to-memory mode. During memory-to-memory transfers, $\overline{\text{EOP}}$ will be output when the TC for channel 1 occurs.<br>The $\overline{\text{EOP}}$ pin is driven by an open driven transistor on-chip, and requires an external pull-up resistor to $V_{CC}$.<br>When an $\overline{\text{EOP}}$ pulse occurs, whether internally or externally generated, the 82C37A will terminate the service, and if autointialize is enabled, the base registers will be written to the current registers of that channel. The mask bit and TC bit in the status word will be set for the currently active channel by $\overline{\text{EOP}}$ unless the channel is programmed fro autointialize. In that case, the mask bit remains clear. |
| A0-A3 | 32–35 | I/O | **ADDRESS:** The four least significant address lines are bidirectional three-state signals. IN the Idle cycle, they are inputs and are used by the 82 C37A to address the control register to be loaded or read. In the Active cycle, they are outputs and provide the lower 4-bits of the output address. |
| A4-A7 | 37–40 | O | **ADDRESS:** The four most significant address lines are three-state outputs and provide 4-bits of address. These lines are enabled only during the DMJA service. |
| HRQ | 10 | O | **HOLD REQUEST:** The Hold Request (HRQ) output is used to request control of the system bus. When a DREQ occurs and the corresponding mask bit is clear, or a software DMA request is made, the 82C37A issues HRQ. The HLDA signal then informs the controller when access to the system busses is permitted. For stand-alone operation where the 82C37A always controls the busses, HRQ may be tied to HLDA. This will result in one SO state before the transfer. |

| Symbol | Pin Number | Type | Description |
|---|---|---|---|
| DACK0-<br>DACK3 | 14,15<br>24,25 | O | **DMA Acknowledge:** DMA acknowledge is used to notify the individual peripherals when one has been granted a DMA cycle. The sense of these lines is programmable. RESET initializes them to active low. |
| AEN | 9 | O | **ADDRESS ENABLE:** Address Enable enables the 8-bit latch containing the upper 8 address bits onto the system address bus. AEN can also be used to disable other system bus drivers during DMA transfers AEN is active high. |
| ADSTB | 8 | O | **ADDRESS STROBE:** This is an active high signal used to control latching of the upper address byte. It will drive directly the strobe input of external transparent octal latches, such as the 82C82, During block operations, ADSTB will only be issued when the upper address byte must be updated, thus speeding operation through elimination of S1 states. ADSTB timing is referenced to the falling edge of the 82C37A clock. |
| $\overline{MEMR}$ | 3 | O | **MEMORY READ:** The Memory Read signal is an active low three-state output used to access data from the selected memory location during a DMA Read or a memory-to-memory transfer. |
| $\overline{MEMW}$ | 4 | O | **MEMORY WRITE:** The memory Write signal is an active low three-state output used to write data to the selected memory location during a DMA Write or a memory-to-memory transfer. |
| NC | 5 | | **NO CONNECT:** Pin 5 is open and should not be tested for continuity. |

The maximum data transfer rates obtainable with the 82C37A are shown in Figure 9.5. Figure 9.6 lists the name and size of the internal registers. The timing and control block derives internal timing from clock input, and generates external control signals. The Priority Encoder block resolves priority contention between DMA channels requesting service simultaneously.

| 82C37A Transfer Type | 5 MHz | 8 MHz | 12.5 MHz | Unit |
|---|---|---|---|---|
| Compressed | 2.50 | 4.00 | 6.25 | MByte/sec |
| Normal I/O | 1.67 | 2.67 | 4.17 | MByte/sec |
| Memory-to-Memory | 0.63 | 1.00 | 1.56 | MByte/sec |

**Figure 9.5:** DMA Transfer Rates

## 9.5 DMA OPERATION

In a system, the 82C37A address and control outputs and data bus pins are basically connected in parallel with the system busses. An external latch is required for the upper address byte. While inactive, the controller's outputs are in a high impedance state. When activated by a DMA request and bus control is relinquished by the host, the 82C37A drives the busses and generates the control signals to perform the data transfer. The operation performed by activating one of the four DMA request inputs has previously been programmed into the controller via the Command, Mode, Address, and Word Count registers. For example, if a block of data is to be transferred from RAM to an I/O device, the starting address of the data is loaded into the 82C37A Current and Base Address registers for a particular channel, and the length of the block is loaded into the channel's Word Count register. The corresponding Mode register is programmed for a memory-to-I/O operation (read transfer), and various options are selected by the Command register and the other Mode register bits. The channel's mask bit is cleared to enable recognition of a DMA request (DREQ). The DREQ can either be a hardware signal or a software command.

Once initiated, the block DMA transfer will proceed as the controller outputs the data address, simultaneous $\overline{MEMR}$ and $\overline{IOW}$ pulses, and selects an I/O device via the DMA acknowledge (DACK) outputs. The data byte flows directly from the RAM to the I/O device. After each byte is transferred, the address is automatically incremented (or decremented) and the word count is decremented. The operation is then repeated for the next byte. The controller stops transferring data when the Word Count register underflows or an external $\overline{EOP}$ is applied.

| Name | Size | Number |
|---|---|---|
| Base Address Registers | 16-Bits | 4 |
| Base Word Count Registers | 16-Bits | 4 |
| Current Address Registers | 16-Bits | 4 |
| Current Word Count Registers | 16-Bits | 4 |
| Temporary Address Register | 16-Bits | 1 |
| Temporary Word Count Register | 16-Bits | 1 |
| Status Register | 8-Bits | 1 |
| Command Register | 8-Bits | 1 |
| Temporary Register | 8-Bits | 1 |
| Mode Register | 6-Bits | 4 |
| Mask Register | 4-Bits | 1 |
| Request Register | 4-Bits | 1 |

**Figure 9.6:** 82C37A Internal Registers

To further understand 82C37A operation, the states generated by each clock cycle must be considered. The DMA controller operates in two major cycles, active and idle. After being programmed, the controller is normally idle until a DMA request occurs on an unmasked channel, or a software request is given. The 82C37A will then request control of the system busses and enter the active cycle. The active cycle is composed of several internal states, depending on what options have been selected and what type of operation has been requested.

The 82C37A can assume seven separate states, each composed of one full clock period. State I (SI) is the idle state. It is entered when the 82C37A has no valid DMA requests pending, at the end of a transfer sequence, or when a Reset or Master Clear has occurred. While in SI, the DMA controller is inactive but may be in the Program Condition (being programmed by the processor).

State 0 (S0) is the first state of a DMA service. The 82C37A has requested a hold but the processor has not yet returned an acknowledge. The 82C37A may still be programmed until it has received HLDA from the CPU. An acknowledge from the CPU will signal the DMA transfer may begin. S1, S2, S3, and S4 are the working state of the DMA service. If more time is needed to complete a transfer than is available with normal timing, wait states (SW) can be inserted between S3 and S4 in normal transfers by the use of the Ready line on the 82C37A. For compressed transfers, wait states can be inserted between S2 and S4. See timing Figures 9.7 and 9.8. Note that the data is transferred directly from the I/O device to memory (or vice versa) with $\overline{IOR}$ and $\overline{MEMW}$ (or $\overline{MEMR}$ and $\overline{IOW}$) being active at the same time. The data is not read into or driven out of the 82C37A in I/O-to-memory or memory-to-I/O DMA transfers.

Memory-to-memory transfers require a read-from and a write to memory to complete each transfer. The states, which resemble the normal working states, use two-digit numbers for identification. Eight states are required for a single transfer. The first four states (S11, S12, S13, S14) are used for the read-from-memory half and the last four state (S21, S22, S23, S24) for the write-to-memory half of the transfer.

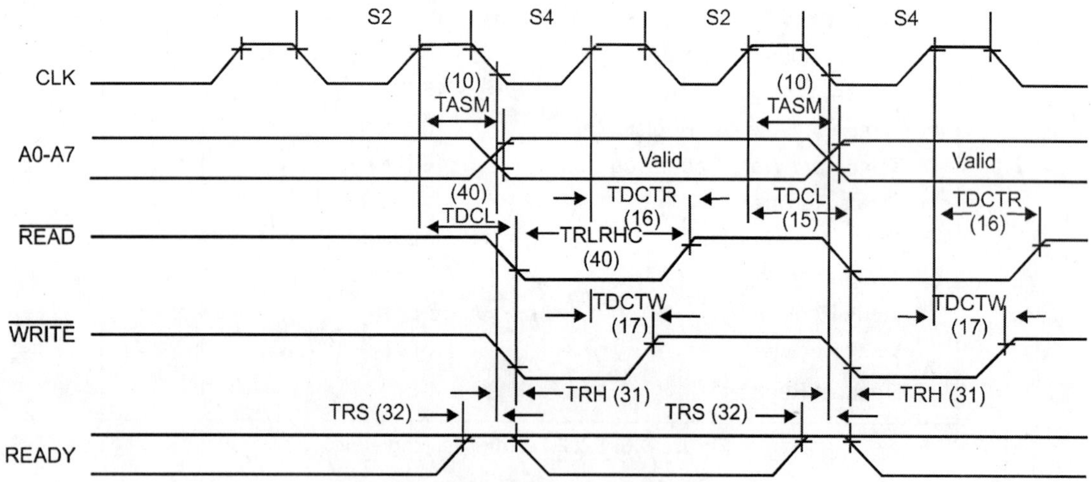


**Figure 9.7:** Compressed Transfer

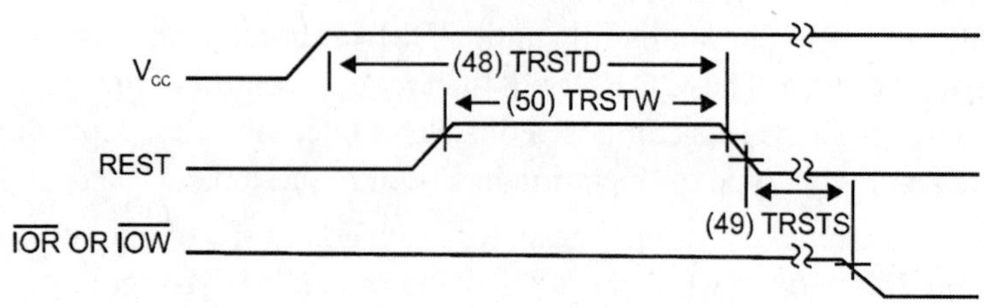


**Figure 9.8:** Reset

### 9.5.1 IDLE Cycle

When no channel is requesting service, the 82C37A will enter the idle cycle and perform "SI" states. In this cycle, the 82C37A will sample the DREQ lines on the falling edge of every clock cycle to determine if any channel is requesting a DMA service.

Note that for standby operation where the clock has been stopped, DMA requests will be ignored. The device will respond to $\overline{CS}$ (chip select), in case of an attempt by the microprocessor to write or read the internal registers of the 82C37A. When $\overline{CS}$ is low and HLDA is low, the 82C37A enters the Program Condition. The CPU can now establish, change or inspect the internal definition of the part by reading from or writing to the internal registers.

The 82C37A may be programmed with the clock stopped, provided that HLDA is low and at least one rising clock edge has occurred after HLDA was driven low, so the controller is in an SI state. Address lines A0-A3 are inputs to the device and select which registers will be read or written. The $\overline{IOR}$ and $\overline{IOW}$ lines are used to select and time the read or write operations. Due to the number and size of the internal registers, an internal flip-flop called the First/Last Flip-Flop is used to generate an additional bit of address. The bit is used to determine the upper or lower byte of the 16-bit Address and Work Count registers.

The flip-flop is reset by Master Clear or RESET. Separate software commands can also set or reset this flip-flop. Special software commands can be executed by the 82C37A in the Program Condition. These commands are decoded as sets of addresses with $\overline{CS}$, $\overline{IOR}$, and $\overline{IOW}$. The commands do not make use of the data bus. Instructions include Set and Clear First/Last Flip-Flop, Master Clear, Clear Mode Register Counter, and Clear Mask Register.

### 9.5.2 Active Cycle

When the 82C37A is in the idle cycle, and a software request or an unmasked channel requests a DMA service, the device will issue HRQ to the microprocessor and enter the Active cycle. It is in this cycle that the DMA service will take place, in one of four modes:

1. **Single Transfer Mode -** In Single Transfer mode, the device is programmed to make one transfer only. The word count will be decremented and the address decremented or incremented following each transfer. When the word count "rolls over" from zero to FFFFH, a terminal count bit in the status register is set, an $\overline{EOP}$ pulse is generated, and the channel will autoinitialize if this option has been selected. If not programmed to autoinitialize, the mask bit will be set, along with the TC bit and $\overline{EOP}$ pulse.

   DREQ must be held active until DACK becomes active. If DREQ is held active throughout the single transfer, HRQ will go inactive and release the bus to the system. It will again go active and, upon receipt of a new HLDA, another single transfer will be performed, unless a higher priority channel takes over. In 8080A, 8085A, 80C88, or 80C86 systems, this will ensure one full machine cycle execution between DMA transfers. Details of timing between the 82C37A and other bus control protocols will depend upon the characteristics of the microprocessor involved.

2. **Block Transfer Mode -** In Block Transfer mode, the device is activated by DREQ or software request and continues making transfers during the service until a TC, caused by word count going to FFFFH, or an external End of Process ( $\overline{EOP}$ ) is encountered. DREQ need only be held active until DACK becomes active. Again, an Auto initialization will occur at the end of the service if the channel has been programmed for that option.
3. **Demand Transfer Mode -** In Demand Transfer mode the device continues making transfers until a TC or external EOP is encountered, or until DREQ goes inactive. Thus, transfer may continue until the I/O device has exhausted its data capacity. After the I/O device has had a chance to catch up, the DMA service is reestablished by means of a DREQ. During the time between services when the microprocessor is allowed to operate the intermediate values of address and word count are stored in the 82C37A Current Address and Current Word Count registers. Higher priority channels may intervene in the demand process, once DREQ has gone inactive. Only an $\overline{EOP}$ can cause an Autoinitialization at the end of service. $\overline{EOP}$ is generated either by TC or by an external signal.
4. **Cascade Mode -** This mode is used to cascade more than one 82C37A for simple system expansion. The HRQ and HLDA signals from the additional 82C37A are connected to the DREQ and DACK signals respectively of a channel for the initial 82C37A.This allows the DMA requests of the additional device to propagate through the priority network circuitry of the preceding device. The priority chain is preserved and the new device must wait for its turn to acknowledge requests. Since the cascade channel of the initial 82C37A is used only for prioritizing the additional device, it does not output an address or control signals of its own. These could conflict with the outputs of the active channel in the added device. The initial 82C37A will respond to DREQ and generate DACK but all other outputs except HRQ will be disabled. An external $\overline{EOP}$ will be ignored by the initial device, but will have the usual effect on the added device. Figure 9.9 shows two additional devices cascaded with an initial device using two of the initial device's channels. This forms a two-level DMA system. More 82C37As could be added at the second level by using the remaining channels of the first level. Additional devices can also be added by cascading into the channels of the second level devices, forming a third level.

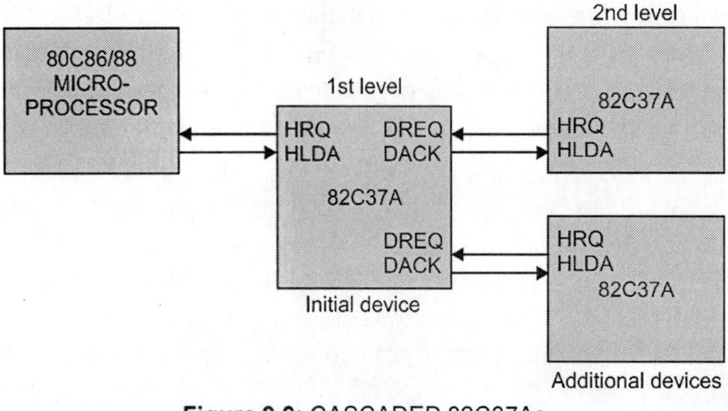


**Figure 9.9:** CASCADED 82C37As

When programming cascaded controllers, start with the first level device (closest to the microprocessor). After RESET, the DACK outputs are programmed to be active low and are held in the high state. If they are used to drive HLDA directly, the second level device(s) cannot be programmed until DACK polarity is selected as active high on the initial device. Also, the initial device's mask bits function normally on cascaded channels, so they may be used to inhibit second level services.

### 9.5.3 Transfer Types

Each of the three active transfer modes can perform three different types of transfers. These are Read, Write and Verify. Write transfers move data from an I/O device to the memory by activating MEMW and IOR. Read transfers move data from memory to an I/O device by activating MEMR and IOW. Verify transfers are pseudo-transfers. The 82C37A operates as in Read or Write transfers generating addresses and responding to $\overline{\text{EOP}}$, etc.; however the memory and I/O control lines all remain inactive. Verify mode is not permitted for memory-to-memory operation. READY is ignored during Verify transfers.

#### *1. Autoinitialize*

By setting bit 4 in the Mode register, a channel may be set up as an Autoinitialize channel. During Autoinitialization, the original values of the Current Address and Current Word Count registers are automatically restored from the Base Address and Base Word Count registers of the channel following $\overline{\text{EOP}}$. The base registers are loaded simultaneously with the current registers by the microprocessor and remain unchanged throughout the DMA service. The mask bit is not set when the channel is in Autoinitialize mode. Following Autoinitialization, the channel is ready to perform another DMA service, without CPU intervention, as soon as a valid DREQ is detected, or software request made.

#### *2. Memory-To-Memory*

To perform block moves of data from one memory address space to another with minimum of program effort and time, the 82C37A includes a memory-to-memory transfer feature. Setting bit 0 in the Command register selects channels 0 and 1 to operate as memory-to-memory transfer channels.

The transfer is initiated by setting the software or hardware DREQ for channel 0. The 82C37A requests a DMA service in the normal manner. After HLDA is true, the device, using four-state transfers in Block Transfer mode, reads data from the memory. The channel 0 Current Address register is the source for the address used and is decremented or incremented in the normal manner. The data byte read from the memory is stored in the 82C37A internal Temporary register. Another four-state transfer moves the data to memory using the address in channel one's Current Address register and incrementing or decrementing it in the normal manner. The channel 1 Current Word Count is decremented. When the word count of channel 1 decrements to FFFFH, a TC is generated causing an $\overline{\text{EOP}}$ output, terminating the service, and setting the channel 1 TC bit in the Status register. The channel 1 mask bit will also be set, unless the channel 1 mode register is programmed for autoinitialization. Channel 0 word count decrementing to FFFFH will not set the channel 0 TC bit in the status register nor generate an $\overline{EOP}$, nor set the channel 0 mask bit in this mode. It will cause an autoinitialization of channel 0, if that option has been selected.

If full Autoinitialization for a memory-to-memory operation is desired, the channel 0 and channel 1 word counts must be set to equal values before the transfer begins. Otherwise, if channel 0 underflows before channel 1, it will autoinitialize and set the data source address back to the beginning of the block. If the channel 1 word count underflows before channel 0, the memory-to-memory DMA service will terminate, and channel 1 will autoinitialize but channel 0 will not. In memory-to-memory mode, Channel 0 may be programmed to retain the same address for all transfers. This allows a single byte to be written to a block of memory. This channel 0 address hold feature is selected by setting bit 1 in the Command register.

The 82C37A will respond to external $\overline{EOP}$ signals during memory-to-memory transfers, but will only relinquish the system busses after the transfer is complete (i.e. after an S24 state). It should be noted that an external $\overline{EOP}$ cannot cause the channel 0 Address and Word Count registers to autoinitialize, even if the Mode register is programmed for autoinitialization. An external $\overline{EOP}$ will autoinitialize the channel 1 registers, if so programmed. Data comparators in block search schemes may use the $\overline{EOP}$ input to terminate the service when a match is found. The timing of memory-to-memory transfers is found in Figure 9.10. Memory-to-memory operations can be detected as an active AEN with no DACK outputs.

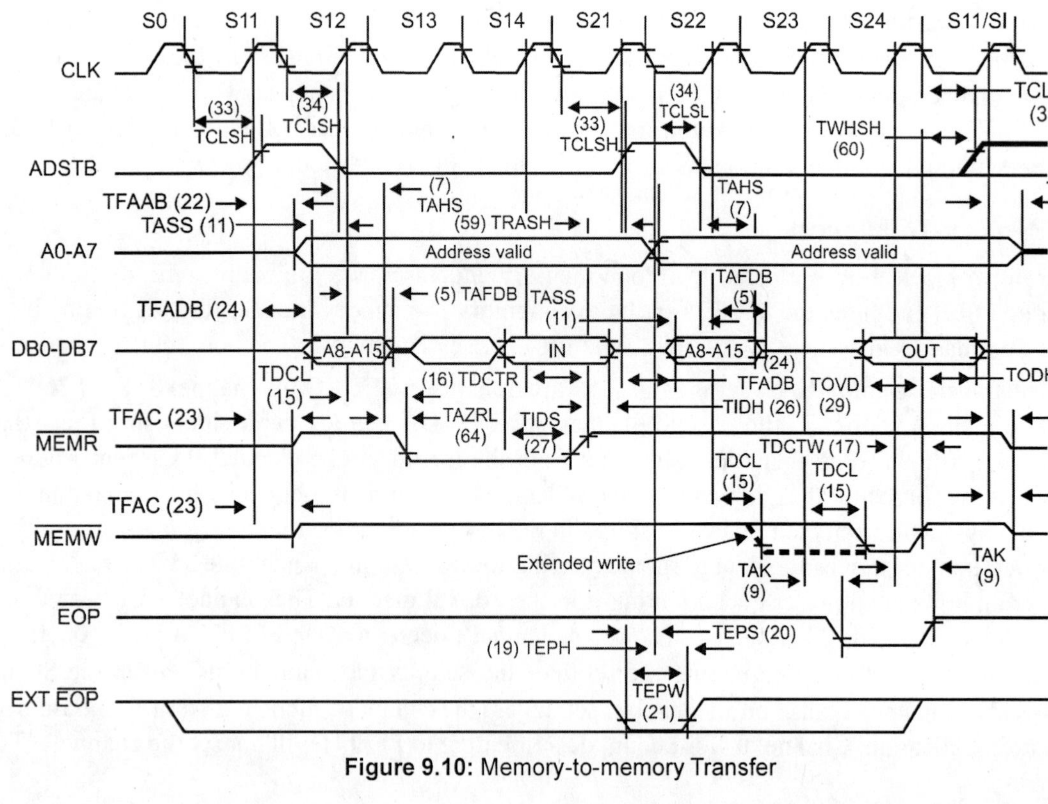


**Figure 9.10:** Memory-to-memory Transfer

### 3. *Priority*

The 82C37A has two types of priority encoding available as software selectable options. The first is Fixed Priority which fixes the channels in priority order based upon the descending value of their numbers. The channel with the lowest priority is 3 followed by 2, 1 and the highest priority channel, 0. After the recognition of any one channel for service, the other channels are prevented from interfering with the service until it is completed. The second scheme is Rotating Priority. The last channel to get service becomes the lowest priority channel with the others rotating accordingly. The next lower channel from the channel serviced has highest priority on the following request. Priority rotates every time control of the system busses is returned to the processor.

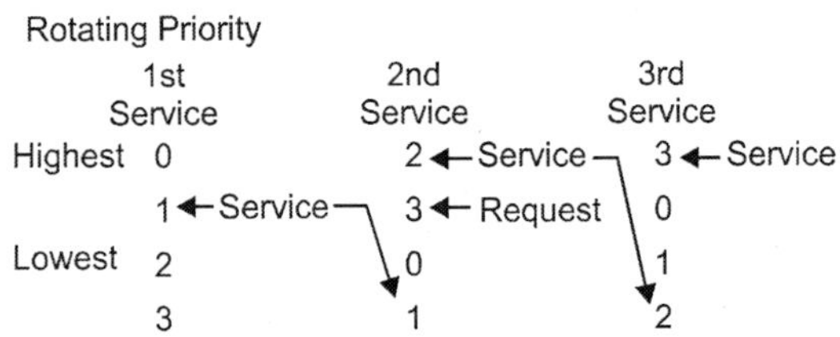


With Rotating Priority in a single chip DMA system, any device requesting service is guaranteed to be recognized after no more than three higher priority services have occurred. This prevents any one channel from monopolizing the system. Regardless of which priority scheme is chosen, priority is evaluated every time a HLDA is returned to the 82C37A.

### 4. *Compressed Timing*

In order to achieve even greater throughput where system characteristics permit, the 82C37A can compress the transfer time to two clock cycles. From Figure 9.11 it can be seen that state S3 is used to extend the access time of the read pulse. By removing state S3, the read pulse width is made equal to the write pulse width and a transfer consists only of state S2 to change the address and state S4 to perform the read/write. S1 states will still occur when A8-A15 need updating (see Address Generation). Timing for compressed transfers is found in Figure 9.12. $\overline{EOP}$ will output in S2 if compressed timing is selected. Compressed timing is not allowed for memory-to-memory transfers.

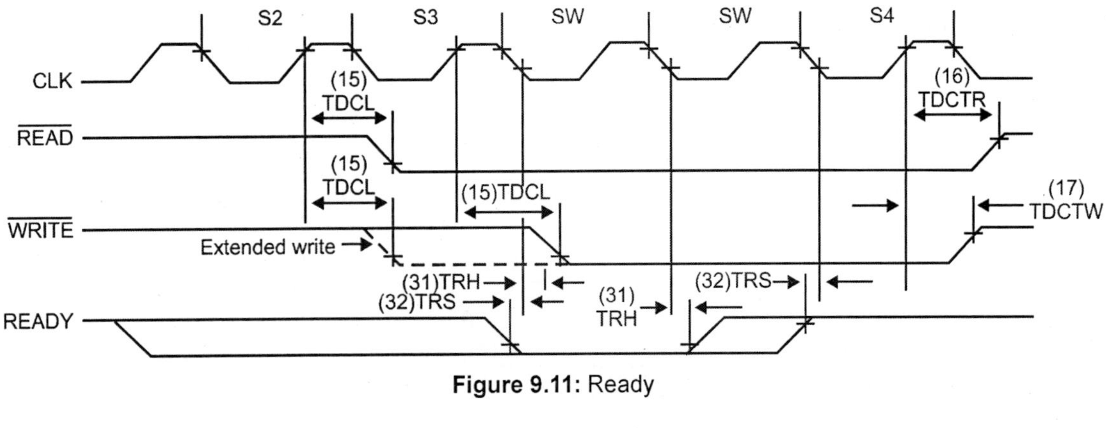


**Figure 9.11:** Ready

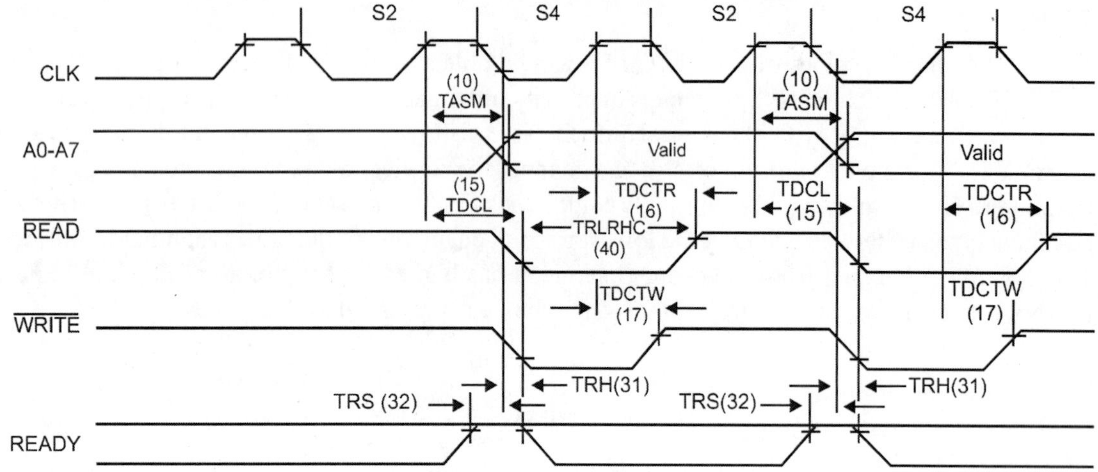


**Figure 9.12:** Compressed Transfer

## 5. *Address Generation*

In order to reduce pin count, the 82C37A multiplexes the eight higher order address bits on the data lines. State S1 is used to output the higher order address bits to an external latch from which they may be placed on the address bus. The falling edge of Address Strobe (ADSTB) is used to load

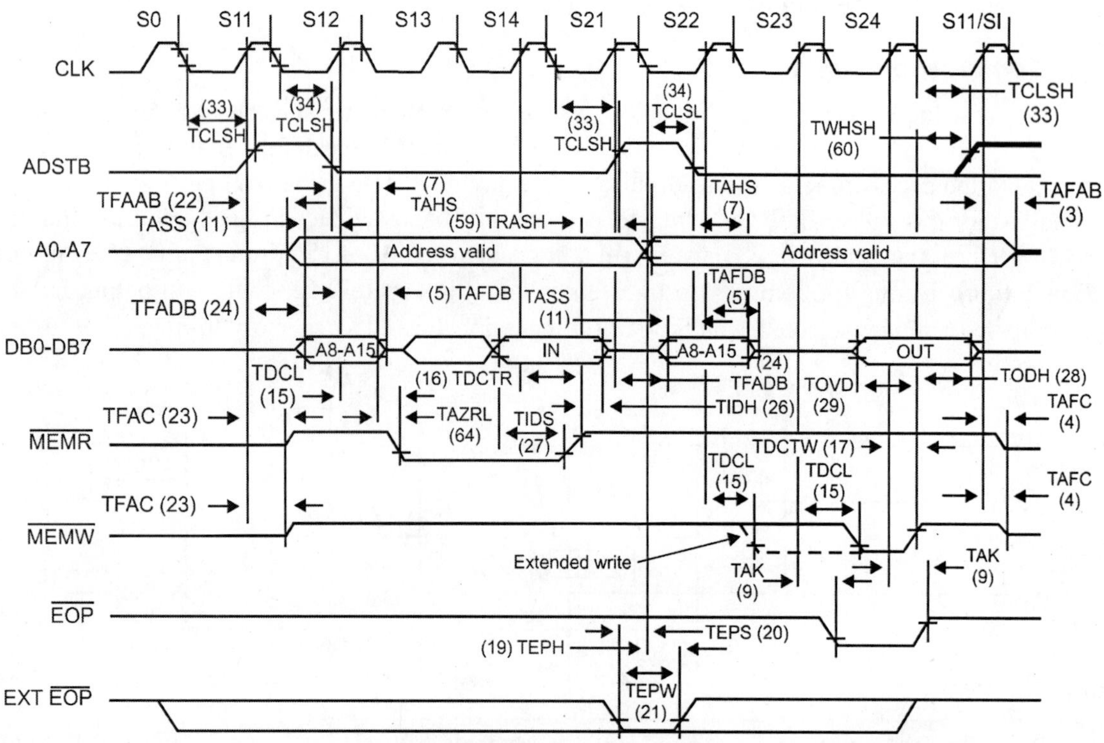


**Figure 9.13:** Memory -To-Memory

these bits from the data lines to the latch. Address Enable (AEN) is used to enable the bits onto the address bus through a three-state enable. The lower order address bits are output by the 82C37A directly. Lines A0-A7 should be connected to the address bus.

Figure 9.13 shows the time relationships between CLK, AEN, ADSTB, DB0-DB7 and A0-A7. During Block and Demand Transfer mode service, which include multiple transfers, the addresses generated will be sequential. For many transfers the data held in the external address latch will remain the same. This data need only change when a carry or borrow from A7 to A8 takes place in the normal sequence of addresses.

To save time and speed transfers, the 82C37A executes S1 states only when updating of A8-A15 in the latch is necessary. This means for long services, S1 states and Address Strobes may occur only once every 256 transfers, a savings of 255 clock cycles for each 256 transfers.

## 9.6 PROGRAMMING THE 8237

To implement the DMA transfer, the 8237 should be initialized by writing into various control registers discussed earlier in the DMA channels and interfacing section. To initialize the 8237, the following steps are necessary.

1. Write a control word in the Mode register that selects the channel and specifies the type of transfer (Read, Write, or Verify) and the DMA mode (block, single-byte, etc.).
2. Write a control word in the Command register that specifies parameters such as priority among four channels, DREQ and DACK active levels, and timing, and enables the 8237.
3. Write the starting address of the data block to be transferred in the channel Memory Address Register (MAR).
4. Write the count (the number of the bytes in the data block) in the channel Count Register.

The 82C37A will accept programming from the host processor anytime that HLDA is inactive, and at least one rising clock edge has occurred after HLDA went low. It is the responsibility of the host to assure that programming and HLDA are mutually exclusive. Note that a problem can occur if a DMA request occurs on an unmasked channel while the 82C37A is being programmed.

For instance, the CPU may be starting to reprogram the two byte Address register of channel 1 when channel 1 receives a DMA request. If the 82C37A is enabled (bit 2 in the Command register is 0), and channel 1 is unmasked, a DMA service will occur after only one byte of the Address register has been reprogrammed. This condition can be avoided by disabling the controller (setting bit 2 in the Command register) or masking the channel before programming any of its registers. Once the programming is complete, the controller can be enabled/unmasked. After power-up it is suggested that all internal locations be loaded with some known value, even if some channels are unused. This will aid in debugging.

### 9.6.1 Register Description

#### *1. Current Address Register*

Each channel has a 16-bit Current Address register. This register holds the value of the address used during DMA transfers. The address is automatically incremented or decremented by one after each transfer and the values of the address are stored in the Current Address register during the transfer. This register is written or read by the microprocessor in successive 8-bit bytes. See Figure 9.16 for programming information. It may also be reinitialized by an autoinitialize back to its original value. Autoinitialize takes place only after an $\overline{\text{EOP}}$. In memory-to-memory mode, the channel 0 Current Address register can be prevented from incrementing or decrementing by setting the address hold bit in the Command register.

### 2. *Current Word Count Register*

Each channel has a 16-bit Current Word Count register. This register determines the number of transfers to be performed. The actual number of transfers will be one more than the number programmed in the Current Word Count register (i.e., programming a count of 100 will result in 101 transfers). The word count is decremented after each transfer. When the value in the register goes from zero to FFFFH, a TC will be generated. This register is loaded or read in successive 8-bit bytes by the microprocessor in the Program Condition. See Figure 9.16 for programming information. Following the end of a DMA service it may also be reinitialized by an Autoinitialization back to its original value. Autoinitialization can occur only when an $\overline{\text{EOP}}$ occurs. If it is not Autoinitialized, this register will have a count of FFFFH after TC.

### 3. *Base Address and Base Word Count Registers*

Each channel has a pair of Base Address and Base Word Count registers. These 16-bit registers store the original value of their associated current registers. During Autoinitialize these values are used to restore the current registers to their original values. The base registers are written simultaneously with their corresponding current register in 8-bit bytes in the Program Condition by the microprocessor. See Figure 9.16 for programming information. These registers cannot be read by the microprocessor.

### 4. *Command Register*

This 8-bit register controls the operation of the 82C37A. It is programmed by the microprocessor and is cleared by RESET or a Master Clear instruction. The following diagram lists the function of the Command register bits. See Figure 9.14 for Read and Write addresses.

### 5. *Mode Register*

Each channel has a 6-bit Mode register associated with it. When the register is being written to by the microprocessor in the Program condition, bits 0 and 1 determine which channel Mode register is to be written. When the processor reads a Mode register, bits 0 and 1 will both be ones. See the following diagram and Figure 9.14 for Mode register functions and addresses.

### 6. *Request Register*

The 82C37A can respond to requests for DMA service which are initiated by software as well as by a DREQ. Each channel has a request bit associated with it in the 4-bit Request register. These are

**Command Register**

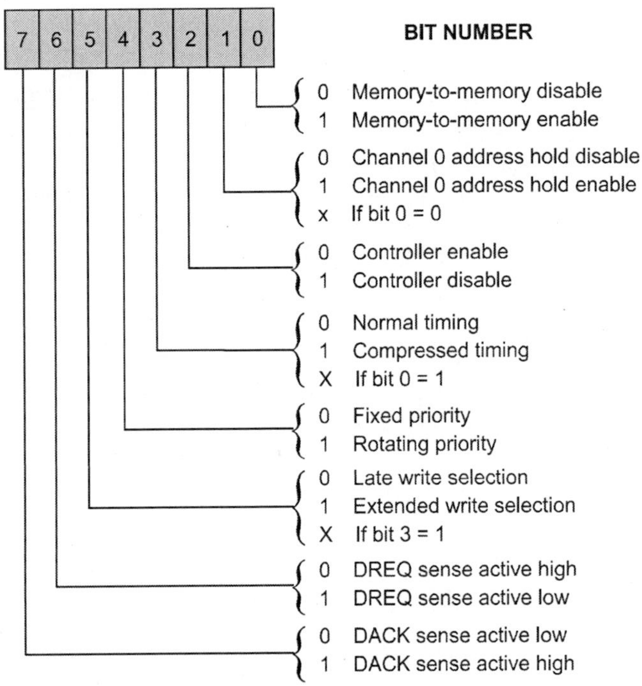


**Mode Register**

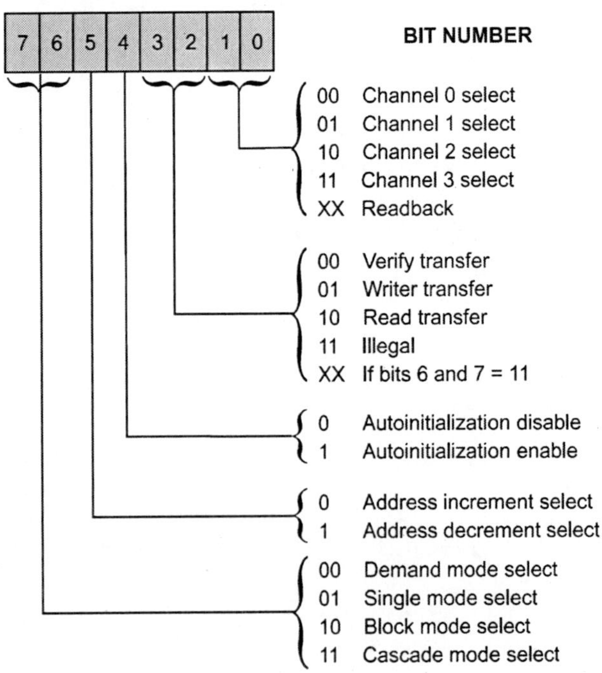

non-maskable and subject to prioritization by the Priority Encoder network. Each register bit is set or reset separately under software control. The entire register is cleared by a Reset or Master Clear instruction. To set or reset a bit, the software loads the proper form of the data word. See Figure 9.14 for register address coding, and the following diagram for Request register format. A software request for DMA operation can be made in block or single modes. For memory-to-memory transfers, the software request for channel 0 should be set. When reading the Request register, bits 4-7 will always read as ones, and bits 0-3 will display the request bits of channels 0-3 respectively.

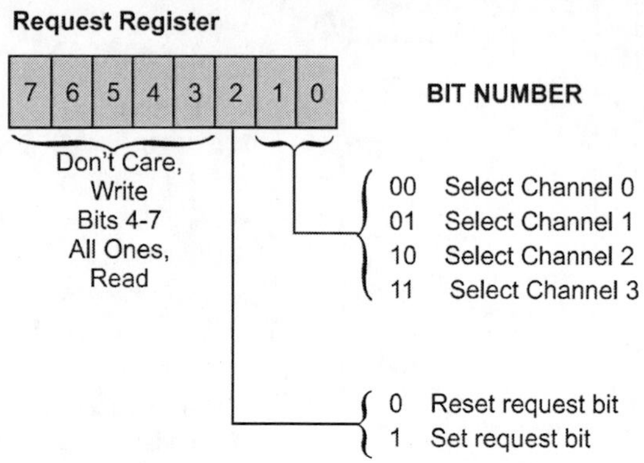


## 7. *Mask Register*

Each channel has associated with it a mask bit which can be set to disable an incoming DREQ. Each mask bit is set when its associated channel produces an $\overline{EOP}$ if the channel is not programmed to Autoinitialize. Each bit of the 4-bit Mask register may also be set or cleared separately or simultaneously under software control. The entire register is also set by a Reset or Master clear. This disables all hardware DMA requests until a Clear Mask Register instruction allows them to occur. The instruction to separately set or clear the mask bits is similar in form to that used with the Request register. Refer to the following diagram and Figure 9.14 for details. When reading the Mask register, bits 4-7 will always read as logical ones, and bits 0-3 will display the mask bits of channels 0-3, respectively. The 4 bits of the Mask register may be cleared simultaneously by using the Clear Mask Register command (see software commands section).

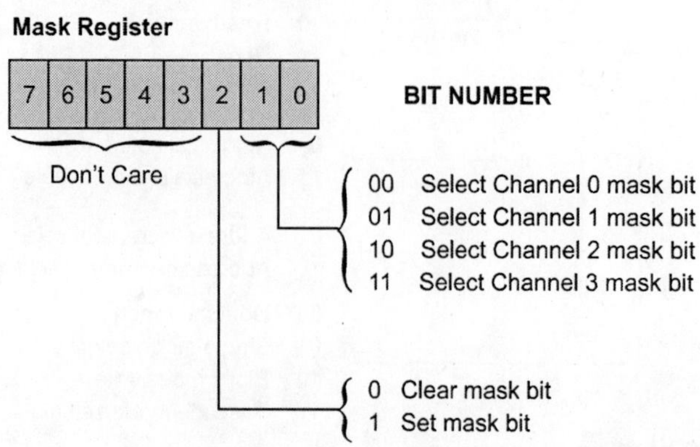

All four bits of the Mask register may also be written with a single command.

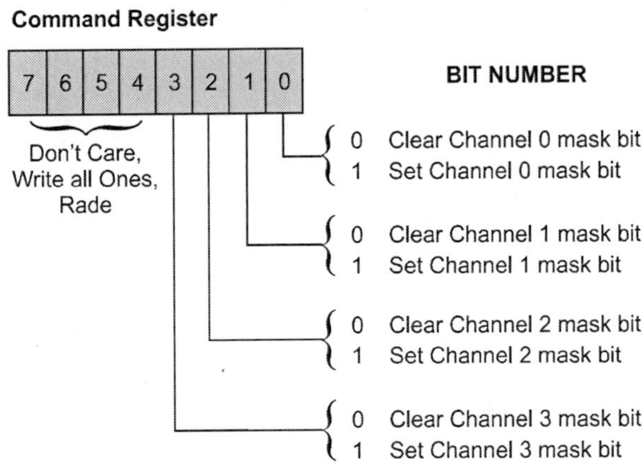


### 8. *Status Register*

The Status register is available to be read out of the 82C37A by the microprocessor. It contains information about the status of the devices at this point. This information includes which channels have reached a terminal count and which channels have pending DMA requests. Bits 0-3 are set every time a TC is reached by that channel or an external EOP is applied. These bits are cleared upon RESET, Master Clear, and on each Status Read. Bits 4-7 are set whenever their corresponding channel is requesting service, regardless of the mask bit state. If the mask bits are set, software can poll the Status register to determine which channels have DREQs, and selectively clear a mask bit, thus allowing user defined service priority. Status bits 4-7 are updated while the clock is high, and latched on the falling edge. Status Bits 4-7 are cleared upon RESET or Master Clear.

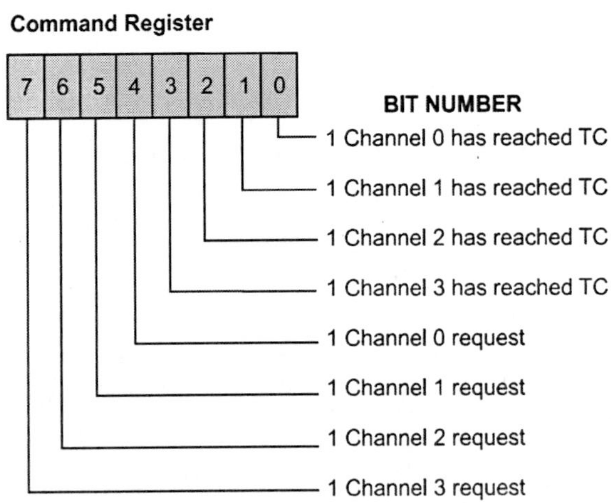


### 9. *Temporary Register*

The Temporary register is used to hold data during memory-to-memory transfers. Following the completion of the transfers, the last byte moved can be read by the microprocessor. The Temporary register always contains the last byte transferred in the previous memory-to-memory operation, unless cleared by a Reset or Master Clear.

| Operation | A3 | A2 | A1 | A0 | IOR | IOW |
|---|---|---|---|---|---|---|
| Read Status Register | 1 | 0 | 0 | 0 | 0 | 1 |
| Write Command Register | 1 | 0 | 0 | 0 | 1 | 0 |
| Read Request Register | 1 | 0 | 0 | 1 | 0 | 1 |
| Write Request Register | 1 | 0 | 0 | 1 | 1 | 0 |
| Read Command Register | 1 | 0 | 1 | 0 | 0 | 1 |
| Write Single Mask Bit | 1 | 0 | 1 | 0 | 1 | 0 |
| Read Mode Register | 1 | 0 | 1 | 1 | 0 | 1 |
| Write Mode Register | 1 | 0 | 1 | 1 | 1 | 0 |
| Set First/Last F/F | 1 | 1 | 0 | 0 | 0 | 1 |
| Clear First/Last F/F | 1 | 1 | 0 | 0 | 1 | 0 |
| Read Temporary Register | 1 | 1 | 0 | 1 | 0 | 1 |
| Master Clear | 1 | 1 | 0 | 1 | 1 | 0 |
| Clear Mode Reg. Counter | 1 | 1 | 1 | 0 | 0 | 1 |
| Clear Mask Register | 1 | 1 | 1 | 0 | 1 | 0 |
| Read All Mask Bits | 1 | 1 | 1 | 1 | 0 | 1 |
| Write All Mask Bits | 1 | 1 | 1 | 1 | 1 | 0 |

**Figure 9.14:** Software Command Codes and Register Codes

### 9.6.2 Software Commands

There are special software commands which can be executed by reading or writing to the 82C37A. These commands do not depend on the specific data pattern on the data bus, but are activated by the I/O operation itself. On read type commands, the data value is not guaranteed. These commands are:

1. **Clear First/Last Flip-Flop:** This command is executed prior to writing or reading new address or word count information to the 82C37A. This command initializes the flip-flop to a known state (low byte first) so that subsequent accesses to register contents by the microprocessor will address upper and lower bytes in the correct sequence.
2. **Set First/Last Flip-Flop:** This command will set the flip-flop to select the high byte first on read and write operations to address and word count registers.
3. **Master Clear:** This software instruction has the same effect as the hardware Reset. The Command, Status, Request, and Temporary registers, and Internal First/Last Flip-Flop and mode register counter are cleared and the Mask register is set. The 82C37A will enter the idle cycle.
4. **Clear Mask Register:** This command clears the mask bits of all four channels, enabling them to accept DMA requests.
5. **Clear Mode Register Counter:** Since only one address location is available for reading the Mode registers, an internal two-bit counter has been included to select Mode registers during read operation. To read the Mode registers, first execute the Clear Mode Register Counter command, then do consecutive reads until the desired channel is read. Read order is channel 0 first, channel 3 last. The lower two bits on all Mode registers will read as ones.

### 9.6.3 External $\overline{EOP}$ Operation

The $\overline{EOP}$ pin is a bidirectional, open drain pin which may be driven by external signals to terminate DMA operation. Because $\overline{EOP}$ is an open drain pin an external pull-up resistor to VCC is required. The value of the external pull-up resistor used should guarantee a rise time of less than 125ns. It is important to note that the 82C37A will not accept external $\overline{EOP}$ signals when it is in a SI (Idle) state. The controller must be active to latch EXT $\overline{EOP}$ .Once latched, the EXT $\overline{EOP}$ will be acted upon during the next S2 state, unless the 82C37A enters an idle state first. In the latter case, the latched $\overline{EOP}$ is cleared. External $\overline{EOP}$ pulses occurring between active DMA transfers in demand mode will not be recognized, since the 82C37A is in an SI state.

| Channel | Register | Operation | Signals | | | | | | | | Data Bus DB0-DB7 |
|---|---|---|---|---|---|---|---|---|---|---|---|
| | | | CS | IOR | IOW | A3 | A2 | A1 | A0 | First/Last Flip-flop State | |
| 0 | Base and current address | Write | 0 | 1 | 0 | 0 | 0 | 0 | 0 | 0 | A1-A7 |
| | | | 0 | 1 | 0 | 0 | 0 | 0 | 0 | 1 | A8-A15 |
| | Current address | Read | 0 | 0 | 1 | 0 | 0 | 0 | 0 | 0 | A0-A7 |
| | | | 0 | 0 | 1 | 0 | 0 | 0 | 0 | 1 | A8-A15 |
| | Base and current word count | Write | 0 | 1 | 0 | 0 | 0 | 0 | 1 | 0 | W0-W7 |
| | | | 0 | 1 | 0 | 0 | 0 | 0 | 1 | 1 | W8-W15 |
| | Current word | Read | 0 | 0 | 1 | 0 | 0 | 0 | 1 | 0 | W0-W7 |
| | | | 0 | 0 | 1 | 0 | 0 | 0 | 1 | 1 | W8-W15 |
| 1 | Base and current address | Wrtie | 0 | 1 | 0 | 0 | 0 | 1 | 0 | 0 | A0-A7 |
| | | | 0 | 1 | 0 | 0 | 0 | 1 | 0 | 1 | A8-A15 |
| | Current address | Read | 0 | 0 | 1 | 0 | 0 | 1 | 0 | 0 | A0-A7 |
| | | | 0 | 0 | 1 | 0 | 0 | 1 | 0 | 1 | A8-A15 |
| | Base and current word count | Write | 0 | 1 | 0 | 0 | 0 | 1 | 1 | 0 | W0-W7 |
| | | | 0 | 1 | 0 | 0 | 0 | 1 | 1 | 1 | W8-W15 |
| | Current word | Read | 0 | 0 | 1 | 0 | 0 | 0 | 1 | 0 | W0-W7 |
| | | | 0 | 0 | 1 | 0 | 0 | 1 | 1 | 0 | W8-W15 |
| 2 | Base and current address | Wrtie | 0 | 1 | 0 | 0 | 1 | 0 | 0 | 0 | A0-A7 |
| | | | 0 | 1 | 0 | 0 | 1 | 0 | 0 | 1 | A8-A15 |
| | Current address | Read | 0 | 0 | 1 | 0 | 1 | 0 | 0 | 0 | A0-A7 |
| | | | 0 | 0 | 1 | 0 | 1 | 0 | 0 | 1 | A8-A15 |
| | Base and current word count | Write | 0 | 1 | 0 | 0 | 1 | 0 | 1 | 0 | W0-W7 |
| | | | 0 | 1 | 0 | 0 | 1 | 0 | 1 | 1 | W8-W15 |
| | Current word | Read | 0 | 0 | 1 | 0 | 1 | 0 | 1 | 0 | W0-W7 |
| | | | 0 | 0 | 1 | 0 | 1 | 0 | 1 | 1 | W8-W15 |
| 3 | Base and current address | Wrtie | 0 | 1 | 0 | 0 | 1 | 1 | 0 | 0 | A0-A7 |
| | | | 0 | 1 | 0 | 0 | 1 | 1 | 0 | 1 | A8-A15 |
| | Current address | Read | 0 | 0 | 1 | 0 | 1 | 1 | 0 | 0 | A0-A7 |
| | | | 0 | 0 | 1 | 0 | 1 | 1 | 0 | 1 | A8-A15 |
| | Base and current word count | Write | 0 | 1 | 0 | 0 | 1 | 1 | 1 | 0 | W0-W7 |
| | | | 0 | 1 | 0 | 0 | 1 | 1 | 1 | 1 | W8-W15 |
| | Current word | Read | 0 | 0 | 1 | 0 | 1 | 1 | 1 | 0 | W0-W7 |
| | | | 0 | 0 | 1 | 0 | 1 | 1 | 1 | 1 | W8-W15 |

**Figure 9.15:** Word Count and Address Register Command Codes

## 9.7 APPLICATION INFORMATION

Figure 9.16 shows an application for a DMA system utilizing the 82C37A DMA controller and the 80C88 Microprocessor. In this application, the 82C37A DMA controller is used to improve system performance by allowing an I/O device to transfer data directly to or from system memory.

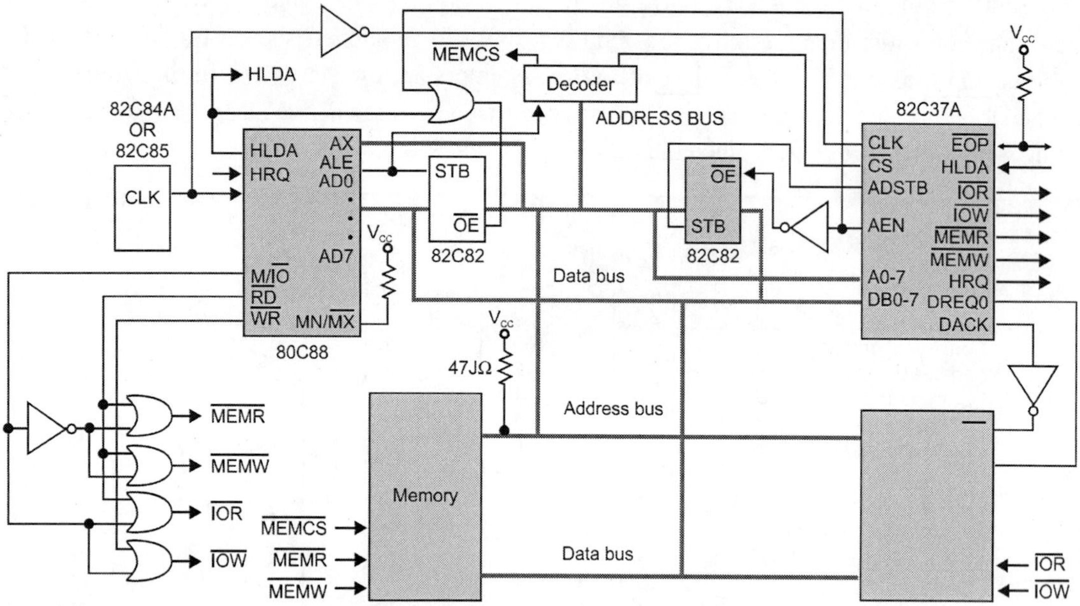


**Figure 9.16:** Application for dma system

### 9.7.1 Components

The system clock is generated by the 82C84A clock driver and is inverted to meet the clock high and low times required by the 82C37A DMA controller. The four OR gates are used to support the 80C88 Microprocessor in minimum mode by producing the control signals used by the processor to access memory or I/O. A decoder is used to generate chip select for the DMA controller and memory. The most significant bits of the address are output on the address/data bus. Therefore, the 82C82 octal latch is used to demultiplex the address. Hold Acknowledge (HLDA) and Address Enable (AEN) are "ORed" together to insure that the DMA controller does not have bus contention with the microprocessor.

### 9.7.2 Operation

A DMA request (DREQ) is generated by the I/O device. After receiving the DMA request, the DMA controller will issue a Hold request (HRQ) to the processor. The system busses are not released to the DMA controller until a Hold Acknowledge signal is returned to the DMA controller from the 80C88 processor. After the Hold Acknowledge has been received, addresses and control signals are generated by the DMA controller to accomplish the DMA transfers. Data is transferred directly from the I/O device to memory (or vice versa) with $\overline{\text{IOR}}$ and $\overline{\text{MEMW}}$ (or $\overline{\text{MEMR}}$ and $\overline{\text{IOW}}$ ) being active. Note that data is not read into or driven out of the DMA controller in I/O-to-memory or memory-to-I/O data transfers.

Figure 9.17 shows an application for a DMA system using the 82C37A DMA controller and the 80C286 Microprocessor. In this application, the system clock comes from the 82C284 clock generator PCLK signal which is inverted to provide proper READY setup and hold times to the DMA controller in an 80C286 system. The Read and Write signals from the DMA controller may be wired directly to the Read/Write control signals from the 82C288 Bus Controller. The octal latch for A8-A15 from the DMA controller's data bus is on the local 80C286 address bus so that memory chip selects may still be generated during DMA transfers. The transceiver on A0- A7 is controlled by AEN and is not necessary, but may be used to drive a heavily loaded system address bus during transfers. The data bus transceivers simply isolate the DMA controller from the local microprocessor bus and allow programming on the upper or lower half of the data bus.

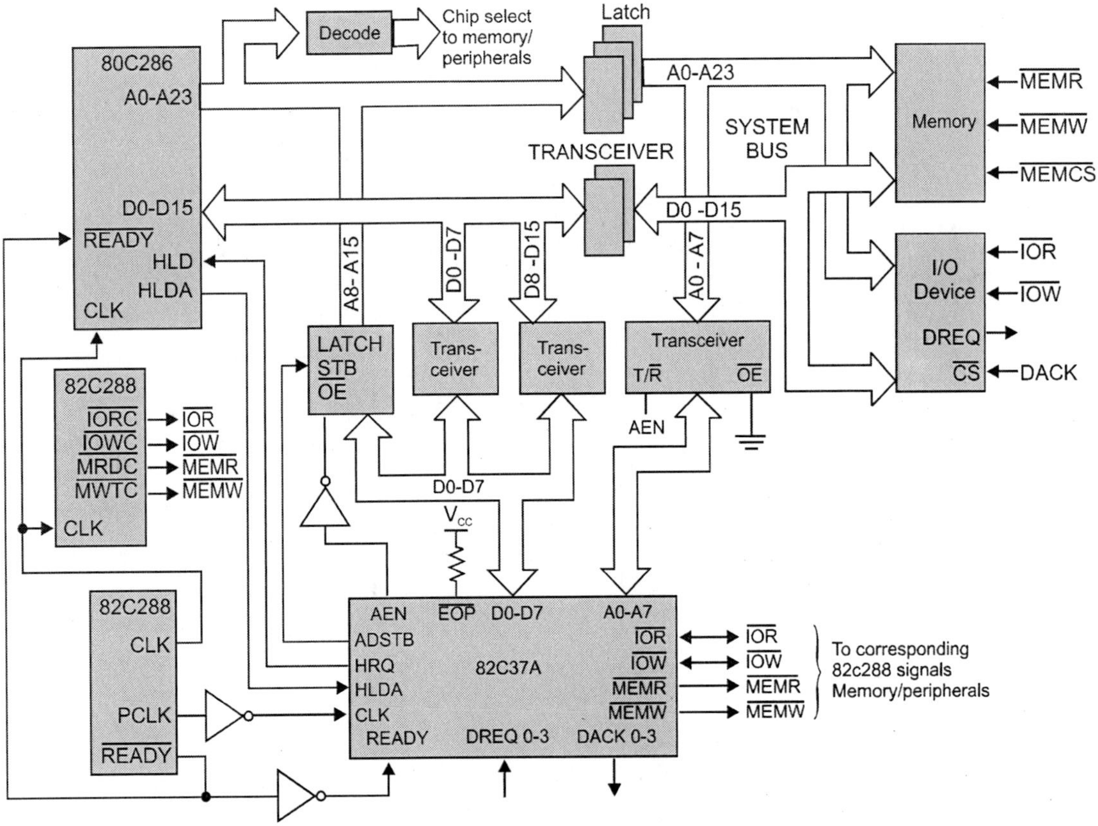


**Figure 9.17:** 80C286 DMA APPLICATION

## QUESTION BANK

Q1. What do you mean by DMA data transfer?

Q2. Explain the DMA controller for 8085.

Q3. Explain the block diagram of 82c37a DMA controller.

Q4. Explain the Pin diagram and description of 8237 DMA Controller.

Q5. Explain the various operations of 8237 DMA.

Q6. Explain the programming of 8237 DMA controller.

Q7. Explain the register description of 8237 DMA.

Q8. Explain the software commands of 8237 DMA.

Q9. What do you mean by external $\overline{EOP}$ operation?

Q10. Discuss the application of DMA system.

CHAPTER 10

# 8255 Programmable Peripheral Interface

## 10.1 THE 8255A PROGRAMMABLE PERIPHERAL INTERFACE

82C55A is a high performance CMOS version of the industry standard 8255A and is manufactured using a self-aligned silicon gate CMOS process. The Intel 8255A is a general purpose programmable I/O device designed for use with Intel microprocessors.

The 8255A is a widely used, programmable, parallel I/O device. It can be programmed to transfer data under various conditions, from simple I/O to interrupt I/O. It is flexible, versatile, and economical (when multiple I/O ports are required), but somewhat complex. It is an important general-purpose I/O device that can be used with almost any microprocessor.

The 8255A has 24 I/O pins that can be grouped primarily in two 8-bit parallel ports: A and B, with the remaining eight bits as port C. The eight bits of port C can be used as individual bits or be grouped inn two 4-bit ports: $C_{UPPER}$ ($C_U$) and $C_{LOWER}$ ($C_L$), as in Figure 10.1(a). The functions of these ports are defined by writing a control word in the control register.

Figure 10.1(b) shows all the functions of the 8255A, classified according to two modes: the Bit Set/Reset (BSR) mode and the I/O mode. The BSR mode is used to set or reset the bits in port C. The I/O mode is further divided into three modes: Mode 0, Mode 1 and Mode 2. In Mode 0, each group of 12 I/O pins may be programmed in sets of 4 to be input or output. Mode 1 is a handshake mode whereby ports A and/or B use bits from port C as handshake signals. In the handshake mode, each group may be programmed to have 8 lines of input or output. Of the remaining 4 pins, 3 are used for handshaking and interrupt control signals.

There are two types of I/O data transfer can be implemented: status check and interrupt. In Mode 2, port A can be set for bidirectional data transfer using handshake signals from port C, and port B can be set up either in Mode 0 or Mode 1.

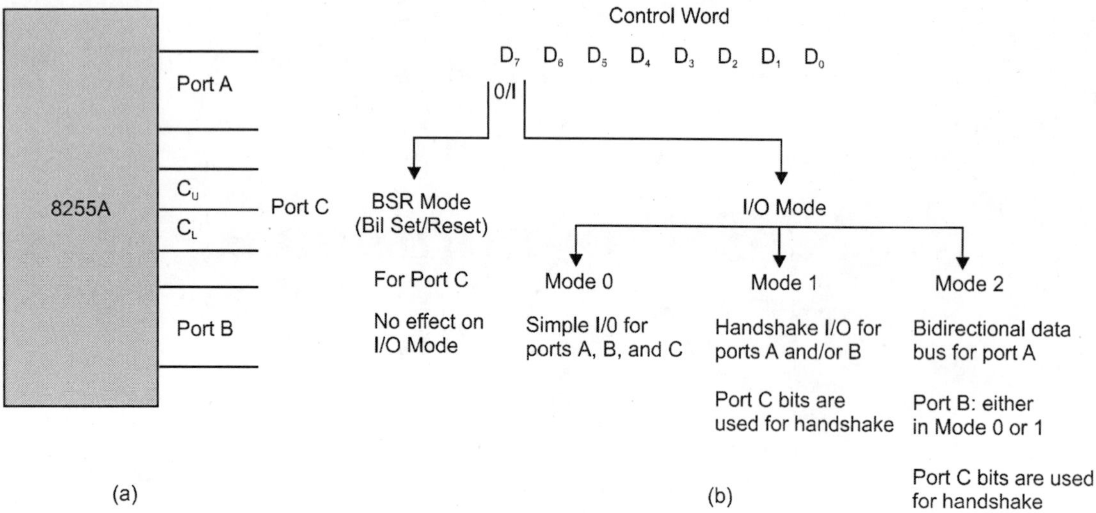


**Figure 10.1:** (a) 8255 I/O Ports (b) Modes of 8255

## 10.2 BLOCK DIAGRAM OF THE 8255A

The block diagram in Figure 10.2(a) shows two 8-bit ports (A and B), two 4-bit ports ($C_U$ and $C_L$), the data bus buffer, and control logic. Figure 10.2(b) shows a simplified but expanded version of the internal structure, including a control register. This block diagram includes all the elements of a programmable device; port C performs functions similar to that of the status register in addition to providing handshake signals.

### 10.2.1 General Description

The 8255A is a programmable peripheral interface (PPI) device designed for use in Intel microcomputer systems. Its function is that of a general purpose I/O component to interface peripheral equipment to the microcomputer system bus. The functional configu-ration of the 8255A is programmed by the system software so that normally no external logic is neces-sary to interface peripheral devices or structures.

### 10.2.2 Data Bus Buffer

This 3-state bidirectional 8-bit buffer is used to inter-face the 8255A to the system data bus. Data is transmitted or received by the buffer upon execution of input or output instructions by the CPU. Control words and status information are also transferred through the data bus buffer.

### 10.2.3 Read/Write And Control Logic

The function of this block is to manage all of the internal and external transfers of both Data and Control or Status words. It accepts inputs from the CPU Address and Control busses and in turn, issues commands to both of the Control Groups.

### 10.2.4 Control Logic

The control section has six lines. Their functions and connections are as follows:

- $\overline{RD}$ (Read): This control signal enables the Read operation. When the signal is low, the MPU reads data from a selected I/O port of the 8255A.
- $\overline{WR}$ (Write): This control signals enables the Write operation. When the signal goes low, the MPU writes into a selected I/O port or the control register.
- RESET (Reset): This is an active high signal; it clears the control register and sets all ports in the input mode.
- $\overline{CS}$, $A_0$, and $A_1$: These are device select signals. $\overline{CS}$ is connected to a decoded address, and $A_0$ and $A_1$ are generally connected to MPU address lines $A_0$ and $A_1$, respectively.

The $\overline{CS}$ signal is the master Chip Select, and $A_0$ and $A_1$ specify one of the I/O ports or the control register as given below:

| $\overline{CS}$ | $A_1$ | $A_0$ | Selected |
|---|---|---|---|
| 0 | 0 | 0 | Port A |
| 0 | 0 | 1 | Port B |
| 0 | 1 | 0 | Port C |
| 0 | 1 | 1 | Control Register |
| 1 | X | X | 8255A is not selected. |

As an example, the port addresses in Figure 10.3(a) are determined by the, $A_0$ and $A_1$ lines. The line goes low when $A_7$=1 and $A_6$ through $A_2$ are at logic 0. When these signals are combined with $A_0$ and $A_1$, the port addresses range from 80H to 83H, as shown in **Figure 10.3(b).**

### 10.2.5 82C55A Basic Operation

| $A_1$ | $A_0$ | $\overline{RD}$ | $\overline{WR}$ | $\overline{CS}$ | Input (Read) Cycle |
|---|---|---|---|---|---|
| 0 | 0 | 0 | 1 | 0 | Port A →data bus |
| 0 | 1 | 0 | 1 | 0 | Port B → data bus |
| 1 | 0 | 0 | 1 | 0 | Port C t→ data bus |
| 1 | 1 | 0 | 1 | 0 | Control word → data bus |
| | | | | | **Output (Write) Cycle** |
| 0 | 0 | 1 | 0 | 0 | Data bus → port A |
| 0 | 1 | 1 | 0 | 0 | Data bus → port B |
| 1 | 0 | 1 | 0 | 0 | Data bus → port C |
| 1 | 1 | 1 | 0 | 0 | Data bus → Control |
| | | | | | **Disable Function** |
| X | X | X | X | 1 | Data bus → Three state |
| X | X | 1 | 1 | 0 | Data bus → Three state |

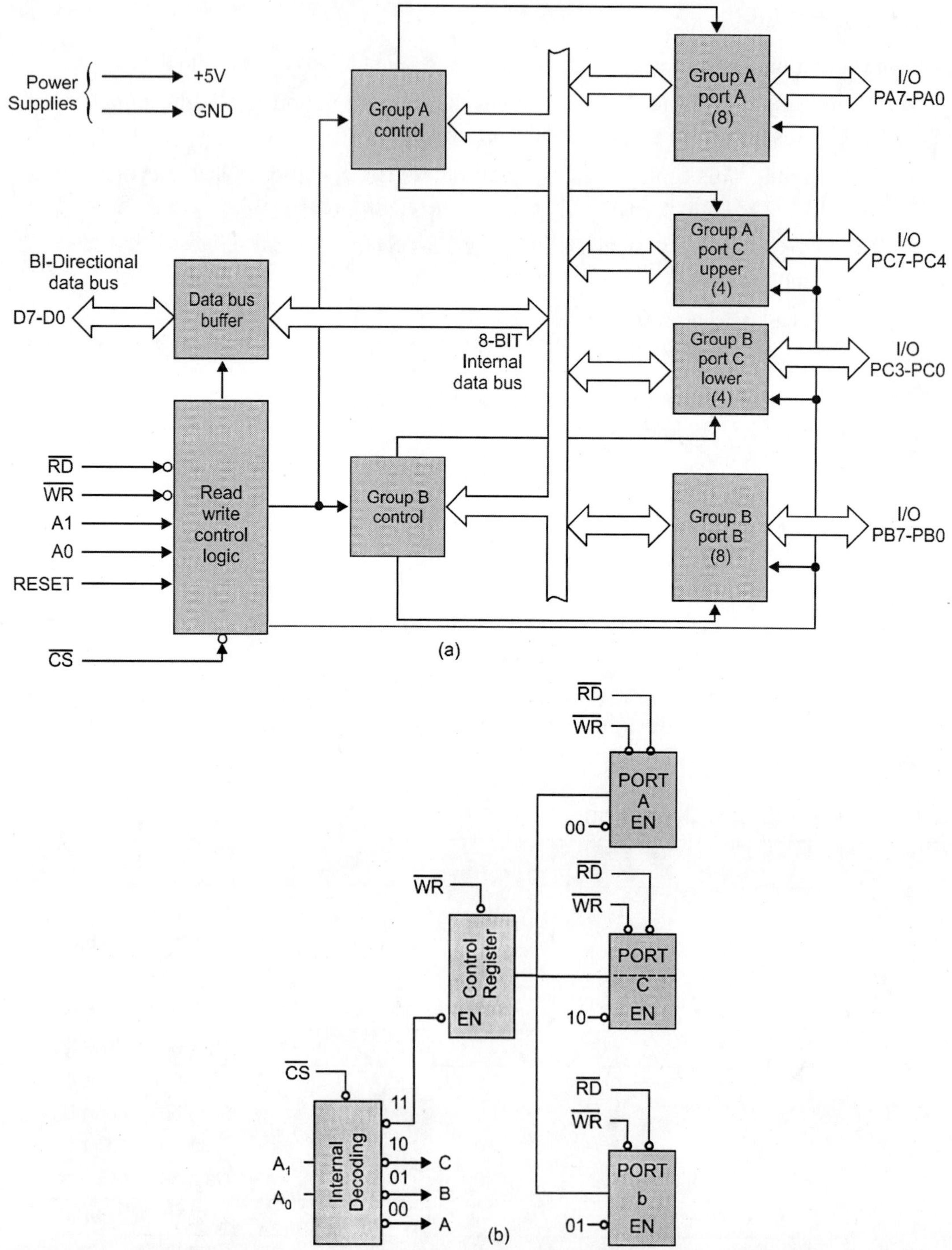


**Figure 10.2:** (a) 82c55a block diagram. Data bus buffer, read/write, group a & b control logic functions (b) Expanded Version of the Control Logic and I/O Ports

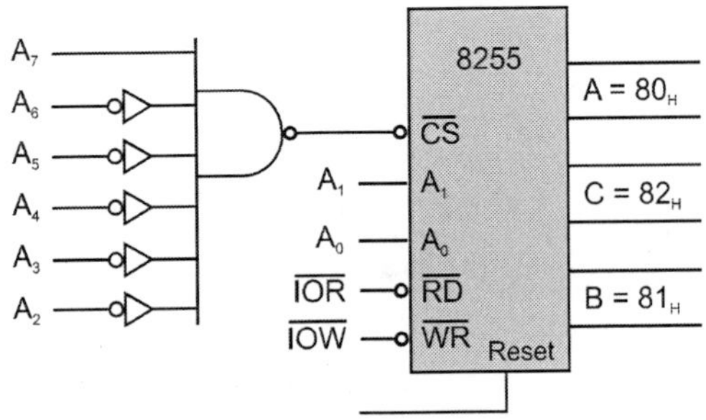


| $\overline{CS}$ | | Hex Address | Port |
|---|---|---|---|
| $A_7$ $A_6$ $A_5$ $A_4$ $A_3$ $A_2$ | A1 A0 | | |
| 1 0 0 0 0 0 | 0 0 | = 80H | A |
| | 0 1 | = 81H | B |
| | 1 0 | = 82H | C |
| | 1 1 | = 83H | Control Register |

**Figure 10.3:** (a) 8255A Chip Select Logic and (b) I/O Port Addresses

## 10.3 CONTROL WORD

The functional configuration of each port is pro-grammed by the systems software. In essence, the CPU "outputs" a control word to the 8255A. The control word contains information such as "mode", "bit set", "bit reset", etc., that initializes the func-tional configuration of the 8255A. Figure **10.2(b)** shows a register called the control register. The contents of this register, called the control word, specify an I/O function for each port. This register can be accessed to write a control word when $A_0$ and $A_1$ are at logic 1, as mentioned previously. The register is not accessible for a Read operation.

Bit $D_7$ of the control register specifies either the I/O function or the Bit Set/Reset function, as classified in Figure 10.1(b). If bit $D_7 = 1$, bits $D_6$-$D_0$ determine I/O functions in various modes, as shown in Figure 10.4. If bit $D_7 = 0$, port C operates in the bit Set/Reset (BSR) mode. The BSR control word does not affect the functions of ports A and B (the BSR mode will be described later).

To communicate with peripherals through the 8255A, three steps are necessary:

1. Determine the addresses of ports A, B and C and of the control register according to the Chip Select logic and address lines $A_0$ and $A_1$.
2. Write a control word in the control register.
3. Write I/O instructions to communicate with peripherals through ports A, B and C.

### 10.3.1 Group A and Group B Controls

Each of the Control blocks (Group A and Group B) accepts "commands" from the Read/Write Control Logic, receives "control words" from the internal data bus and issues the proper commands to its as-sociated ports.

Control Group A—Port A and Port C upper ($C_7$-$C_4$)

Control Group B—Port B and Port C lower ($C_3$-$C_0$)

The Control Word Register can only be written into. No Read operation of the Control Word Register is allowed.

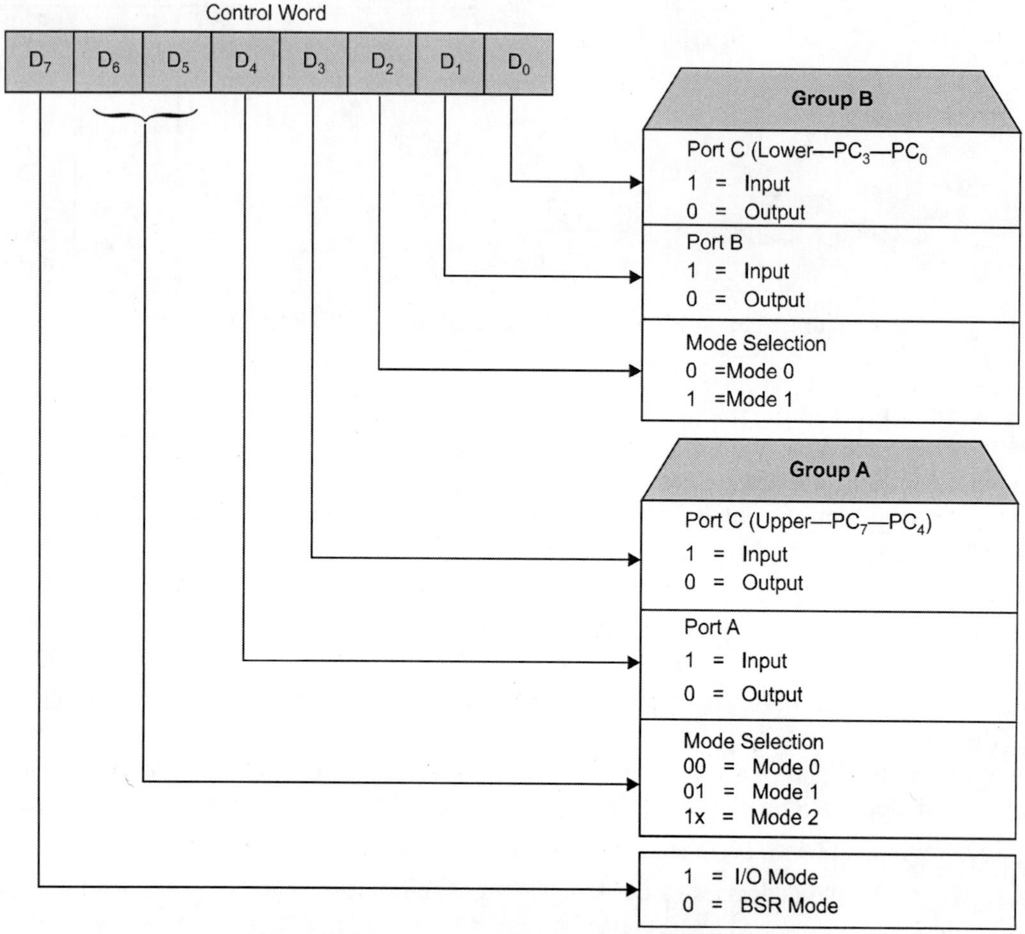


**Figure 10.4:** 8255A Control Word Format for I/O Mode

### 10.3.2 Ports A, B, AND C

The 82C55A contains three 8-bit ports (A, B, and C). All can be configured to a wide variety of functional characteristics by the system software but each has its own special Features or "personality" to further enhances the power and flexibility of the 82C55A.

**Port A:** One 8-bit data output latch/buffer and one 8-bit data input latch. Both "pull-up" and "pull-down" bus-hold devices are present on Port A (Figure 10.5a).

**Port B:** One 8-bit data input/output latch/buffer and one 8-bit data input buffer (Figure 10.5b).

**Port C:** One 8-bit data output latch/buffer and one 8-bit data input buffer (no latch for input). This port can be divided into two 4-bit ports under the mode control. Each 4-bit port contains a 4-bit latch and it can be used for the control signal output and status signal inputs in conjunction with ports A and B (Figure 10.5b).

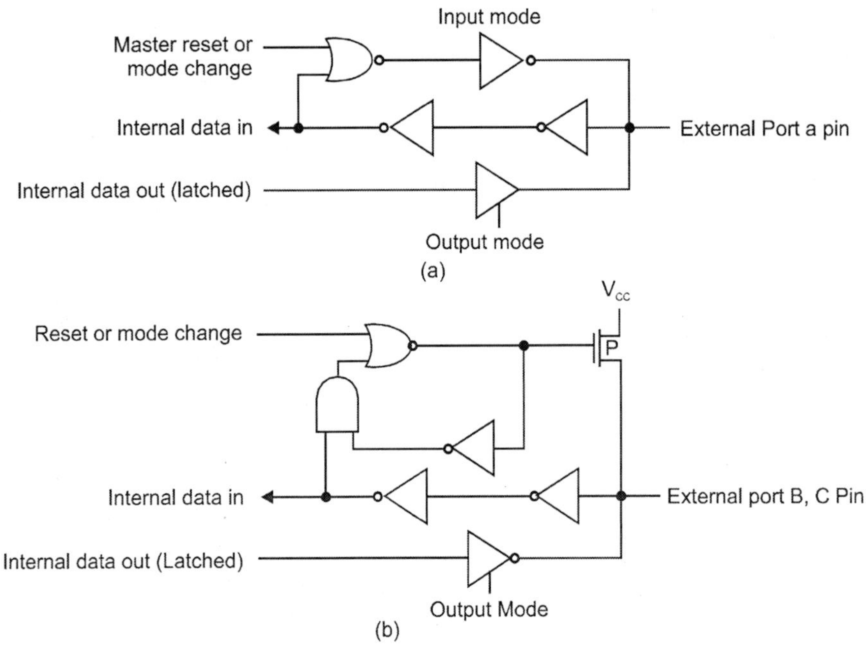


**Figure 10.5:** Bus-hold configuration (a)Port a bus-hold configuration (b) Port b and c bus-hold configuration

Examples of the various modes are given in the next section.

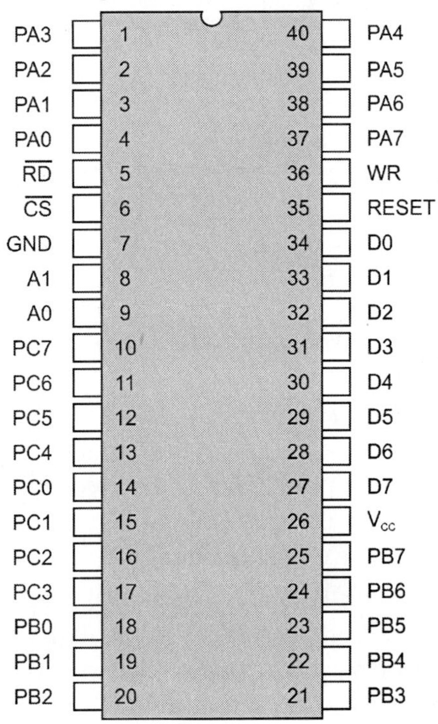

## 10.4 Pin Description of 8255

| Symbol | Pin number | Type | Description |
|---|---|---|---|
| $V_{CC}$ | 26 | | $V_{CC}$: The +5V power supply pin. A.0.1mF capacitor between pins 26 and 7 is recommended for decoupling. |
| GND | 7 | | Ground |
| D0-D7 | 27-34 | I/O | **Data Bus:** The data bus lines are bidirectional three-state pins connected to the system data bus. |
| Reset | 35 | I | Reset: A high on this input clears the control register and all ports (A, B, C)are set to the input mode with the 'Bus Hold' circuitry turned on. |
| $\overline{CS}$ | 6 | I | **Chip select** is an active low input used to enable the 82C55A onto the Data Bus for CPU communications. |
| $\overline{RD}$ | 5 | I | **Read** is an active low input control signal used by the CPU to read status information or data via the data bus. |
| $\overline{WR}$ | 36 | I | **Write** is an active low input control signal used by the CPU to load control words and data into the 82C55A. |
| AD–A1 | 8,9 | I | **Address:** These input signals, in conjunction with the $\overline{RD}$ and $\overline{WR}$ inputs, control the selection of one of the three ports or the controlword register. A1 and A1 are normally connected to the least significant bits of the Address Bus A0, A1. |
| PA0–PA7 | 1–4, 37–40 | I/O | **Port A:** 8-bit input and output port. Both bus hold high and bus hold low circuitry are present on this port. |
| PB0–PB7 | 18–25 | I/O | **Port B:** 8-bit input and output port. Bus hold high circuitry is present on this port. |
| PC0–PC7 | 10–17 | I/O | **Port C:** 8-bit input and output port. Bus hold circuitry is present on this port. |

## 10.5 8255A OPERATIONAL DESCRIPTION

### 10.5.1 Mode Selection

There are three basic modes of operation that can be selected by the system software

Mode 0—Basic Input/Output
Mode 1—Strobed Input/Output
Mode 2—Bi-Directional Bu

When the reset input goes "high" all ports will be set to the input mode (i.e., all 24 lines will be in the high impedance state). After the reset is removed the 8255A can remain in the input mode with no addi-tional initialization required. During the execution of the system program any of the other modes may be selected using a single output instruction. This al-lows a single 8255A to service a variety of peripheral devices with a simple software maintenance routine.

The modes for Port A and Port B can be separately defined, while Port C is divided into two portions as required by the Port A and Port B definitions. All of the output registers, including the status flip-flops, will be reset whenever the mode is changed. Modes may be combined so that their functional definition can be "tailored" to almost any I/O structure.

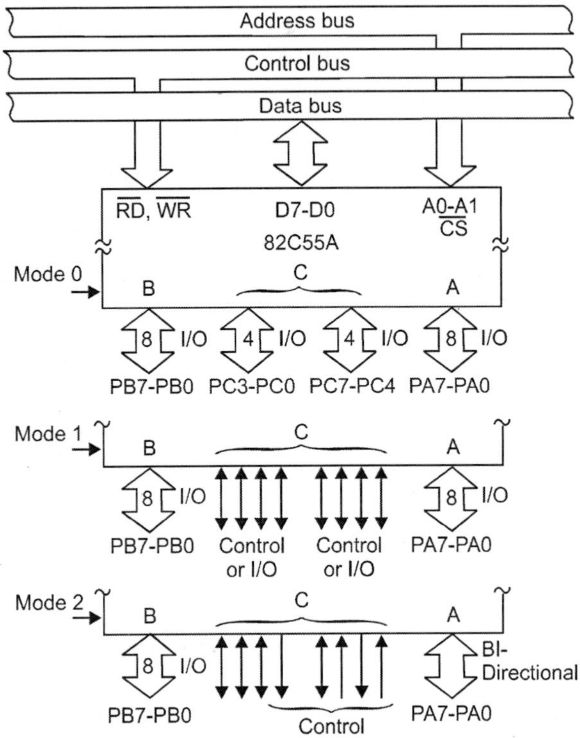


**Figure 10.7:** Basic mode definitions and bus interface

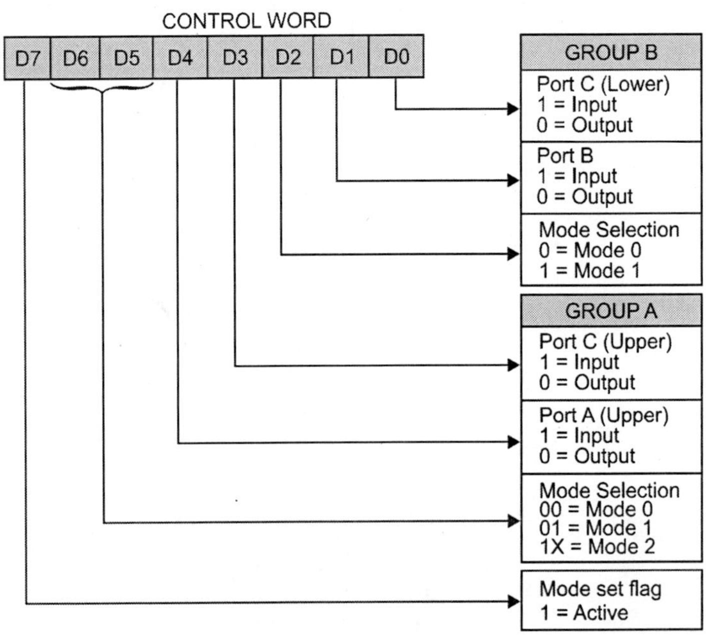


**Figure 10.8:** Mode definition format

For instance; Group B can be programmed in Mode 0 to monitor simple switch closings or display computa-tional results, Group A could be programmed in Mode 1 to monitor a keyboard or tape reader on an interrupt-driven basis.

The mode definitions and possible mode combina-tions may seem confusing at first but after a cursory review of the complete device operation a simple, logical I/O approach will surface. The design of the 8255A has taken into account things such as effi-cient PC board layout, control signal definition vs PC layout and complete functional flexibility to support almost any peripheral device with no external logic. Such design represents the maximum use of the available pins.

### 10.5.2 BSR (Bit Set/Reset) Mode

The BSR mode is concerned only with the eight bits of port C, which can be set or reset by writing an appropriate control word in the control register. A control word with bit $D_7 = 0$ is recognized as a BSR control word, and it does not alter any previously transmitted control word with bit $D_7 = 1$; thus the I/O operations of ports A and B are not affected by a BSR control word. In the BSR mode, individual bits of port C can be used for applications such as an on/off switch.

#### *1. BSR Control Word*

This control word, when written in the control register, sets or resets one bit at a time, as specified in Figure 10.9.

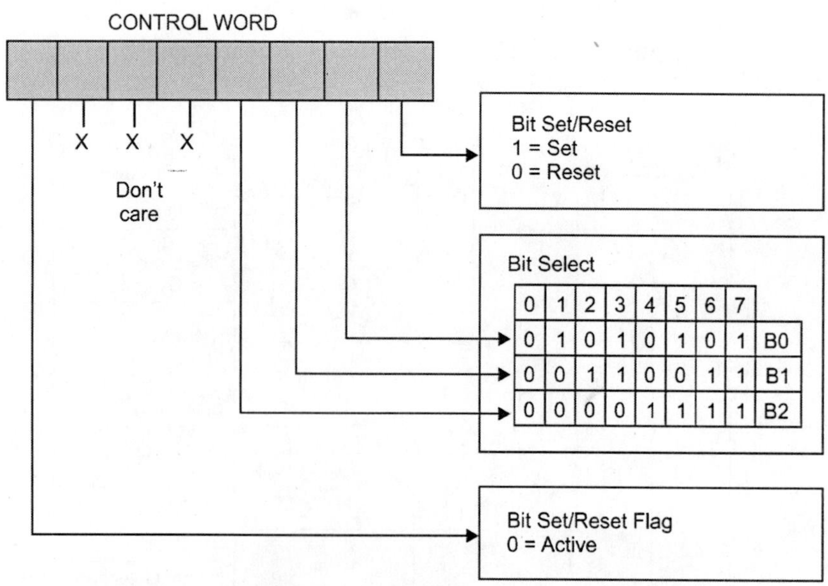


**Figure 10.9:** 8255A Control Word Format in the BSR Mode

Write a BSR control word subroutine to set bits $PC_7$ and $PC_3$ and reset them after 10 ms. Use the schematic in Figure 10.3 and assume that a delay subroutine is available.

### 2. BSR Control Words

| | | $D_7$ | $D_6$ | $D_5$ | $D_4$ | $D_3$ | $D_2$ | $D_1$ | $D_0$ | | |
|---|---|---|---|---|---|---|---|---|---|---|---|
| To set bit $PC_7$ | = | 0 | 0 | 0 | 0 | 1 | 1 | 1 | 1 | = | 0FH |
| To reset bit $PC_7$ | = | 0 | 0 | 0 | 0 | 1 | 1 | 1 | 0 | = | 0EH |
| To set bit $PC_3$ | = | 0 | 0 | 0 | 0 | 0 | 1 | 1 | 1 | = | 07H |
| To reset bit $PC_3$ | = | 0 | 0 | 0 | 0 | 0 | 1 | 1 | 0 | = | 06H |

### 3. Port Address

Control register address = 83H; refer to Figure 10.3(b).

### 4. Subroutine

```
BSR:    MVI A,0FH     ;Load byte in accumulator to set PC7
        OUT 83H       ;Set PC7 = 1
        MVI A,07H     ;Load byte in accumulator to set PC3
        OUT 83H       ;Set PC3 = 1
        CALL DELAY    ;This is a 10-ms delay
        MVI A,06H     ;Load accumulator with the byte to reset PC3
        OUT 83H       ;Reset PC7
        MVI A,0EH     ;Load accumulator with the byte to reset PC7
        OUT 83H       ;Rest PC7
        RET
```

From an analysis of the above routine, the following points can be noted:

1. TO set/reset bits in port C, a control word is written in the control register and not in port C.
2. A BSR control word affects only one bit in port C.
3. The BSR control word does not affect the I/O mode.

### 5. Interrupt Control Functions

When the 8255A is programmed to operate in mode 1 or mode 2, control signals are provided that can be used as interrupt request inputs to the CPU. The in-terrupt request signals, generated from port C, can be inhibited or enabled by setting or resetting the associated INTE flip-flop, using the bit set/reset function of port C.

This function allows the Programmer to disallow or allow a specific I/O device to interrupt the CPU with-out affecting any other device in the interrupt struc-ture.

**INTE flip-flop definition:** (BIT-SET)—INTE is set—Interrupt enable (BIT-RESET)—INTE is RESET—Interrupt disable

**Note:** All Mask Flip-Flops Are Automatically Reset During Mode Selection And Device Reset.

## 10.6 OPERATING MODES

### 10.6.1 Mode 0: Simple Input or Output

In this mode, ports A and B are used as two simple 8-bit I/O ports and ports and port C as two 4-bit ports. Each port (or half-port, in case of C) can be programmed to function as simply an input port or an output port. No "handshaking" is required; data is simply written to or read from a specified port.

#### *Mode 0 Basic Functional Definitions:*

1. Two 8-bit ports and two 4-bit ports.
2. Any port can be input or output.
3. Outputs are latched.
4. Inputs are not latched.
5. 16 different Input/Output configurations are pos-sible in this Mode.

**Table 5.3:** Port configuration in mode 0

| A | | B | | Group A | | Group B | | |
|---|---|---|---|---|---|---|---|---|
| $D_4$ | $D_3$ | $D_1$ | $D_0$ | Port A | Port C (upper) | # | Port B | Port C (lower) |
| 0 | 0 | 0 | 0 | Output | Output | 0 | Output | Output |
| 0 | 0 | 0 | 1 | Output | Output | 1 | Output | Input |
| 0 | 0 | 1 | 0 | Output | Output | 2 | Input | Output |
| 0 | 0 | 1 | 1 | Output | Output | 3 | Input | Input |
| 0 | 1 | 0 | 0 | Output | Input | 4 | Output | Output |
| 0 | 1 | 0 | 1 | Output | Input | 5 | Output | Input |
| 0 | 1 | 1 | 0 | Output | Input | 6 | Input | Output |
| 0 | 1 | 1 | 1 | Output | Input | 7 | Input | Input |
| 1 | 0 | 0 | 0 | Input | Output | 8 | Output | Output |
| 1 | 0 | 0 | 1 | Input | Output | 9 | Output | Input |
| 1 | 0 | 1 | 0 | Input | Output | 10 | Input | Output |
| 1 | 0 | 1 | 1 | Input | Output | 11 | Input | Input |
| 1 | 1 | 0 | 0 | Input | Input | 12 | Output | Output |
| 1 | 1 | 0 | 1 | Input | Input | 13 | Output | Input |
| 1 | 1 | 1 | 0 | Input | Input | 14 | Input | Output |
| 1 | 1 | 1 | 1 | Input | Input | 15 | Input | Input |

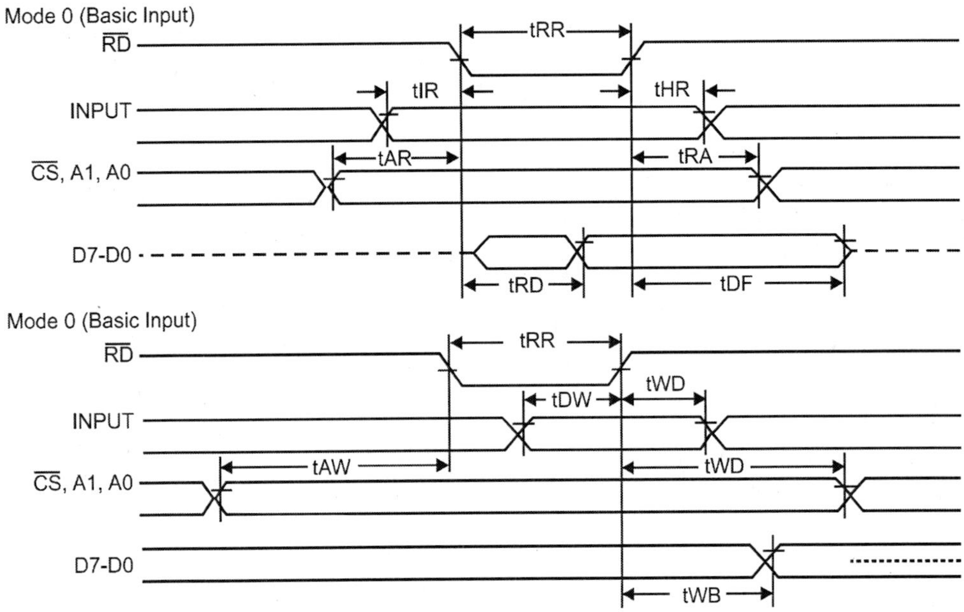


**Figure 10.10**

## Example

1. Identify the port addresses in Figure 10.11.

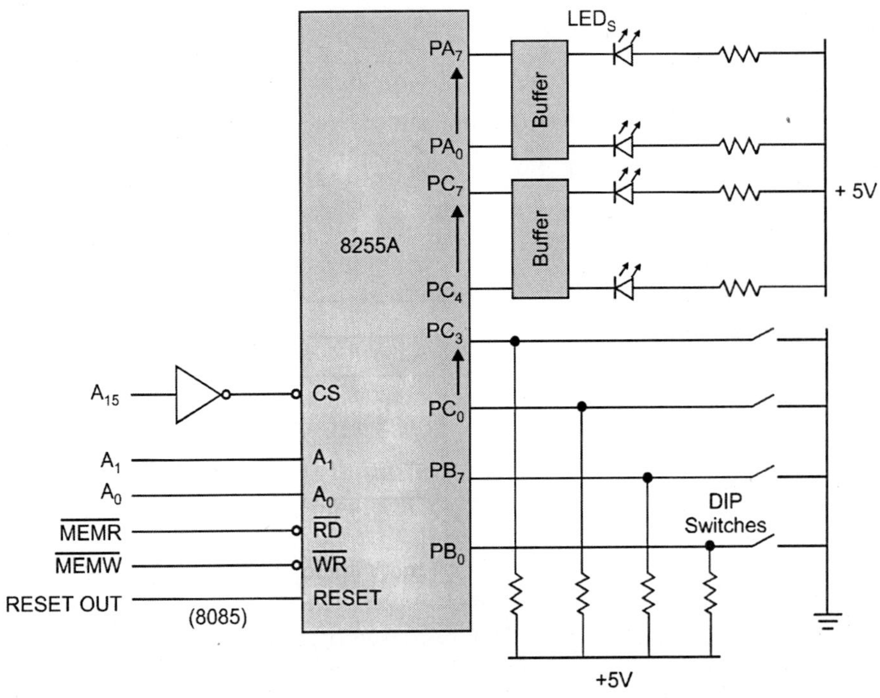


**Figure 10.11:** Interfacing 8255A I/O Ports in Mode 0.

2. Identify the Mode 0 control word to configure port A and port $C_U$ as output ports and port B and port $C_L$ as input ports.
3. Write a program to read the DIP switches and display the reading from port B at port A and from port $C_L$ at port $C_U$.

**Solution:**

1. **PORT ADDRESSES:** This is a memory-mapped I/O; when the address line $A_{15}$ is high, the Chip Select line is enabled. Assuming all don't care lines are at logic 0, the port addresses are as follows:

| | | |
|---|---|---|
| Port A | = | 8000H ($A_1 = 0, A_0 = 0$) |
| Port B | = | 8001H ($A_1 = 0, A_0 = 1$) |
| Port C | = | 8002H ($A_1 = 1, A_0 = 0$) |
| Control Register | = | 8003H ($A_1 = 1, A_0 = 1$) |

2. **CONTROL WORD**

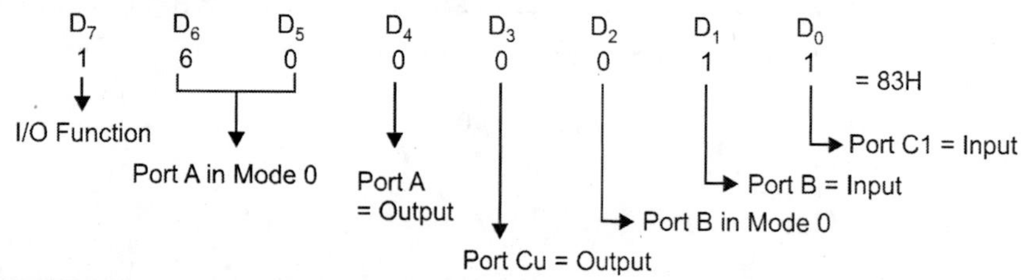


3. **PROGRAM**

```
MVI A,83H     ; Load accumulator with the control word
STA 8003H     ; Write word in the control register to initialize the ports
LDA 8001H     ; Read switches at port B
STA 8000H     ; Display the reading at port A
LDA 8002H     ; Read switches at port C
ANI 0FH       ; Mask the upper four bits of port C; these bits are not input data
RLC           ; Rotate and place data in the upper half of the accumulator
RLC
RLC
RLC
STA 8002H     ; Display data at port CU
HLT
```

4. **PROGRAM DESCRIPTION:** The circuit is designed for memory-mapped I/O; therefore, the instructions are written as if all the 8255A ports are memory locations.

The ports are initialized by placing the control word 83H in the control register. The instruction STA and LDA are equivalent to the instruction OUT and IN, respectively.

In this example, the low four bits of port C are configured as input and the high four bits are configured as output: even though port C has one address for both halves $C_U$ and $C_L$ (8002H).

Read and Write operations are differentiated by the control signals $\overline{MEMR}$ and $\overline{MEMW}$. When the MPU reads port C (e.g., LDA 8002H), it receives eight bits in the accumulator. However, the high-order bits ($D_7$–$D_4$) must be ignored because the input data bits are in $PC_3$–$PC_0$. To display these bits at the upper half of port C, bits ($PC_3$–$PC_0$) must be shifted to $PC_7$–$PC_4$.

### 10.6.2 Mode 1: Input or Output with Handshake

This functional configuration provides a means for transferring I/O data to or from a specified port in conjunction with strobes or "hand shaking" signals. In mode 1, port A and port B use the lines on port C to generate or accept these "hand shaking" signals. Handshake signals are exchanged between the MPU and peripherals prior to data transfer.

#### *Mode 1: Basic Function Definitions*

1. Two ports (A and B) function as 8-bit I/O ports. They can be configured either as input or output ports.
2. Each port uses three lines from port C as handshake signals. The remaining two lines of port C can be used for simple I/O functions.
3. Input and output data are latched.
4. Interrupt logic is supported.

In the 8255A, the specific lines from port C used for handshake signals vary according to the I/O function of a port. Therefore, input and output functions in Mode 1 are discussed separately.

#### *1. Mode 1: Input Control Signals*

Figure 10.12(a) shows the associated control signals used for handshaking when ports A and B are configured as input ports. Port A uses the upper three signals: $PC_3$, $PC_4$ and $PC_5$. Port B uses the lower three signals: $PC_2$, $PC_1$ and $PC_0$. The functions of these signals are as follows:

- **$\overline{STB}$ (Strobe Input):** A "low" on this input loads data into the input latch. This signal (active low) is generated by a peripheral device to indicate that it has transmitted a byte of data. The 8255A, in response to $\overline{STB}$, generates IBF and INTR, as shown in Figure 10.13.
- **IBF (Input Buffer Full):** This signal is an acknowledgment by the 8255A to indicate that the input latch has received the data byte. This is reset when the MPU reads the data (Figure 10.13).
- **INTR (Interrupt Request):** This is an output signal that may be used to interrupt the MPU. This signal is generated if, IBF, and INTE (Internal flip-flop) are all at logic 1. This is reset by the falling edge of the  signal (Figure 10.13).

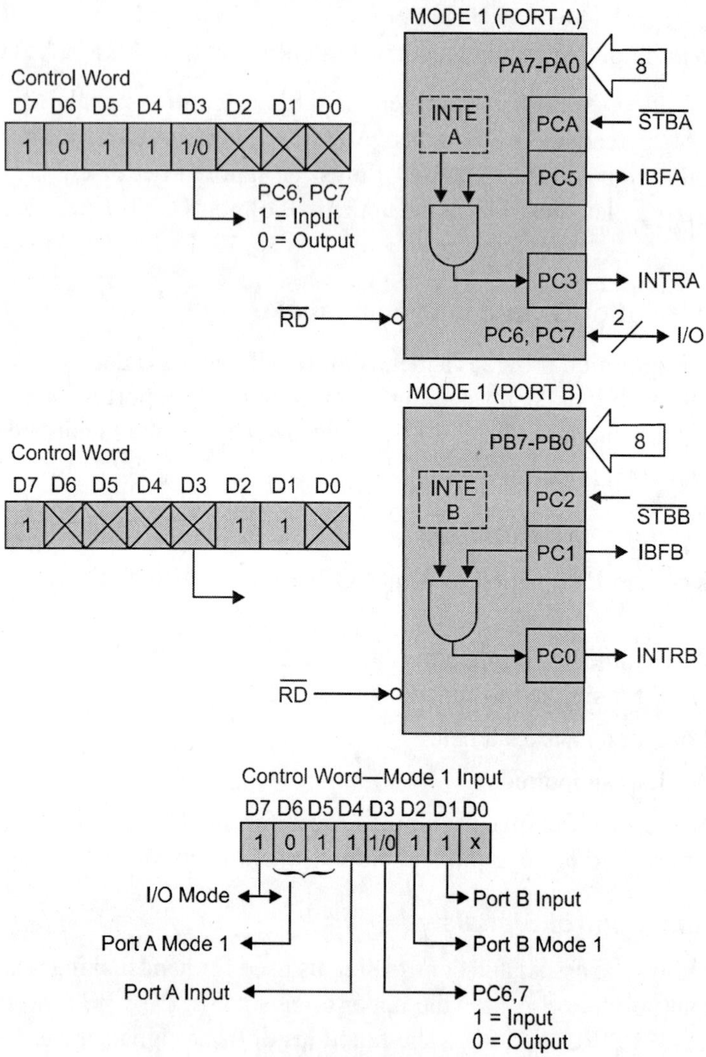


Status word-mode 1 input

| $D_7$ | $D_6$ | $D_5$ | $D_4$ | $D_3$ | $D_2$ | $D_1$ | $D_0$ |
|---|---|---|---|---|---|---|---|
| I/O | I/O | $IBF_A$ | $INTE_A$ | $INTR_A$ | $INTE_B$ | $IBF_B$ | $INTR_B$ |

**Figure 10.12:** 8255A Mode 1: Input Configuration

- **INTE (Interrupt Enable):** This is an internal flip-flop used to enable or disable the generation of the INTR signal. The two flip-flops $INTE_A$ and $INTE_B$ are set/reset using the BSR mode. The $INTE_A$ is enabled or disabled through $PC_4$, and $INTE_B$ is enabled or disabled through $PC_2$.

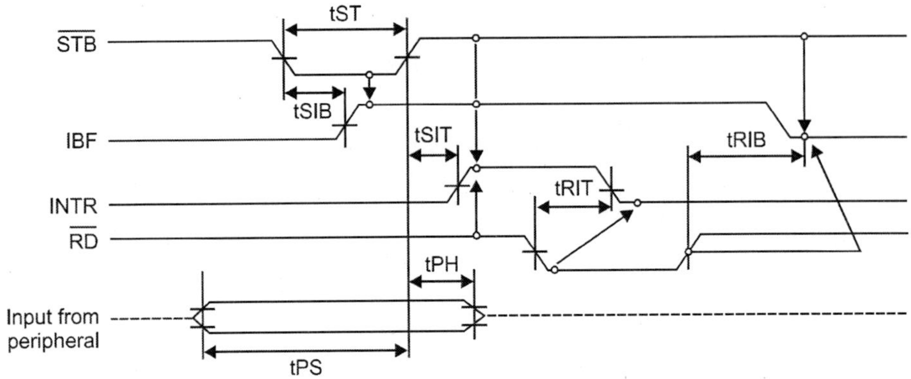


**Figure 10.13:** 8255A Mode 1: Timing Waveforms for Strobed Input (with Handshake)

## 2. *Control and Status Words*

Figure 10.12(b) uses control words derived from Figure 10.4 to set up port A and port B as input ports in Mode 1. Similarly, Figure 10.12(c) also shows the status word, which will be placed in the accumulator if port C is read.

## 3. *Programming The 8255A In Mode 1*

The 8255A can be programmed to function using either status check I/O or interrupt I/O. Figure 10.14(a) shows a flowchart for the status check I/O. In this flowchart, the MPU continues to check data status through the IBF line until goes high. This is a simplified flowchart; however, it does not show how to handle data transfer if two ports are being used. The technique is similar to that of Mode 0 combined with the BSR mode. The disadvantage of the status check I/O with handshake is that the MPU is tied up in the loop.

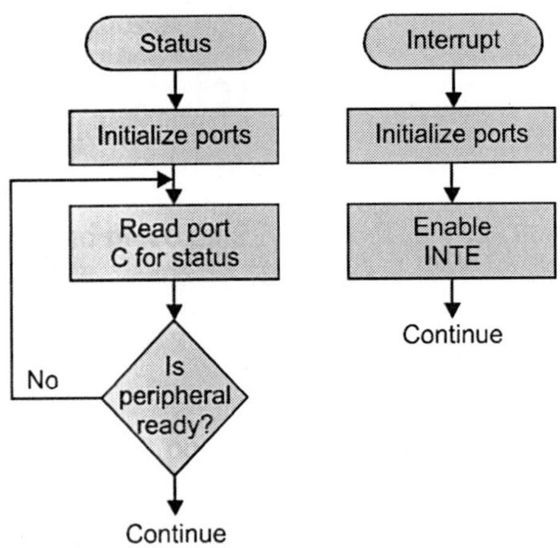


**Figure 10.14:** Flowcharts: (a) Status Check I/O (b) Interrupt I/O

The flowchart in Figure 10.14(b) shows the steps required for the interrupt I/O, assuming that vectored interrupts are available. The confusing step in the interrupt I/O is to set INTE either for port A or port B. Figure 10.12(a) shows that the $\overline{STB}$ signal is connected to pin $PC_4$ and the $INTE_A$ is set or reset in the BSR mode and the BSR control word has no effect when ports A and B are set in Mode 1.

In case the INTR line is used to implement the interrupt, it may be necessary to read the status of $INTR_A$ and $INTR_B$ to identify the port requesting an interrupt service and to determine the priority through software, if necessary.

### 4. *Mode 1: Output Control Signals*

Figure 10.15 shows the control signals when ports A and B are configured as output ports. These signals are defined as follows:

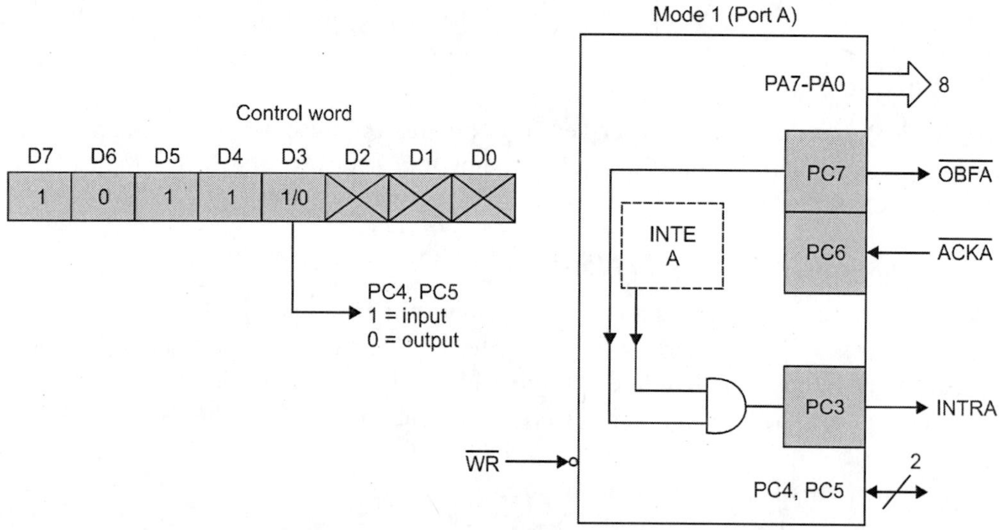


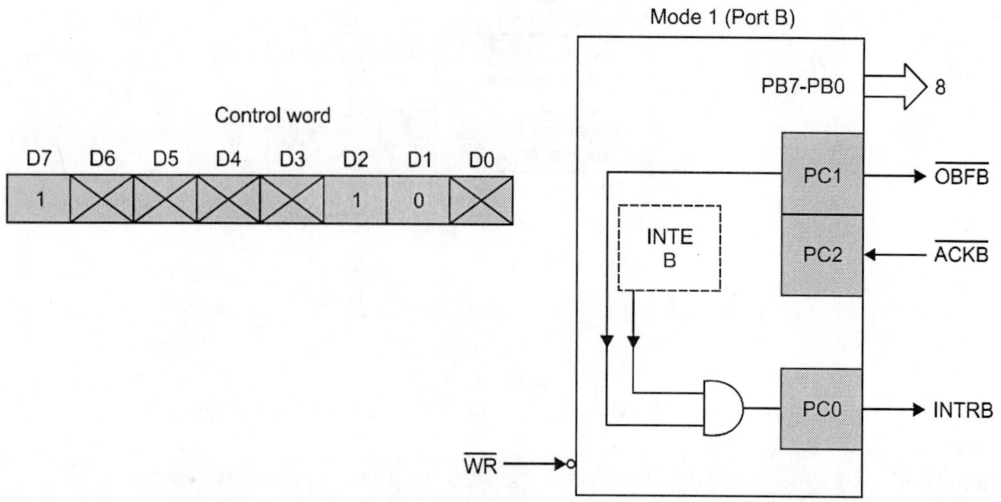

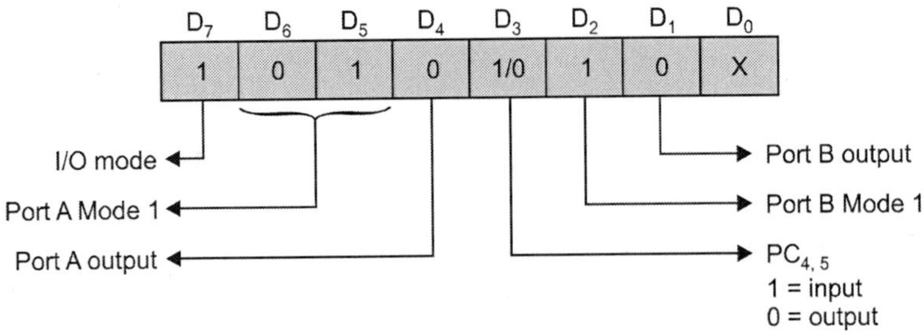


Status word

| $D_7$ | $D_6$ | $D_5$ | $D_4$ | $D_3$ | $D_2$ | $D_1$ | $D_0$ |
|---|---|---|---|---|---|---|---|
| $\overline{OBF}_A$ | $INTE_A$ | 1/0 | 1/0 | $INTR_A$ | $INTE_B$ | $\overline{OBF}_B$ | $INTR_B$ |

**Figure 10.15:** 8255A Mode 1: Output Configuration

- **OBF (Output Buffer Full):** This is an output signal that goes low when the MPU writes data into the output latch of the 8255A. This signal indicates to an output peripheral that new data are ready to be read (Figure 10.16). It goes high again after the 8255A receives an $\overline{ACK}$ from the peripheral.
- **$\overline{ACK}$ (Acknowledge):** A "low" on this input informs the 82C55A that the data from Port A or Port B is ready to be accepted. In essence, a response from the peripheral device indicating that it is ready to accept data (Figure 10.16).
- **INTR (Interrupt Request):** This is an output signal, and it is set by the rising edge of the $\overline{ACK}$ signal. This signal can be used to interrupt the MPU to request the next data byte for output. The INTR is set when $\overline{OBF}$ , $\overline{ACK}$, and INTE are all one (Figure 10.16) and reset by the falling edge of $\overline{WR}$.
- **INTE (Interrupt Enable):** This is an internal flip-flop to a port and needs to be set to generate the INTR signal. The two flip-flops $INTE_A$ and $INTE_B$ are controlled by bits $PC_6$ and $PC_2$ respectively, through the BSR mode.
- **$PC_{4,5}$:** These two lines can be set up either as input or output.

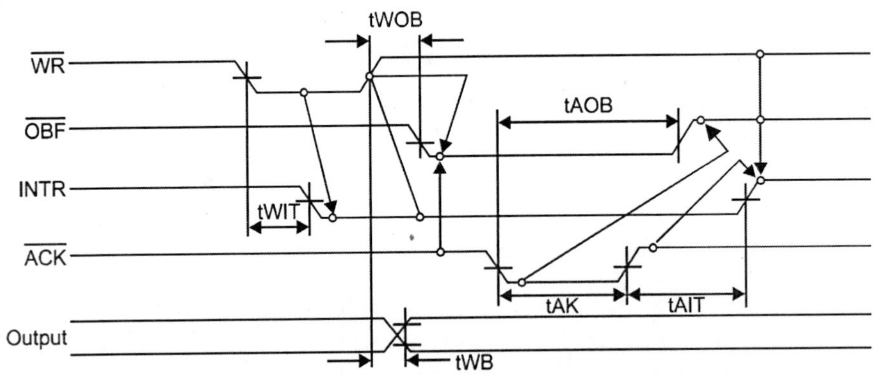


**Figure 10.16:** 8255 Mode 1: Timing Waveforms for Strobed (with Handshake) Output

### *5. Control and Status Words*

Figure 10.15(b) shows the control word used to set up ports A and B as output ports in Mode 1. Similarly, Figure 10.15(c) also shows the status word, which will be placed in the accumulator if port C is read.

### *6. Combinations of Mode 1*

Port A and Port B can be individually defined as in-put or output in MODE 1 to support a wide variety of strobed I/O applications.

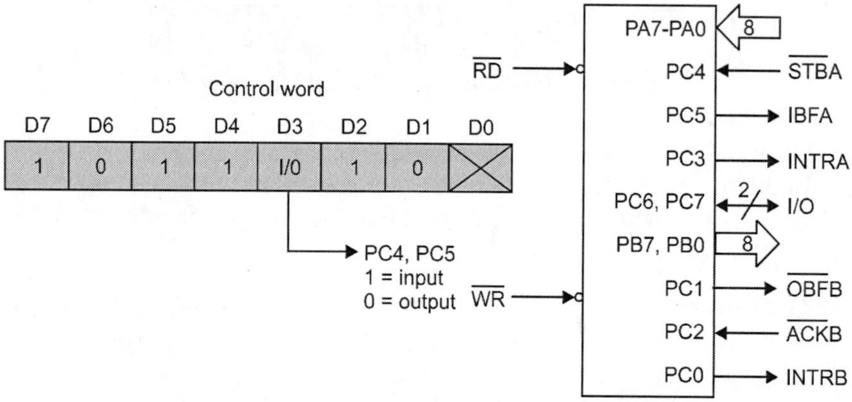


Port A - (Strobed input)
Port B - (Strobed output)

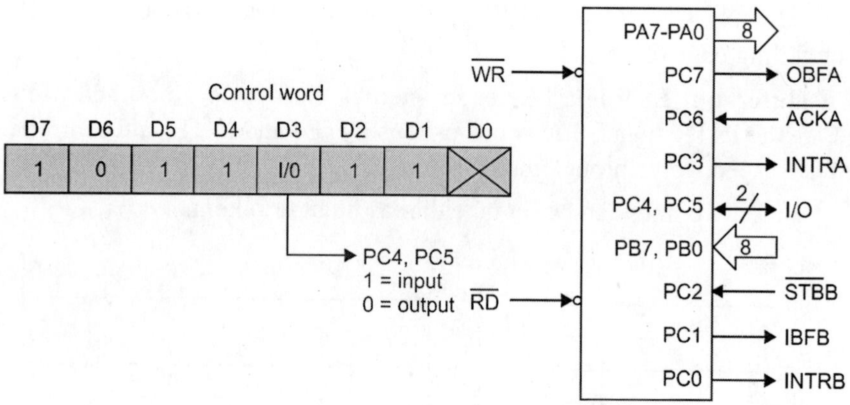


Port A - (Strobed output)
Port B - (Strobed input)

### 10.6.3 Mode 2: Bidirectional Data Transfer

This functional configuration provides a means for com-municating with a peripheral device or structure on a single 8-bit bus for both transmitting and receiving data (bidirectional bus I/O). "Handshaking" signals are provided to maintain proper bus flow discipline in a similar manner to MODE 1. Interrupt generation and enable/disable functions are also available.

This mode is used primarily in applications such as data transfer between two computers or floppy disk controller interface. In this mode, port A can be configured as the bidirectional port and port B either in Mode 0 or Mode 1. Port A uses five signals from port C as handshake signals for data transfer. The remaining three signals from port C can be used either as simple I/O or as handshake for port B.

### *Mode 2 Basic Functional Definitions*

- Used in Group A only.
- One 8-bit, bi-directional bus Port (Port A) and a 5- bit control Port (Port C).
- Both inputs and outputs are latched.
- The 5-bit control port (Port C) is used for control and status for the 8-bit, bi-directional bus port (Port A).

1. **INTR (Interrupt Request).** A high on this output can be used to interrupt the CPU for both input or output operations.
2. **OUTPUT OPERATIONS**

   **$\overline{\text{OBF}}$ (Output Buffer Full):** The $\overline{\text{OBF}}$ output will go "low" to indicate that the CPU has written data out to port A.

   $\overline{\text{ACK}}$ **(Acknowledge).** A "low" on this input enables the tri-state output buffer of port A to send out the data. Otherwise, the output buffer will be in the high impedance state.

   **INTE 1 (The INTE Flip-Flop Associated with $\overline{\text{OBF}}$).** Controlled by bit set/reset of $PC_6$.
3. **INPUT OPERATIONS**

   **$\overline{\text{STB}}$ (Strobe Input):** A "low" on this input loads data into the input latch.

   **IBF (Input Buffer Full F/F):** A "high" on this output indicates that data has been loaded into the input latch.

   **INTE 2 (The INTE Flip-Flop Associated with IBF):** Controlled by bit set/reset of $PC_4$.

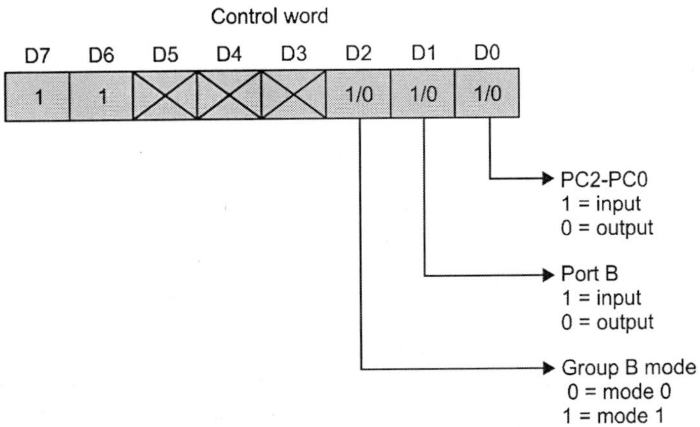


**Figure 10.17:** Mode Control Word

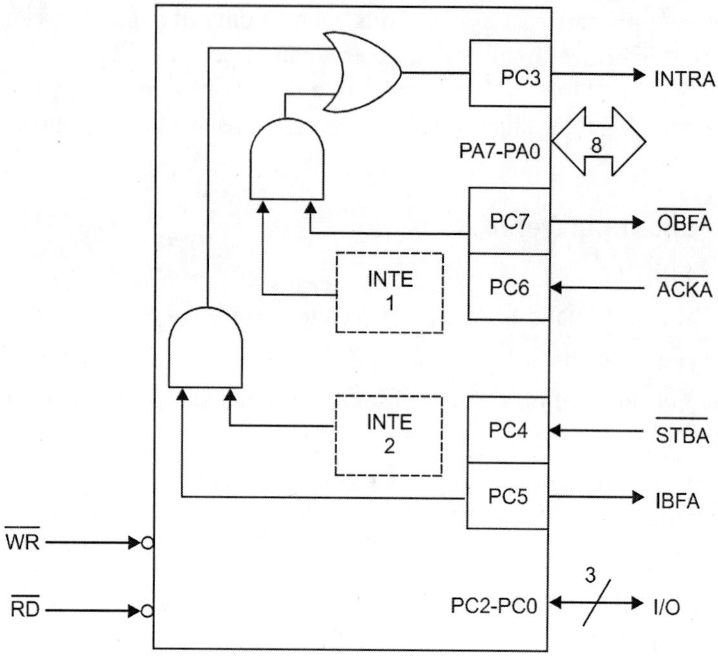


**Figure 10.18:** Mode 2

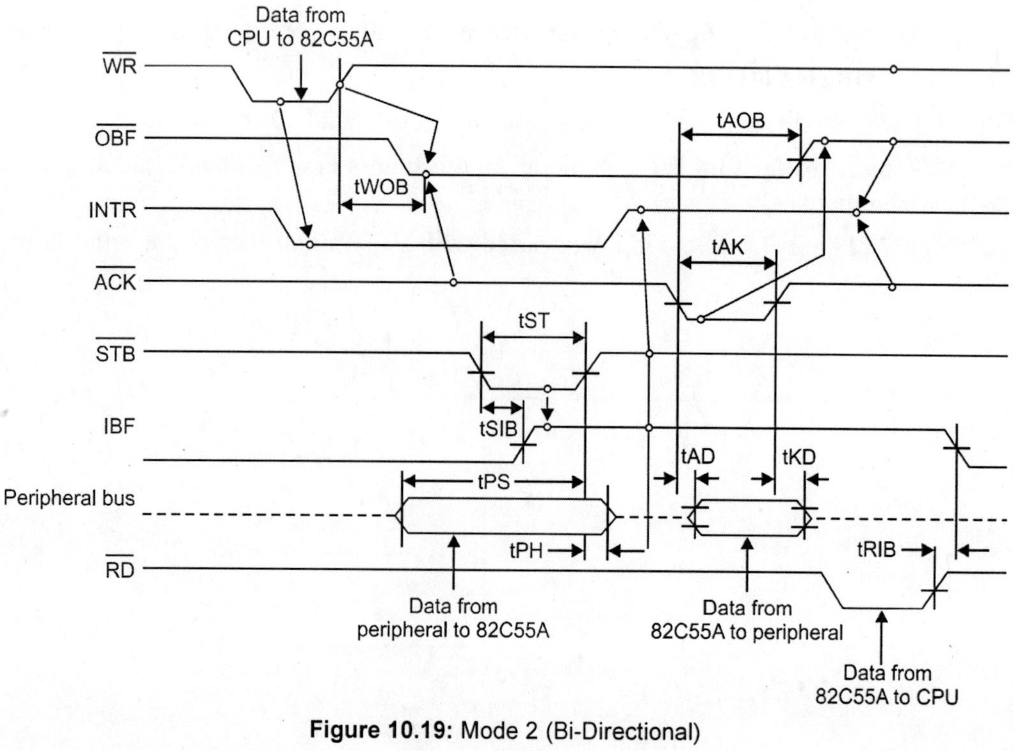


**Figure 10.19:** Mode 2 (Bi-Directional)

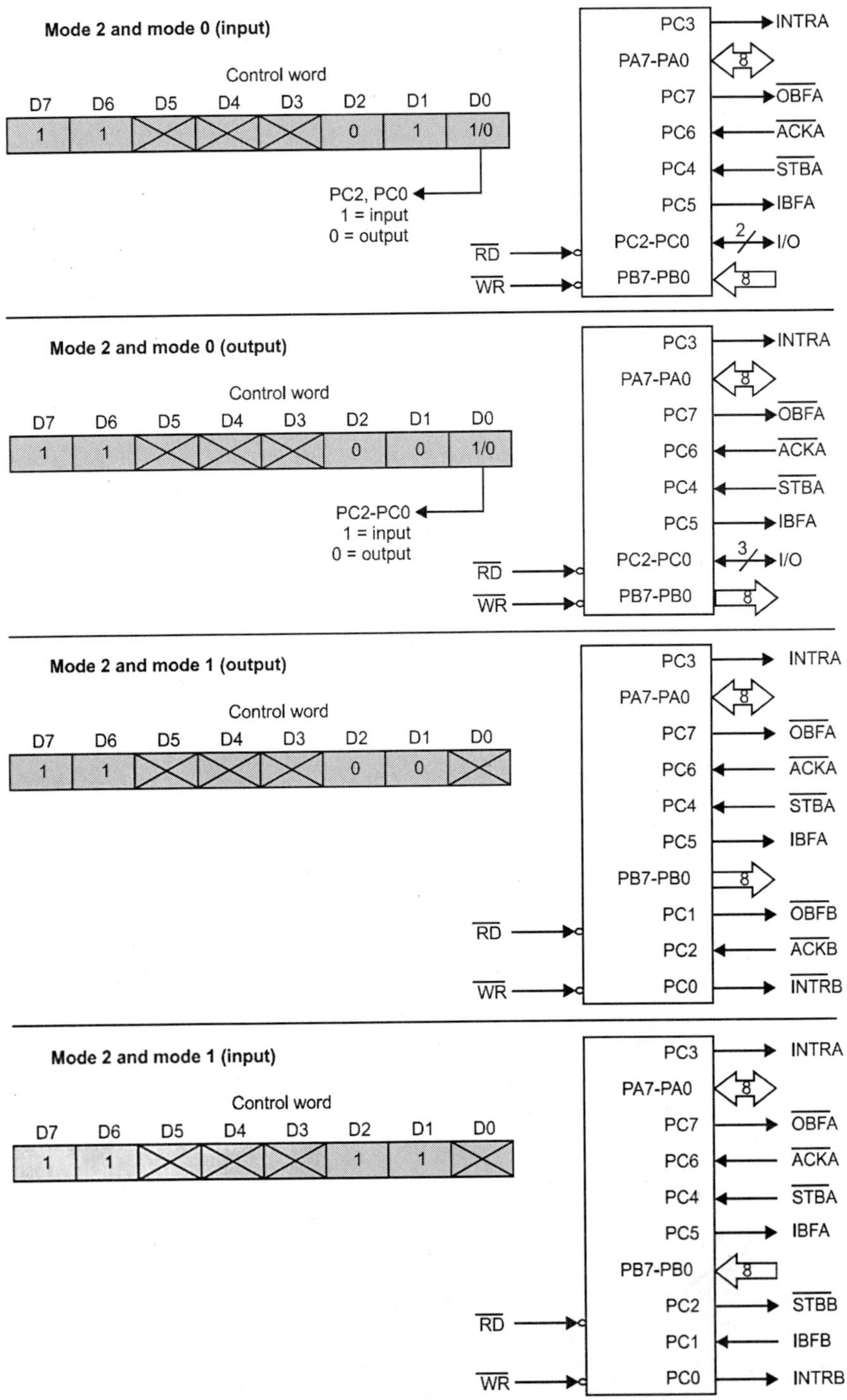


**Figure 10.20:** Mode 2 Combinations

| Mode Definition Summary | | | | | |
|---|---|---|---|---|---|
| | Mode 0 | | Mode 1 | | Mode 2 |
| | In | Out | In | Out | Group A only |
| PA0 | In | Out | In | Out | ↔ |
| PA1 | In | Out | In | Out | ↔ |
| PA2 | In | Out | In | Out | ↔ |
| PA3 | In | Out | In | Out | ↔ |
| PA4 | In | Out | In | Out | ↔ |
| PA5 | In | Out | In | Out | ↔ |
| PA6 | In | Out | In | Out | ↔ |
| PA7 | In | Out | In | Out | ↔ |
| PB0 | In | Out | In | Out | Mode 0 or Mode 1 Only |
| PB1 | In | Out | In | Out | |
| PB2 | In | Out | In | Out | |
| PB3 | In | Out | In | Out | |
| PB4 | In | Out | In | Out | |
| PB5 | In | Out | In | Out | |
| PB6 | In | Out | In | Out | |
| PB7 | In | Out | In | Out | |
| PC0 | In | Out | INTRB | INTRB | I/O |
| PC1 | In | Out | IBFB | $\overline{\text{OBF}}$ B | I/O |
| PC2 | In | Out | $\overline{\text{STB}}$ B | ACKB | I/O |
| PC3 | In | Out | INTRA | INTA | INTRA |
| PC4 | In | Out | $\overline{\text{STB}}$ A | I/O | $\overline{\text{STB}}$ A |
| PC5 | In | Out | IBFA | I/O | IBFA |
| PC6 | In | Out | I/O | $\overline{\text{ACK}}$ A | $\overline{\text{ACK}}$ A |
| PC7 | In | Out | I/O | $\overline{\text{OBF}}$ A | $\overline{\text{OBF}}$ A |

## 10.7 SPECIAL MODE COMBINATION CONSIDERATIONS

There are several combinations of modes possible. For any combination, some or all of Port C lines are used for control or status. The remaining bits are either inputs or outputs as defined by a "Set Mode" command. During a read of Port C, the state of all the Port C lines, except the $\overline{\text{ACK}}$ and $\overline{\text{STB}}$ lines, will be placed on the databus. In place of the $\overline{\text{ACK}}$ and $\overline{\text{STB}}$ line states, flag status will appear on the data bus in the PC2, PC4, and PC6 bit positions as illustrated by Figure 10.23. Through a "Write Port C" command, only the Port C pins programmed as outputs in a Mode 0 group can be written. No other pins can be affected by a "Write Port C" command, nor can the interrupt enable flags be accessed. To write to any Port C output programmed as an output in Mode 1 group or to change an interrupt enable flag, the "Set/Reset Port C Bit" command must be used. With a "Set/Reset Port Cea Bit" command, any Port C line programmed as an output (including IBF and $\overline{\text{OBF}}$) can be written, or an interrupt enable flag can be either set or reset.

Port C lines programmed as inputs, including $\overline{ACK}$ and $\overline{STB}$ lines, associated with Port C fare not affected by a "Set/Reset Port C Bit" command. Writing to the corresponding Port C bit positions of the $\overline{ACK}$ and $\overline{STB}$ lines with the "Set Reset Port C Bit" command will affect the Group A and Group B interrupt enable flags, as illustrated in Figure 10.23.

Input configuration

| $D_7$ | $D_6$ | $D_5$ | $D_4$ | $D_3$ | $D_2$ | $D_1$ | $D_0$ |
|---|---|---|---|---|---|---|---|
| I/O | I/O | IBFA | INTEA | INTRA | INTEB | IBFB | INTRB |
| Group A | | | | | Group B | | |

Output configuration

| $D_7$ | $D_6$ | $D_5$ | $D_4$ | $D_3$ | $D_2$ | $D_1$ | $D_0$ |
|---|---|---|---|---|---|---|---|
| OBFA | INTEA | I/O | I/O | INTRA | INTEB | OBFB | INTRB |
| Group A | | | | | Group B | | |

**Figure 10.21:** Mode 1 Status Word Format

| $D_7$ | $D_6$ | $D_5$ | $D_4$ | $D_3$ | $D_2$ | $D_1$ | $D_0$ |
|---|---|---|---|---|---|---|---|
| OBFA | INTE1 | IBFA | INTE2 | INTRA | X | X | X |
| Group A | | | | | Group B (Defined by mode 0 or mode 1 selection) | | |

**Figure 10.22:** Mode 2 Status Word Format

### 10.7.1 Current Drive Capability

Any output on Port A, B or C can sink or source 2.5mA. This feature allows the 82C55A to directly drive Darlington type drivers and high-voltage displays that require such sink or source current.

### 10.7.2 Reading Port C Status

In Mode 0, Port C transfers data to or from the peripheral device. When the 82C55A is programmed to function in Modes 1 or 2, Port C generates or accepts "hand shaking" signals with the peripheral device. Reading the contents of Port C allows the programmer to test or verify the "status" of each peripheral device and change the program flow accordingly.

There is not special instruction to read the status information from Port C. A normal read operation of Port C is executed to perform this function (Figures 10.21 and 10.22).

| Interrupt enable flag | Position | Alternate port C pin signal (Mode) |
|---|---|---|
| INTE B | PC2 | $\overline{ACK}$ B (Output Mode 1)<br>or $\overline{STB}$ B (Input Mode 1) |
| INTE A2 | PC4 | $\overline{STB}$ A (Input Mode 1 or Mode 2) |
| INTE A1 | PC6 | $\overline{ACK}$ A (Output Mode 1 or Mode 2) |

**Figure 10.23:** Interrupt Enable Flags In Modes 1 and 2

## 10.8 APPLICATIONS OF THE 82C55A

The 82C55A is a very powerful tool for interfacing peripheral equipment to the microcomputer system. It represents the optimum use of available pins and flexible enough to interface almost any I/O device without the need for additional external logic.

Each peripheral device in a microcomputer system usually has a "service routine" associated with it. The routine manages the software interface between the device and the CPU. The functional definition of the 82C55A is programmed by the I/O service routine and becomes an extension of the system software.

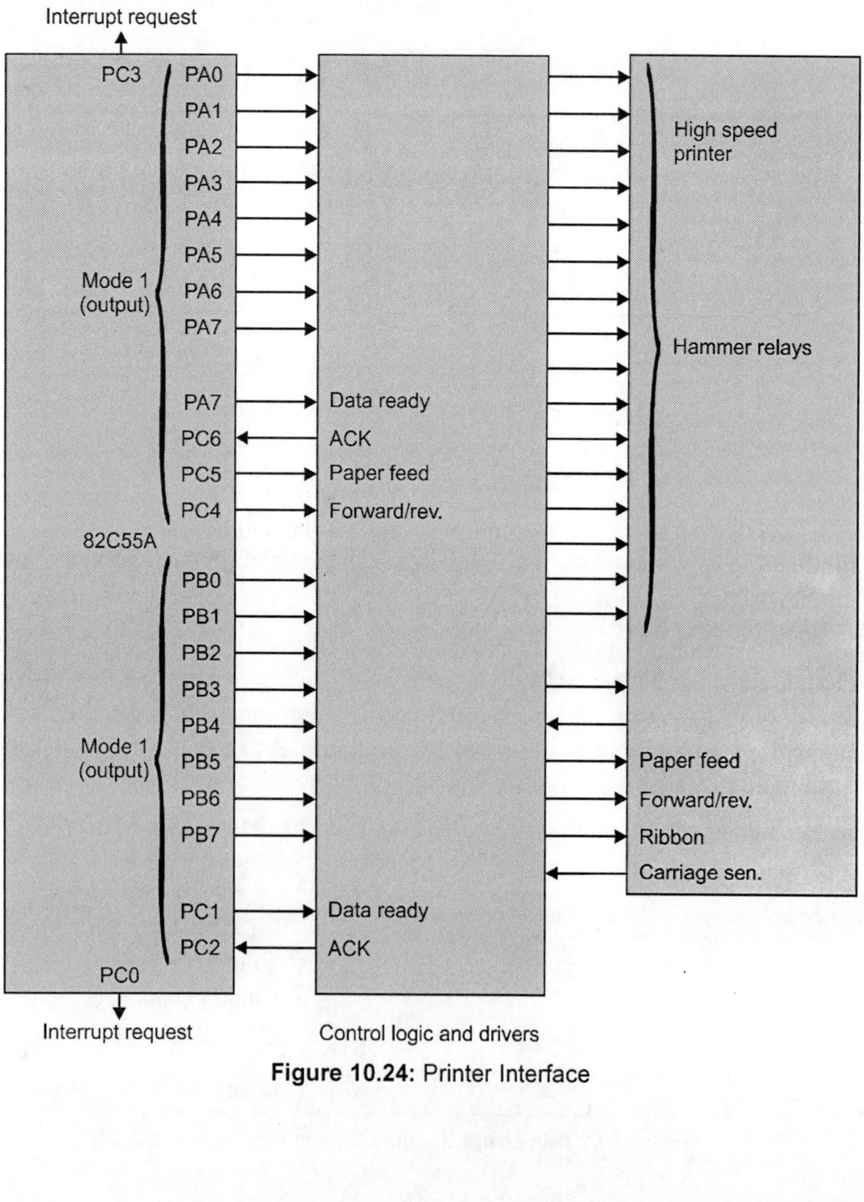


**Figure 10.24:** Printer Interface

By examining the I/O devices interface characteristics for both data transfer and timing, and matching this information to the examples and tables in the detailed operational description, a control word can easily be developed to initialize the 82C55A to exactly "fit" the application. Figures 10.24 through 10.27 present a few examples of typical applications of the 82C55A.

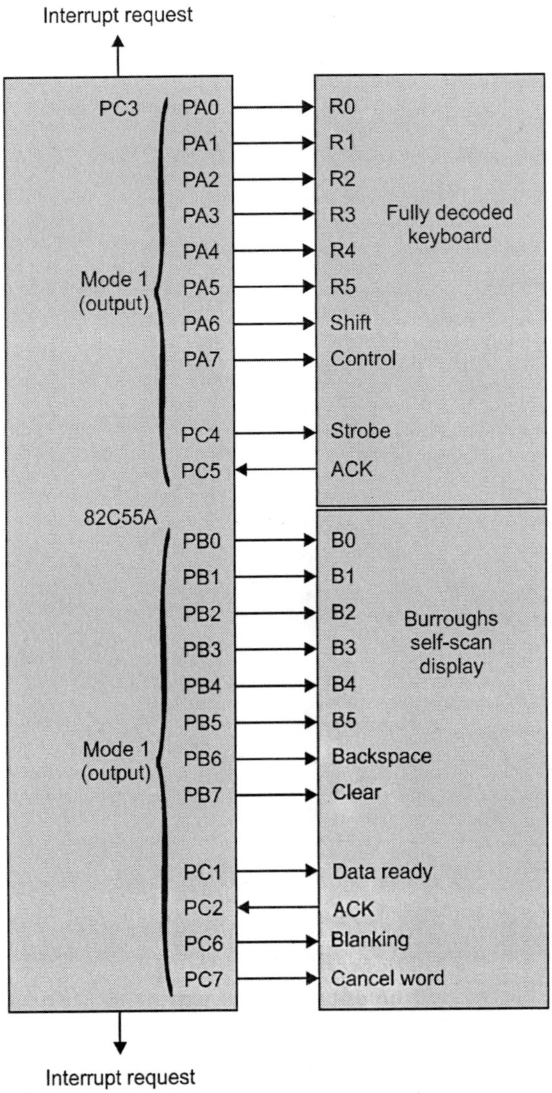


**Figure 10.25:** Keyboard And Display Interface

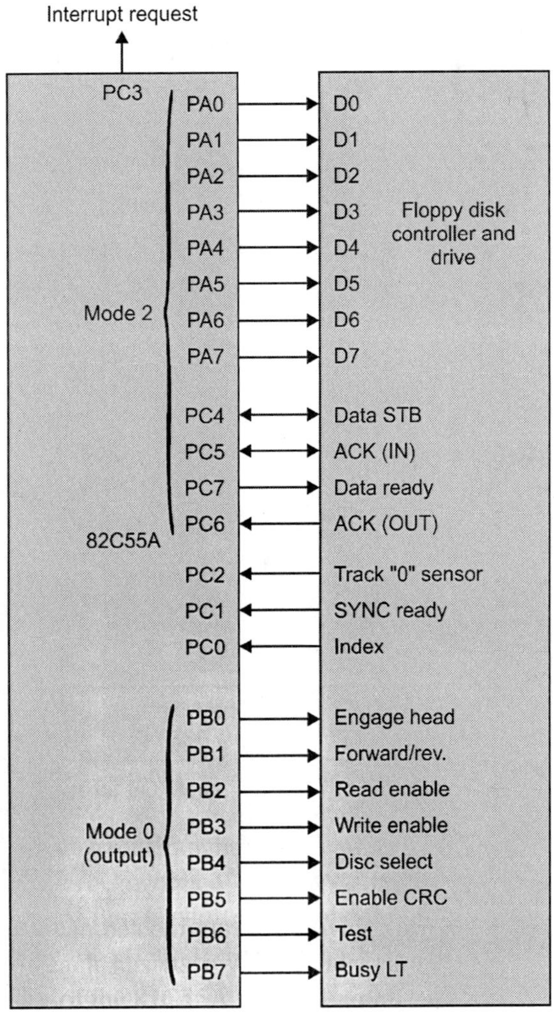


**Figure 10.26:** Basic Floppy Disc Interface

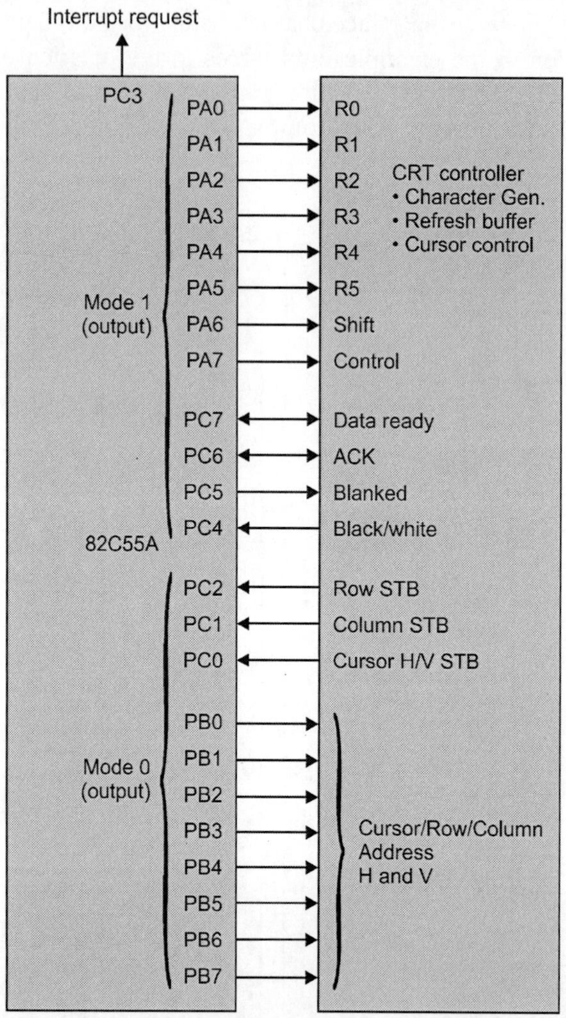


**Figure 10.27:** Basic CRT Controller Interface

## QUESTION BANK

Q1. What do you mean by 8255A programmable peripheral interface?

Q2. Explain the block diagram of the 8255A.

Q3. Explain the control word of the 8255A.

Q4. Explain the pin diagram and description of 8255A PPI.

Q5. Explain the operational description of 8255A.

Q6. Explain the different operating modes of 8255A.

Q7. Explain the special mode combinations.

Q8. Explain the various applications of the 8255A.

CHAPTER 11

# 8253/8254 Programmable Timer/Counter

## 11.1 TIMER

Timer is a very common and useful peripheral. It is used to generate events at specific times or measures the duration of specific events which are external to the processor. It is a programmable device, i.e. the time period can be adjusted by writing specific bit patterns to some of the registers called timer-control registers.

## 11.2 COUNTER

A counter is a more general version of the timer. It is used to count events in the form of pulses which is fed to it.

Figure 11.1(a) shows the block diagram of a simple timer. This has a 16-bit up counter which increments with each input clock pulse. Thus the output value Cnt represents the number of pulses since the counter was last reset to zero. An additional output top indicates when the terminal count has been reached. It may go high for a predetermined time as set by the programmable control word inside the timer unit. The count can be loaded by the external program.

Figure 11.1(b) provides the structure of another timer where a multiplexer is used to choose between an internal clock or external clock. The mode bit when set or reset decided the selection. For internal clock (Clk) it behaves like the timer in Figure 11.1(a). For the external count in (*cnt_in*) it just counts the number of occurrences.

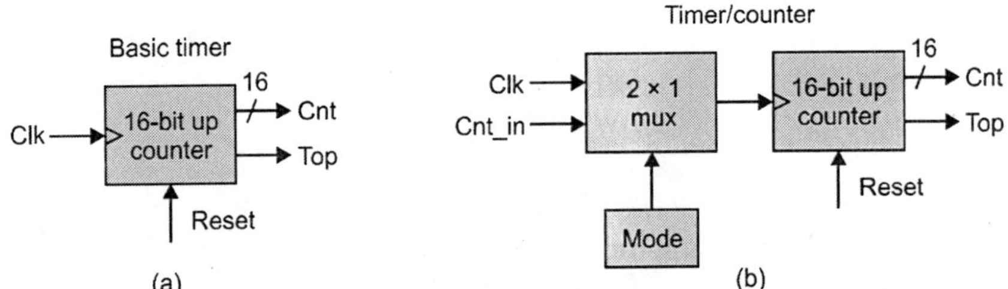


**Figure 11.1**

Figure11.1(c) shows a timer with the terminal count. This can generate an event if a particular interval of time has been elapsed. The counter restarts after every terminal count.

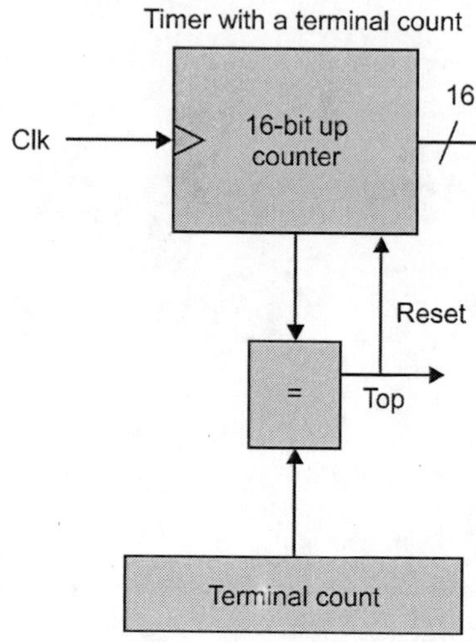


**Figure 11.1(c)**

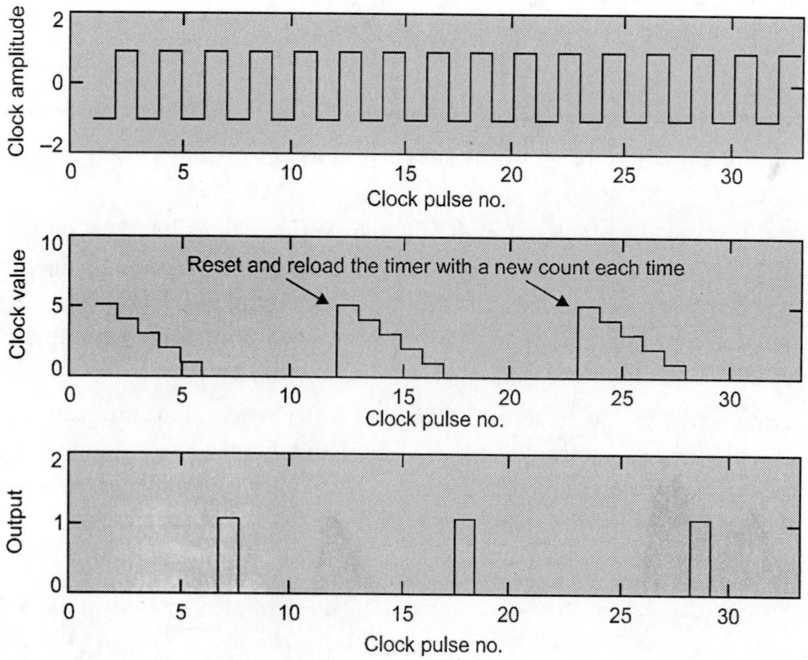


**Figure 11.2:** The Timer Count and Output. The timer is in count-down mode. In every clock pulse the count is decremented by 1. When the count value reaches zero the output of the counter i.e. TOP goes high for a predetermined time. The counter has to be loaded with a new or previous value of the count by external program or it can be loaded automatically every time the count reaches zero.

## 11.3 NECESSITY AND INTRODUCTION

The 8253/54 solves one of most common problem in any microcomputer system, the generation of accurate time delays under software control. Instead of setting up timing loops in system software, the programmer configures the 8253/54 to match his requirements, initializes one of the counters of the 8253/54 with the desired quantity, then upon command the 8253/54 will count out the delay and interrupt the CPU when it has completed its tasks. It is easy to see that the software overhead is minimum and that multiple delays can be easily be maintained by assignment of priority levels.

The 8253/54 includes three identical 16 bit counters that can operate independently. To operate a counter, a 16-bit count is loaded in its register and, on command; it begins to decrement the count until it reaches 0. At the end of the count, it generates a pulse that can be used to interrupt the CPU. The counter can count either in binary or BCD. In addition, a count can be read by the CPU while the counter is decrementing. In this chapter, we are going to study two timer ICs 8253 and 8254. The 8254 is a superset of 8253. The functioning of these two ICs are almost similar along with the pin configuration. Only the differences are:

| 8253 | 8254 |
|---|---|
| 1. Operating frequency 0–2.6 MHz. | 1. Operating frequency 0–10 MHz. |
| 2. Uses N-MOS technology. | 2. Uses H-MOS technology |
| 3. Read-Back command not available. | 3. Read-Back command availble. |
| 4. Reads and writes of the same counter can not be interleaved. | 4. Reads and writes of the same counter can be interleaved. |

In this chapter, where the things are common 8253/54 is mentioned and 8254 is mentioned specifically for giving only information about 8254.

### 11.3.1 Features

1. Three independent 16-bit down counters.
2. 8254 can handle inputs from DC to 10 MHz (5MHz 8254-5 8MHz 8254 10MHz 8254-2) where as 8253 can operate upto 2.6 MHz.
3. Three counters are identical presettable, and can be programmed for either binary or BCD count.
4. Counter can be programmed in six different modes.
5. Compatible with all Intel and most other microprocessors.
6. 8254 has powerful command called READ BACK command which allows the user to check the count value, programmed mode and current mode and current status of the counter.

## 11.4 THE PROGRAMMABLE INTERVAL TIMER 8253

The 8253 is a programmable Interval timer/counter specifically designed for use with the Intel

Micro computer systems. Its function is that of a general purpose, multi-timing element that can be treated as an array of I/O ports in the system software.

The 8253 solves one of the most common problems any microcomputer system. The generation of accurate time delay under software control. Instead of selling up timing loops in systems software. The programmer configures the 8253 to match his requirements. Initializes one of the counters of the 8253 with the desired quantity, then upon command the 8253 will count-out the delay and interrupt the CPU when it has completed its tasks. It is easy to see that the software overhead is minimal and that multiple delays can easily be maintained by assignment of priority levels.

Other counter/timer functions that are non-delay in nature but also common to most microcomputers can be implemented with the 8253.

- Programmable Rate Generator
- Event Counter
- Binary Rate Multiplier
- Real Time Clock
- Digital One-Shot
- Complex Motor Controller

For processors where the timer unit is not internal the programmable interval timer can be used. Figure 11.3 shows the signals for 8253 programmable interval timer.

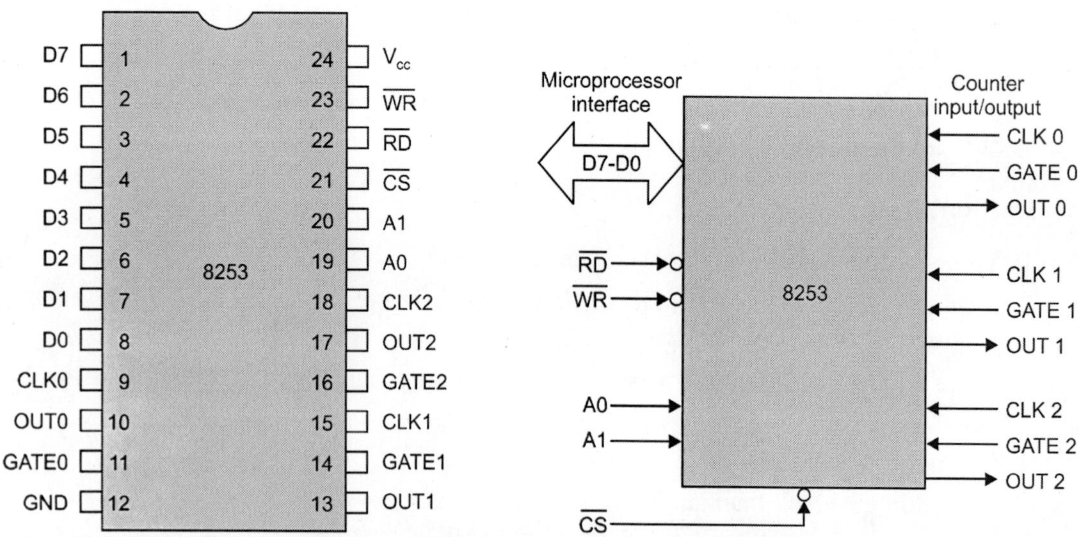


**Figure 11.3:** The pin configuration of the timer

Figure 11.4 shows the internal block diagram. There are three separate counter units controlled by configuration register (Fig.11.5).

Each counter has two inputs, clock and gate and one output. The clock is signal that helps in counting by decrementing a preloaded value in the respective counter register. The gate serves as an enable input. If the gate is maintained low the counting is disabled. The timing diagram explains in detail about the various modes of operation of the timer.

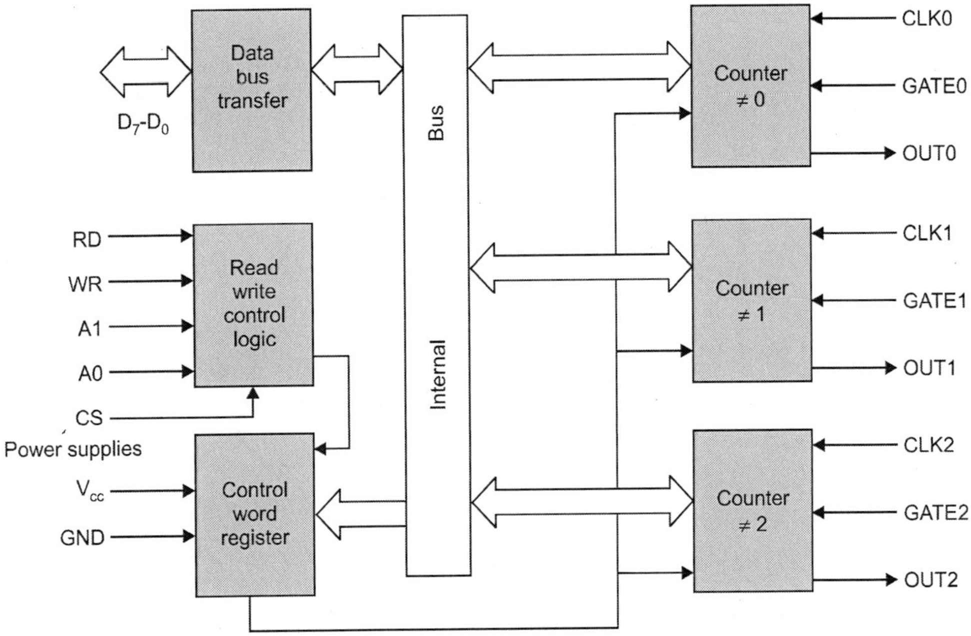


**Figure 11.4:** The internal block diagram of 8253 Table The address map

### 1. Data Bus Buffer

This 3-state, be directional,8-bit buffer is used to interface the 8253 to the systems data bus. Data is transmitted or received by the buffer upon execution of Input and Output CPU instructions. The Data Bus Buffer has three basic functions.

1. Programming the MODES of the 8253
2. Loading the count registers
3. Reading the count values

### 2. Read/Write Logic

The Read/Write Logic accepts inputs from the system bus and in turn generates control signals for overall device operation. It is enabled by CS so that no operation can occur to change the function unless the device has been selected by the system logic.

### 3. RD (Read)

A "low" on this input informs the 8253 that the CPU is inputting data in the form of a counters value.

### 4. *WR (Write)*

A "low" on this input informs the 8253 that the CPU is outputting data in the form of mode information or loading counters.

### 5. *A0, A1*

These inputs are normally connected to the address bus. Their function is to select one of the three counters to be operated on and to address the control word register for mode selection.

### 6. *CS (Chip Select)*

A 'low' on this input enables the 8253. No reading or writing will occur unless the device is selected. The CS Input has no elect upon the actual operation of the counters.

| $\overline{CS}$ | A1 | A0 | Port |
|---|---|---|---|
| 0 | 0 | 0 | Counter 0 |
| 0 | 0 | 1 | Counter 1 |
| 0 | 1 | 0 | Counter 2 |
| 0 | 1 | 1 | Counter register |

## 11.4.1 Control Word Register

The Control Word Register is selected when A0, A1 are 11 then accepts information from the data bus buffer and stores it in a register. The information stored in this register controls the operational NODE of each counter, selection of binary or BCD counting and the loading of each count register. The Control Word Register can only be written into no. read operation of its contents is available.

**Counter #0,**

**Counter #1,**

**Counter #2**

These three functional blocks are identical in operation so only a single Counter will be described. Each Counter consists of a single 16.bit,pre-sellable DOWN counter. The counter can operate in either binary or BCD and its input, gate and output are configured by the selection of MODES stored in the Control Word Register.

The counters are fully independent and each can have separate Mode configuration and counting operation, binary or BCD. Also there are special features in the control word that handle the loading lf the count value so that software overhead can be minimized for these functions.

The reading of the contents of each counter is available to the programmer with simple READ operations for event counting applications and special commands and logic are included in the 8253 so that the contents of each counter can be read "on the fly" without having to inhibit the clock input.

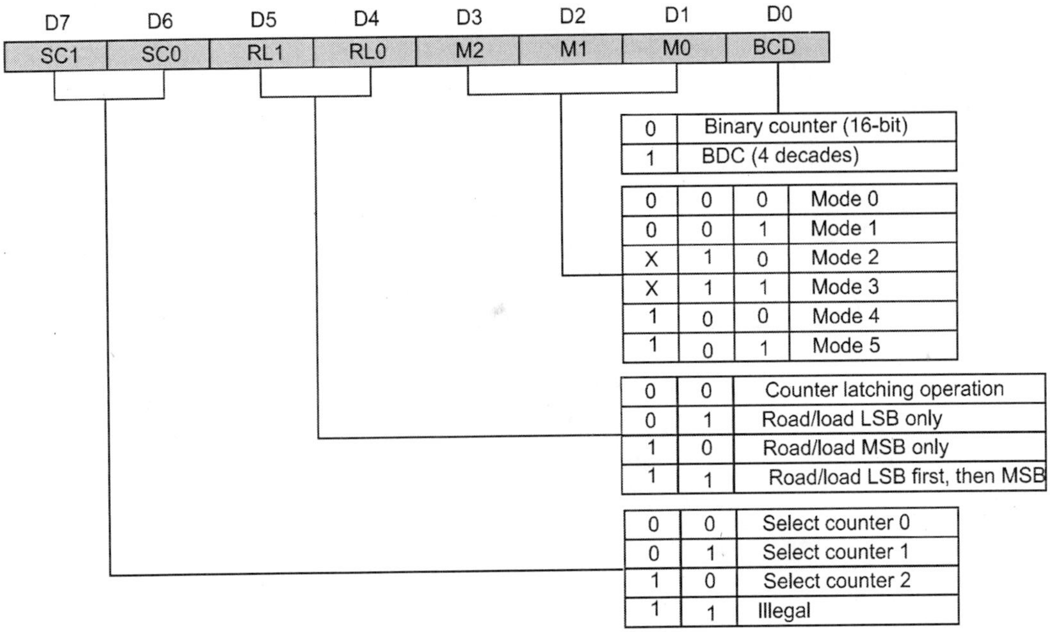


**Figure 11.5:** Control Register

### 11.4.2 8253 Systems Interface

The 8253 is a component of the Intel-Microcomputer Systems and Interfaces in the same manner as all other peripherals of the family. It is treated by the systems software as an array of peripheral I/O ports: three are counters and the fourth is a control register for MODE programming.

Basically, the select Inputs A0, A1 connect to the A0, A1 connect to the A0, A1 address bus signals of the CPU. The CS can be derived directly from the address bus using a linear select method or it can be connected to the output of a decoder, such as an Intel 8205 for larger systems.

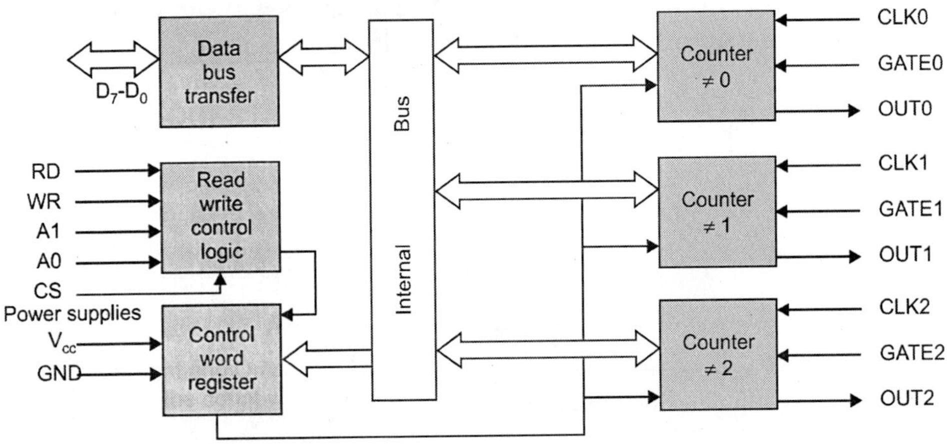


**Figure 11.6:** Block Diagram Showing Control Word and Register and Counter Functions

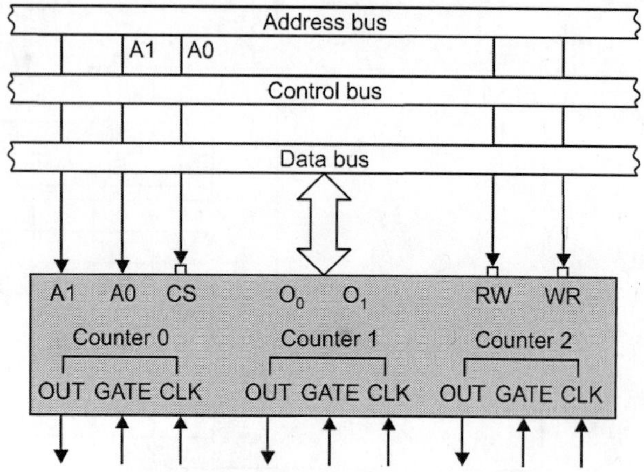


**Figure 11.7:** 8253 System Interface

### 11.4.3 8253 Operating Modes

1. Mode 0 Interrupt on terminal count
2. Mode 1 Programmable one shot
3. Mode 2 Rate Generator
4. Mode 3 Square wave rate Generator
5. Mode 4 Software triggered strobe
6. Mode 5 Hardware triggered strobe

#### *1. Mode 0 Interrupt on terminal count*

The output goes high after the terminal count is reached. The counter stops if the Gate is low. (Figure 11.8(a) & (b)). The timer count register is loaded with a count (say 6) when the WR line is made low by the processor. The counter unit starts counting down with each clock pulse. The output goes high when the register value reaches zero. In the mean time if the GATE is made low [Figure 11.8(b)] the count is suspended at the value(3) till the GATE is enabled again.

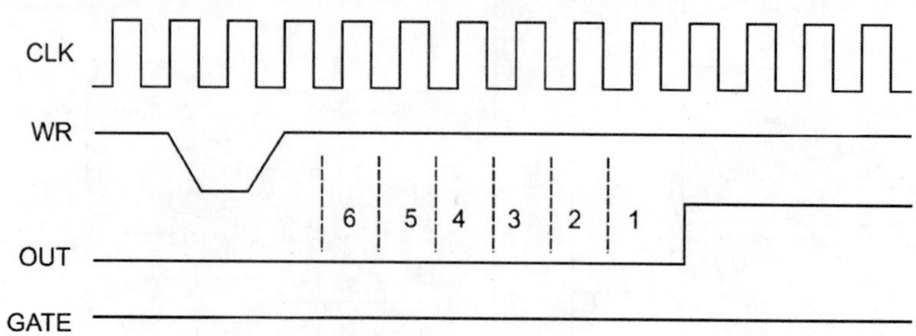


**Figure 11.8** (a) Mode 0 count when Gate is high (enabled)

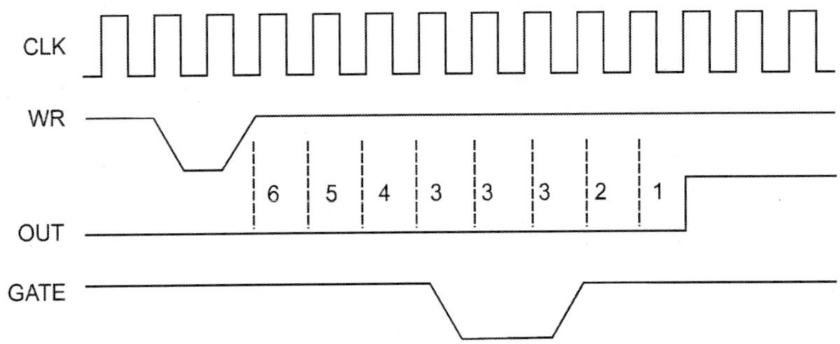


**Figure 11.8** (b) Mode 0 count when Gate is low temporarily (disabled)

### *2. Mode 1 Programmable mono-shot*

The output goes low with the Gate pulse for a predetermined period depending on the counter. The counter is disabled if the GATE pulse goes momentarily low.

The counter register is loaded with a count value as in the previous case (say 5) [Figure 11.9 (a)]. The output responds to the GATE input and goes low for period that equals the count down period of the register (5 clock pulses in this period). By changing the value of this count the duration of the output pulse can be changed. If the GATE becomes low before the count down is completed then the counter will be suspended at that state as long as GATE is low (Figure 11.9 (b)). Thus it works as a mono-shot.

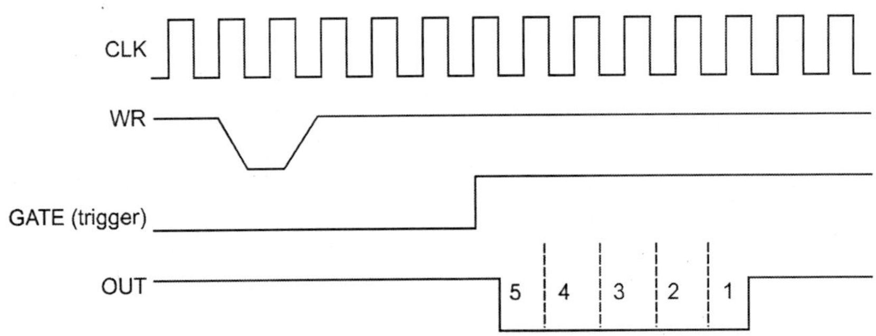


**Figure 11.9** (a) Mode 1 gate goes high. The outputs go low for the period depending on the count

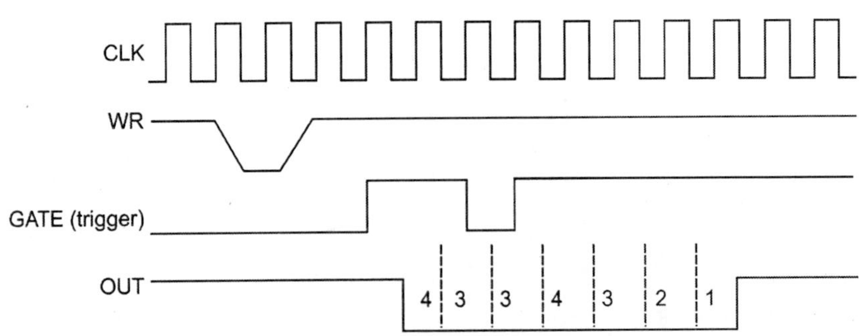


**Figure 11.9** (b) Mode 1 the gate pulse is disabled momentarily causing the counter to stop

### 3. Mode 2 Programmable Rate Generator

Figure 11.10(a) and (b) shows the waveforms corresponding the Timer operation in this mode. In this mode it operates as a rate generator. The output goes high for a period that equals the time of count down of the count register (3 in this case). The output goes low exactly for one clock period before it becomes high again. This is a periodic operation.

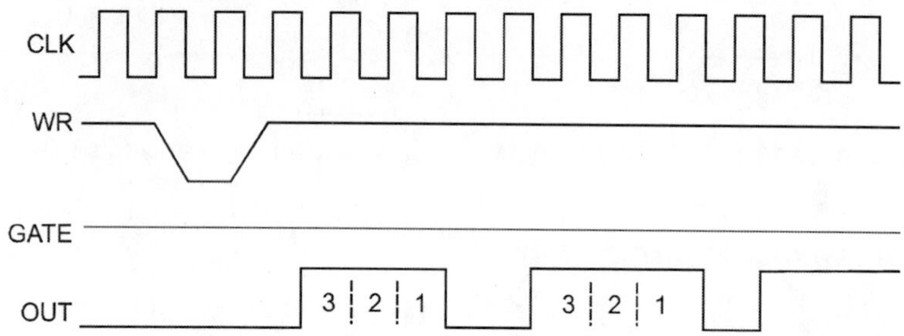


**Figure 11.10:** (a) Mode 2 Operation when the GATE is kept high

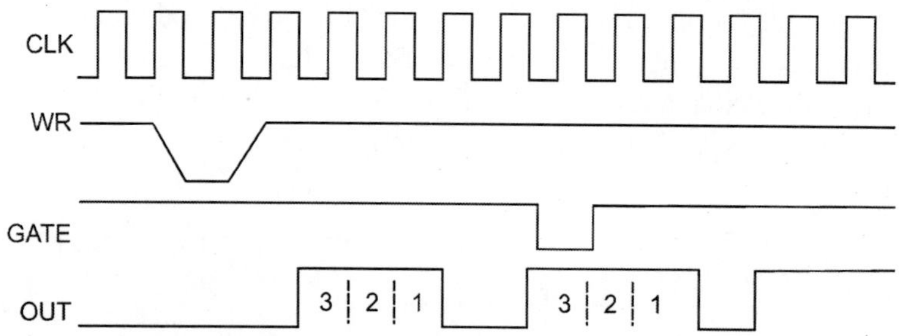


**Figure 11.10:** (b) Mode 2 operation when the GATE is disabled momentarily.

### 4. Mode 3 Programmable Square Wave Rate Generator

It is similar to Mode 2 but the output high and low period is symmetrical. The output goes high after the count is loaded and it remains high for period which equals the count down period of the counter register. The output subsequently goes low for an equal period and hence generates a symmetrical square wave unlike Mode 2. The GATE has no role here. **(Figure 11.11).**

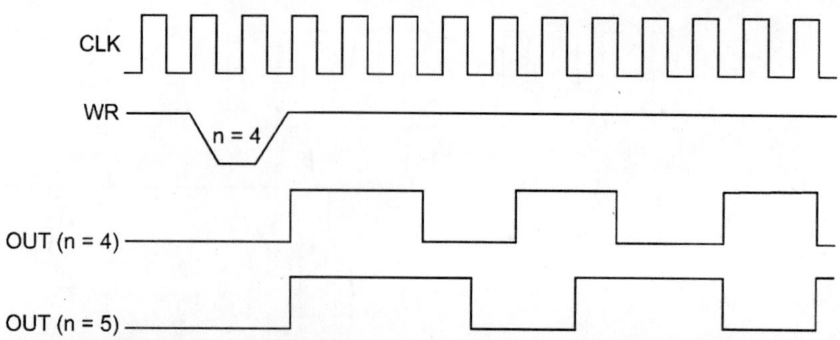


**Figure 11.11:** Mode3 Operation: Square Wave generator

### 5. Mode 4 Software Triggered Strobe

In this mode after the count is loaded by the processor the count down starts. The output goes low for one clock period after the count down is complete. The count down can be suspended by making the GATE low [Figure 11.12(a) (b)]. This is also called a software triggered strobe as the count down is initiated by a program.

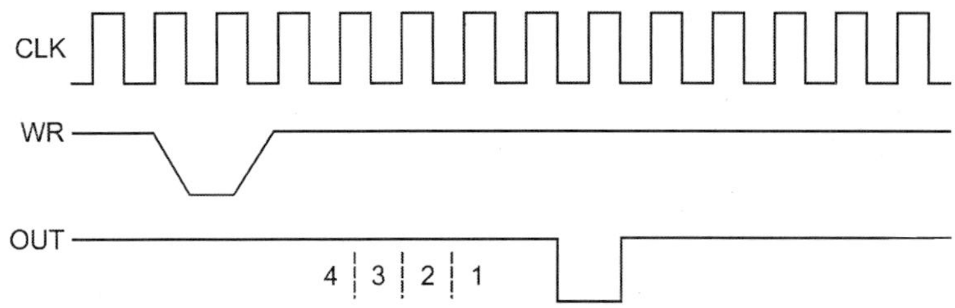


**Figure 11.12:** (a) Mode 4 Software Triggered Strobe when GATE is high

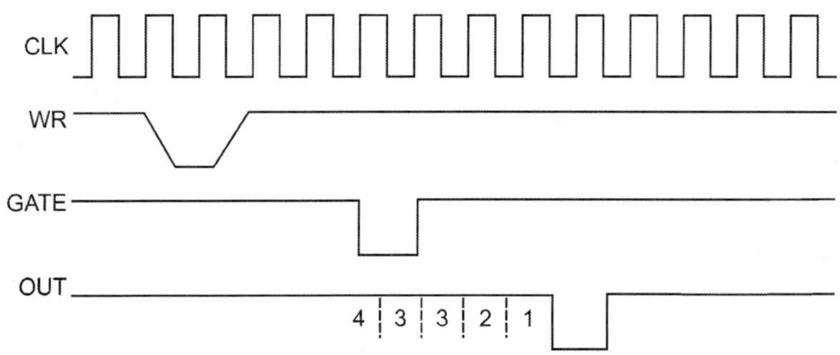


**Figure 11.12:** (b) Mode 4 Software Triggered Strobe when GATE is momentarily low

### 6. Mode 5 Hardware Triggered Strobe

The count is loaded by the processor but the count down is initiated by the GATE pulse. The transition from low to high of the GATE pulse enables count down. The output goes low for one clock period after the count down is complete (Figure 11.13).

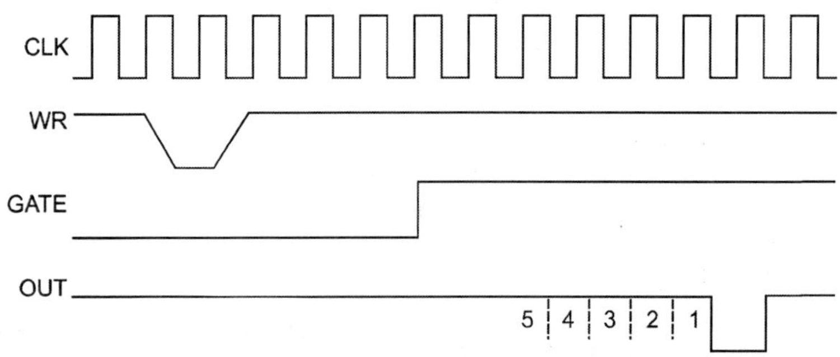


**Figure 11.13:** Mode 5 Hardware triggered strobe

| Modes/ Signal Status | Low or Going Low | Rising | High |
|---|---|---|---|
| 0 | Disables counting | ——— | Enables counting |
| 1 | | 1. Initiates Counting<br>2. Resets output after next clock | ——— |
| 2 | 1. Disables Counting<br>2. Sets output immediately high | 1. Reloads counter<br>2. Initiates counting | Enables counting |
| 3 | 1. Disables counting<br>2. Sales output immediately high | Initiates counting | Enables counting |
| 4 | Disables counting | ——— | Enables counting |
| 5 | | Initiates counting | ——— |

**Figure 11.14:** Gate Pin Operations Summary

## 11.5 8254 PROGRAMMABLE INTERVAL TIMER

The 8254 programmable interval timer/counter is functionally similar to the software-designed counters and timers. It generates accurate time delays and can be used for applications such as a real-time click, an event counter, a digital one-shot, a square-wave generator, and a complex waveform generator.

The 8254 includes three identical 16-bit counters that can operate independently in any one of the six modes (to be described later). It is packaged in a 24-pin DIP and requires a single +5 V power supply. To operate a counter, a 16-bit count is loaded in its register and, on command, begins to decrement the count until it reaches 0. At the end of the count, it generates a pulse that can be used to interrupt the MPU. The counter can count either in binary or BCD. In addition, a count can be read by the MPU while the counter is decrementing.

The 8254 is an upgraded version of the 8253, and they are pin-compatible. The features of these two devices are almost identical except that

- The 8254 can operate with higher clock frequency range (DC to 8 $MH_Z$ and 10 $MH_Z$ for 8254-2), and the 8253 can operate with clock frequency from DC to 2 $MH_Z$.
- The 8254 includes a Status Read-Back Command that can latch the count and the status of the counters.

Figure 11.16 shows the block diagram of 8253/54. It includes three counters, a data bus buffer, Read/Write control logic, and a control register. Each counter has two input signals CLOCK and GATE and one output signal OUT

### 11.5.1 Pin Diagram of 8254

| Pin Names | | | |
|---|---|---|---|
| $D_7$–$D_0$ | Data bus (8 Bit) | CLK N | Counter Clock Inputs |
| Gate N | Counter Gate Inputs | Out N | Counter Outputs |
| RD | Read Counter | WR | Wrtie Command or Data |
| CS | Chip Select | $A_0$–$A_1$ | Counter Select |
| $V_{CC}$ | +5 Volts | GND | Ground |

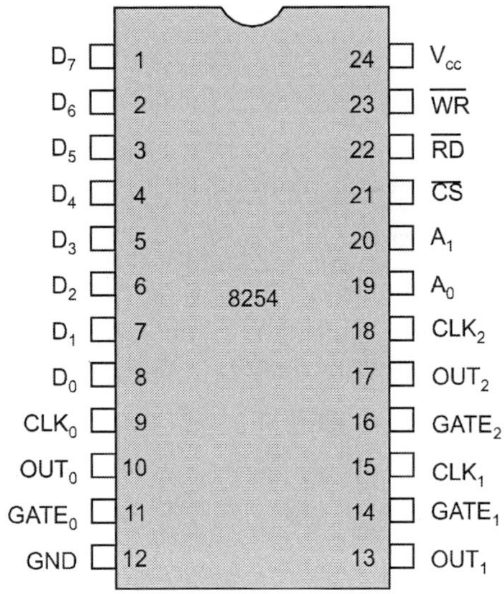


**Figure 11.15:** Pin diagram of 8254

## 11.5.2 Block Diagram of 8254

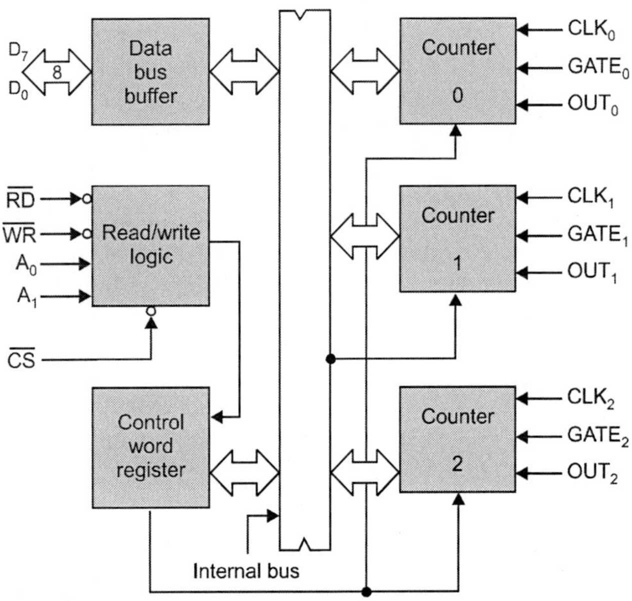


**Figure 11.16:** Block diagram of 8254

### *1. Data Bus Buffer*

This tri-state, bi-directional, 8-bit buffer is used to interface the 8253/54 to the system data bus. The Data bus buffer has three basic functions.

1. Programming the modes of 8253/54.

2. Loading the count registers.
3. Reading the count values.

### 2. Read/Write Logic

The Read/Write logic has five signals : $\overline{RD}$, $\overline{WR}$, $\overline{CS}$ and the address lines A0 and A1. In the peripheral I/O mode, the RD, and WR signals are connected to $\overline{IOR}$ and $\overline{IOW}$, respectively. In memory-mapped I/O, these are connected to $\overline{MEMR}$ and $\overline{MEMW}$. Address lines A0 and A1 of the CPU are usually connected to lines A0 and A1 of the 8253/54, and is tied to a decoded address. The control word register and counters are selected according to the signals on lines A0 and A1.

| $A_1$ | $A_0$ | Selection |
|---|---|---|
| 0 | 0 | Counter 0 |
| 0 | 1 | Counter 1 |
| 1 | 0 | Counter 2 |
| 1 | 1 | Control word register |

### 3. Control Word Register

This register is accessed when lines A0 and A1 are at logic 1. It is used to write a command word which specifies the counter to be used (binary or BCD), its mode, and either a read or write operation.

### 4. Counters

These three functional blocks are identical in operation. Each counter consists of a single, 16 bit, pre-settable, down counter. The counter can operate in either binary or BCD and its input, gate and output are configured by the selection of modes stored in the control word register. The counters are fully independent. The programmer can read the contents of any of the three counters without disturbing the actual count in process.

## 11.5.3 Operational Description

The complete functional definition of the 8253/54 is programmed by the system software. Once programmed, the 8253/54 is ready to perform whatever timing tasks it is assigned to accomplish.

### 1. Programming the 8253/54

Each counter of the 8253/54 is individually programmed by writing a control word into the control word register ($A_0$ - $A_1$ = 11). The Figure 11.17 shows the control word format. Bits $SC_1$ and $SC_0$ select the counter, bits $RW_1$ and $RW_0$ select the read, write or latch command, bits $M_2$, $M_1$ and $M_0$ select the mode of operation and bit BCD decides whether it is a BCD counter or binary counter.

### 2. Write Operation

1. Write a control word into control register.

2. Load the low-order byte of a count in the counter register.
3. Load the high-order byte of count in the counter register.

| $D_7$ | $D_6$ | $D_5$ | $D_4$ | $D_3$ | $D_2$ | $D_1$ | $D_0$ |
|---|---|---|---|---|---|---|---|
| $SC_1$ | $SC_0$ | $RW_1$ | $RW_0$ | $M_2$ | $M_1$ | $M_0$ | BCD |

**SC - Select counter**

| $SC_1$ | $SC_0$ | |
|---|---|---|
| 0 | 0 | Select counter 0 |
| 0 | 1 | Select counter 1 |
| 1 | 0 | Select counter 2 |
| 1 | 1 | Illegal for 8253<br>Read-Back command for 8254<br>(See Read operations) |

**M - Mode**

| $M_2$ | $M_1$ | $M_0$ | |
|---|---|---|---|
| 0 | 0 | 0 | Mode 0 |
| 0 | 0 | 1 | Mode 1 |
| X | 1 | 0 | Mode 2 |
| X | 1 | 1 | Mode 3 |
| 1 | 0 | 0 | Mode 4 |
| 1 | 0 | 1 | Mode 5 |

**RW - Read/Write**

| $RW_1$ | $RW_0$ | |
|---|---|---|
| 0 | 0 | Couner latch command<br>(see Read operations) |
| 0 | 1 | Read/Write least significant byte only |
| 1 | 0 | Read/Write most significant byte only |
| 1 | 1 | Read/write least significant byte first, then most significant byte |

**BCD**

| | |
|---|---|
| 0 | Binary counter 16-bits |
| 1 | Binary coded decimal (BCD) Counter (4 Decades) |

Note : Don't care bits (X) should be 0 to ensure compatibility with future Intel products.

**Figure 11.17:** Control word format

### 3. *Read Operation*

In some applications, especially in event counters, it is necessary to read the value of the count in process. This can be done by three possible methods:

1. **Simple Read**: It involves reading a count after inhibiting the counter by controlling the gate input or the clock input of the selected counter, and two I/O read operations are performed by the CPU. The first I/O operation reads the low-order byte, and the second I/O operation reads the high order byte.
2. **Counter Latch Command**: In the second method, an appropriate control word is written into the control register to latch a count in the output latch, and two I/O read operations are performed by the CPU. The first I/O operation reads the low-order byte, and the second I/O operation reads the high order byte.
3. **Read-Back Command (Available only for 8254)**: The third method uses the Read-Back command. This command allows the user to check the count value, programmed Mode, and current status of the OUT pin and Null count flag of the selected counter(s). Figure 11.18 shows the format of the control word register for Read-Back command.

$A_0, A_1 = 11\ \overline{CS} = 0\ \overline{RD} = 1\ \overline{WR} = 0$

| $D_7$ | $D_6$ | $D_5$ | $D_4$ | $D_3$ | $D_2$ | $D_1$ | $D_0$ |
|---|---|---|---|---|---|---|---|
| 1 | 1 | $\overline{COUNT}$ | $\overline{STATUS}$ | $CNT_2$ | $CNT_1$ | $CNT_0$ | 0 |

$D_5$ : 0 = Latch count of selected counter(s)

$D_4$ : 0 = Latch status of selected counter(s)

$D_3$ : 1 = Select counter 2

$D_2$ : 1 = Select counter 1

$D_1$ : 1 = Select counter 0

$D_0$ : Reserved for future expansion: must be 0.

**Figure 11.18:** Control word register for read-back command

The Read-Back command may be used to latch multiple counter output latches by setting the $\overline{COUNT}$ bit $D_5 = 0$ and selecting the desired counter (s). Each counter's latch count in held until it is read (or the counter is reprogrammed). That counter is automatically unlatched when read.

### *Other Features of Read - Back Command (Available only for 8254)*

The Read-Back command may also be used to latch status information of selected counter(s) by setting STATUS bit $D_4 = 0$. The contents of the counter must be latched before reading. The status of a counter is then accessed by a read from that counter. The Figure 11.19 shows the counter status format.

Bit $D_5$–$D_0$ contains the counter's programmed mode exactly as written in the last mode control word. Bit $D_7$ contains the current status of the output pin. In 8254, it is not possible to read count from the counter, if the count is not loaded into the counting element (CE). The Bit $D_6$ indicates whether the counting element has count or not. If $D_6 = 0$, counting element has count otherwise null count.

| $D_7$ | $D_6$ | $D_5$ | $D_4$ | $D_3$ | $D_2$ | $D_1$ | $D_0$ |
|---|---|---|---|---|---|---|---|
| Output | Null count | $RW_1$ | $RW_0$ | $M_2$ | $M_1$ | $M_0$ | BCD |

**Figure 11.19:** Counter status format

**Interleaved Read and Write:** Another feature of the 8254 is that reads and writes of the same counter may be interleaved. For example, if the counter is programmed for the two byte counts, the following sequence is valid.

1. Read least significant byte.
2. Write new least significant byte.
3. Read most significant byte.
4. Write new most significant byte.

## 11.6 MODE DEFINITION

### 11.6.1 Mode 0: Interrupt on Terminal Count

**(a) Normal Operation**

1. The output will be initially low after the mode set operation.
2. After the count is loaded into the selected count Register the output will remain low and the counter will count.
3. When the terminal count is reached the output will go high and remain high until the selected count is reloaded.

**(b) Gate Disable**

1. Gate = 1 enables counting.
2. Gate = 0 disables counting.

**Note:** Gate has no effect on OUT.

**(c) New Count:** If a new count is written to the counter, it will be loaded on the next CLK pulse and counting will continue from the new count.

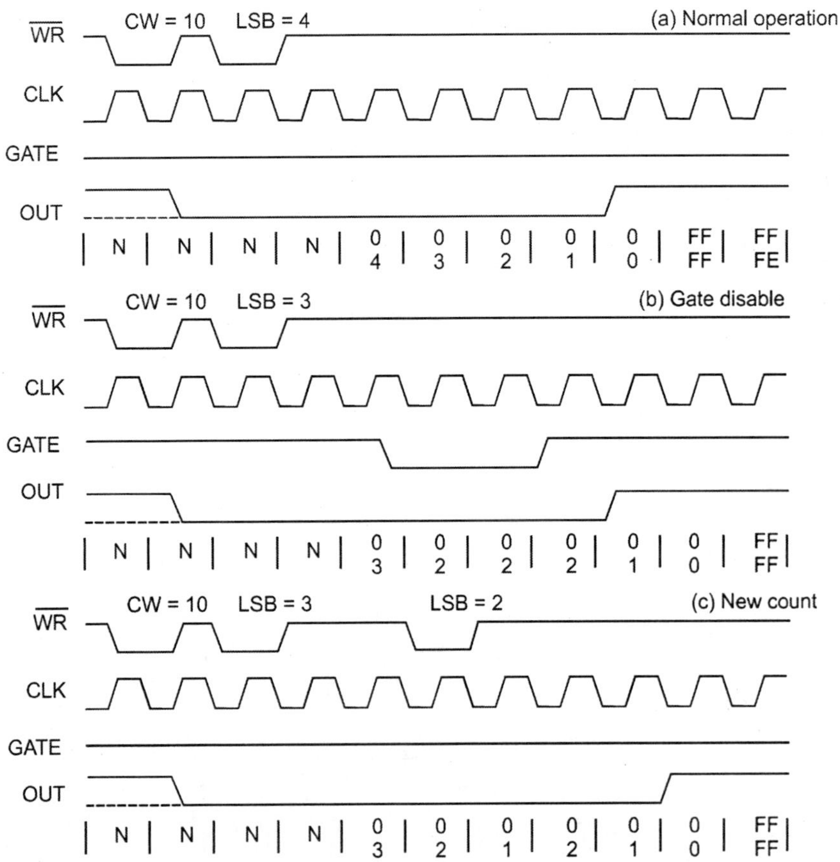


**Figure 11.20:** Mode 0 interrupt on terminal count

**In case of two byte count:**

1. Writing the first byte disables counting.
2. Writing the second byte loads the new count on the next CLK pulse and counting will continue from the new count.

### 11.6.2 Mode 1: Hardware Retriggerable One-shot

**(a) Normal operation**

1. The output will be initially high
2. The output will go low on the CLK pulse following the rising edge at the gate input.
3. The output will go high on the terminal count and remain high until the next rising edge at the gate input.

**(b) Retriggering:** The one shot is retriggerable, hence the output will remain low for the full count after any rising edge of the gate input.

**(c) New count:** If the counter is loaded during one shot pulse, the current one shot is not affected unless the counter is retriggered. If retriggered, the counter is loaded with the new count and the one-shot pulse continues until the new count expires.

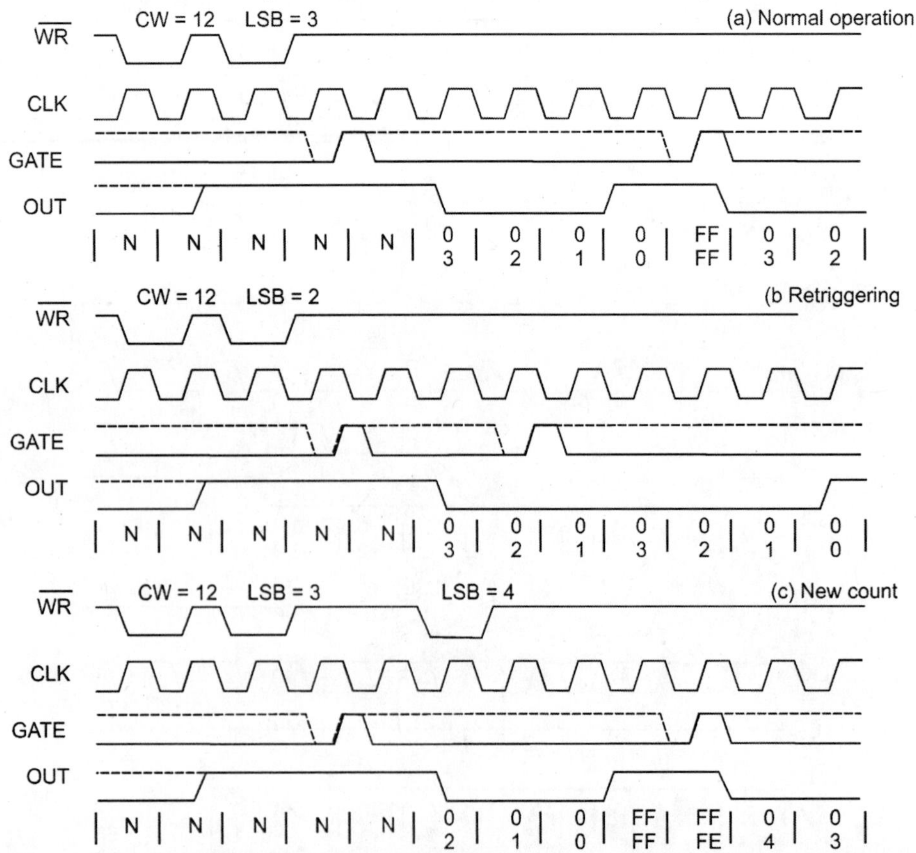


**Figure 11.21:** Mode 1 hardware retriggerable one-shot

### 11.6.3 Mode 2: Rate Generator

This mode functions like a divide by-N counter.

**(a) Normal Operation**

1. The output will be initially high.
2. The output will go low for one clock pulse before the terminal count.
3. The output then goes high, the counter reloads the initial count and the process is repeated.
4. The period from one output pulse to the next equals the number of input counts in the count register.

**(b) Gate Disable**

1. If Gate = 1 it enables a counting otherwise it disables counting (Gate = 0).
2. If Gate goes low during an low output pulse, output is set immediately high.

A trigger reloads the count and the normal sequence is repeated.

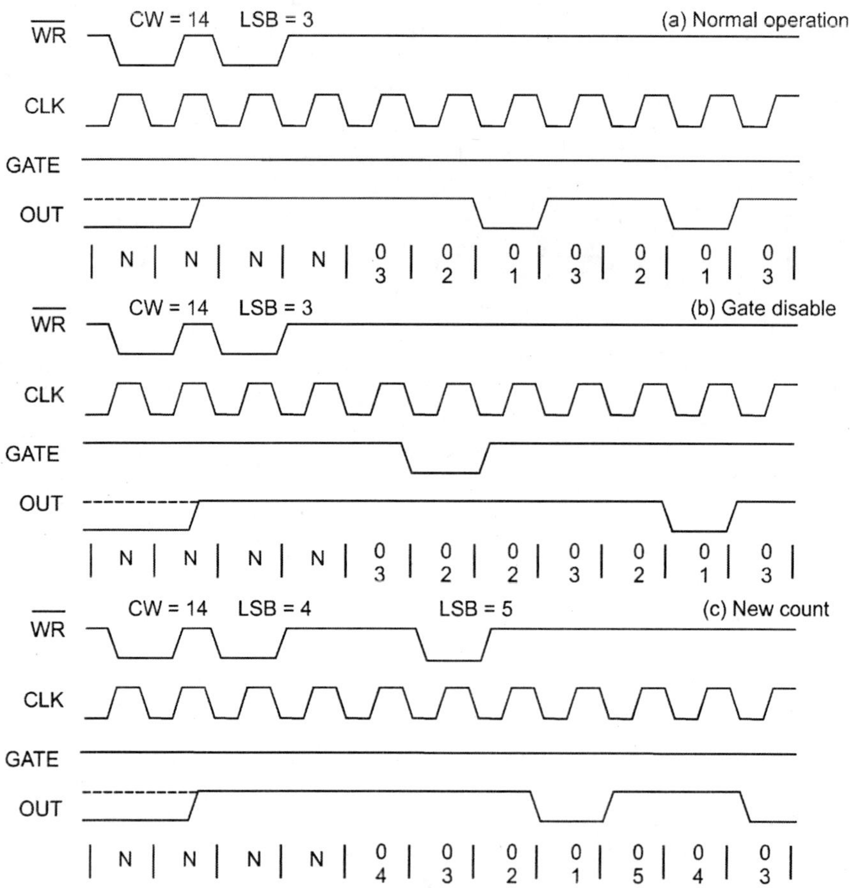


**Figure 11.22:** Mode 2 rate generator

(c) **New count:** The current counting sequence does not affect when the new count is written. If a trigger is received after writing a new count but before the end of the current period, the new count will be loaded with the new count on the next CLK pulse and counting will continue from the new count. Otherwise, the new count will be loaded at the end of the current counting cycle.

**Note:** In mode 2, a count of 1 is illegal.

### 11.6.4 Mode 3: Square Wave Rate Generator

**(a) Normal operation**

1. Initially output is high.
2. For even count, counter is decremented by 2 on the falling edge of each clock pulse. When the counter reaches terminal count, the state of the output is changed and the counter is reloaded with the full count and the whole process is repeated.
3. If the count is odd and the output is high the first clock pulse (after the count is loaded) decrements the count by 1. Subsequent clock pulses decrement the clock by 2. After

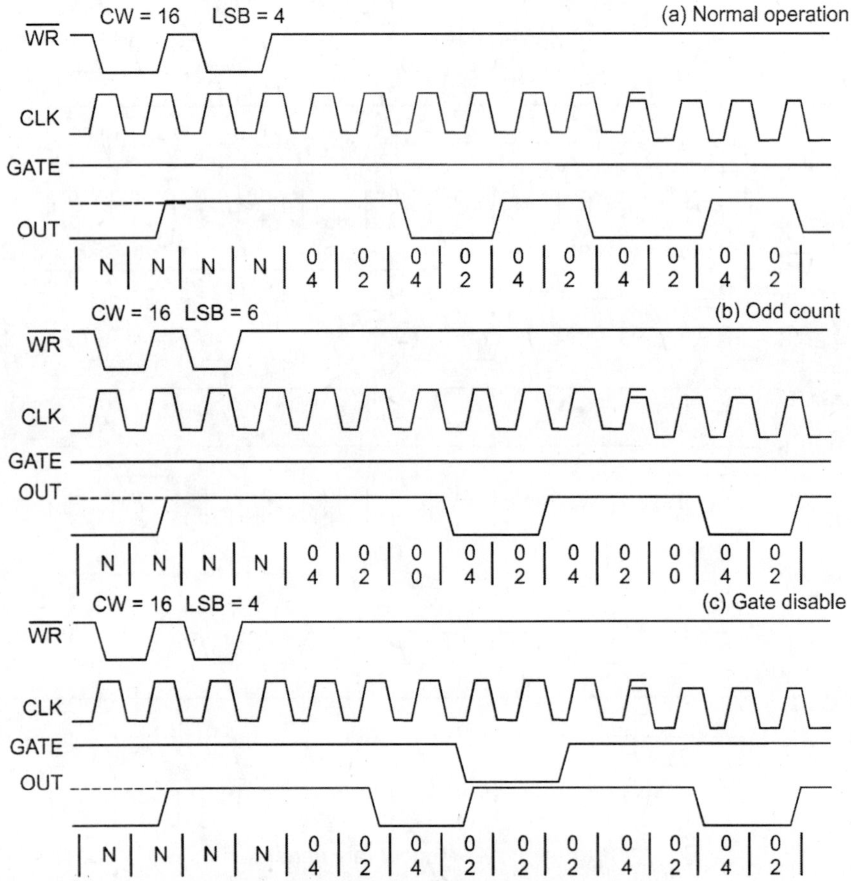


**Figure 11.23:** Mode 3 square wave mode

timeout, the output goes low and the full count is reloaded. The first clock pulse (following the reload) decrements the count by 3 and subsequent clock pulse decrement the count by two. Then the whole process is repeated. In this way, if the count is odd, the output will be high for (n+1)/2 counts and low for (n-1)/2 counts.

**(b) Gate Disable:** If Gate is 1 counting is enabled otherwise it is disabled. If Gate goes low while output is low, output is set high immediately. After this, When Gate goes high, the counter is loaded with the initial count on the next clock pulse and the sequence is repeated.

**(c) New Count:** The current counting sequence does not affect when the new count is written. If a trigger is received after writing a new count but before the end of the current half-cycle of the square wave, the counter will be loaded with the new count on the next CLK pulse and counting will continue from the new count. Otherwise, the new count will be loaded at end of the current half-cycle.

### 11.6.5 Mode 4: Software Triggered Strobe

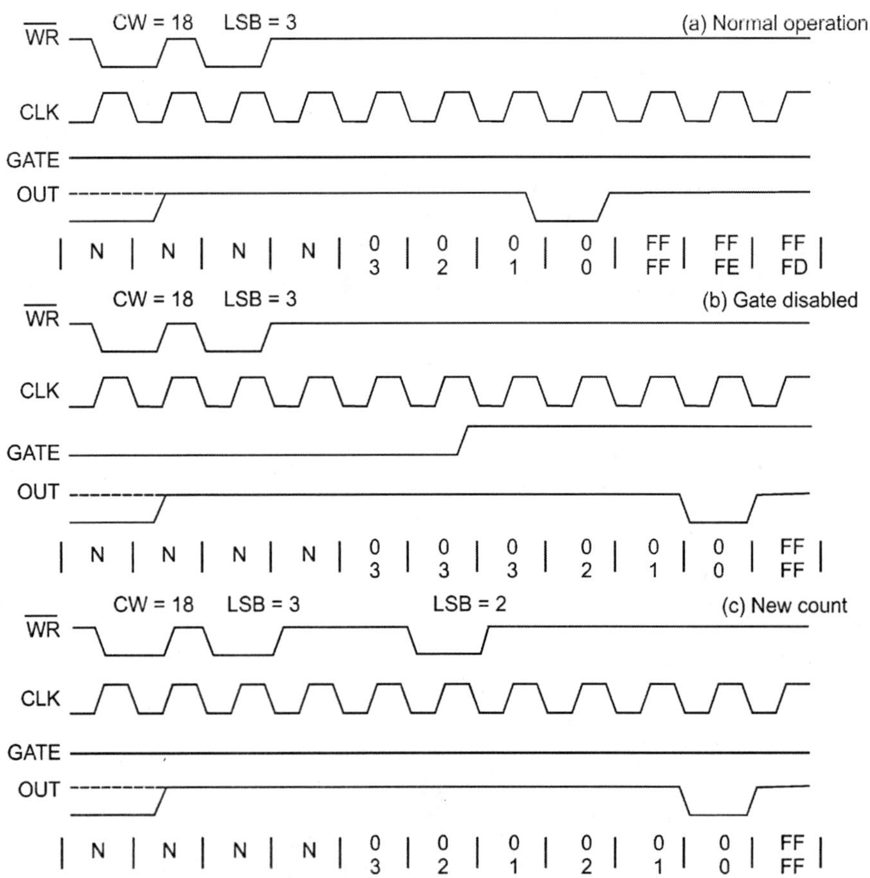


**Figure 11.24:** Mode 4 software triggered strobe

**(a) Normal operation**

1. The output will be initially high
2. The output will go low for one CLK pulse after the terminal count (TC).

**(b) Gate Disable:** If Gate is one the counting is enabled otherwise it is disabled. The Gate has no effect on the output.

**(c) New count:** If a new count is written during counting, it will be loaded on the next CLK pulse and counting will continue from the new count. If the count is two byte then

1. Writing the first byte has no effect on counting.
2. Writing the second byte allows the new count to be loaded on the next CLK pulse.

### 11.6.6 Mode 5: Hardware Triggered Strobe (Retriggerable)

**(a) Normal operation**

1. The output will be initially high.

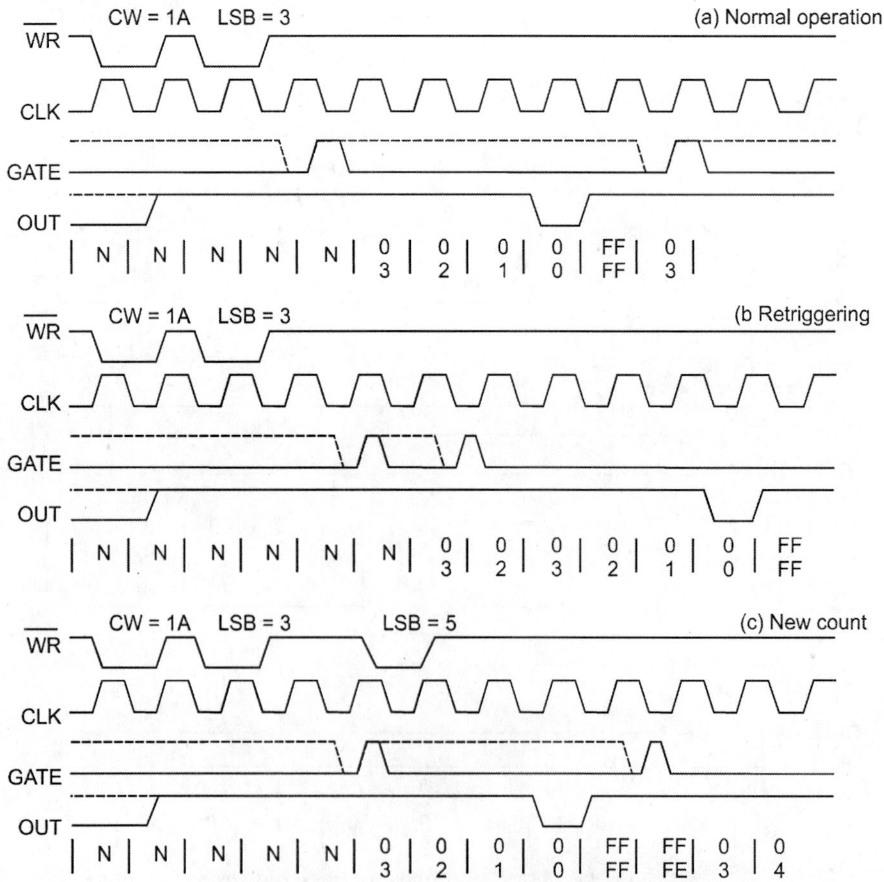


**Figure 11.25:** Mode 5 hardware triggered strobe

2. The counting is triggered by the rising edge of the Gate.
3. The output will go low for one CLK pulse after the terminal count (TC).

**(b) Retriggering:** If the triggering occurs on the Gate input during the counting, the initial count is loaded on the next CLK pulse and the counting will be continued until the terminal count is reached.

**(c) New count:** If a new count is written during counting, the current counting sequence will not be affected. If the trigger occurs after the new count is written but before the terminal count, the counter will be loaded with the new count on the next CLK pulse and counting will continue from there.

| Signal status modes | Low or going low | Rising | High |
|---|---|---|---|
| 0 | Disable counting | — | Enables counting |
| 1 | — | (i) Initiates counting<br>(ii) Resets output after next clock | — |
| 2 | (i) Disables counting<br>(ii) Sets output immediately high | Initiates counting | Enables countings |
| 3 | (i) Disables counting<br>(ii) Sets output immediately high | Initiates counting | Enable countings |
| 4 | Disables counting | — | Enables counting |
| 5 | — | Initiates counting | — |

## 11.7 ILLUSTRATION: THE 8254 AS A COUNTER

### Problem Statement

1. Identify the port addresses of the control register and counter 2 in Figure 11.26.
2. Write a subroutine to initialize counter 2 in Mode 0 with a count of $50{,}000_{10}$. The subroutine should also include reading counts on the fly; when the count reaches zero, it should return to the main program.
3. Write a main program to display seconds by calling the subroutine as many times as necessary.

1. **Port Addresses** The Chip Select is enabled when $A_7 = 1$ (see Figure 11.26), and the control register is selected when $A_1$ and $A_0 = 1$. Similarly, counter 2 is selected when $A_1 = 1$ and $A_0 = 0$. Assuming that the unused address lines $A_6$ to $A_2$ are at logic 0, the port addresses will be as follows:

   Control Register = 83H

   Counter 2 = 82H

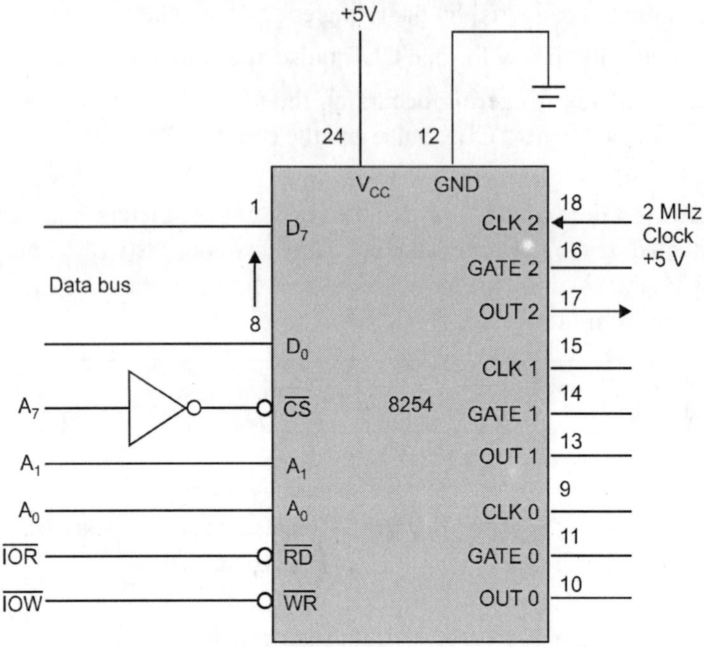


**Figure 11.26:** Schematic Interfacing The 8254

2. **Subroutine Counter** To initialize the 8254 for counter 2 in mode 0, the following control word is necessary (See figure 11.24):

   Control word (Load operation)

| $D_7$ | $D_6$ | $D_5$ | $D_4$ | $D_3$ | $D_2$ | $D_1$ | $D_0$ = 09H | |
|---|---|---|---|---|---|---|---|---|
| 1 | 0 | 1 | 1 | | 0 | 0 | 0 | 0 |
| Select counter 2 | | Load 16-bit count | | | Mode 0 | | Count in binary | |

Control word for Latching Bits $D_5$ and $D_4$ - should be 0 = 80H

**Subroutine**

```
Counter:  MVI A, B0H         ;Control word to initialize counter 2
          OUT 83H            ;Write in then control register
          MVI A, LOBYTE      ;Low-order byte of the count 50000
          OUT 82H            ;Load counter 2 with the low-order byte
          MVI A, HIBYTE      ;High-order byte of the count 50000
          OUT 82H            ;Load counter 2 with the high-order byte
          READ: MVI A, 80H   ;Control word to latch a count
          OUT 83H            ;Write in the control register
          IN 82H             ;Read low-order byte
          MOV D, A           ;Store low-order byte in register D
```

```
        IN 82H             ;Read high-order byte
        ORA D              ;OR low-and high-order bytes to set Z flag
        JNZ READ           ;If counter "" 0, go back to read next count
        RET
```

**Subroutine Description:** The subroutine has two segments. In the first segment, counter 2 is initialized by writing a control word in the control register and a 16-bit count specified as LOBYTE and HIBYTE in the counter register. The hexadecimal value equivalent to $50000_{10}$ must be calculated.

In the second segment (beginning at READ), a control word is writing into the control register to sample a count, and the 16-bit count is read by performing to input operations. The reading of the counter is repeated until the counter reaches 0; the Zero flag is checked-by ORing the low-and high-order bytes.

3. **Main Program** The subroutine COUNTER provides 25 ms (50000 x0.5 us Clock) delay; if this routine is called 40 times, the total delay will be one second.

**Program**

```
        LXI SP, STACK      ;Initialize stack pointer
        MVI B, 00H         ;Clear register B to save number of seconds
SECOND: MVI C, 28H         ;Set up register C to count 40 (decimal)
WAIT:   CALL COUNTER       ;Wait for 25 ms
        DCR C
        JNZ WAIT           ;Is this one second? If not, go back and wait
        MVI A, B
        ADI 01             ;Add one second
        DAA
        OUT PORTI
        MOV B, A           ;Save seconds
        JMP SECOND         ;Go back and start counting the next second
```

## Program Description

The main program initializes the stack pointer; loads register C with the count of 28H ($40_{10}$) and sets up the WAIT loop. The loop calls the COUNTER subroutine 40 times to generate a one-second delay. At the end of the loop, it increments the seconds in register B, decimal-adjusts the byte, and display seconds. The sequence is repeated until register B reaches $99_{BCD}$. After the $99^{th}$ second, register B is cleared and the clock sequence is repeated.

This program is just to demonstrate the Read and Write operations of the 8254; this clock design does not take into account the errors causes by the delay in executing the program instructions. A better way of designing a real-time clock is to interrupt the MPU at the end of count.

## QUESTION BANK

Q1. What do you mean by timer and counter?

Q2. Explain the programmable interval timer 8253 in detail with the help of block diagram, its pin diagram and description.

Q3. Explain the control word register and system interface of 8253 programmable interval timer.

Q4. Explain the different operating modes of 8253 timer.

Q5. Explain the programmable interval timer 8254 in detail with the help of block diagram, its pin diagram and description.

Q6. Explain the operational description 8253/8254.

Q7. Explain the different operating modes of 8254 timer.

Q8. Explain the 8254 as a counter.

CHAPTER 12

# 8259 Programmable Interrupt Controller

## 12.1 8259A PROGRAMMABLE INTERRUPT CONTROLLER

The 82C59A is a high performance CMOS Priority Interrupt Controller manufactured using an advanced 2μm CMOS process. The 82C59A is designed to relieve the system CPU from the task of polling in a multilevel priority system. The high speed and industry standard configuration of the 82C59A make it compatible with microprocessors such as 80C286, 80286, 80C86/88, 8086/88, 8080/85 and NSC800. The 82C59A can handle up to eight vectored priority interrupting sources and is cascadable to 64 without additional circuitry. Individual interrupting sources can be masked or prioritized to allow custom system configuration. Two modes of operation make the 82C59A compatible with both 8080/85 and 80C86/88/286 formats. Static CMOS circuit design ensures low operating power.

The 8259A is a programmable interrupt controller designed to work with Intel microprocessor 8085, 8086 and 8088. The 8259A interrupt controller can

1. Manage eight interrupts according to the instructions written into its control registers. This is equivalent to providing eight interrupt pins on the processor in place of one INTR (8085) pin.
2. Vector an interrupt request anywhere in the memory map. However, all eight interrupts are spaced at the interval of either four or eight locations. This eliminates the major drawback of the 8085 interrupts in which all interrupts are vectored to memory locations on page 00H.
3. Resolve eight levels of interrupt priorities in a variety of modes, such as fully nested mode, automatic rotation mode, and specific rotation mode (to be explained later).
4. Mask each interrupt request individually.
5. Read the status of pending interrupts, in-service interrupts, and masked interrupts.
6. Be set up to accept either the level-triggered or the edge-triggered interrupt request.
7. Be expanded to 64 priority levels by cascading additional 8259As.

8. Be set up to work with either the 8085 microprocessor mode or the 8086/8088 microprocessor mode.

The 8259A is upward-compatible with its predecessor, the 8259. The main difference between the two is that the 8259A can be used with Intel's 8086/88 16-bit microprocessors. It also includes additional features such as the level-triggered mode, buffered mode, and automatic-end-of-interrupt mode. To simplify the explanation of the 8259A, illustrative examples will not include the cascade mode or the 8086/88 mode and will be limited to modes commonly used with the 8085.

## 12.2 INTERRUPTS IN MICROCOMPUTER SYSTEMS

Microcomputer system design requires that I/O devices such as keyboards, displays, sensors and other components receive servicing in an efficient manner so that large amounts of the total system tasks can be assumed by the microcomputer with little or no effect on throughput. The most common method of servicing such devices is the Polled approach. This is where the processor must test each device in sequence and in effect "ask" each one if it needs servicing. It is easy to see that a large portion of the main program is looping through this continuous polling cycle and that such a method would have a serious, detrimental effect on system throughput, thus, limiting the tasks that could be assumed by the microcomputer and reducing the cost effectiveness of using such devices.

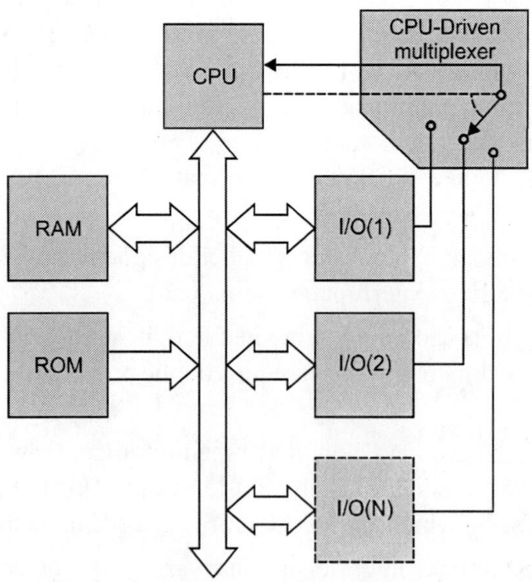


**Figure 12.1:** Polled method

A more desirable method would be one that would allow the microprocessor to be executing its main program and only stop to service peripheral devices when it is told to do so by the device itself. In effect, the method would provide an external asynchronous input that would inform the processor that it should complete whatever instruction that is currently being executed and fetch a new routine that will service the requesting device. Once this servicing is complete, however, the processor would resume exactly where it left off. This is the Interrupt-driven method. It is easy to see that system throughput would drastically increase, and thus, more tasks could be assumed by the microcomputer to further enhance its cost effectiveness.

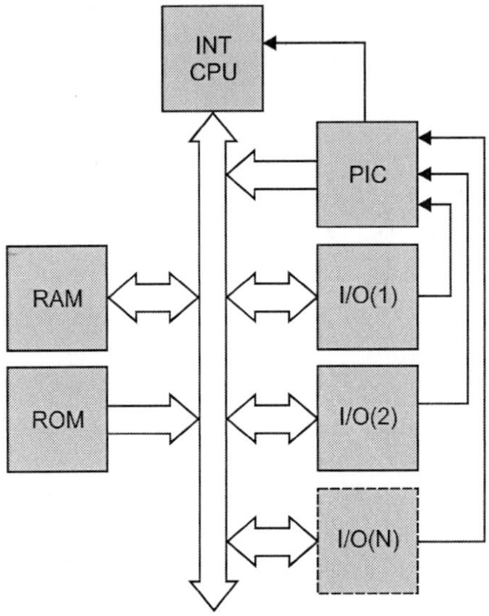


**Figure 12.2:** Interrupt method

The Programmable Interrupt Controller (PIC) functions as an overall manager in an Interrupt-Driven system. It accepts requests from the peripheral equipment, determines which of the incoming requests is of the highest importance (priority), ascertains whether the incoming request has a higher priority value than the level currently being serviced, and issues an interrupt to the CPU based on this determination. Each peripheral device or structure usually has a special program or "routine" that is associated with its specific functional or operational requirements; this is referred to as a "service routine". The PIC, after issuing an interrupt to the CPU, must somehow input information into the CPU that can "point" the Program Counter to the service routine associated with the requesting device. This "pointer" is an address in a vectoring table and will often be referred to, in this document, as vectoring data.

## 12.3 BLOCK DIAGRAM OF THE 8259A

The 82C59A is a device specifically designed for use in real time, interrupt driven microcomputer systems. It manages eight levels of requests and has built-in features for expandability to other 82C59A's (up to 64 levels). It is programmed by system software as an I/O peripheral. A selection of priority modes is available to the programmer so that the manner in which the requests are processed by the 82C59A can be configured to match system requirements. The priority modes can be changed or reconfigured dynamically at any time during main program operation. This means that the complete interrupt structure can be defined as required, based on the total system environment.

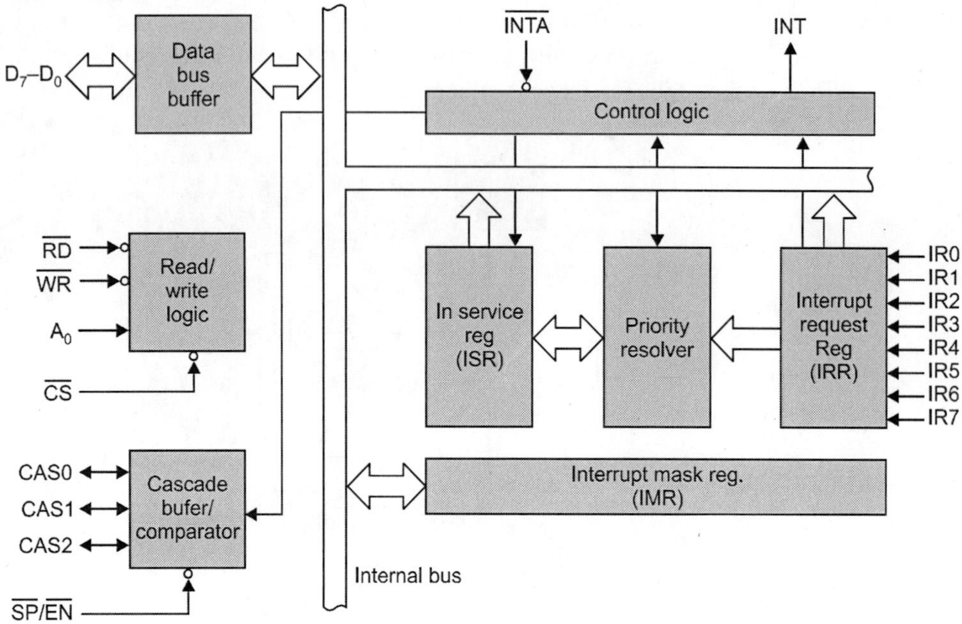


**Figure 12.3:** 8259a functional diagram

### 1. Interrupt Request Register (IRR) and In-service Register (ISR)

The interrupts at the IR input lines are handled by two registers in cascade, the Interrupt Request Register (IRR) and the In- Service Register (ISR). The IRR is used to indicate all the interrupt levels which are requesting service, and the ISR is used to store all the interrupt levels which are currently being serviced.

### 2. Priority Resolver

This logic block determines the priorities of the bits set in the IRR. The highest priority is selected and strobed into the corresponding bit of the ISR during the $\overline{\text{INTA}}$ sequence.

### 3. Interrupt Mask Register (IMR)

The IMR stores the bits which disable the interrupt lines to be masked. The IMR operates on the output of the IRR. Masking of a higher priority input will not affect the interrupt request lines of lower priority.

### 4. Interrupt (INT)

This output goes directly to the CPU interrupt input. The VOH level on this line is designed to be fully compatible with the 8080, 8085, 8086/88, 80C86/88, 80286, and 80C286 input levels.

### 5. Interrupt Acknowledge ($\overline{\text{INTA}}$)

$\overline{\text{INTA}}$ pulses will cause the 82C59A to release vectoring information onto the data bus. The format of this data depends on the system mode (µPM) of the 82C59A.

### 6. Data Bus Buffer

This 3-state, bidirectional 8-bit buffer is used to interface the 82C59A to the System Data Bus. Control words and status information are transferred through the Data Bus Buffer.

### 7. Read/Write Control Logic

The function of this block is to accept output commands from the CPU. It contains the Initialization Command Word (ICW) registers and Operation Command Word (OCW) registers which store the various control formats for device operation. This function block also allows the status of the 82C59A to be transferred onto the Data Bus.

### 8. Chip Select ($\overline{\text{CS}}$)

A LOW on this input enables the 82C59A. No reading or writing of the device will occur unless the device is selected.

### 9. Write ($\overline{\text{WR}}$)

A LOW on this input enables the CPU to write control words (ICWs and OCWs) to the 82C59A.

### 10. Read ($\overline{\text{RD}}$)

A LOW on this input enables the 82C59A to send the status of the Interrupt Request Register (IRR), In-Service Register (ISR), the Interrupt Mask Register (IMR), or the interrupt level (in the poll mode) onto the Data Bus.

### 11. A0

This input signal is used in conjunction with $\overline{\text{WR}}$ and $\overline{\text{RD}}$ signals to write commands into the various command registers, as well as to read the various status registers of the chip. This line can be tied directly to one of the system address lines.

### 12. The Cascade Buffer/Comparator

This function block stores and compares the IDs of all 82C59As used in the system. The associated three I/O pins (CAS0 - 2) are outputs when the 82C59A is used as a master and are inputs when the 82C59A is used as a slave. As a master, the 82C59A sends the ID of the interrupting slave device onto the CAS0 - 2 lines. The slave, thus selected will send its preprogrammed subroutine address onto the Data Bus during the next one or two consecutive $\overline{\text{INTA}}$ pulses.

## 13. Interrupt Sequence

The powerful features of the 82C59A in a microcomputer system are its programmability and the interrupt routine addressing capability. The latter allows direct or indirect jumping to the specified interrupt routine requested without any polling of the interrupting devices. The normal sequence of events during an interrupt depends on the type of CPU being used.

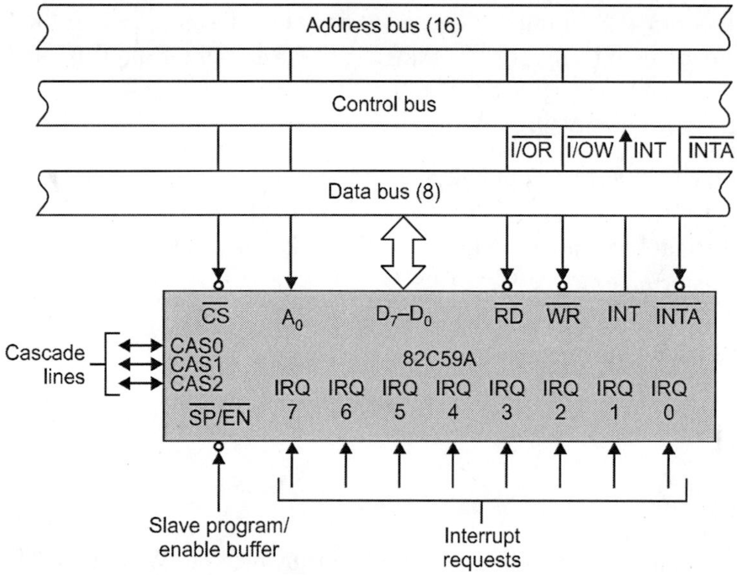


**Figure 12.4:** 8259A Standard System Bus Interface

**These events occur in an 8080/8085 system**

1. One or more of the INTERRUPT REQUEST lines (IR0–IR7) are raised high, setting the corresponding IRR bit(s).
2. The 82C59A evaluates those requests in the priority resolver and sends an interrupt (INT) to the CPU, if appropriate.
3. The CPU acknowledges the INT and responds with an $\overline{\text{INTA}}$ pulse.
4. Upon receiving an $\overline{\text{INTA}}$ from the CPU group, the highest priority ISR bit is set, and the corresponding IRR bit is reset. The 82C59A will also release a CALL instruction code (11001101) onto the 8-bit data bus through D0–D7.
5. This CALL instruction will initiate two additional $\overline{\text{INTA}}$ pulses to be sent to 82C59A from the CPU group.
6. These two $\overline{\text{INTA}}$ pulses allow the 82C59A to release its preprogrammed subroutine address onto the data bus. The lower 8-bit address is released at the first $\overline{\text{INTA}}$ pulse and the higher 8-bit address is released at the second $\overline{\text{INTA}}$ pulse.

7. This completes the 3-byte CALL instruction released by the 82C59A. In the AEOI mode, the ISR bit is reset at the end of the third $\overline{\text{INTA}}$ pulse. Otherwise, the ISR bit remains set until an appropriate EOI command is issued at the end of the interrupt sequence. The events occurring in an 80C86/88/286 system are the same until step 4.
8. The 82C59A does not drive the data bus during the first $\overline{\text{INTA}}$ pulse.
9. The 80C86/88/286 CPU will initiate a second $\overline{\text{INTA}}$ pulse. During this $\overline{\text{INTA}}$ pulse, the appropriate ISR bit is set and the corresponding bit in the IRR is reset. The 82C59A outputs the 8-bit pointer onto the data bus to be read by the CPU.
10. This completes the interrupt cycle. In the AEOI mode, the ISR bit is reset at the end of the second $\overline{\text{INTA}}$ pulse. Otherwise, the ISR bit remains set until an appropriate EOI command is issued at the end of the interrupt subroutine. If no interrupt request is present at step 4 of either sequence (i.e., the request was too short in duration), the 82C59A will issue an interrupt level
11. If a slave is programmed on IR bit 7, the CAS lines remain inactive and vector addresses are output from the master 82C59A.

## 12.4 PIN DIAGRAM OF 8259A

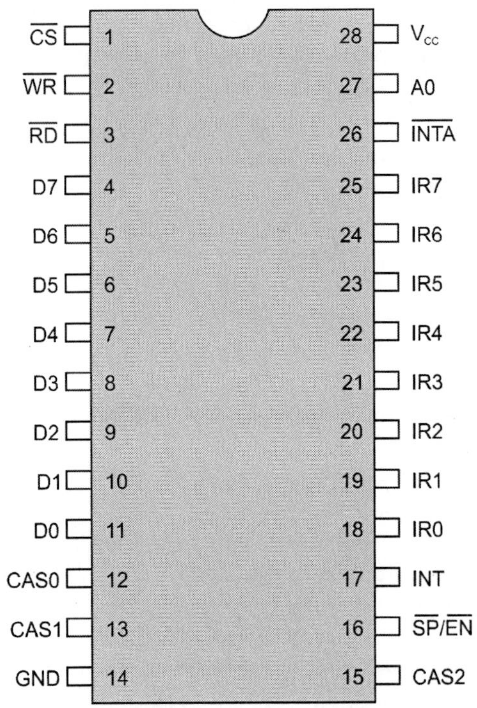


| Pin | Description |
|---|---|
| D7–D0 | Data Bus (Bidirectional) |
| $\overline{\text{RD}}$ | Read Input |
| $\overline{\text{WR}}$ | Write Input |
| A0 | Command Select Address |
| $\overline{\text{CS}}$ | Chip Select |
| CAS2–CAS0 | Cascade Lines |
| $\overline{\text{SP}}/\overline{\text{EN}}$ | Slave Program Input Enable |
| INT | Interrupt Output |
| $\overline{\text{INTA}}$ | Interrupt Acknowledge Input |
| IR0–IR7 | Interrupt Request Inputs |

**Figure 12.5:** Pin diagram of 8259A

### 12.4.1 Pin Description of 8259A

| Symbol | Type | Description |
|---|---|---|
| $V_{CC}$ | I | $V_{CC}$. is the + 5v power supply pin. A 0.1 mF capacitor between pins 31 and 20 is recommended fro decoupling. |
| GND | I | Ground |
| $\overline{CS}$ | I | **CHIP SELECT:** A low on this pin enables $\overline{RD}$ and $\overline{WR}$ communications between the CPU and the 82C59A $\overline{INTA}$ functions are independent of $\overline{CS}$. |
| $\overline{WR}$ | I | **Write:** A low on this pin when $\overline{CS}$ is low enable the 82C59A to accept command words from the CPU |
| $\overline{RD}$ | I | **Read:** A low on this pin when $\overline{CS}$ is low enables the 82C59A to release status onto the data bus for the CPU |
| D7–D0 | I/O | **Bidirectional Data Bus:** Control, status and interrup-vector information is transferred via this bus. |
| CAS0–CAS2 | I/O | **Cascade Lines:** The CAS lines from a private 82C59A bus tp cpmtrp; a ,i;to[;e 82C59A strictire. These pins are outputs for a master 82C59A and inputs for a siave 82C59A. |
| $\overline{SP}/\overline{EN}$ | I/O | **Slave Program/Enable Buffer:** This is a dual function pin. When in the Buffered Mode it can be used as an ouput to control buffer transceivers ($\overline{EN}$).When Not in the Buffered Mode it is used as an input to designate a master ($\overline{SP}$ =1) or Slave ($\overline{SP}$ = 0). |
| INT | O | **Interrupt:** This pin goes high whenever a valide interrupt request is asserted. It is used to interrupt the CPU, thus, it is connected to the CPU's interrupt pin. |
| IR0–IR7 | I | **Interrupt Requiests:** Asynchronous inputs. An interrupt request is executed by raising an IR input (low to high), and holding it high until it is acknowledged (Edge Triggered Mode), or just by a high level on an IR input (Level Triggered Mode). Internal pull-up resistors are implemented on IR0–7. |
| $\overline{INTA}$ | I | **Interrupt Acknowledge:** This pin is used to enable 82C59A interrupt-vector data onto the data bus by a sequence of interrupt acknowledge pulses issued by the CPU. |
| A0 | I | **Address Line:** This pin acts in conjuction with the $\overline{CS}$, $\overline{WR}$ and $\overline{RD}$ pins. It is used by teh 82C59A to decipher various Command Words the CPU writes and status the CPU wishes to read. It is typically connected to the CPU A0 address line (A1 for 80C86/88/286). |

## 12.5 INTERRUPT OPERATION

To implement interrupts, the Interrupt Enable flip-flop in the microprocessor should be enabled by writing the EI instruction, and the 8259A should be initialized by writing control words in the control register. The 8259A requires two types of control words: Initialization Command Words (ICWs) and Operational Command Words (OCWs). The ICWs are used to set up the proper conditions and specify RST vector addresses. The OCWs are used to perform functions such as masking interrupts, setting up status-read operations, etc. After the 8259A is initialized, the following sequence of events occurs when one or more interrupt request lines go high:

1. The IRR stores the requests.
2. The priority resolver checks three registers: the IRR for interrupt requests, the IMR for masking bits, and the ISR for the interrupt request being served. It resolves the priority and sets the INT high when appropriate.
3. The MPUS acknowledges the interrupt by sending $\overline{INTA}$.
4. After the $\overline{INTA}$ is received, the appropriate priority bit in the ISR is set to indicate which interrupt level is being served, and the corresponding bit in the IRR is reset to indicate that the request is accepted. Then, the opcode for the CALL instruction is placed on the data bus.
5. When the MPU decodes the CALL instruction, it places two more $\overline{INTA}$ signals on the data bus.
6. When the 8259A receives the second $\overline{INTA}$, it places the low-order byte of the CALL address on the data bus. At the third $\overline{INTA}$, it places the high-order byte on the data bus. The CALL address is the vector memory location for the interrupt; this address is placed in the control register during the initialization.
7. During the third $\overline{INTA}$ pulse, the ISR bit is reset either automatically (Automatic-End-of-Interrupt — AEOI) or by a command word that must be issued at the end of the service routine (End-of-Interrupt — EOI). This option is determined by the initialization command word (ICW).
8. The program sequence is transferred to the memory location specified by the CALL instruction.

### 12.5.1 Interrupt Sequence Outputs

#### *1. 8085 Interrupt Response Mode*

This sequence is timed by three $\overline{INTA}$ pulses. During the first $\overline{INTA}$ pulse, the CALL opcode is enabled onto the data bus. First Interrupt Vector Byte Data: Hex CD

| | $D_7$ | $D_6$ | $D_5$ | $D_4$ | $D_3$ | $D_2$ | $D_1$ | $D_0$ |
|---|---|---|---|---|---|---|---|---|
| Call Code | 1 | 1 | 0 | 0 | 1 | 1 | 0 | 1 |

During the second $\overline{INTA}$ pulse, the lower address of the appropriate service routine is enabled onto the data bus. When interval = 4 bits, A5–A7 are programmed, while A0–A4 are automatically inserted by the 82C59A. When interval = 8, only A6 and A7 are programmed, while A0–A5 are automatically inserted.

| IR | Interval = 4 | | | | | | | |
|---|---|---|---|---|---|---|---|---|
| | D7 | D6 | D5 | D4 | D3 | D2 | D1 | D0 |
| 7 | A7 | A6 | A5 | 1 | 1 | 1 | 0 | 0 |
| 6 | A7 | A6 | A5 | 1 | 1 | 0 | 0 | 0 |
| 5 | A7 | A6 | A5 | 1 | 0 | 1 | 0 | 0 |
| 4 | A7 | A6 | A5 | 1 | 0 | 0 | 0 | 0 |
| 3 | A7 | A6 | A5 | 0 | 1 | 1 | 0 | 0 |
| 2 | A7 | A6 | A5 | 0 | 0 | 1 | 0 | 0 |
| 1 | A7 | A6 | A5 | 0 | 0 | 1 | 0 | 0 |
| 0 | A7 | A6 | A5 | 0 | 0 | 0 | 0 | 0 |

| IR | Interval = 8 | | | | | | | |
|---|---|---|---|---|---|---|---|---|
| | D7 | D6 | D5 | D4 | D3 | D2 | D1 | D0 |
| 7 | A7 | A6 | 1 | 1 | 1 | 0 | 0 | 0 |
| 6 | A7 | A6 | 1 | 1 | 0 | 0 | 0 | 0 |
| 5 | A7 | A6 | 1 | 0 | 1 | 0 | 0 | 0 |
| 4 | A7 | A6 | 1 | 0 | 0 | 0 | 0 | 0 |
| 3 | A7 | A6 | 0 | 1 | 1 | 0 | 0 | 0 |
| 2 | A7 | A6 | 0 | 1 | 0 | 0 | 0 | 0 |
| 1 | A7 | A6 | 0 | 0 | 1 | 0 | 0 | 0 |
| 0 | A7 | A6 | 0 | 0 | 0 | 0 | 0 | 0 |

During the third $\overline{\text{INTA}}$ pulse, the higher address of the appropriate service routine, which was programmed as byte 2 of the initialization sequence (A8 - A15), is enabled onto the bus.

Content of Third Interrupt Vector Byte

| $D_7$ | $D_6$ | $D_5$ | $D_4$ | $D_3$ | $D_2$ | $D_1$ | $D_0$ |
|---|---|---|---|---|---|---|---|
| A15 | A14 | A13 | A12 | A11 | A10 | A9 | A8 |

### 2. *8086, 8088, 80286 Interrupt Response Mode*

8086/88/286 mode is similar to 8080/85 mode except that only two Interrupt Acknowledge cycles are issued by the processor and no CALL opcode is sent to the processor. The first interrupt acknowledge cycle is similar to that of 8080/85 systems in that the 82C59A uses it to internally freeze the state of the interrupts for priority resolution and, as a master, it issues the interrupt code on the cascade lines. On this first cycle, it does not issue any data to the processor and leaves its data bus buffers disabled. On the second interrupt acknowledge cycle in the 86/88/286 mode, the master (or slave if so programmed) will send a byte of data to the processor with the acknowledged interrupt code composed as follows (note the state of the ADI mode control is ignored and A5 - A11 are unused in the 86/88/286 mode).

| | Content of Interrupt Vector Byte for 80C86/88/288 System Mode | | | | | | | |
|---|---|---|---|---|---|---|---|---|
| | D7 | D6 | D5 | D4 | D3 | D2 | D1 | D0 |
| IR7 | T7 | T6 | T5 | T4 | T3 | 1 | 1 | 1 |
| IR6 | T7 | T6 | T5 | T4 | T3 | 1 | 1 | 0 |
| IR5 | T7 | T6 | T5 | T4 | T3 | 1 | 0 | 1 |
| IR4 | T7 | T6 | T5 | T4 | T3 | 1 | 0 | 0 |
| IR3 | T7 | T6 | T5 | T4 | T3 | 0 | 1 | 1 |
| IR2 | T7 | T6 | T5 | T4 | T3 | 0 | 1 | 0 |
| IR1 | T7 | T6 | T5 | T4 | T3 | 0 | 0 | 1 |
| IR0 | T7 | T6 | T5 | T4 | T3 | 0 | 0 | 0 |

### 12.5.2 Priority Modes and Other Features

Many types of priority modes are available under software control in the 8259A, and they can be changed dynamically during the program by writing appropriate command words. Commonly used priority modes are discussed below.

1. **Fully Nested Mode:** This is a general-purpose mode in which all IRs (Interrupt Requests) are arranged from highest to lowest, with $IR_0$ as the highest and $IR_7$ as the lowest.

   In addition, any IR can be assigned the highest priority in this mode; the priority sequence will then begin at that IR. In the example below, $IR_4$ has the highest priority, and $IR_3$ has the lowest priority:

   | $IR_0$ | $IR_1$ | $IR_2$ | $IR_3$ | $IR_4$ | $IR_5$ | $IR_6$ | $IR_7$ |
   |---|---|---|---|---|---|---|---|
   | 4 | 5 | 6 | 7 | 0 | 1 | 2 | 3 |
   | | | | ↑ Lowest Priority | ↑ Highest Priority | | | |

2. **Automatic Rotation Mode:** In this mode, a device, after being serviced, receives the lowest priority. Assuming that the $IR_2$ has just been serviced, It will receive the seventh priority, as shown below:

   | $IR_0$ | $IR_1$ | $IR_2$ | $IR_3$ | $IR_4$ | $IR_5$ | $IR_6$ | $IR_7$ |
   |---|---|---|---|---|---|---|---|
   | 1 | 6 | 7 | 0 | 1 | 2 | 3 | 4 |

3. **Specific Rotation Mode:** This mode is similar to the automatic rotation mode, except that the user can select any IR for the lowest priority, thus fixing all other priorities.

### 12.5.3 End of Interrupt

After the completion of an interrupt service, the corresponding ISR bit needs to be reset to update the information in the ISR. This is called the End-of-Interrupt (EOI) command. It can be issued in three formats:

1. **Nonspecific EOR Command:** When this command is set to the 8259A, it resets the highest priority ISR bit.
2. **Specific EOI command:** This command specifies which ISR bit to reset.
3. **Automatic EOI:** In this mode, no command is necessary. During the third INTA, the ISR bit is reset. The major drawback with this mode is that the ISR does not have information on which IR is being serviced. Thus, any IR can interrupt the service routine, irrespective of its priority, if the Interrupt Enable flip-flop is set.

## 12.6 PROGRAMMING THE 8259A

As mentioned before, the 8259A requires two types of command words: Initialization Command Words (ICWs) and Operational Command Words (OCWs). The 8259A can be initialized with four ICWs; the first two are essential, and the other two are operational based on the modes being used. These words must be issued in a given sequence. Once initialized, the 8259A can be set up to operate in various modes by using three different OCWs; however, they no longer need be issued in a specific sequence.

Figure 12.6 shows the bit specification of the first two ICWs. The $ICW_1$, shown in Figure 12.6(a), specifies

1. Single or multiple 8259As in the system.
2. 4- or 8-bit interval between the interrupt vector locations.
3. The address bits $A_7$-$A_5$ of the CALL instruction; the rest are supplied by the 8259A, as shown in Figure 12.6(b).

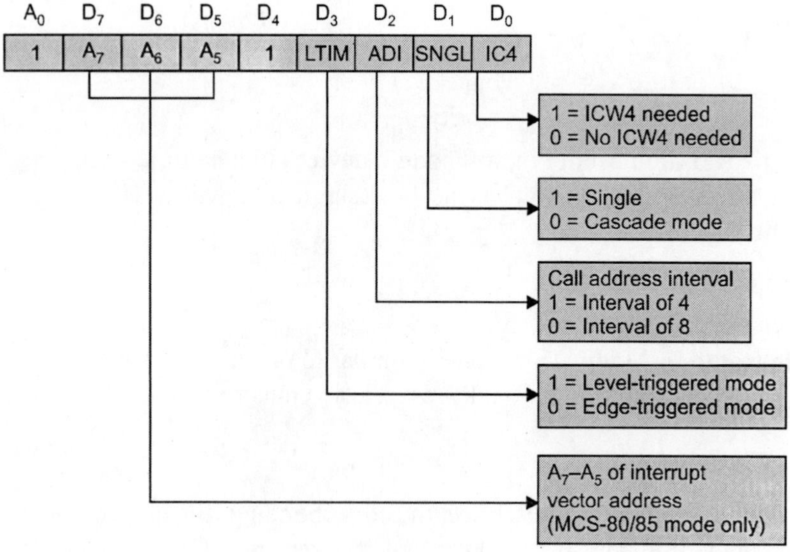


(a)

| IR | Interval = 4 | | | | | | | |
|---|---|---|---|---|---|---|---|---|
| | $D_7$ | $D_6$ | $D_5$ | $D_4$ | $D_3$ | $D_2$ | $D_1$ | $D_0$ |
| 7 | $A_7$ | $A_6$ | $A_5$ | 1 | 1 | 1 | 0 | 0 |
| 6 | $A_7$ | $A_6$ | $A_5$ | 1 | 1 | 0 | 0 | 0 |
| 5 | $A_7$ | $A_6$ | $A_5$ | 1 | 0 | 1 | 0 | 0 |
| 4 | $A_7$ | $A_6$ | $A_5$ | 1 | 0 | 0 | 0 | 0 |
| 3 | $A_7$ | $A_6$ | $A_5$ | 0 | 1 | 1 | 0 | 0 |
| 2 | $A_7$ | $A_6$ | $A_5$ | 0 | 1 | 0 | 0 | 0 |
| 1 | $A_7$ | $A_6$ | $A_5$ | 0 | 0 | 1 | 0 | 0 |
| 0 | $A_7$ | $A_6$ | $A_5$ | 0 | 0 | 0 | 0 | 0 |

| IR | Interval = 8 | | | | | | | |
|---|---|---|---|---|---|---|---|---|
| | $D_7$ | $D_6$ | $D_5$ | $D_4$ | $D_3$ | $D_2$ | $D_1$ | $D_0$ |
| 7 | $A_7$ | $A_6$ | 1 | 1 | 1 | 0 | 0 | 0 |
| 6 | $A_7$ | $A_6$ | 1 | 1 | 0 | 0 | 0 | 0 |
| 5 | $A_7$ | $A_6$ | 1 | 0 | 1 | 0 | 0 | 0 |
| 4 | $A_7$ | $A_6$ | 1 | 0 | 0 | 0 | 0 | 0 |
| 3 | $A_7$ | $A_6$ | 0 | 1 | 1 | 0 | 0 | 0 |
| 2 | $A_7$ | $A_6$ | 0 | 1 | 0 | 0 | 0 | 0 |
| 1 | $A_7$ | $A_6$ | 0 | 0 | 1 | 0 | 0 | 0 |
| 0 | $A_7$ | $A_6$ | 0 | 0 | 0 | 0 | 0 | 0 |

(b)

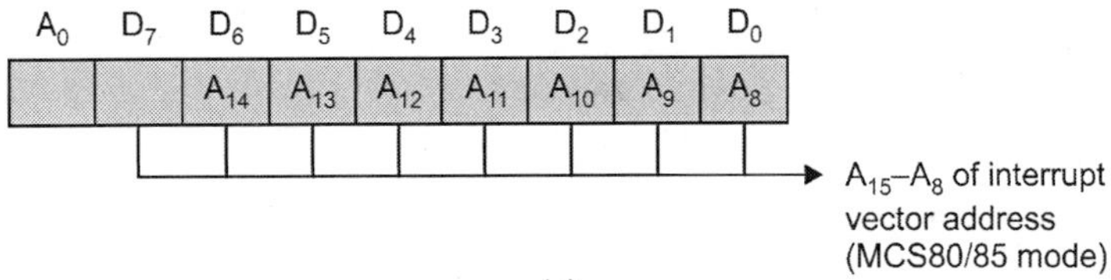


(c)

**Figure 12.6:** Initialization Command Words for the 8259A

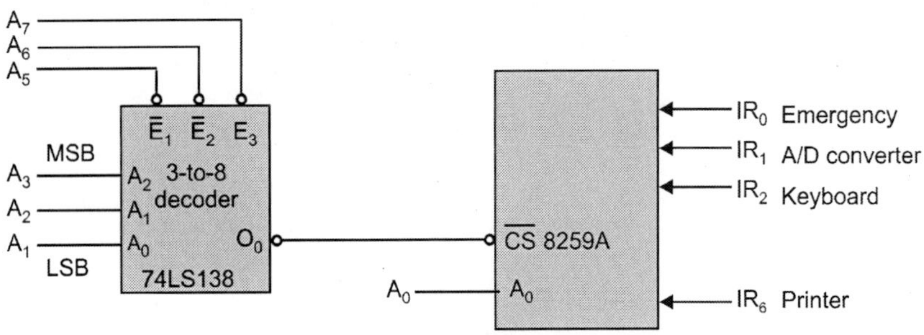


**Figure 12.7:** Schematic of an Interrupt system using the 8259A

The $ICW_2$ of **Figure 12.6(c)** specifies the high-order byte of the CALL instruction.

### *Examples*

Figure 12.7 shows the schematic of an interrupt-driven system using the 8259A. For sources are connected to the IR lines of the 8259A: Emergency Signal, Keyboard, A/D Converter, and Printer. Of these, the Emergency Signal has the highest priority and the Printer has the lowest priority.

**The 82C59A accepts two types of command words generated by the CPU:**

1. **Initialization Command Words (ICWs):** Before normal operation can begin, each 82C59A in the system must be brought to a starting point - by a sequence of 2 to 4 bytes timed by pulses.
2. **Operation Command Words (OCWs):** These are the command words which command the 82C59A to operate in various interrupt modes.

**Among these modes are:**

(a) Fully nested mode.

(b) Rotating priority mode.

(c) Special mask mode.

(d) Polled mode.

The OCWs can be written into the 82C59A anytime after initialization.

### 12.6.1 Initialization Command Words (ICWs)

Whenever a command is issued with A0 = 0 and D4 = 1, this is interpreted as Initialization Command Word 1 (ICW1). ICW1 starts the initialization sequence during which the following automatically occur:

(a) The edge sense circuit is reset, which means that following initialization, an interrupt request (IR) input must make a low-to-high transition to generate an interrupt.

(b) The Interrupt Mask Register is cleared.

(c) IR7 input is assigned priority 7.

(d) Special Mask Mode is cleared and Status Read is set to IRR.

(e) If IC4 = 0, then all functions selected in ICW4 are set to zero. (Non-Buffered mode (see note), no Auto-EOI, 8080/85 system).

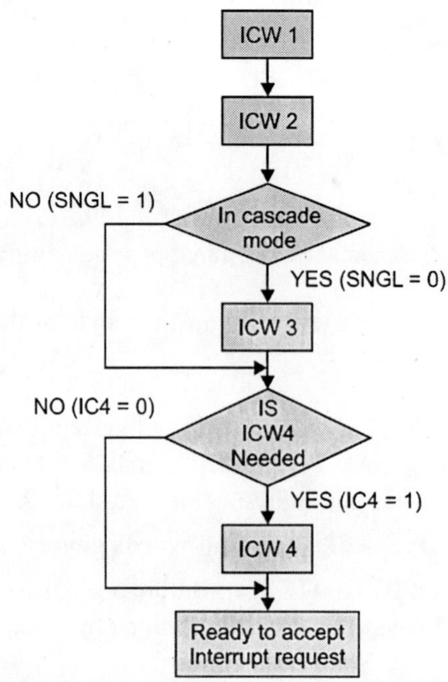


**Figure 12.8:** 82C59A Initialization Sequence

## 1. *Initialization Command Words 1 and 2 (ICW1, LCW2)*

(a) **A5 - A15:** Page starting address of service routines. In an 8080/85 system the 8 request levels will generate CALLS to 8 locations equally spaced in memory. These can be programmed to be spaced at intervals of 4 or 8 memory locations, thus, the 8 routines will occupy a page of 32 or 64 bytes, respectively. The address format is 2 bytes long (A0–A15). When the routine interval is 4, A0–A4 are automatically inserted by the 82C59A, while A5–A15 are programmed externally. When the routine interval is 8, A0–A5 are automatically inserted by the 82C59A while A6–A15 are programmed externally. The 8-byte interval will maintain compatibility with current software, while the 4-byte interval is best for a compact jump table. In an 80C86/88/286 system, A15–A11 are inserted in the five most significant bits of the vectoring byte and the 82C59A sets the three least significant bits according to the interrupt level. A10–A5 are ignored and ADI (Address interval) has no effect.

(b) **LTIM:** If LTIM = 1, then the 82C59A will operate in the level interrupt mode. Edge detect logic on the interrupt inputs will be disabled.

(c) **ADI:** ALL address interval. ADI = 1 then interval = 4; ADI = 0 then interval = 8.

(d) **SNGL:** Single. Means that this is the only 82C59A in the system. If SNGL = 1, no ICW3 will be issued.

(e) **IC4:** If this bit is set - ICW4 has to be issued. If ICW4 is not needed, set IC4 = 0.

## 2. *Initialization Command Word 3 (ICW3)*

This word is read only when there is more than one 82C59A in the system and cascading is used, in which case SNGL = 0. It will load the 8-bit slave register. The functions of this register are:

(a) **In the master mode** (either when  = 1, or in buffered mode when M/S = 1 in ICW4) a "1" is set for each slave in the bit corresponding to the appropriate IR line for the slave. The master then will release byte 1 of the call sequence (for 8080/85 system) and will enable the corresponding slave to release bytes 2 and 3 (for 80C86/88/ 286, only byte 2) through the cascade lines.

(c) **In the slave mode** (either when  = 0, or if BUF = 1 and M/S = 0 in ICW4), bits 2 - 0 identify the slave. The slave compares its cascade input with these bits and if they are equal, bytes 2 and 3 of the call sequence (or just byte 2 for 80C86/88/286) are released by it on the Data Bus.

## 3. *Initialization Command Word 4 (ICW4)*

(a) **SFNM:** If SFNM = 1, the special fully nested mode is programmed.

(b) **BUF:** If BUF = 1, the buffered mode is programmed. In buffered mode, / becomes an enable output and the master/slave determination is by M/S.

(c) **M/S:** If buffered mode is selected: M/S = 1 means the 82C59A is programmed to be a master, M/S = 0 means the 82C59A is programmed to be a slave. If BUF = 0, M/S has no function.

(d) **AEOI:** If AEOI = 1, the automatic end of interrupt mode is programmed.

(e) **µPM:** Microprocessor mode: µPM = 0 sets the 82C59A for 8080/85 system operation, µPM = 1 sets the 82C59A for 80C86/88/286 system operation.

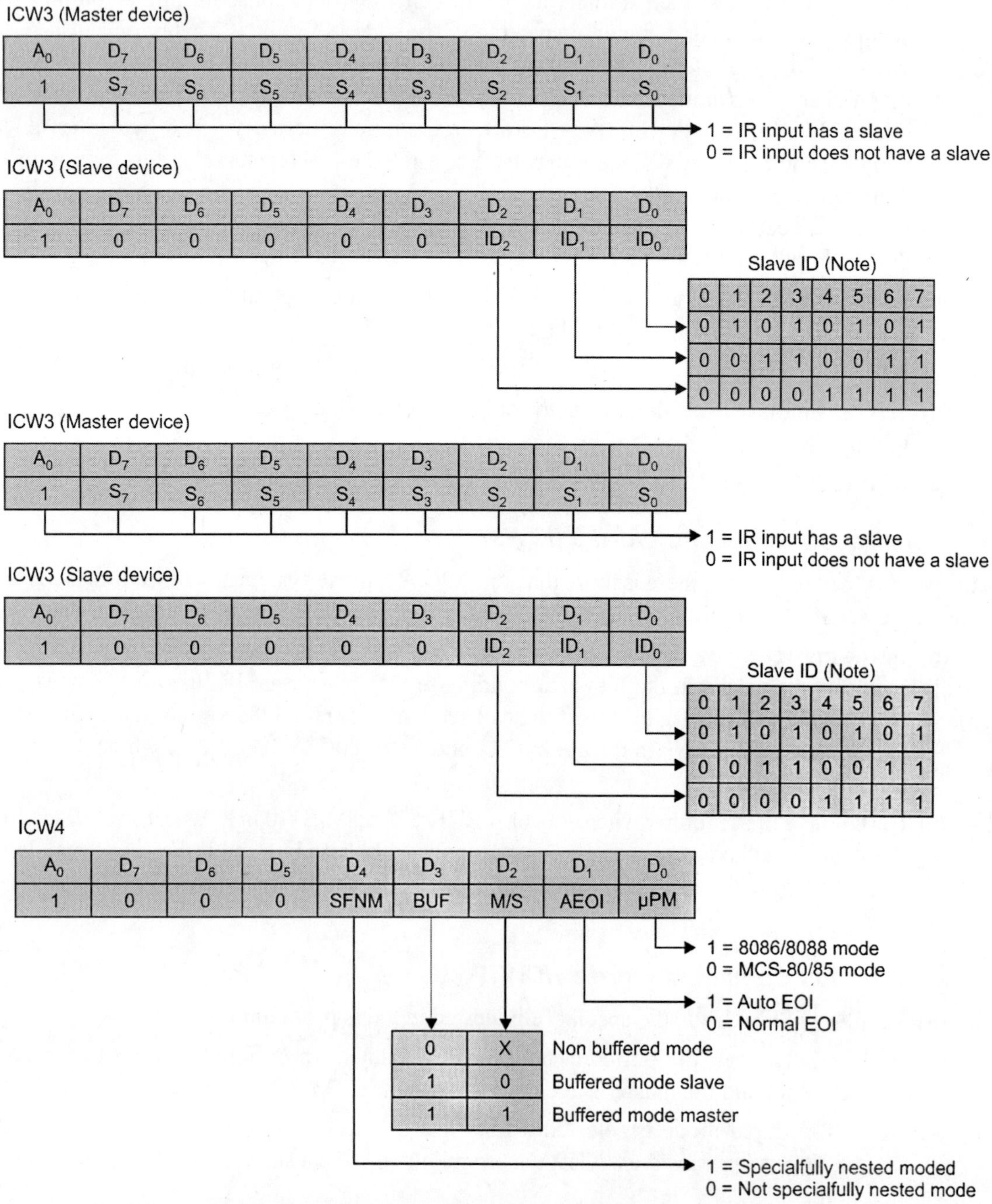


**Figure 12.9:** 82C59A Initialization Command Word Format

### 12.6.2 Operation Command Words (OCWs)

After the Initialization Command Words (ICWs) are programmed into the 82C59A, the device is ready to accept interrupt requests at its input lines. However, during the 82C59A operation, a selection of algorithms can command the 82C59A to operate in various modes through the Operation Command Words (OCWs).

Operation Command Words (OCWs)

| A0 | D7 | D6 | D5 | D4 | D3 | D2 | D1 | D0 |
|---|---|---|---|---|---|---|---|---|
| OCW1 | | | | | | | | |
| 1 | M7 | M6 | M5 | M4 | M3 | M2 | M1 | M0 |
| OCW2 | | | | | | | | |
| 0 | R | SL | EOI | 0 | 0 | L2 | L1 | L0 |
| OCW3 | | | | | | | | |
| 0 | 0 | ESMM | SMM | 0 | 1 | P | RR | RS |

#### 1. Operation Command Word 1 (OCW1)

OCW1 sets and clears the mask bits in the Interrupt Mask Register (IMR) M7–M0 represent the eight mask bits. M = 1 indicates the channel is masked (inhibited), M = 0 indicates the channel is enabled.

#### 2. Operation Command Word 2 (OCW2)

(a) **R, SL, EOI** - These three bits control the Rotate and End of Interrupt modes and combinations of the two. A chart of these combinations can be found on the Operation Command Word Format.

(a) **L2, L1, L0** - These bits determine the interrupt level acted upon when the SL bit is active.

#### 3. Operation Command Word 3 (OCW3)

(a) **ESMM (Enable Special Mask Mode):** When this bit is set to 1 it enables the SMM bit to set or reset the Special Mask Mode. When ESMM = 0, the SMM bit becomes a "don't care".

(a) **SMM (Special Mask Mode):** If ESMM = 1 and SMM = 1, the 82C59A will enter Special Mask Mode. If ESMM = 1 and SMM = 0, the 82C59A will revert to normal mask mode. When ESMM = 0, SMM has no effect.

### 12.6.3 Fully Nested Mode

This mode is entered after initialization unless another mode is programmed. The interrupt requests are ordered in priority from 0 through 7 (0 highest). When an interrupt is acknowledged the highest priority request is determined and its vector placed on the bus. Additionally, a bit of the Interrupt Service register (IS0–7) is set. This bit remains set until the microprocessor issues an End of Interrupt (EOI) command immediately before returning from the service routine, or if the AEOI (Automatic End of Interrupt) bit is set, until the trailing edge of the last $\overline{\text{INTA}}$. While the IS bit is set, all further interrupts of the same or lower priority are inhibited, while higher levels will generate an interrupt (which will be acknowledged only if the microprocessor internal interrupt enable flip-flop has been re-enabled through software). After the initialization sequence, IR0 has the highest priority and IR7 the lowest. Priorities can be changed, as will be explained in the rotating priority mode or via the set priority command.

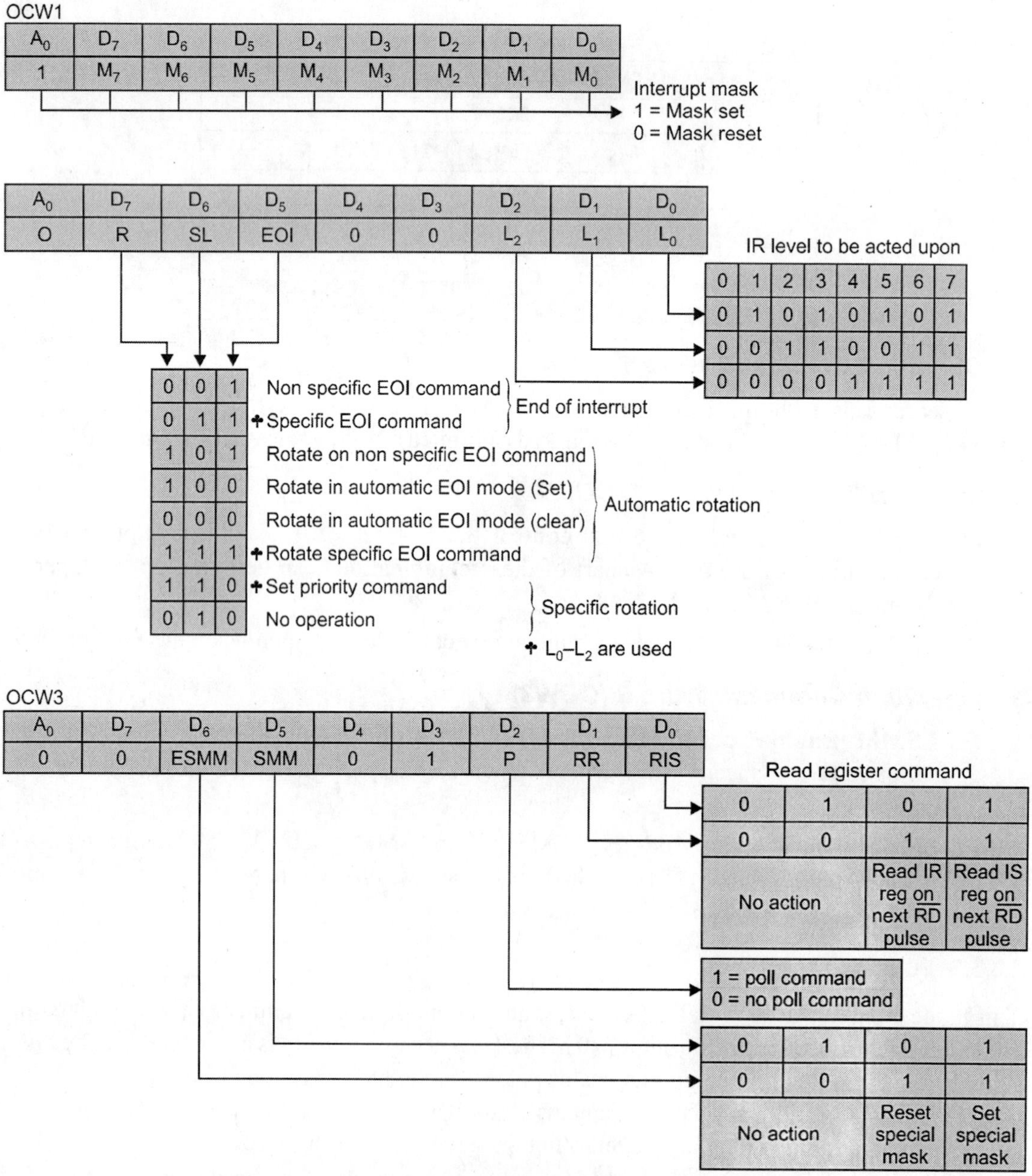


**Figure 12.10:** 82C59A Operation Command Word Format

### 12.6.4 End of Interrupt (EOI)

The In-Service (IS) bit can be reset either automatically following the trailing edge of the last in sequence $\overline{\text{INTA}}$ pulse (when AEOI bit in ICW1 is set) or by a command word that must be issued to the 82C59A before returning from a service routine (EOI Command). An EOI command must be issued twice if servicing a slave in the Cascade mode, once for the master and once for the corresponding slave. There are two forms of EOI command: Specific and Non- Specific. When the 82C59A is operated in modes which preserve the fully nested structure, it can determine which IS bit to reset on EOI. When a Non-Specific command is issued the 82C59A will automatically reset the highest IS bit of those that are set, since in the fully nested mode the highest IS level was necessarily the last level acknowledged and serviced. A non-specific EOI can be issued with OCW2 (EOI = 1, SL = 0, R = 0).

When a mode is used which may disturb the fully nested structure, the 82C59A may no longer be able to determine the last level acknowledged. In this case a Specific End of Interrupt must be issued which includes as part of the command the IS level to be reset. A specific EOI can be issued with OCW2 (EOI = 1, SL = 1, R = 0, and L0 - L2 is the binary level of the IS bit to be reset). An IRR bit that is masked by an IMR bit will not be cleared by a non-specific EOI if the 82C59A is in the Special Mask Mode.

### 12.6.5 Automatic End of Interrupt (AEOI) Mode

If AEOI = 1 in ICW4, then the 82C59A will operate in AEOI mode continuously until reprogrammed by ICW4. In this mode the 82C59A will automatically perform non-specific EOI operations at the trailing edge of the last interrupt acknowledge pulse (third pulse in 8080/85, second in 80C86/88/286). Note that from a system standpoint, this mode should be used only when a nested multilevel interrupt structure is not required within a single 82C59A.

### 12.6.6 Automatic Rotation (Equal Priority Devices)

In some applications there are a number of interrupting devices of equal priority. In this mode a device, after being serviced, receives the lowest priority, so a device requesting an interrupt will have to wait, in the worst case until each of 7 other devices are serviced at most once. For example, if the priority and "in service" status is:

Before Rotate (IR4 the highest priority requiring service)

| | IS7 | IS6 | IS5 | IS4 | IS3 | IS2 | IS1 | IS0 |
|---|---|---|---|---|---|---|---|---|
| "IS" status | 0 | 1 | 0 | 1 | 0 | 0 | 0 | 0 |
| Priority status | 7 | 6 | 5 | 4 | 3 | 2 | 1 | 0 |

Lowest (IS7) … Highest (IS0)

After Rotate (IR4 was serviced, all other priorities rotated correspondingly)

| | IS7 | IS6 | IS5 | IS4 | IS3 | IS2 | IS1 | IS0 |
|---|---|---|---|---|---|---|---|---|
| "IS" status | 0 | 1 | 0 | 0 | 0 | 0 | 0 | 0 |
| Priority status | 2 | 1 | 0 | 7 | 6 | 5 | 4 | 3 |
| | ↑ Highest | | | | | | | ↑ Lowest |

There are two ways to accomplish Automatic Rotation using OCW2, the Rotation on Non-Specific EOI Command (R = 1, SL = 0, EOI = 1) and the Rotate in Automatic EOI Mode which is set by (R = 1, SL = 0, EOI = 0) and cleared by (R = 0, SL = 0, EOI = 0).

### 12.6.7 Specific Rotation (Specific Priority)

The programmer can change priorities by programming the lowest priority and thus, fixing all other priorities; i.e., if IR5 is programmed as the lowest priority device, then IR6 will have the highest one.

The Set Priority command is issued in OCW2 where: R = 1, SL = 1, L0 - L2 is the binary priority level code of the lowest priority device.

Observe that in this mode internal status is updated by software control during OCW2. However, it is independent of the End of Interrupt (EOI) command (also executed by OCW2). Priority changes can be executed during an EOI command by using the Rotate on Specific EOI command in OCW2 (R = 1, SL = 1, EOI = 1, and L0–L2 = IR level to receive lowest priority).

#### *1. Interrupt Masks*

Each Interrupt Request input can be masked individually by the Interrupt Mask Register (IMR) programmed through OCW1. Each bit in the IMR masks one interrupt channel if it is set (1). Bit 0 masks IR0, Bit 1 masks IR1 and so forth. Masking an IR channel does not affect the operation of other channels.

#### *2. Special Mask Mode*

Some applications may require an interrupt service routine to dynamically alter the system priority structure during its execution under software control. For example, the routine may wish to inhibit lower priority requests for a portion of its execution but enable some of them for another portion. The difficulty here is that if an Interrupt Request is acknowledged and an End of Interrupt command did not reset its IS bit (i.e., while executing a service routine), the 82C59A would have inhibited all lower priority requests with no easy way for the routine to enable them. That is where the Special Mask Mode comes in. In the Special Mask Mode, when a mask bit is set in OCW1, it inhibits further interrupts at that level and enables interrupts from all other levels (lower as well as higher) that are not masked.

Thus, any interrupts may be selectively enabled by loading the mask register. The Special Mask Mode is set by OCW3 where: ESMM = 1, SMM = 1, and cleared where ESMM = 1, SMM = 0.

### 3. *Poll Command*

In this mode, the INT output is not used or the microprocessor internal Interrupt Enable flip flop is reset, disabling its interrupt input. Service to devices is achieved by software using a Poll command.

The Poll command is issued by setting P = 1 in OCW3. The 82C59A treats the next $\overline{RD}$ pulse to the 82C59A (i.e. $\overline{RD}$, = 0, $\overline{CS}$ = 0) as an interrupt acknowledge, sets the appropriate IS bit if there is a request, and reads the priority level. Interrupt is frozen from $\overline{WR}$ to $\overline{RD}$. The word enabled onto the data bus during $\overline{RD}$ is:

| $D_7$ | $D_6$ | $D_5$ | $D_4$ | $D_3$ | $D_2$ | $D_1$ | $D_0$ |
|---|---|---|---|---|---|---|---|
| I | – | – | – | – | W2 | W1 | W0 |

**W0 – W2:** Binary code of the highest priority level requesting service.

**I:** Equal to a "1" if there is an interrupt.

This mode is useful if there is a routine command common to several levels so that the $\overline{INTA}$ sequence is not needed (saves ROM space). Another application is to use the poll mode to expand the number of priority levels to more than 64.

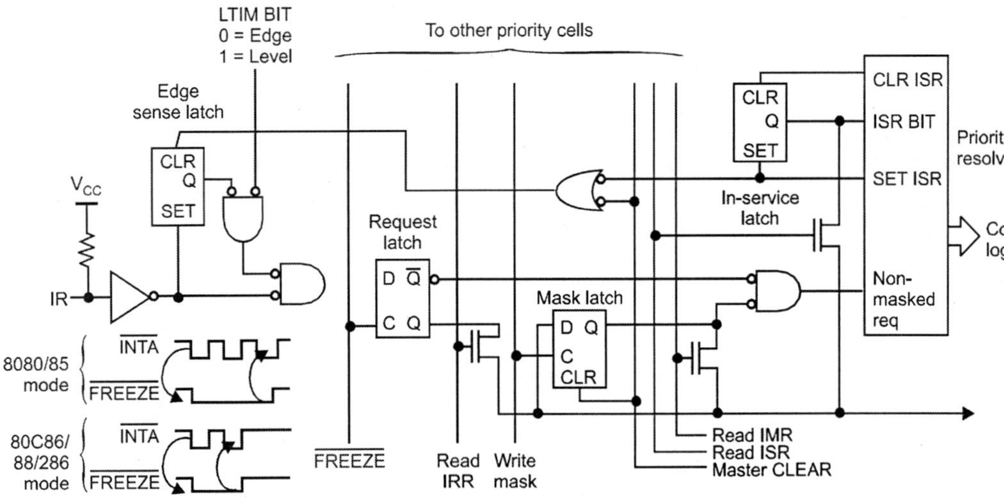


**Figure 12.11:** Diagram of poll mode

## 12.6.8 READING THE 8259A STATUS

The input status of several internal registers can be read to update the user information on the system. The following registers can be read via OCW3 (IRR and ISR) or OCW1 (IMR).

1. **Interrupt Request Register (IRR):** 8-bit register which contains the levels requesting an interrupt to be acknowledged. The highest request level is reset from the IRR when an interrupt is acknowledged. IRR is not affected by IMR.

2. **In-service Register (ISR):** 8-bit register which contains the priority levels that are being serviced. The ISR is updated when an End of Interrupt Command is issued.
3. **Interrupt Mask Register:** 8-bit register which contains the interrupt request lines which are masked.

The IRR can be read when, prior to the $\overline{RD}$ pulse, a Read Register Command is issued with OCW3 (RR = 1, RIS = 0). The ISR can be read when, prior to the RD pulse, a Read Register Command is issued with OCW3 (RR = 1, RIS = 1). There is no need to write an OCW3 before every status read operation, as long as the status read corresponds with the previous one: i.e., the 82C59A "remembers" whether the IRR or ISR has been previously selected by the OCW3. This is not true when poll is used. In the poll mode, the 82C59A treats the $\overline{RD}$ following a "poll write" operation as an $\overline{INTA}$. After initialization, the 82C59A is set to IRR.

For reading the IMR, no OCW3 is needed. The output data bus will contain the IMR whenever $\overline{RD}$ is active and A0 = 1 (OCW1). Polling overrides status read when P = 1, RR = 1 in OCW3.

### 12.6.9 Edge and Level Triggered Modes

This mode is programmed using bit 3 in ICW1. If LTIM = "0", an interrupt request will be recognized by a low to high transition on an IR input. The IR input can remain high without generating another interrupt.

If LTIM = "1", an interrupt request will be recognized by a "high" level on an IR input, and there is no need for an edge detection. The interrupt request must be removed before the EOI command is issued or the CPU interrupt is enabled to prevent a second interrupt from occurring.

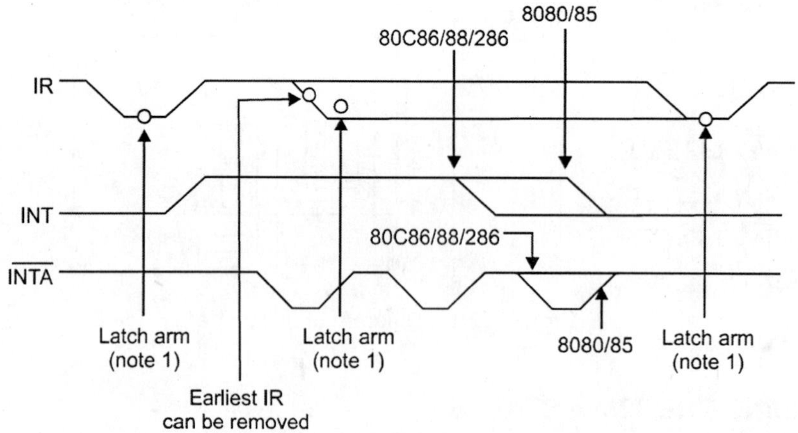


**Figure 12.12:** Timing diagram of edge and level triggered modes

The priority cell diagram shows a conceptual circuit of the level sensitive and edge sensitive input circuitry of the 82C59A. Be sure to note that the request latch is a transparent D type latch. In both the edge and level triggered modes the IR inputs must remain high until after the falling edge of the first . If the IR input goes low before this time a DEFAULT IR7 will occur when the CPU acknowledges the interrupt. This can be a useful safeguard for detecting interrupts caused by spurious noise glitches on the IR inputs.

To implement this feature the IR7 routine is used for "clean up" simply executing a return instruction, thus, ignoring the interrupt. If IR7 is needed for other purposes a default IR7 can still be detected by reading the ISR. A normal IR7 interrupt will set the corresponding ISR bit, a default IR7 won't. If a default IR7 routine occurs during a normal IR7 routine, however, the ISR will remain set.

In this case it is necessary to keep track of whether or not the IR7 routine was previously entered. If another IR7 occurs it is a default. In power sensitive applications, it is advisable to place the 82C59A in the edge-triggered mode with the IR lines normally high. This will minimize the current through the internal pull-up resistors on the IR pins.

### *1. The Special Fully Nested Mode*

This mode will be used in the case of a big system where cascading is used, and the priority has to be conserved within each slave. In this case the special fully nested mode will be programmed to the master (using ICW4).

**This mode is similar to the normal nested mode with the following exceptions:**

(a) When an interrupt request from a certain slave is in service, this slave is not locked out from the master's priority logic and further interrupt requests from higher priority IRs within the slave will be recognized by the master and will initiate interrupts to the processor. (In the normal nested mode a slave is masked out when its request is in service and no higher requests from the same slave can be serviced.

(b) When exiting the Interrupt Service routine the software has to check whether the interrupt serviced was the only one from that slave. This is done by sending a non-specific End of Interrupt (EOI) command to the slave and then reading its In-Service register and checking for zero. If it is empty, a non-specified EOI can be sent to the master, too. If not, no EOI should be sent.

### *2. Buffered Mode*

When the 82C59A is used in a large system where bus driving buffers are required on the data bus and the cascading mode is used, there exists the problem of enabling buffers The buffered mode will structure the 82C59A to send an enable signal on $\overline{SP}/\overline{EN}$ to enable the buffers. In this mode, whenever the 82C59A's data bus outputs are enabled, the $\overline{SP}/\overline{EN}$ output becomes active. This modification forces the use of software programming to determine whether the 82C59A is a master or a slave. Bit 3 in ICW4 programs the buffered mode, and bit 2 in ICW4 determines whether it is a master or a slave.

### *3. Cascade Mode*

The 8259A can be easily interconnected in a system of one master with up to eight slaves to handle up to 64 priority levels. The master controls the slaves through the 3 line cascade bus (CAS2 - 0).

The cascade bus acts like chip selects to the slaves during the $\overline{INTA}$ sequence. In a cascade configuration, the slave interrupt outputs (INT) are connected to the master interrupt request inputs. When a slave request line is activated and afterwards acknowledged, the master will enable the corresponding slave to release the device routine address during bytes 2 and 3 of $\overline{INTA}$. (Byte 2 only for 8086/88/286).

The cascade bus lines are normally low and will contain the slave address code from the leading edge of the first $\overline{INTA}$ pulse to the trailing edge of the last $\overline{INTA}$ pulse. Each 82C59A in the system must follow a separate initialization sequence and can be programmed to work in a different mode. An EOI command must be issued twice: once for the master and once for the corresponding slave. Chip select decoding is required to activate each 82C59A.

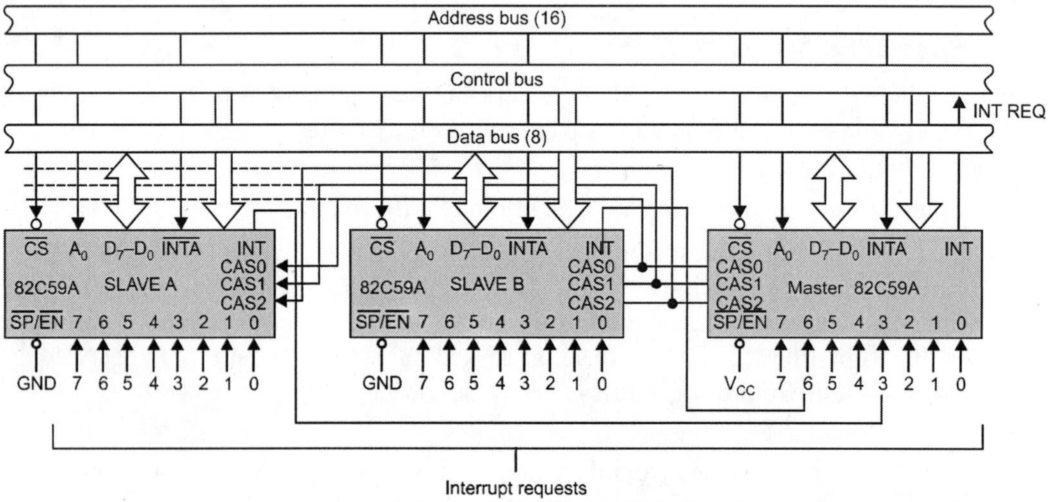


**Figure 12.13:** Cascading the 8259A

## QUESTION BANK

Q1. What do you mean by 8259A programmable interrupt controller?

Q2. Explain the concept of interrupts in microcomputer systems.

Q3. Explain the block diagram of the 8259A programmable interrupt controller.

Q4. Explain the pin diagram and description of of 8259A programmable interrupt controller.

Q5. Explain the various interrupt operations.

Q6. Explain the programming of the 8259A programmable interrupt controller.

Q7. Explain the different operating modes of the 8259A programmable interrupt controller.

C H A P T E R 13

# 8251 USART

## 13.1 HARDWARE-CONTROLLED SERIAL I/O USING PROGRAMMABLE CHIPS

The hardware approach to serial I/O incorporates the same basic principles and requirements necessary for the software approach. The functions performed separately under software control must be combined in one chip. These functions and requirements for the software approach are summarized below.

1. An input port and an output port of required for interfacing.
2. In data transmission, the MPU converts a parallel word into a stream of serial bits.
3. In data reception, the MPU converts serial bits into a parallel word.
4. Data transfer is synchronized between the MPU and slow-responding peripherals through time delays.

An integrated circuit called USART (Universal Synchronous/Asynchronous Receiver/Transmitter) incorporates all the features described above on the chip, as well as many more sophisticated features used for serial data communication. It is a programmable device; i.e., its functions and specifications for serial I/O can be determined by writing instructions in its internal registers. The Intel 825IA USART is device widely used in serial I/O; the device and its applications are described in the next section.

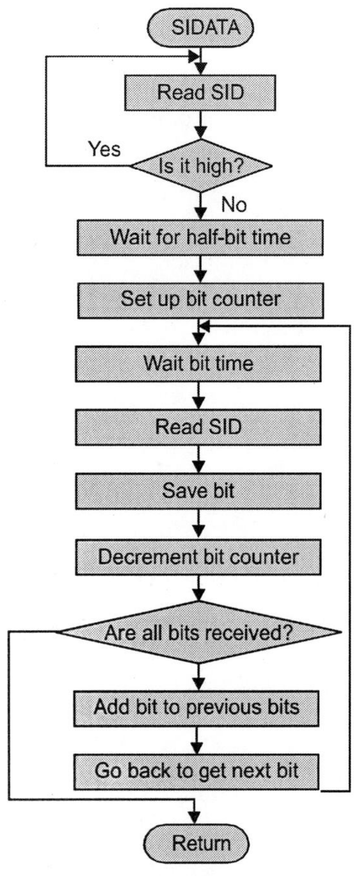


**Figure 13.1:** Flowchart of Data Reception using SID Line

## 13.2 THE 8251A PROGRAMMABLE COMMUNICATION INTERFACE

The 82C51A is a USART (Universal Synchronous Asynchronous Receiver Transmitter) for serial data communication.

As a peripheral device of a microcomputer system, the 82C51A receives parallel data from the CPU and transmits serial data after conversion. This device also receives serial data from the outside and transmits parallel data to the CPU after conversion. The 82C51A configures a fully static circuit using silicon gate CMOS technology. Therefore, it operates on extremely low power at 100 mA (max) of standby current by suspending all operations.

### 13.2.1 FEATURES OF 8251A

1. Wide power supply voltage range from 3 V to 6 V
2. Wide temperature range from –40°C to 85°C
3. Synchronous communication upto 64 Kbaud
4. Asynchronous communication upto 38.4 Kbaud
5. Transmitting/receiving operations under double buffered configuration.
6. Error detection (parity, overrun and framing)
7. 28-pin Plastic DIP (DIP28-P-600-2.54): (Product name: MSM82C51A-2RS)
8. 28-pin Plastic QFJ (QFJ28-P-S450-1.27): (Product name: MSM82C51A-2JS)
9. 32-pin Plastic SSOP(SSOP32-P-430-1.00-K): (Product name: MSM82C51A-2GS-K)

The 825IA is a programmable chip designed for synchronous and asynchronous serial data communication, packaged in a 28-pin DIP. The 825I A is the enhanced version of its predecessor, the 8251, and is compatible with the 8251. Figure 13.2 shows the block diagram of the 8251A. It includes five sections. Read/write Control logic. Transmitter, Receiver, Data Bus Buffer, and Modem Control.

The control logic interfaces the chip with the MPU, determines the functions of the chip according to the control word in its register (to be explained below), and monitors the data flow. The transmitter section converts a parallel word received from the MPU into serial bits and transmits them over the TxD line to a peripheral. The receiver section receives serial bits from a peripheral, converts a parallel word, and transfers the word to the MPU. The modem control is used to establish data communication through modems over telephone lines. The 8251A is a complex device, capable of performing various functions. For the sake of clarity, this chapter focuses only on the asynchronous mode of serial I/O and excludes any discussion of the synchronous mode and the modem control. The asynchronous mode is ofter used for data communication between the MPU and serial peripherals such as terminals and floppy disks.

Figure 13.3 shows an expanded version of the 8251A block diagram. The block diagram shows all the elements of a programmable chip; it includes the interfacing signals, the control register, and the status register. The functions of various blocks are described below.

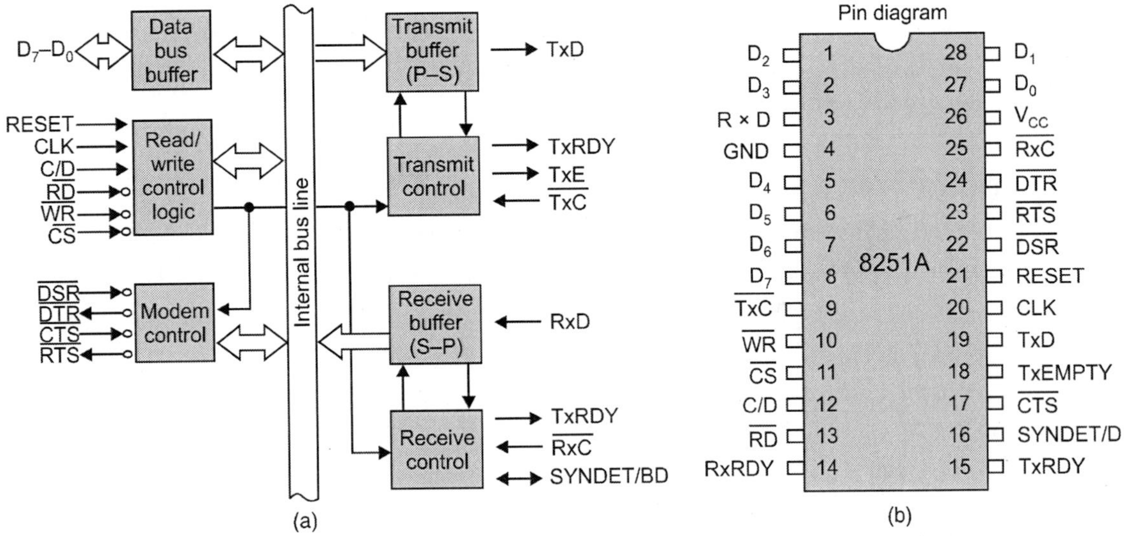


**Figure 13.2:** The 8251A: (a) Block Diagram, (b) Pin Configuration and Description

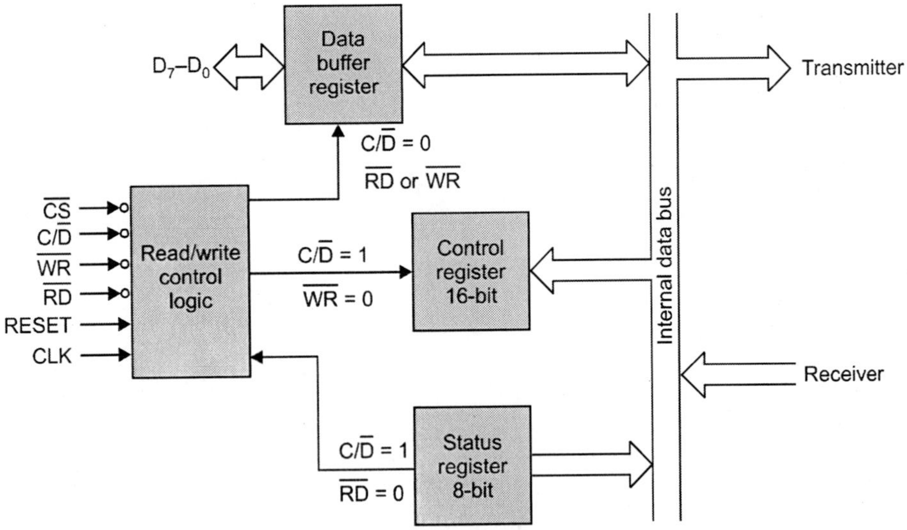


**Figure 13.3:** The 8251A: Expanded Block Diagram of Control Logic and Register

### 13.2.2 Read/Write Control Logic and Register

This section includes R/W control logic, six input signals, control logic, and three buffer registers: data register, control register and status register. The input signals to the control logic are as follows.

### 13.2.3 Input Signals

- **$\overline{\text{CS}}$-Chip Select:** When this signal goes low, the 8251A is selected by the MPU for communication. This is usually connected to a decoded address bus.

- $\overline{CD}$ **-Control/Data:** When this signal is high, the control register or the status register is addressed; when it is low, the data buffer is addressed. The control register and the status register are differentiated by $\overline{WR}$ and $\overline{RD}$ signals, respectively.
- $\overline{WR}$ **-Write:** When this signal goes low, the MPU either writes in the control register or sends output to the data buffer. This is connected to $\overline{IOW}$ or $\overline{MEMW}$
- $\overline{RD}$ **-Read:** When this signal goes low, the MPU either reads a status from the status register or accepts (inputs) data from the data buffer. This is connected to either $\overline{IOR}$ or $\overline{MEMR}$.
- **RESET-Reset:** A high on this input resets the 8251A and forces it into the idle mode.
- **CLK-Clock:** This is the clock input, usually connected to the system clock. This clock does not control either the transmission or the reception rate. The clock is necessary for communication with the microprocessor.

The 82C51A functional configuration is programmed by software. Operation between the 82C51A and a CPU is executed by program control. Table 13.1 shows the operation between a CPU and the device.

**Table 13.1: Operation between 82C51A and CPU**

| $\overline{CS}$ | $\overline{CD}$ | $\overline{RD}$ | $\overline{WR}$ | |
|---|---|---|---|---|
| 1 | X | X | X | Data Bus 3-State |
| 0 | X | 1 | 1 | Data Bus 3-State |
| 0 | 1 | 0 | 1 | Status → CPU |
| 0 | 1 | 1 | 0 | Control Word ← CPU |
| 0 | 0 | 0 | 1 | Data → CPU |
| 0 | 0 | 1 | 0 | Data ← CPU |

It is necessary to execute a function-setting sequence after resetting the 82C51A. Figure 13.4 shows the function-setting sequence.

If the function was set, the device is ready to receive a command, thus enabling the transfer of data by setting a necessary command, reading a status and reading/writing data.

**Control Register:** This 16-bit register for a control word consists of two independent bytes: the first byte is called the mode instruction (word) and the second byte is called the command instruction (word). This register can be accessed as an output port when the $\overline{CD}$ pin is high.

**Status Register:** This input register checks the ready status of a peripheral. This register is addressed as an input port when the $\overline{CD}$ pin is high; it has the same port address as the control register.

**Data Buffer:** This bidirectional register can be addressed as an input port and an output port when the C/pin is low. Table 13.2 summarizes all the interfacing and control signals.

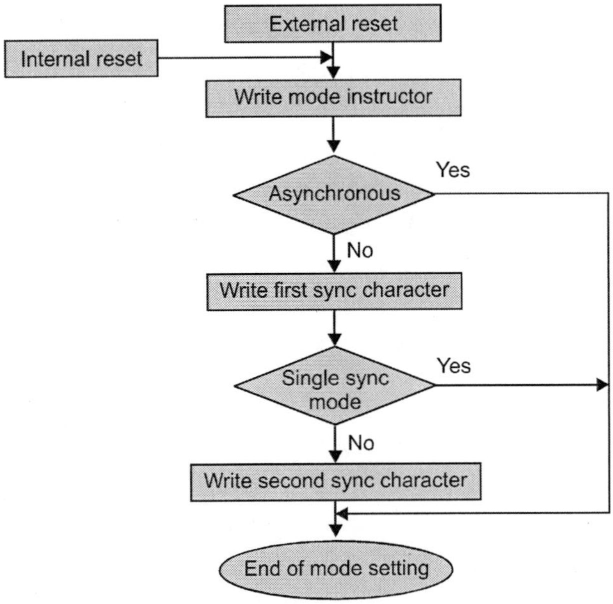


**Figure 13.4:** Function-setting Sequence (Mode Instruction Sequence)

### 13.2.4 Transmitter Section

The transmitter accepts parallel data from the MPU and converts them into serial data. It has two registers: a buffer register to hold eight bits and an output register to convert eight bits into a stream of serial bits (Figure 13.5). The MPU writes a byte in the buffer register; whenever the output register is empty, the contents of the buffer register are transferred to the output register. This section transmits data on the TxD pin with the appropriate framing bits (Start and Stop). Three output signals and one input signal are associated with the transmitter section.

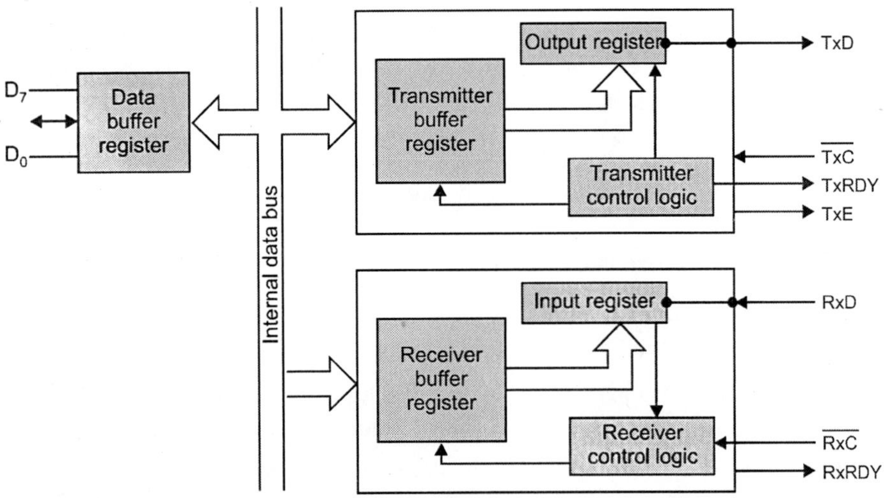


**Figure 13.5:** The 8251A: Expanded Block Diagram of Transmitter and Receiver Sections

- **TxD-Transmit Data:** Serial bits are transmitted on this line.
- **$\overline{TxC}$ -Transmitter Clock:** This input signal control the rate at which bits are transmitted by the USART. The clock frequency can be 1, or 64 times the baud.
- **TxRDY-Transmitter Ready:** This is an output signal. When it is high, it indicates that the buffer register is empty and the USART is ready to accept a byte. It can be used either to interrupt the MPU or to indicate the status. This signal is reset when a data byte is loaded into the buffer.
- **TxE-Transmitter Empty:** This is an output signal, Logic 1 on this line indicates that the output register is empty. This signal is reset when a byte is transferred from the buffer to the output registers.

### 13.2.5 Receiver Section

The receiver accepts serial data on the RxD line from a peripheral and converts them into parallel data. The section has two registers: the receiver input register and buffer register (Figure 13.5).

When the RxD line goes low, the control logic assumes it is a Start bit, waits for half a bit time, and samples the line again. If the line is still low, the input register accepts the following bits, forms a character, and loads it into the buffer register. Subsequently, the parallel byte is transferred to the MPU when requested. In the asynchronous mode, two parallel byte is transferred to the MPU when requested. In the asynchronous mode, two input signals and one output signal are necessary, as described below.

- **RxD-Receive Data:** Bits are received serially on this line and converted into a parallel byte in the receiver input register.
- **$\overline{RxC}$ -Receiver Clock:** This is a clock signal that controls the rate at which bits are received by the USART. In the asynchronous mode, the clock can be set to 1, 16, or 64 times the baud.
- **RxRDY-Receiver Ready:** This is an output signal. It goes high when the USART has a character in the buffer register and is ready to transfer it to the MPU. This line can be used either to indicate the status or to interrupt the MPU.

## 13.3 PIN DIAGRAM OF 8251USART

### Pin Description of 8251USART

1. **D0 to D7 (I/O terminal):** This is bidirectional data bus which receive control words and transmits data from the CPU and sends status words and received data to CPU.
2. **RESET (Input terminal):** A "High" on this input forces the MSM82C51A-2 into "reset status." The device waits for the writing of "mode instruction." The min. reset width is six clock inputs during the operating status of CLK.

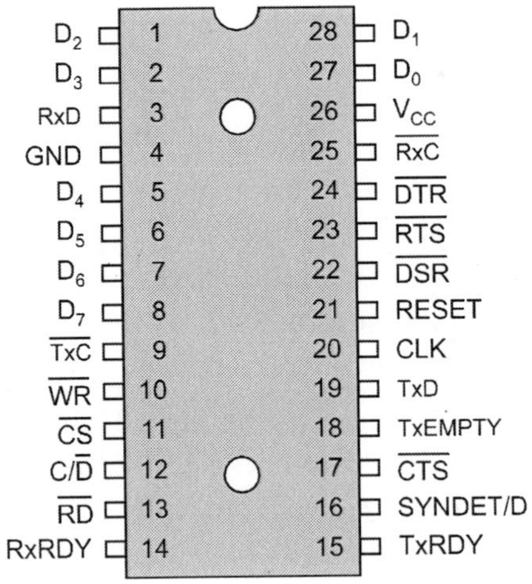


**Figure 13.6:** pin diagram of 8251 USART

3. **CLK (Input terminal):** CLK signal is used to generate internal device timing. CLK signal is independent of $\overline{RxC}$ or $\overline{TxC}$. However, the frequency of CLK must be greater than 30 times the $\overline{RxC}$ and $\overline{TxC}$ at Synchronous mode and Asynchronous "x1" mode, and must be greater than 5 times at Asynchronous "x16" and "x64" mode.
4. **$\overline{WR}$ (Input terminal):** This is the "active low" input terminal which receives a signal for writing transmit data and control words from the CPU into the MSM82C51A-2.
5. **$\overline{RD}$ (Input terminal):** This is the "active low" input terminal which receives a signal for reading receive data and status words from the MSM82C51A-2.
6. **$C\overline{D}$ (Input terminal):** This is an input terminal which receives a signal for selecting data or command words and status words when the MSM82C51A-2 is accessed by the CPU.

   If $C\overline{D}$ = low, data will be accessed.

   If $C\overline{D}$ = high, command word or status word will be accessed.
7. **$\overline{CS}$ (Input terminal):** This is the "active low" input terminal which selects the MSM82C51A-2 at low level when the CPU accesses.

   **Note:** The device won't be in "standby status"; only setting $\overline{CS}$ = High.
8. **TXD (output terminal):** This is an output terminal for transmitting data from which serial-converted data is sent out. The device is in "mark status" (high level) after resetting or during a status when transmit is disabled. It is also possible to set the device in "break status" (low level) by a command.
9. **TXRDY (output terminal):** This is an output terminal which indicates that the MSM82C51A-2 is ready to accept a transmitted data character. But the terminal is always at low level if $\overline{CTS}$ = high or the device was set in "TX disable status" by a command.

**Note:** TXRDY status word indicates that transmit data character is receivable, Regardless of $\overline{CTS}$ or command.

If the CPU writes a data character, TXRDY will be reset by the leading edge or $\overline{WR}$ signal.

10. **TXEMPTY (Output terminal):** This is an output terminal which indicates that the MSM82C51A-2 has transmitted all the characters and had no data character. In "synchronous mode," the terminal is at high level, if transmit data characters are no longer remaining and sync characters are automatically transmitted. If the CPU writes a data character, TXEMPTY will be reset by the leading edge of $\overline{WR}$ signal.

    **Note:** As the transmitter is disabled by setting "High" or command, data written before disable will be sent out. Then TXD and TXEMPTY will be "High". Even if a data is written after disable, that data is not sent out and TXE will be "High".After the transmitter is enabled, it sent out.

11. **$\overline{TXC}$ (Input terminal):** This is a clock input signal which determines the transfer speed of transmitted data. In "synchronous mode," the baud rate will be the same as the frequency of $\overline{TXC}$. In "asynchronous mode", it is possible to select the baud rate factor by mode instruction. It can be 1, 1/16 or 1/64 the $\overline{TXC}$. The falling edge of $\overline{TXC}$ sifts the serial data out of the MSM82C51A-2.
12. **RXD (input terminal):** This is a terminal which receives serial data.
13. **RXRDY (Output terminal):** This is a terminal which indicates that the MSM82C51A-2 contains a character that is ready to READ. If the CPU reads a data character, RXRDY will be reset by the leading edge of $\overline{RD}$ signal. Unless the CPU reads a data character before the next one is received completely, the preceding data will be lost. In such a case, an overrun error flag status word will be set.
14. **$\overline{RXC}$ (Input terminal):** This is a clock input signal which determines the transfer speed of received data. In "synchronous mode," the baud rate is the same as the frequency of $\overline{RXC}$. In "asynchronous mode," it is possible to select the baud rate factor by mode instruction. It can be 1, 1/16, 1/64 the $\overline{RXC}$.
15. **SYNDET/BD (Input or output terminal):** This is a terminal whose function changes according to mode. In "internal synchronous mode." this terminal is at high level, if sync characters are received and synchronized. If a status word is read, the terminal will be reset. In "external synchronous mode, "this is an input terminal. A "High" on this input forces the MSM82C51A-2 to start receiving data characters. In "asynchronous mode," this is an output terminal which generates "high level" output upon the detection of a "break" character if receiver data contains a "low-level" space between the stop bits of two continuous characters. The terminal will be reset, if RXD is at high level. After Reset is active, the terminal will be output at low level.
16. **$\overline{DSR}$ (Input terminal):** This is an input port for MODEM interface. The input status of the terminal can be recognized by the CPU reading status words.

17. $\overline{\textbf{DTR}}$ **(Output terminal):** This is an output port for MODEM interface. It is possible to set the status of $\overline{\text{DTR}}$ by a command.

18. $\overline{\textbf{CTS}}$ **(Input terminal):** This is an input terminal for MODEM interface which is used for controlling a transmit circuit. The terminal controls data transmission if the device is set in "TX Enable" status by a command. Data is transmitable if the terminal is at low level.

19. $\overline{\textbf{RTS}}$ **(Output terminal):** This is an output port for MODEM interface. It is possible to set the status $\overline{\text{RTS}}$ by a command.

## 13.4 PROGRAMMING THE 8251A

The complete functional definition of the 8251A is programmed by the system's software. A set of con-trol words must be sent out by the CPU to initialize the 8251A to support the desired communications format. These control words will program the: Baud Rate, Character Length, Number of Stop Bits, Synchronous or Asynchronous Op-Eration, Even/Odd/Off Parity, etc. In the Synchronous Mode, options are also provided to se-lect either internal or external character synchroni-zation.

Once programmed, the 8251 A is ready to perform its communication functions. The TxRDY output is raised "high" to signal the CPU that the 8251A is ready to receive a data character from the CPU. This output (TxRDY) is reset automatically when the CPU writes a character into the 8251A. On the other hand, the 8251A receives serial data from the MO-DEM or I/O device. Upon receiving an entire charac-ter, the RxRDY output is raised "high" to signal the CPU that the 8251A has a complete character ready for the CPU to fetch. RxRDY is reset automatically upon the CPU data read operation.

The 8251A cannot begin transmission until the Tx Enable (Transmitter Enable) bit is set in the Com-mand Instruction and it has received a Clear To Send ($\overline{\text{CTS}}$) input. The TxD output will be held in the marking state upon Reset.

Prior to starting data transmission or reception, the 8251A must be loaded with a set of control words generated by the CPU. These control signals define the complete functional definition of the 8251A and must immediately follow a Reset operation (internal or external).

### 13.4.1 Control Words

There are two types of control word:

1. Mode instruction (setting of function)
2. Command (setting of operation)

#### *1. Mode Instruction*

Mode instruction is used for setting the function of the MSM82C51A-2. Mode instruction will be in "wait for write" at either internal reset or external reset. That is, the writing of a control word after resetting will be recognized as a "mode instruction." Items set by mode instruction are as follows:

- Synchronous/asynchronous mode
- Stop bit length (asynchronous mode)
- Character length
- Parity bit
- Baud rate factor (asynchronous mode)
- Internal/external synchronization (synchronous mode)
- Number of synchronous characters (Synchronous mode)

The bit configuration of mode instruction is shown in Figures 13.7 and 13.8. In the case of synchronous mode, it is necessary to write one-or two byte sync characters.

If sync characters were written, a function will be set because the writing of sync characters constitutes part of mode instruction.

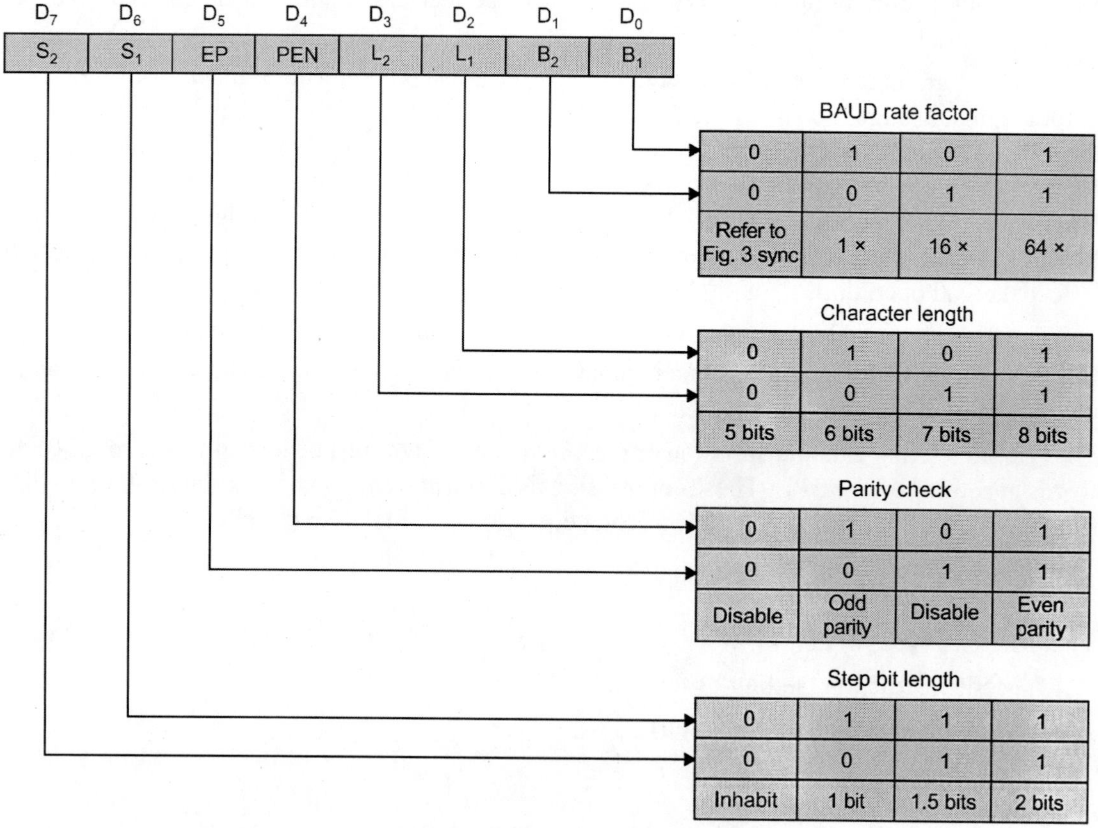


**Figure 13.7:** Bit Configuration of Mode Instruction (Asynchronous)

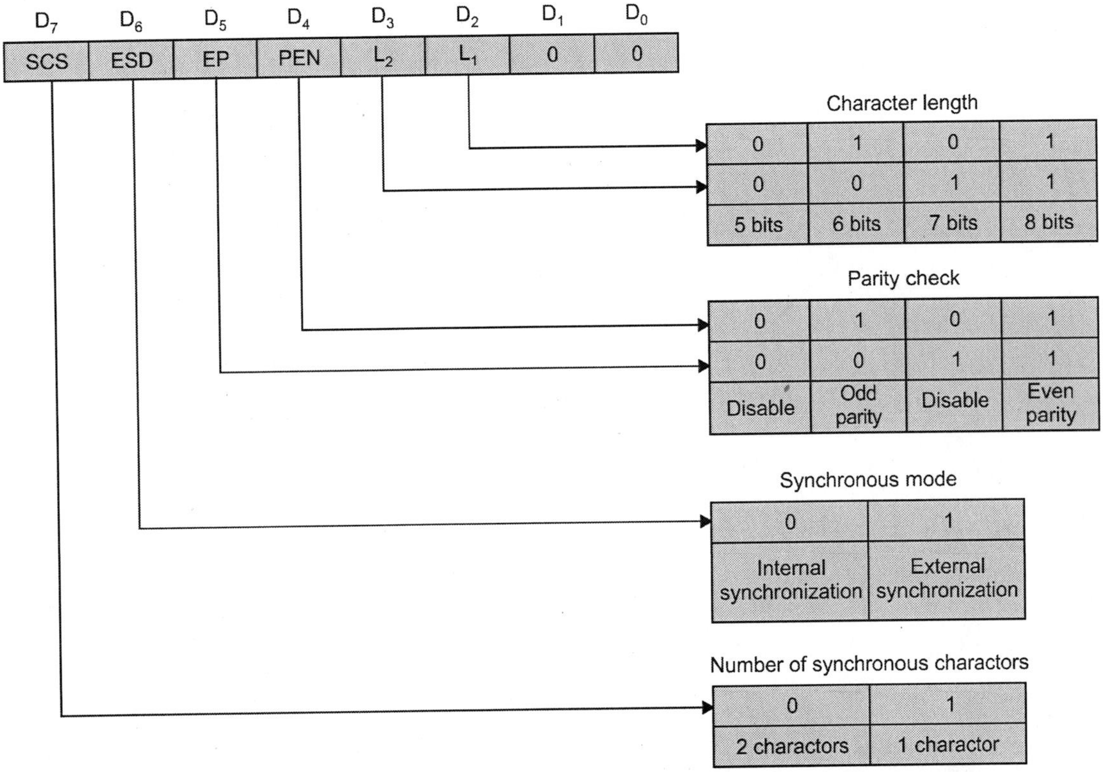


**Figure 13.8:** Bit Configuration of Mode Instruction (Synchronous)

## 2. *COMMAND*

Command is used for setting the operation of the MSM82C51A-2. It is possible to write a command whenever necessary after writing a mode instruction and sync characters. Items to be set by command are as follows:

- Transmit Enable/Disable
- Receive Enable/Disable
- $\overline{\text{DTR}}$, $\overline{\text{RTS}}$ Output of data.
- Resetting of error flag.
- Sending to break characters
- Internal resetting
- Hunt mode (synchronous mode)

The bit configuration of a command is shown in Figure 13.9.

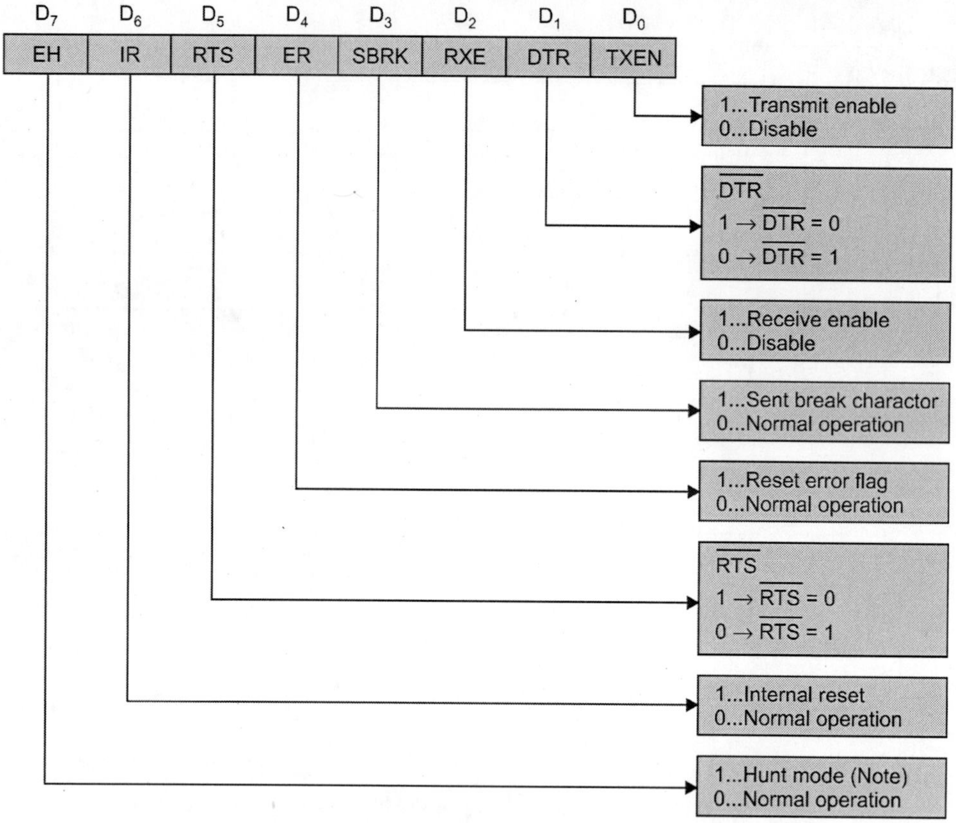


**Figure 13.9:** Bit Configuration of Command

### 13.4.2 Status Word

It is possible to see the internal status of MSM82C51A-2 by reading a status word. The bit configuration of status word is shown in Figure 13.10.

#### *1. STANDBY STATUS*

It is possible to put the MSM82C51A-2 in "standby status" When the following conditions have been satisfied the MSM82C51A-2 is in "standby status."

1. Terminal is fixed at Vcc level.
2. Input pins other , D0 to D7, , and C/ are fixed at Vcc or GND level (including SYNDET in external synchronous mode).

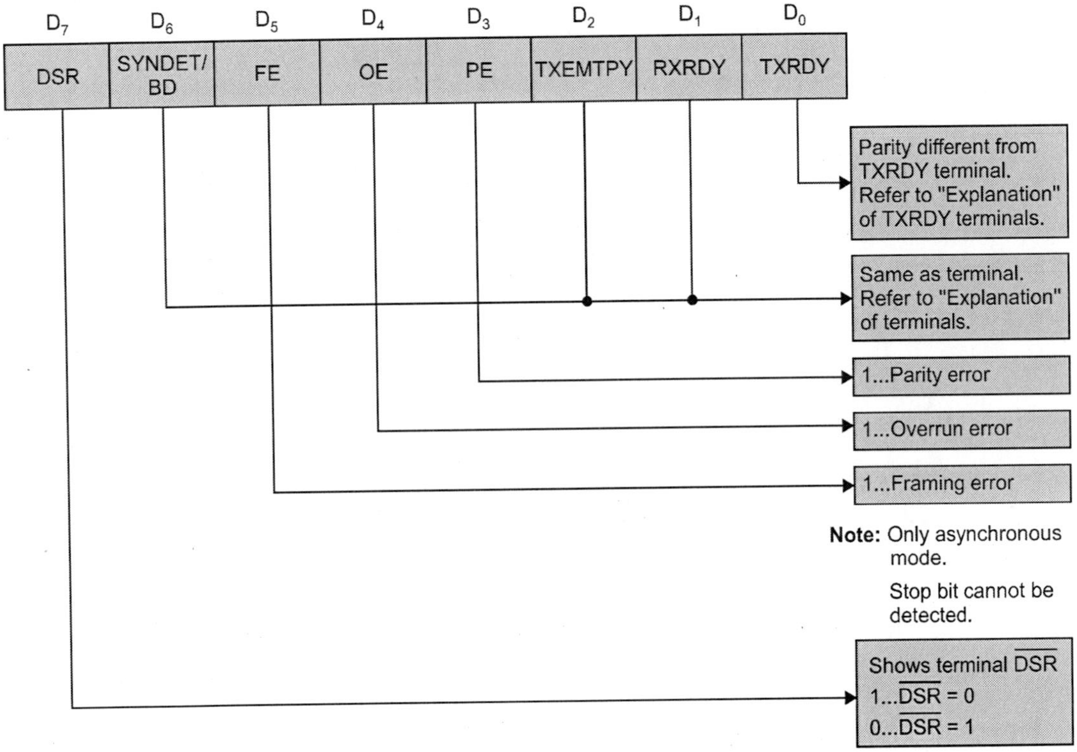


**Figure 13.10:** Bit Configuration of Status Word

## 13.5 INITIALIZING THE 8251A

To implement serial communication, the MPU must inform the 8251A of all details, such as mode, baud, Stop bits, parity, etc. Therefore, prior to data transfer, a set of control words must be loaded into the 16-bit control register of the 8251A. In addition, the MPU must check the readiness of a peripheral by reading the status register. The control words are divided into tow formats: mode words and command words. This mode word specifies the general characteristics of operation (such as baud, parity, number of Stop bits), the command word enables data transmission and/or reception, and the status word provides the information concerning register status and transmission errors. Figure shows the definitions of these words.

To initialize the 8251A in the asynchronous mode, a certain sequence of control words must be followed. After a Reset operation (system Reset or through instruction), a mode word must be written in the control register followed by a command word. Any control word written into the control register immediately after a mode word will be interpreted as a command word; that means a command word can be changed anytime during the operation. However, the 8251A should be reset prior to writing a new mode word, and it can be reset by using the Internal Reset bit ($D_6$) in the command word.

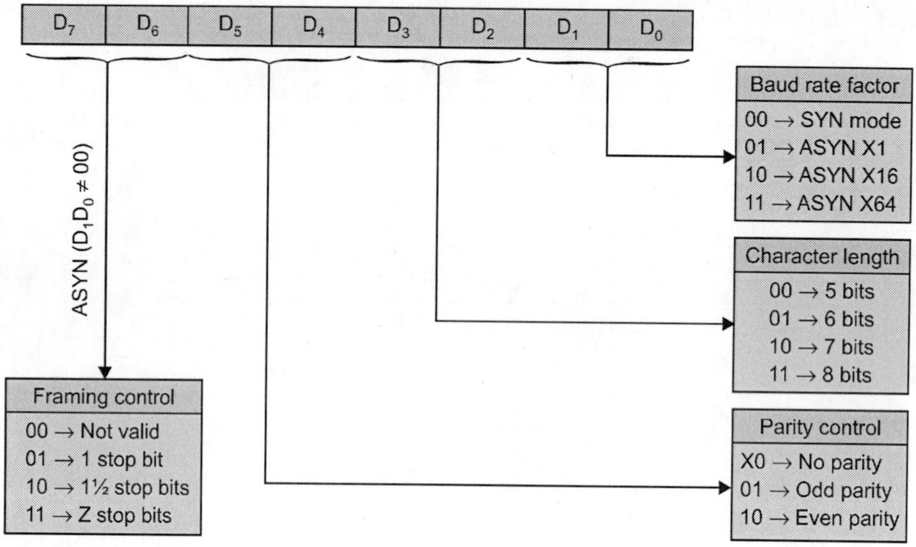


(a)

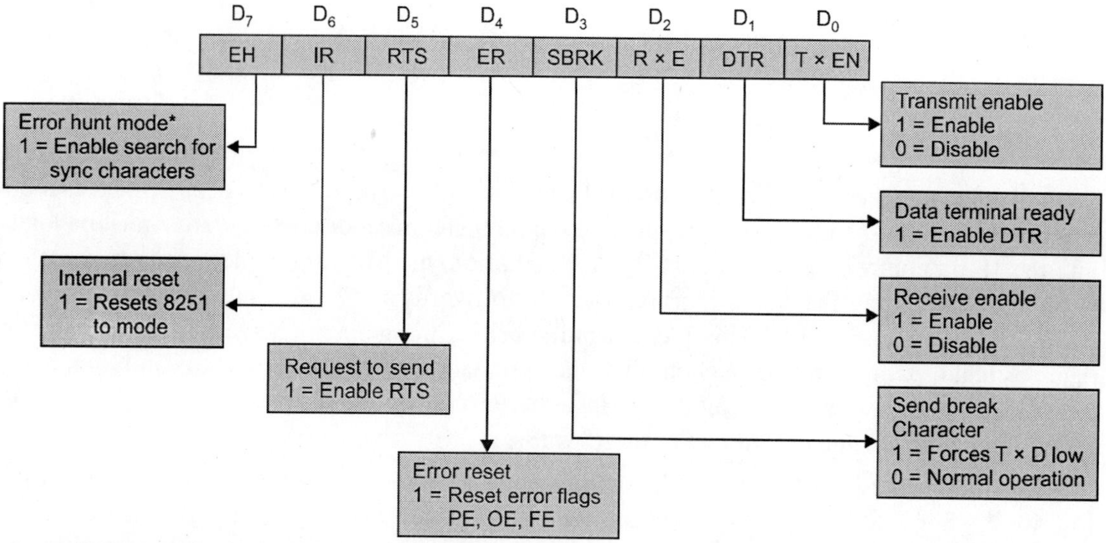


*(Has no effect in async mode)

**Note:** Error reset must be performed whenever R × Enable and enter hunt are programmed.

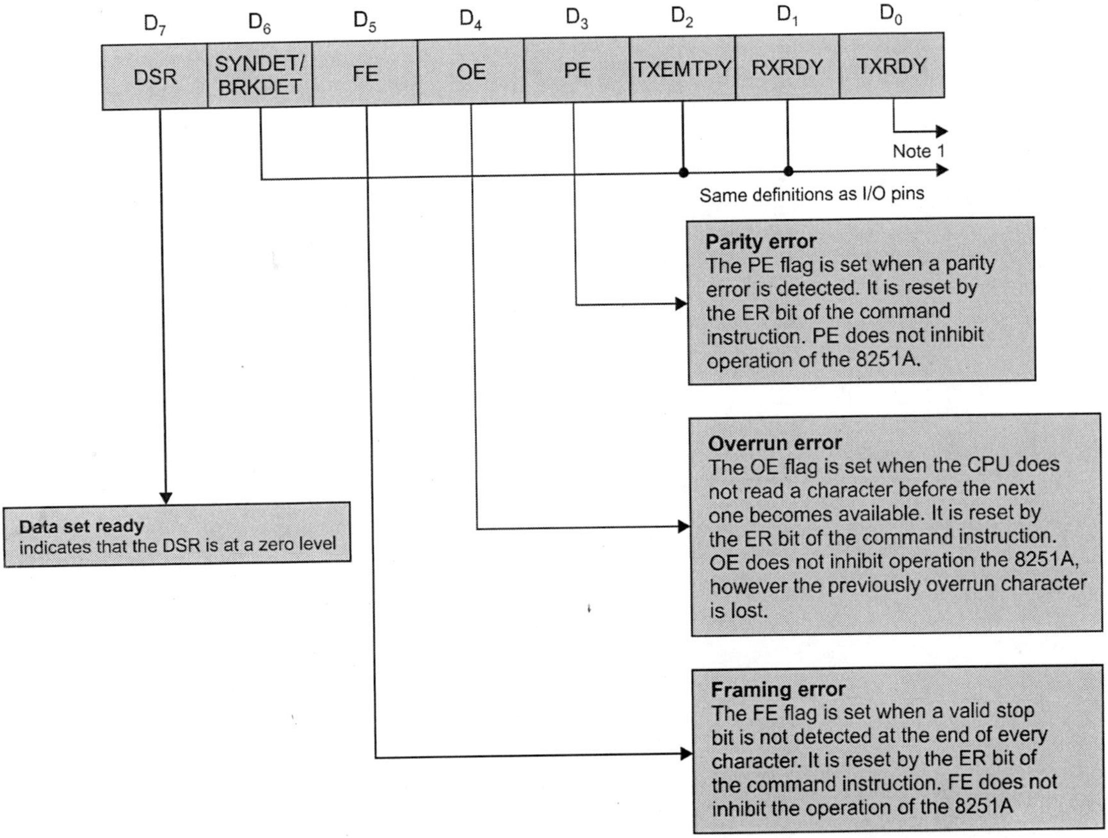


**Figure 13.11:** (a) Mode Word Format (b) Command Word Format and (c) Status Word Format

## 13.6 ILLUSTRATION: INTERFACING AN RS-232 TERMINAL USING THE 8251A

### Problem Statement

1. Identify the port addresses of the control register, the status register, and the data register in Figure 13.12
2. Explain the RS-232 signals and the operations of the line driver (MC 1488) and the line receiver (MC) 1489) shown in Figure 13.12
3. Specify the initialization instructions and the status word to transmit characters with the following parameters if the transmitter clock frequency (TxC) is 153.6 kHz.
   - Asynchronous mode with 9600 baud
   - Character length = seven bits and two Stop bits
   - No parity check
4. Write instructions to initialize the 8251A, to read the status word, and set up a loop unitl the transmitter (TxRDY) is ready.

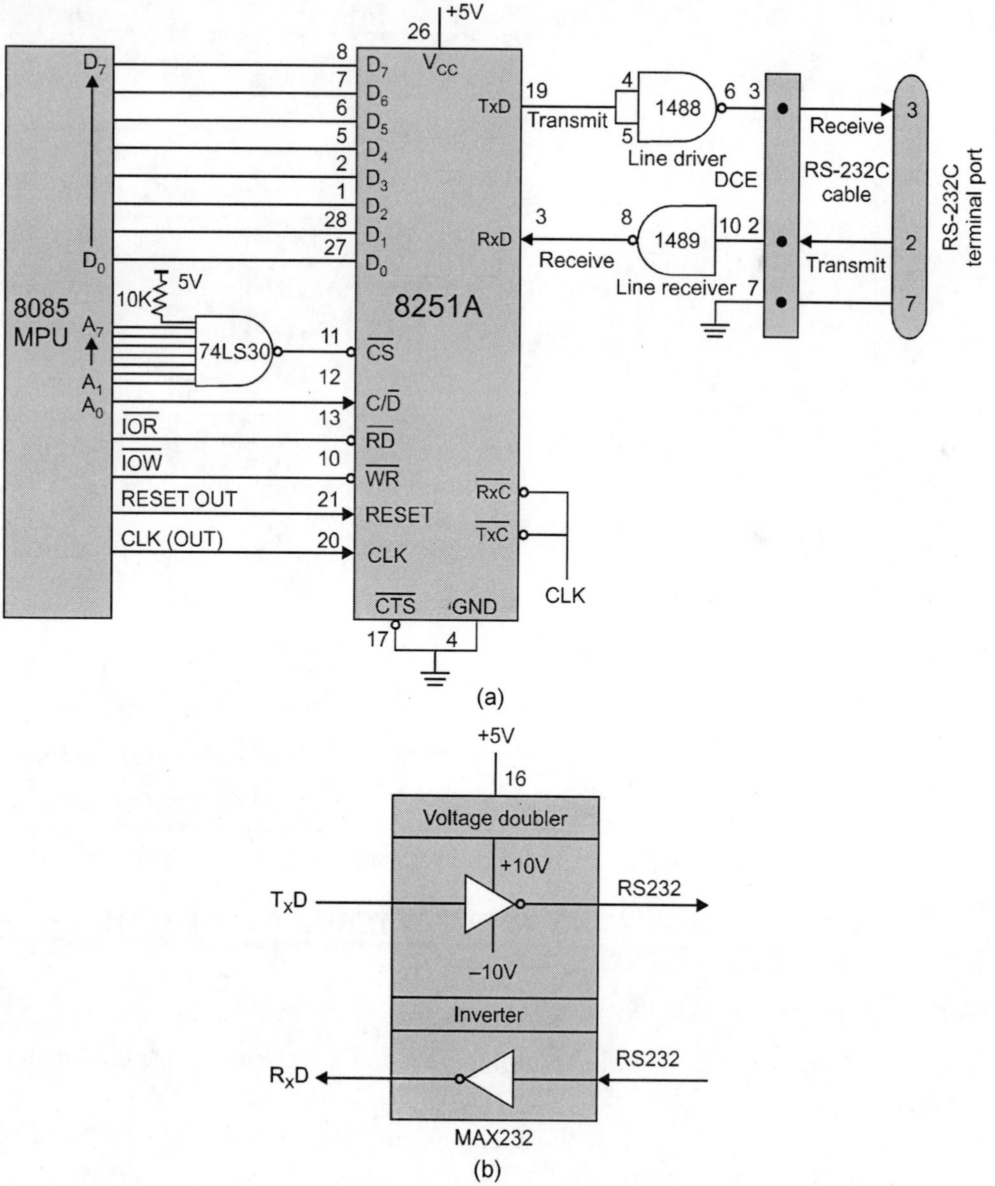


**Figure 13.12** (a) Schematic of Interfacing an RS-232 Terminal with an 8085 System Using the 8215A and (b) AX 232 Logic Diagram

### *1. Port Addresses*

(a) The Chip Select line of the 8251A is enabled when the address lines $A_7$ through $A_1$ are at logic 1. To select the control register or the status register, the C/line should be high, which means that address line $A_0$ should be 1. Therefore, the port address of the control register and the status register = FFH.

The control register is an output port and the status register is an input port; they are identified by $\overline{WR}$ and $\overline{RD}$ signals, even if their port addresses are the same.

(a) The data register is selected when the $C\overline{D}$ line goes low; thus, $A_0$ should be low. The port address of the data register = FEH. The register is bidirectional, and the same address is used to receive or transmit data. The input and output functions are identified by $\overline{RD}$ and $\overline{WR}$ signals.

## 2. *RS-232C Signals, Line Drivers, and Line Receivers*

Figure 13.12 shows that three RS-232 signals–TxD, RxD and Ground – are being used for serial communication between the CRT terminal and the 8085 system. The terminal transmits data on pin 2 and receives on pin 3; on the other hand, the 8085 system receives on pin 2 and transmits on pin 3 using the 8251A. Therefore, the terminal is connected as the DTE and the system plays the role of the DCE; the 8251A is part of the 8085 system.

Data transmitted over the TxD line (pin 19 of the 8251A) are at the TTL logic level. These bits are converted to RS-232 voltage levels and negative logic by line driver MC 1488. Data received by the 8251A over the RxD line (pin 3) should be at the TTL logic level. Therefore, the RS-232 signals at pin 2 of the connector are converted to the positive TTL logic levels by line receiver MC 1489. The line driver and receiver are described here briefly.

**Line Driver: MC 1488:** This is a quad line driver that converts TTL input levels to a maximum +15 $V_{DC}$ output signal. Typically, it is used with a ±12 V power supply. For logic 0 input (< 0.8 $V_{DC}$) the output is around +10 V, and for logic 1 input (> +2.4 $V_{DC}$) the output is around -10 V; thus, the positive true logic is converted into negative true logic for RS-232C signals. The internal circuit of the MC 1488 functions much like a comparator. For an input lower than the threshold voltage, the output approaches positive power supply voltage, and for an input higher than the threshold voltage, the output approaches negative power supply voltage.

**Line Receiver: MC 1489:** This is a quad line receiver that converts high voltage signals (+15 V) into TTL logic levels. Output voltages usually range from 0.2 V (low) to 4.0 V (high). The internal circuit functions as an on/off transistor. When the transistor base has a negative input voltage, the transistor is turned off and the collector voltage (the output of the MC 1489) is high. When the transistor base has a positive input voltage, the transistor is driven into saturation to 0.2 V.

**RS-232 Drivers/Receivers:** MAX 232 One of the drawbacks of the line driver MC 1488 is that it requires additional voltage ±12 vv from a power supply. Typically, such voltages are unavailable in systems compatible with the TTL logic. This difficulty can be resolved by using a specialized integrated device, MAX 232 . It includes drivers and receivers (two each), and it generates ± 10V internally by using a voltage doubler abd inverter circuits. The MAX 232 can replace the 1488 and 1489 and climinate the need for ± 12 v power supplies.

## 3. *Initialization*

The control words necessary for the given specifications are as follows:

- **Mode Word (Figure 13.11(a)):** In a mode word, bits $D_1$ and $D_0$ can specify a baud factor that divides the clock frequency (TxC) to provide three different transmission rates. In this illustration, TxC is sixteen times the specified baud.
- **Command Word (Figure 13.11(b)):** In this command word, bit $D_0$ enables the transmitter, bit $D_4$ ignores any errors, and bit $D_6$ prevents reset of the 825IA; all other bits are don't care. In this illustration, bit $D_4$ also can assume don't care logic level.
- **Status word (Figure 13.11(c)):** The MPU should check bit $D_0$ before transferring a character to the 8251A; it indicates the status of the pin TxRDY. When a byte is transferred from the transmitter buffer to the output register, bit $D_0$ is set to 1, and it is reset when the MPU loads the next byte in the buffer. Bit $D_2$ (TxEMPTY) indicates the status of the output register; this bit usually is not used except in applications such as half duplex mode.

### 4. *Initialization Instructions*

```
SETUP:  MVI A,CAH          ;Load the mode word
        OUR FFH            ;Write mode word in control register
        MVI A, 11H         ;Load the command word to enable transmitter
        OUT FFH            ;Enable the transmitter
        STATUS: IN FFH     ;Read status register
        ANI 01H            ;Mask all bits except D0
        JZ STATUS          ;If D0 = 0, the transmitter buffer is full; go back
                           ; and wait
```

Figure 13.13 shows the contents of various registers after the initialization. Mode word CAH and command word 11H are loaded in the control register by the OUT FFH instructions. The MPU checks the transmitter status by reading the status register and examining bit $D_0$ (Transmitter Ready).

When a character is transferred from the buffer register to the output (parallel-to-serial) register, bit $D_0$ is set to 1, indicating to the MPU that the transmitter is ready to accept the next character.

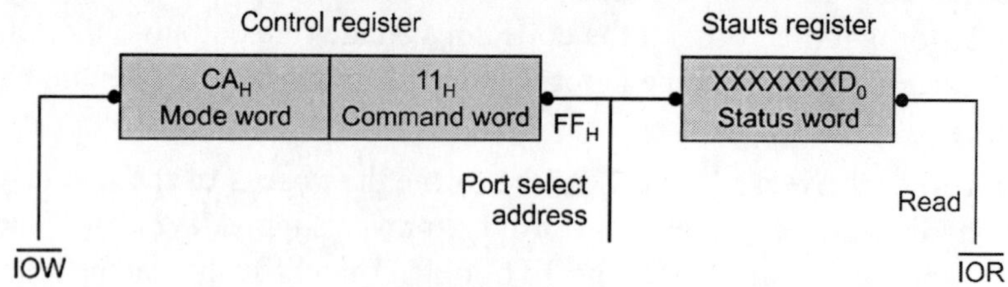


**Figure 13.13:** Control registers contents after the initialization

## QUESTION BANK

Q1. What do you mean by 8251A programmable communication interface?

Q2. Explain the block diagram of 8251A programmable communication interface.

Q3. Explain the pin diagram and description of 8251A programmable communication interface.

Q4. Explain the programming of 8251A.

Q5. Explain the control words of 8251A.

Q6. Explain the initialization of the 8251A.

Q7. Explain the interfacing of an RS-232 terminal using the 8251A.

CHAPTER 14

# RS232C

## 14.1 SYNCHRONOUS AND ASYNCHRONOUS COMMUNICATIONS

There are two basic types of serial communications, synchronous and asynchronous. With synchronous communications, the two devices initially synchronize themselves to each other, and then continually send characters to stay in sync. Even when data is not really being sent, a constant flow of bits allows each device to know where the other is at any given time. That is, each character that is sent is either actual data or an idle character. Synchronous communications allows faster data transfer rates than asynchronous methods, because additional bits to mark the beginning and end of each data byte are not required. The serial ports on IBM-style PCs are asynchronous devices and therefore only support asynchronous serial communications.

Asynchronous means "no synchronization", and thus does not require sending and receiving idle characters. However, the beginning and end of each byte of data must be identified by start and stop bits. The start bit indicates when the data byte is about to begin and the stop bit signals when it ends. The requirement to send these additional two bits causes asynchronous communication to be slightly slower than synchronous however it has the advantage that the processor does not have to deal with the additional idle characters.

An asynchronous line that is idle is identified with a value of 1 (also called a mark state). By using this value to indicate that no data is currently being sent, the devices are able to distinguish between an idle state and a disconnected line. When a character is about to be transmitted, a start bit is sent. A start bit has a value of 0 (also called a space state). Thus, when the line switches from a value of 1 to a value of 0, the receiver is alerted that a data character is about to be sent.

## 14.2 STANDARDS IN SERIAL I/O

The serial I/O technique is commonly used to interface terminals, printers, and modems. These peripherals and computers are designed and manufactured by various manufacturers. Therefore, a common understanding must exist, among various manufacturing and user groups, which can ensure compatibility among different equipment. When this understanding is defined and generally accepted in industry (and by users), it is known as a standard. A standard is normally defined by a professional organization (such as IEEE-Institute of Electrical and Electronics Engineers); however, occasionally, a widespread practice can become a de facto standard. A standard may include such items as assignment of pin positions for signals, voltage levels, speed of data transfer, length of cables, and mechanical specifications.

In serial I/O, data can be transmitted as either current or voltage. Typically, 20 mA (or 60 mA) current loops are used in teletype equipment. When a teletype is marking or at logic 1, current flows; when it is at logic 0 (or Space), the current flow is interrupted. The advantage of the current loop method is that signals are relatively noise-free and are suitable for transmission over a distance.

When data are transmitted as voltage, the commonly used standard is known as RS-232C. It is defined in reference to Data Terminal Equipment (DTE) and Data Communication Equipment (DCE) – terminal and modem – as shown in Figure 14.1(a); however, its voltage levels are not compatible with TTL logic levels. The rate of data transmission in RS-232C is restricted to a maximum of 20 kbaud and a distance of 50 ft. For high-speed data transmission, two new standards – RS-422A and RS – 423A – have been developed in recent years; however, they are not yet widely used.

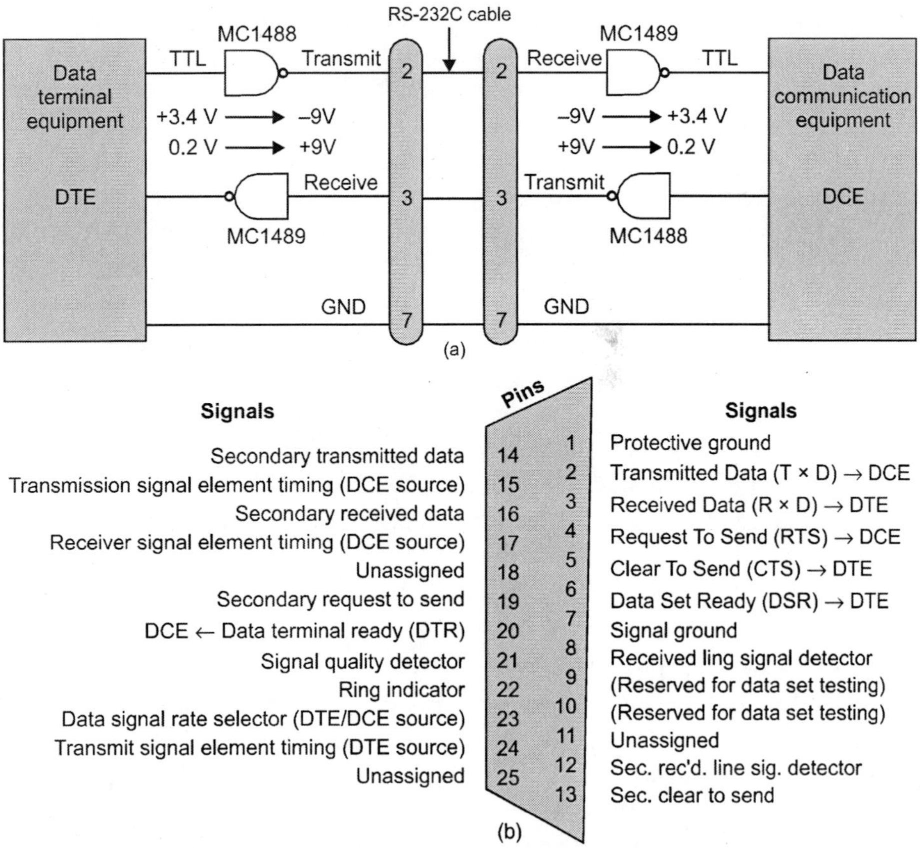


**Figure 14.1:** (a) Minimum Configurations of RS-232C Signals and Voltage Levels and (b) RS-232C Signal Definitions and Pin Assignments.

To appreciate the difficulties and confusion in this standard, one has to examine its historical background. The RS-232 standards was developed during the initial days of computer timesharing, long before the existence of TTL logic, and its primary focus was to have compatibility between a terminal and a modem. However, the same standard is now being used for communications between computers and peripherals, and the roles of a data terminal and a modem have become ambiguous. Should a computer be considered a terminal or a modem? The answer is that it can be either. Therefore, the lines used for transmission and reception will differ, depending on the manufacturer's role-definition of its equipment.

## 14.3 RS-232C

Figure 14.1 (b) shows the RS-232C 25-pin connector and its signals. The signals are divided into four groups: data signals, control signals, timing signals, and grounds. For data lines, the voltage level +3 V to +15 V is defined as logic 0; from -3 V to -15 V is defined as logic 1 (normally, voltage levels are ±12V). This is negative true logic. However, other signals (control and timing) are compatible with the TTL level. Because of incompatibility of the data lines with the TTL logic, voltage translators, called line drivers and line receivers, are required to interface TTL logic with the RS-232 signals, as shown in Figure 14.1(a).The line driver, MC1488, converts logic 1 into approximately -9V and logic 0 into +9 V, as shown in Figure 14.1 (a). Before it is received by the DCE, it is again converted by the line receiver, MC1489, into TTL-compatible logic. This raises the question: If the received signal is to be converted back to the TTL level, what is the reason, in the first place, to convert the transmitted signals to the higher level? The primary reason is that the standard was defined before the TTL levels came into existence; before 1960, most equipment was designed to handle voltages. The other reason is that this standard provides a higher level of noise margin—from -3 V to +3 V.

The minimum interface between a computer and a peripheral requires three lines: pins 2, 3 and 7, as shown in Figure 14.1 (a). These lines are defined in relation to the DTE; the terminal transmits on pin 2 and receives on pin3. On the other hand, the DCE transmits on pin 3 and receives on pin 2. Now the dilemma is: How does a manufacturer define the role of its equipment? For example, the user may connect its microcomputer to a serial printer configured as a DTE. Therefore, to remain compatible with the defined signals of the RS-232C, the RS-232 cable must be reconfigured as shown in Figure 14.2 (b). In Figure 14.2, the microcomputer is defined as a DTE, and it can be connected to the modem, defined as a DCE, without any modification in the RS-232 cable, as shown in Figure 14.2 (a). However, when it is connected to the printer, the transmit and the receives lines must be crossed as shown in Figure 14.2 (b); this is shown as a until-modem connection.

Typically, data transmission with handshake requires eight lines, listed in Table 14.1. Specific functions of handshake lines differ in different peripherals and, therefore should be referred to in the manufacturers' manuals

For high-speed transmission, the standards RS-422A and RS-423A are used. These standards use differential amplifiers to reject noise levels and can transmit data at higher speed with longer cable. The RS-422A allows a maximum speed of 10 kbaud for a 40-ft distance and 100 kbaud for 4000 ft. The RS-423A is limited to 100 kbaud for a 30-ft distance and 10 kbaud for 300 ft. See Table 14.1 for comparison of the three standards: RS-232C, RS-422A, and RS-432A.

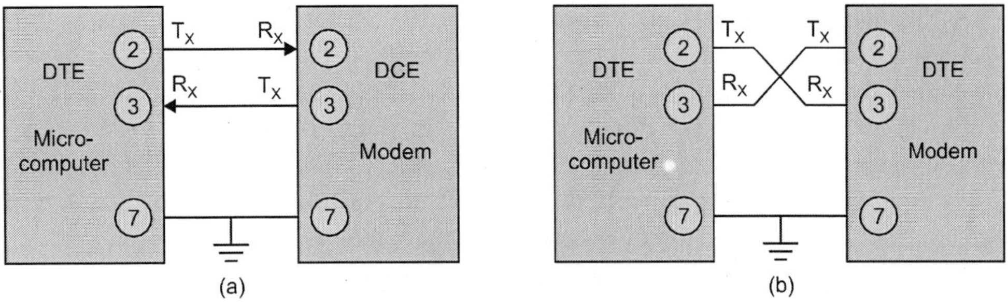


**Figure 14.2:** (a) RS-232 Connections: DTE to DCE and (b) DTE to DTE

| Pin No. | Signals[a] | | Functions |
|---|---|---|---|
| 2 | Transmitted Data | TxD | Output: transmits data from DTE to DCE |
| 3 | Received Data | RxD | Input: DTE receives data from DCE |
| 4 | Request to Send | RTS | General-purpose output from DTE |
| 5 | Clear to Send | CTS | General-purpose input to DTE; can be used as a handshake signal |
| 6 | Data Set Ready | DSR | General-purpose input to dTE; can be used to indicate that DCE is ready |
| 7 | Signal Ground | GND | Common reference between DTE and DCE |
| 8 | Data Carrier Detect | DCD | Generally used by DTE to disable data reception |
| 20 | Data Terminal Ready | DTR | Output; generally used to indicate that DTE is ready |

**Table 14.1:** Comparison of the three standards: RS-232C, RS-422A, and RS-432A.

| Specifications | RS-232C | RS-422A | RS-423A |
|---|---|---|---|
| Speed | 20 kbaud | 10 Mbaud at 40 ft<br>100 kbaud at 4000 ft | 100 kbaud at 30 ft<br>1 kbaud at 4000 ft |
| Distance | 50 ft | 4000 ft | 4000 ft |
| Logic 0 | > +3 to +25 V | B > A[a] | +4 to +6 V |
| Logic 1 | < –3 to –25 V | B < A | –4 to –6 V |
| Receiver Input Voltage | ± 15 V | ± 7 V | ± 12 V |

In telecommunications, RS-232 (Recommended Standard 232) is a standard for serial binary data signals connecting between a DTE (Data Terminal Equipment) and a DCE (Data Circuit-terminating Equipment). It is commonly used in computer serial ports.

RS-232 stands for Recommend Standard number 232 and C is the latest revision of the standard. The serial ports on most computers use a subset of the RS-232C standard. The full RS-232C standard specifies a 25-pin "D" connector of which 22 pins are used. Most of these pins are not needed for normal PC communications, and indeed, most new PCs are equipped with male D type connectors having only 9 pins.

In RS-232, user data is sent as a time-series of bits. Both synchronous and asynchronous transmissions are supported by the standard. In addition to the data circuits, the standard defines a number of control circuits used to manage the connection between the DTE and DCE. Each data or control circuit only operates in one direction that is, signaling from a DTE to the attached DCE or the reverse. Since transmit data and receive data are separate circuits, the interface can operate in a full duplex manner, supporting concurrent data flow in both directions. The standard does not define character framing within the data stream, or character encoding.

The RS232 standard describes a communication method where information is sent bit by bit on a physical channel. The information must be broken up in data words. The length of a data word is variable. On PC's a length between 5 and 8 bits can be selected. This length is the net to information length of each word. For proper transfer additional bits are added for synchronization and error checking purposes. It is important, that the transmitter and receiver use the same number of bits. Otherwise, the data word may be misinterpreted, or not recognized at all.

With synchronous communication, a clock or trigger signal must be present which indicates the beginning of each transfer. The absence of a clock signal makes an asynchronous communication channel cheaper to operate. Fewer lines are necessary in the cable. A disadvantage is that the receiver can start at the wrong moment receiving the information. Resynchronization is then needed which costs time. All data received in the resynchronization period is lost. Another disadvantage is that extra bits are needed in the data stream to indicate the start and end of useful information. These extra bits take up bandwidth.

Data bits are sent with a predefined frequency, the baud rate. Both the transmitter and receiver must be programmed to use the same bit frequency. After the first bit is received, the receiver calculates at which moments the other data bits will be received. It will check the line voltage levels at those moments.

With RS232, the line voltage level can have two states. The on state is also known as mark, the off state as space. No other line states are possible. When the line is idle, it is kept in the mark state.

RS-232 standards (EIA-232) are defined by EIA/TIA (Electronic Industries Alliance / Telecommunications Industry Association). RS-232 defines both the physical and electrical characteristics of the interface. RS-232 is practically identical to ITU V.24 (signal description and names) and V.28 (electrical). RS232 is an Active LOW voltage driven interfaces and operates at +12V to –12V where:

Signal = 0 (LOW) > +3.0V (SPACE)

Signal = 1 (HIGH) < –3.0V (MARK)

## 14.4 DTE (PC) AND DCE (MODEM)

In serial communications the terminal end (PC) is called the Data Terminal Equipment (DTE) and the modem is called the Data Communications Equipment (DCE) as shown in the diagram below.

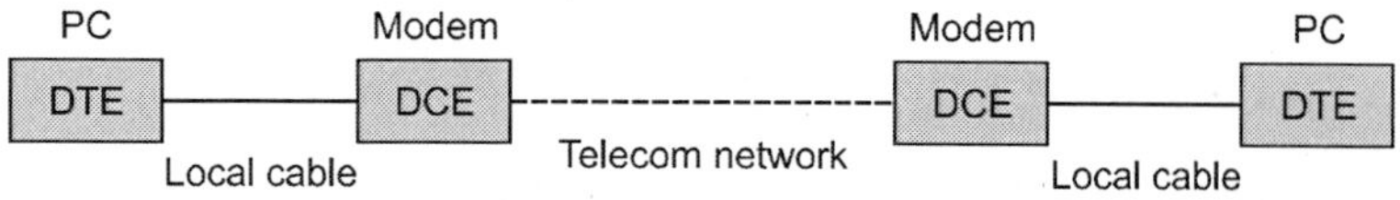


**Figure 14.3:** Serial Communications with a modem

RS-232 signals have a direction (in or out) depending on whether they are with respect to a DTE or a DCE. In all the pinout diagrams below the signal direction is with respect to the DTE (PC).

## 14.5 NULL MODEM CONNECTIONS

When PCs are connected back-to-back each is acting as a DTE (there is no DCE in this case) and consequently certain signals have to be looped in the connection to satisfy any input signal requirement. This is called a NULL (no) modem configuration. For example, when the DTE raises Request to Send (RTS) it expects Clear to Send (CTS) from the DCE. Since there is no DCE to raise CTS, the outgoing RTS signal is looped in the NULL modem cable to the incoming CTS to satisfy the DTE's need for this signal. This is shown in the diagram below.

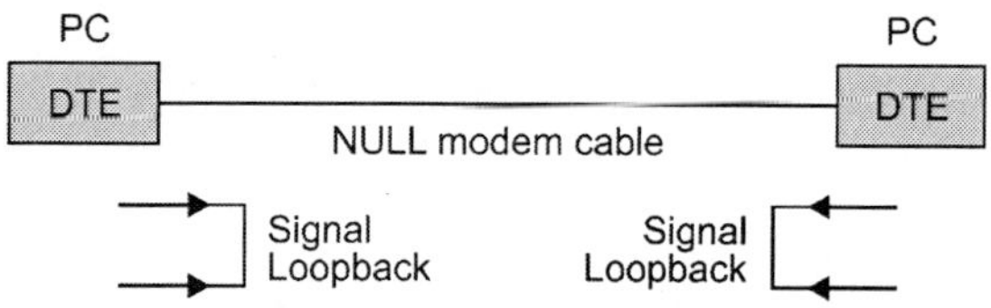


**Figure 14.4:** Serial Communications with a NULL modem configuration

## 14.6 DB25 MALE AND FEMALE PIN NUMBERING

These diagrams show the male (grey background) and female (black background) pin numbering for DB25 sub-miniature connectors. Generally Pin 1 is marked on the front of the connector right next to the pin - though you may need a magnifying glass to read it. Some manufacturers mark each pin number on the plastic housing at the rear of the connector. The male connector has the pins sticking out!

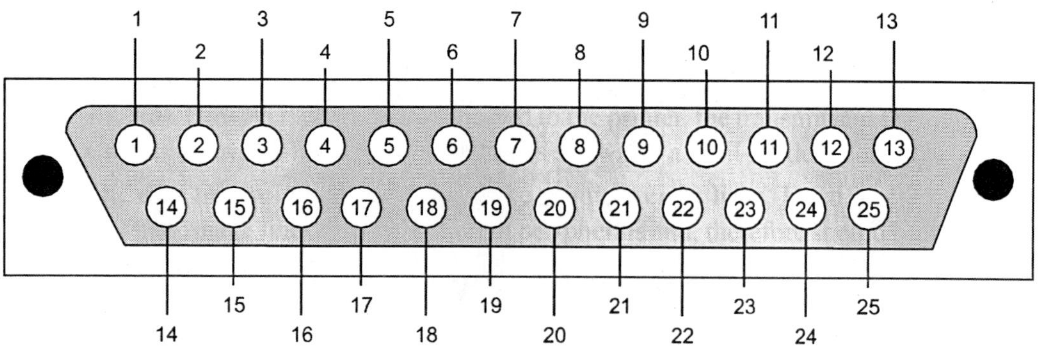


(a) DB25: View looking into male connector

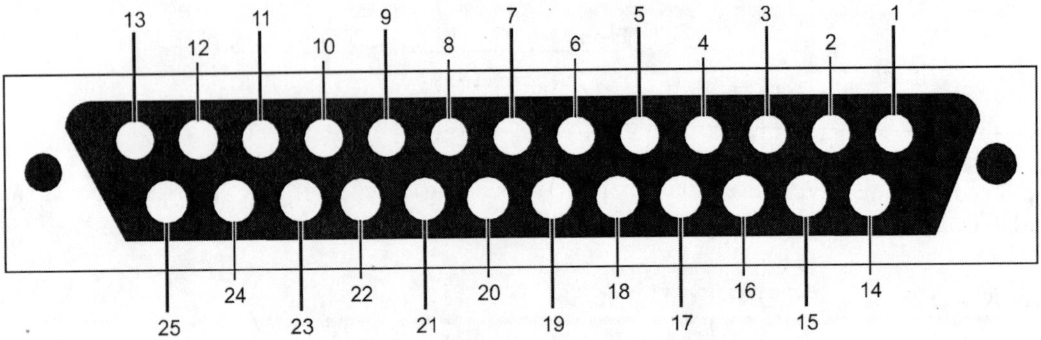


(b) DB25: View looking into female connector

**Figure 14.5:** (a) DB25 MALE AND (b) FEMALE

## 14.7 PIN DIAGRAM OF RS-232C

The use of each pin including methods for spoofing signals is described in our Signal/pin primer. The RS-232 DB25 connector is capable of supporting two separate connections - each with its own optional clock when used in Synchronous mode or Bit-Synchronous mode. If you are using the interface purely for Asynchronous communications then you only need those marked with (ASYNC) below or you can use even fewer (if you understand what is happening). The column marked Dir shows the signal direction with respect to the DTE.

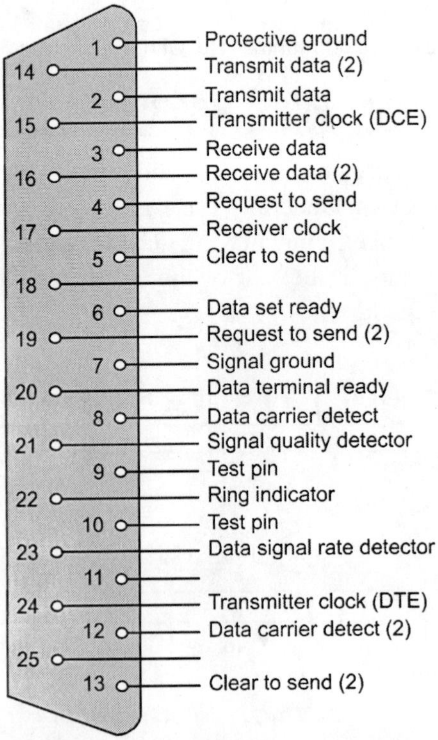


**Figure 14.6:** Pin diagram of RS-232C

The following table lists commonly-used rs-232 signals and pin assignments:

| Pin No. | Description | |
|---|---|---|
| 1 | Protective/shielded ground | |
| 2 | TD | Transmit Data (a.k.a TxD, Tx) (ASYNC) |
| 3 | RD | Receive Data (a.k.a RxD, Rx) (ASYNC) |
| 4 | RTS | Request To Send (ASYNC) |
| 5 | CTS | Clear To Send (ASYNC) |
| 6 | DSR | Data Set Ready (ASYNC) |
| 7 | SGND | Signal Ground |
| 8 | CD | Carrier Detect (a.k.a DCD). |
| 9 | Reserved for data set testing. | |
| 10 | Reserved for data set testing. | |
| 11 | Unassigned | |
| 12 | SDCD | Secondary Carrier Detect. Only needed if second channel being used. |
| 13 | SCTS | Secondary Clear to send. Only needed if second channel being used. |
| 14 | STD | Secondary Transmit Data. Only needed if second channel being used. |
| 15 | DB | Transmit Clock (TCLK, TxCLK). Synchronous use only. |
| 16 | SRD | Secondary Receive Data. Only needed if second channel being used. |
| 17 | DD | Receive Clock (a.k.a. RCLK). Synchronous use only. |
| 18 | LL | Local Loopback |
| 19 | SRTS | Secondary Request to Send. Only needed if second channel being used. |
| 20 | DTR | Data Terminal Ready. (ASYNC) |
| 21 | RL/SQ | Signal Quality Detector/Remote loopback. |
| 22 | RI Ring Indicator. DCE (Modem) raises when incoming call detected used for auto answer applications. | |
| 23 | CH/CI | Signal Rate selector. |
| 24 | DA | Auxiliary Clock (a.k.a. ACLK). Secondary Channel only. |
| 25 | Unassigned | |

**Some commonly-used signals are:**

1. **Transmitted Data (TxD):** Data sent from DTE to DCE.
2. **Received Data (RxD):** Data sent from DCE to DTE.
3. **Request To Send (RTS):** Asserted (set to logic 0, positive voltage) by DTE to prepare DCE to receive data. This may require action on the part of the DCE, e.g. transmitting a carrier or reversing the direction of a half-duplex channel. For the modern usage of "RTS/CTS handshaking," see the section of that name.
4. **Ready To Receive (RTR):** Asserted by DTE to indicate to DCE that DTE is ready to receive data. If in use, this signal appears on the pin that would otherwise be used for Request To

Send, and the DCE assumes that RTS is always asserted; see RTS/CTS handshaking for details.

5. **Clear To Send (CTS):** Asserted by DCE to acknowledge RTS and allow DTE to transmit. This signaling was originally used with half-duplex modems and by slave terminals on multidrop lines: The DTE would raise RTS to indicate that it had data to send, and the modem would raise CTS to indicate that transmission was possible. For the modern usage of "RTS/CTS handshaking," see the section of that name.
6. **Data Terminal Ready (DTR):** Asserted by DTE to indicate that it is ready to be connected. If the DCE is a modem, this may "wake up" the modem, bringing it out of a power saving mode. This behavior is seen quite often in modern PSTN and GSM modems. When this signal is de-asserted, the modem may return to its standby mode, immediately hanging up any calls in progress.
7. **Data Set Ready (DSR):** Asserted by DCE to indicate the DCE is powered on and is ready to receive commands or data for transmission from the DTE. For example, if the DCE is a modem, DSR is asserted as soon as the modem is ready to receive dialing or other commands; DSR is not dependent on the connection to the remote DCE (see Data Carrier Detect for that function). If the DCE is not a modem (e.g. a null modem cable or other equipment), this signal should be permanently asserted (set to 0), possibly by a jumper to another signal.
8. **Data Carrier Detect (DCD):** Asserted by DCE when a connection has been established with remote equipment.
9. **Ring Indicator (RI):** Asserted by DCE when it detects a ring signal from the telephone line.

**Table 14.2: RS-232 Serial Port Pinout**

| Name | pin | Signal | Function |
|---|---|---|---|
| AA | 1 | PG | **Protective Ground:** This line is connected to the chassis ground of the GPIB-232CV. Since the GPIB-232CV chassis ground is not connected to earth ground, pin 1 should be connected on both serial devices |
| BA | 2 | TxD | **Transmit Data:** This line carries serial data from the GPIB-232CV to the serial host |
| BB | 3 | RxD | **Receive Data:** This line caries serial data fron the serial host to the GPIB-232CV |
| CA | 4 | RTS | **Request to Send:** This signal line is driven by the GPIB-232CV and when asserted indicates that the GPIB-232CV is ready to accept serial data. The GPIB-232CV unasserts RTS when it is no longer ready to accept serial data because of a buffer full condition |
| CB | 5 | CTS | **Clear to Send:** This signal line is asserted by the serial host and sensed by the GPIB-232CV. When asserted, it indicates that the serial host is ready to accept serial data. When unasserted, it indicates that data transmission should be disabled. |
| AB | 7 | SG | **Signal Ground:** This line establishes a reference point for all interface voltages. |
| CD | 20 | DTR | **Data Terminal Ready:** This signal line is asserted by the GPIB-232CV to signal that it has been powered on, and is ready to operate. |

| Table 14.3: RS-232C Interface Signals | | | |
|---|---|---|---|
| **Pin** | **Description** | **Pin** | **Description** |
| 1 | Protective Ground | 2 | Transmitted Data |
| 3 | Received Data | 4 | Request to Send |
| 5 | Clear to Send | 6 | Data Set Ready |
| 7 | Signal Ground (Common Return) | 8 | Received Line Signal Detector |
| 9 | (Recerved for Data Set Testing) | 10 | (Reserved for data Set Testing) |
| 11 | Unassigned | 12 | Sec. Rec'd. Line Sig. Detector |
| 13 | Sec. Clear to Send | 14 | Secondary Transmitted Data |
| 15 | Transmission Signal Element Timing (DCE Source) | 16 | Secondary Received Data |
| 17 | Reciver Signal Element Timing | 18 | Unassigned |
| 19 | Secondary Request to Send | 20 | Data Terminal Ready |
| 21 | Signal Quality Detector | 22 | Ring Indicator |
| 23 | Data Signal Rate Selector (DTE/DCE Source) | 24 | Transmit Signal Element Timing (DTE Source) |
| 25 | Unassigned | | |

### *1. Electrical Characteristics*

The RS-232-C specifies the signaling rate between the DTE and DCE, and a digital signal is used on all interchange circuits.

The RS-232 standard specifies that logic "1" is to be sent as a voltage in the range -15 to -5 V and that logic "0" is to sent as a voltage in the range +5 to +15 V.

The standard specifies that voltages of at least 3 V in amplitude will always be recognized correctly at the receiver according to their polarity, so that appreciable attenuation along the line can be tolerated. The transfer rate is rated > 20 kbps and a distance of < 15m. Greater distance and data rates are possible with good design, but it is reasonable to assume that these limits apply in practice as well as in theory.

The load impedance of the terminator side of the interface must be between 3000 and 7000 ohms, and not more than 2500pF.

Table 14.4, summarizes the functional specifications of the most important circuits.

| Table 14.4: RS-232C Circuit Definitions | | |
|---|---|---|
| **Name** | **Direction to** | **Function** |
| **Data Signals** | | |
| Transmited Data (BA) | DCE | Data Generated by DTE |
| Received Data (BB) | DTE | Data Received by DTE |
| **Timing Signals** | | |
| Transmitter Signal Element Timing (DA) | DCE | Clocking signal, transitions to ON and off occur at center of each signal element |

| Name | Direction to | Function |
|---|---|---|
| Transmitter Signal Element Timing (DB) | DTE | Clocking signal, as above; both leads relate to signals on BA |
| Receiver signal Element Timing (DD) | DTE | Clocking signal, as above, for circuit BB |
| **Control Signals** | | |
| Request to Send (CA) | DCE | DTE wishes to transmit |
| Clear to Senc (CB) | DTE | DCE is ready to transmit; response torequest to send |
| Data Set Ready (CC) | DTE | DCE is ready to operate |
| Data Terminal Ready (CD) | DCE | DTE is ready to operate |
| Ring Indicator (CE) | DTE | Indicates that DEC is receiving a ringing signal on the communication channel |
| Carrier Detect (CF) | DTE | Indicates that DCE is receiving a carrier signal |
| Signal Quality Detector (CG) | DTE | Asserted when there is reason to believe there is an error in the received data |
| Data Signal Rate Selector (CH) | DCE | Asserted to select the higher of two possible data rates |
| Data Signal Rate Selector (CI) | DTE | Asserted to select the higher of two possible data rates |
| **Ground** | | |
| Protective Ground (AA) | NA | Attached to machine frame and possible external grounds |
| Signal Ground (AB) | NA | Establishes common ground reference for all circuits |

### 2. Range

The RS-232-C standard specifies that the maximum length of cable between the transmitter and receiver should not exceed 100 feet, Although in practice many systems are used in which the distance between transmitter and receiver exceeds this rather low figure. The limited range of the RS-232C standard is one of its major shortcomings compared with other standards which offer greater ranges within their specifications.

One reason why the range of the RS-232C standard is limited is the need to charge and discharge the capacitance of the cable connecting the transmitter and receiver.

### 3. Mechanical Characteristics

The connector for the RS-232-C is a 25 pin connector with a specific arrangement of wires. In theory, a 25 wire cable could be used to connect the Data Terminal Equipment (DTE) to the Data Communication Equipment (DCE). The DTE is a device that is acting as a data source, data sink, or both, e.g. a terminal, peripheral or computer.

The DCE is a device that provides the functions required to establish, maintain, and terminate a data-transmission connecting, as well as the signal conversion, and coding required for